EXPLORING
Biology
IN THE LABORATORY
CORE CONCEPTS
SECOND EDITION

Murray P. Pendarvis
Southeastern Louisiana University

John L. Crawley

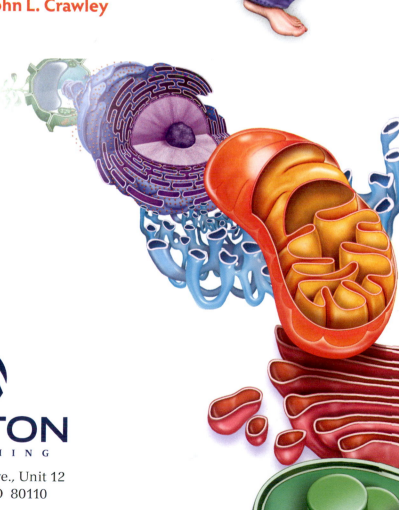

MORTON
P U B L I S H I N G

925 W. Kenyon Ave., Unit 12
Englewood, CO 80110

morton-pub.com

Book Team

President and CEO	David M. Ferguson
Senior Acquisitions Editor	Marta R. Pentecost
Supervising Editor	Adam W. Jones
Developmental Editor	Sarah D. Thomas
Editorial Project Managers	Rayna S. Bailey, Trina Lambert
Production Manager	Will Kelley
Production Associates	Joanne Saliger, Amy Stevens
Indexer	Carolyn Acheson
Cover and Illustrations	Imagineeringart.com, Inc.

Printed in the United States of America

10 9 8 7 6 5 4 3 2

ISBN-10: 1-61731-900-7

ISBN-13: 978-1-61731-900-6

Library of Congress Control Number: 2018956709

Why I Teach

"Ya'll BIOLOGIZE out there" are the final words of wisdom that I often shout to my biology classes as they bolt for the door after class. Many of my students listen and return to my next class with photographs, fossils, leaves, and critters as well as myriad questions and stories related to their exploration of life. I enjoy and look forward to sharing their enthusiasm and excitement about biology.

For me, introductory biology is about more than fulfilling a lab requirement for graduation. This class might be my one opportunity to inspire a student to pursue a career in biology research or education. Falling short of that, it is my 14-week chance to create enthusiasm for students to explore some of life's mysteries. When the light bulb goes off for future math or business majors, I hope the excitement of discovery will lead them to whichever path life takes them.

Thinking back over my teaching career at both the high school and the university levels, I have immersed myself in biology. It has become a passion, a hobby, and a way of life. Many of my fondest memories stem from my experiences with students in the classroom, in the laboratory, and in the field. I have the honor of teaching the most intriguing subject in academia and working with young people. I have dedicated my life to quality biology education.

How I Teach

This laboratory manual is based upon my vision of biology education and the way I teach biology. The exercises have been designed to be safe, interesting, and meaningful. I have arranged the chapters in an order that builds on previous chapters. In addition, I have organized the chapters to reflect variations on a theme. This subtle but important idea—that all living things share a commonality of purpose and a similarity in design—is vital to understanding the unifying concepts of biology. Throughout the manual, many activities emphasize the unity of all living things and evolutionary forces that have resulted in (and continue to act on) the diversity we see before us today.

Chapters and exercises have been condensed from the full edition of *Exploring Biology in the Laboratory*, 3rd edition, to create this *Core Concepts* version. This streamlined edition better fits a one-semester, nonmajors course, although the book can still be used in a two-semester course or a majors course. Each chapter contains a set of five "Just Wondering" pre-laboratory questions to be answered by students prior to coming to lab. Each review section at the end of the chapter contains a series of biology myths for the students to research and "bust." This *Core Concepts* 2nd edition also contains a new chapter as well as new photography and illustrations. As with the full edition, interesting facts and figures are sprinkled throughout the manual to capture students' imaginations and stimulate thought beyond the laboratory. Also included are inspirational and insightful quotes from influential scientists and other well-known historical figures. Whether they are profound or humorous, they all emphasize the pivotal role of scientific study in shaping humankind's understanding of the world around us.

Changes to *Exploring Biology in the Laboratory: Core Concepts*, 2e

Changes throughout:

- All chapters feature new, revised, and updated Check Your Understanding and Review Quiz questions.
- Many chapters received new tables for recording procedural data and experiment results; other original tables were revised and updated.
- Updated art program including over 150 new and revised figures.
- Revised materials lists.
- Use of materials more specific in procedure steps.
- New and revised Check Your Understanding questions (including multiple choice, true and false, short answer, etc.).
- New and revised Review Quiz sections (including multiple choice, true and false, short answer, etc.).
- New and revised "Did you know…" boxes.

Changes by chapter:

- Chapter 2 is new and includes three new exercises concerning scientific notation and the metric system.
- Chapter 3, Exercise 3.3 features an expanded discussion of hydrolysis.
- Chapter 3, Exercise 3.5 now has plain Greek yogurt included in the substances tested in Procedure 1.
- Chapter 8 includes new text regarding botanists of note and discussion of codominance. Chapter 8, Exercise 8.6 uses the term "unifactorial" instead of "monogenic," and discussions of autosomal dominant, X-linked trait, antiparallel have been added to this edition.
- Chapter 10 includes the discussion of evolutionary medicine.

- Chapter 11 no longer discusses viruses. Exercise 11.1 now features text on hyperthermophiles, psychrophiles, and alkaliphiles.
- Chapter 12, Exercise 12.1, Procedure 1 "Observing Spirogyra and Volvox" has been split into two separate experiments in this edition, and discussion of Euglenozoans has been condensed.
- Chapter 12, Exercise 12.3, Procedure 1 has new steps (8 and 9). Discussion of flagellates has been revised. Examination of dinoflagellates has replaced gonyaulax catenella. Examinations of *Vorticella* sp., *Stentor* sp., and *Paramecium* conjugation are new to this edition.
- Chapter 13 no longer includes parenchyma, collenchyma, and sclerenchyma tissues, and Exercise 11.1 has been removed (includes Procedure 1 Plant Tissues).
- Chapter 14 no longer includes phylum Ginkgophyta.
- Chapter 16, Exercise 16.1 now includes discussion of hypogynous, epigynous, and perigynous.
- Chapter 17, Exercise 17.1, Procedure 2 has a wet mount component.
- Chapter 18 has a new Hints & Tips box.
- Chapter 19, Exercise 19.1, Procedure 2 now includes *Dugesia* observation.
- Chapter 19, Exercise 19.2, Procedure 2 was formerly part of Procedure 1.

- Chapter 19, Exercise 19.3 now discusses classes of Phylum Mollusca, and text listing examples of gastropods has been removed.
- Chapter 22, Exercise 22.1 includes new text regarding tuberosities.
- Chapter 23, Exercise 23.5 includes new text discussing receptors, olfaction, gustation, tactile, baroreceptors, thermoreceptors, nociceptors, and proprioceptors.
- Chapter 23, Exercise 23.7, Procedure 2 Equilibrium Test is new to this edition.
- Chapter 24, Exercise 24.1 now longer includes the "Observing the Heart" procedure. Exercise 24.2, Procedure 2 "Listening to the Lungs" is new to this edition.
- Chapter 25, Exercise 25.3 no longer includes the "Oil and Brine" procedure.

If you have any questions or comments about the manual, or suggestions for how it might be improved in future editions, please contact us at exploringbiology@morton-pub.com. Please enjoy this edition of *Exploring Biology in the Laboratory: Core Concepts* and BIOLOGIZE!

—*Murray Paton "Pat" Pendarvis*

About the Authors

Murray Paton "Pat" Pendarvis, a resident of Walker, Louisiana, is currently an assistant professor of biological sciences at Southeastern Louisiana University. He teaches general biology for majors and nonmajors, honors biology, medical terminology, anatomy and physiology, the history of biology, evolutionary biology, and biophotography. In addition, he is an adjunct professor of biology at Our Lady of the Lake College in Baton Rouge, teaching general biology, environmental science, paleontology, evolution, medical genetics, and the history of medicine. Prior to teaching at the university level, he taught science at Doyle and Walker High Schools in Livingston Parish, LA.

As a result of his dedication to quality science education, he has received a number of awards, including the Presidential Award for Excellence in Mathematics and Science Education and the National Association of Biology Teachers Outstanding Teacher Award. In addition to teaching, Pat is an author, writing biology texts for McGraw-Hill and Morton Publishing.

One of Pat's passions is the science and art of biophotography. He has published many photographs in several biology and ecology texts. Pat earned a B.S. degree in biology education, a M.S. degree in zoology from Southeastern Louisiana University, and a Ph.D. in science education from the University of Southern Mississippi.

John L. Crawley currently resides in Provo, Utah. He received his degree in zoology from Brigham Young University in 1988. While working as a researcher for the National Forest Service and Utah Division of Wildlife Resources in the early 1990s, John was invited to work on his first project for Morton Publishing, *A Photographic Atlas for the Anatomy and Physiology Laboratory*. *Exploring Biology in the Laboratory: Core Concepts*, is John's sixth title with Morton Publishing.

John has spent much of his life taking pictures. His photography has allowed him to travel widely, and his photos have appeared in national ads, magazines, and publications. He has worked for groups such as Delta Airlines, *National Geographic*, the U.S. Bureau of Land Management, and many others. His projects with Morton Publishing have been a great fit for his passion for photography and the biological sciences.

Acknowledgments

Many professionals have assisted in the preparation of *Exploring Biology in the Laboratory: Core Concepts*, and have shared our enthusiasm about its value for students of biology. For their help on the second edition we are appreciative of Dr. Barry Ferguson, Dr. Byron Adams at Brigham Young University, and Dr. Samuel R. Rushforth and Dr. Robert Robbins at Utah Valley University.

We appreciate all of the feedback we received from biology laboratory professionals during the creation of the book. We are particularly indebted to the following individuals for their detailed suggestions for improvement:

- Elizabeth R. Abboud, Tulane University, Department of Cell and Molecular Biology
- Warner Bair, Lone Star College–Cy Fair
- Kimberly Beck, Midwestern State University
- Karen E. Braley, Daytona State College
- Emily B. Carlisle, Pearl River Community College
- J. Larry Dew, University of New Orleans
- Amy L.S. Donovan, Columbia Basin College
- Fleur Ferro, Community College of Denver
- Margi G. Flood, University of North Georgia
- Thomas E. Johnson, Tidewater Community College–Chesapeake
- Troy A. Ladine, East Texas Baptist University
- Stephanie Loveless, Danville Area Community College
- Carroll J. Mann, Florida State College at Jacksonville
- Margaret McMichael, Baton Rouge Community College
- Melinda A. Miller, Pearl River Community College
- Thomas Pitzer, Florida International University
- Michele L. Pruyn, Plymouth State University
- Catherine B. Purzycki, M.S., University of the Sciences in Philadelphia
- Peggy Rolfsen, Cincinnati State Technical and Community College
- Dr. Lori Rose, Hill College
- Michelle Spaulding, Purdue University Northwest
- Rachel Wiechman, West Liberty University

Additional comments and suggestions for improvements to the second edition were contributed by the following reviewers:

- Alexie McKee, Portland Community College
- Kristen Miller, University of Georgia
- Christine Simmons, Southern Illinois University Edwardsville
- Amy G. Waites, Central Alabama Community College

Any outstanding errors are the sole responsibility of the authors. We welcome any feedback you might have after using this book.

We gratefully acknowledge the assistance of Imagineering Media Services, Inc., for the art throughout the book. We are indebted to the founder of Morton Publishing, the late Douglas Morton. Additional thanks for the opportunity, encouragement, and support to prepare this lab manual as well as for the organization, copyediting, layout, and proofing go to Morton's President and CEO David Ferguson; the editorial staff: Marta Pentecost, Rayna Bailey, Trina Lambert, Sarah Thomas, and Adam Jones; and the production staff: Will Kelley, Joanne Saliger, and Amy Stevens.

Many of the photographs of living plants and animals in all editions were made possible because of the cooperation and generosity of the San Diego Zoo, San Diego Wild Animal Park, Sea World (San Diego, CA); Hogle Zoo (Salt Lake City, UT); San Francisco Zoo; Aquatica (Orem, UT); Avery Island, LA; and Tickfaw State Park (Springfield, LA). We are especially appreciative to the professional biologists at these fine institutions. Thanks to Satsuma Seafood (Satsuma, LA) for letting us photograph crayfish (crawfish) specimens.

Thank you to Sarah Jean Rayner for her contributions to Chapters 5 through 8 and Chapter 10.

The authors would like to thank their families and friends as well as countless students through the years for their support during this endeavor.

A heartfelt thanks to Shaun Paton Pendarvis (1982–2009) for his strength, courage, and inspiration while battling cystic fibrosis.

Photo Credits

All photos are courtesy of John L. Crawley or Murray P. Pendarvis unless noted here.

Chapter 1 Chapter opener, Martin Shields/ Science Source; Figure 1.3C, Dante Fenolio/Science Source

Chapter 2 Chapter opener, Eric Cohen/Science Source

Chapter 3 Chapter opener, Andrew Lambert Photography/Science Source; Figure 3.3, SPL/Science Source

Chapter 4 Chapter opener, Rafe Swan/Science Source; Figure 4.3, Centers for Disease Control and Prevention (CDC); Figure 4.4, Dartmouth Electron Microscope Facility; Figure 4.5 Leica, Inc.; Figure 4.10A, *A Photographic Atlas for the Microbiology Laboratory*, 3rd ed., by Michael J. Leboffe and Burton E. Pierce, © 2005 Morton Publishing; Figure 4.14A, CDC; Figure 4.15C, Protist Image Database; Figure 4.21B, Spike Walker/Science Source

Chapter 5 Chapter opener, Laguna Design/Science Source; Figure 5.6, Biophoto Associates/Science Source

Chapter 7 Chapter opener, Eric Cohen/Science Source

Chapter 8 Chapter opener, Michael Lustbader/Science Source; Figures 8.1 and 8.9, *A Photographic Atlas for the Anatomy & Physiology Laboratory*, 8th ed., by Morton and Crawley, © 2015 Morton Publishing

Chapter 9 Chapter opener, Alfred Pasieka/Science Source; Figure 9.1, James Gillray (1756–1815), public domain; Figure 9.2, Patrick Dumas/Science Source; Figure 9.3, Pascal Goetgheluck/Science Source; Figure 9.4 Tek Images/Science Source; Figure 9.8, Sidney Paget, www.bestofsherlock.com, public domain

Chapter 10 Figure 10.6C, Géry Parent

Chapter 11 Figure 11.2D, Dr. Lance Liotta Laboratory; 11.2E, Centers for Disease Control (CDC); Figure 11.3A, National Institutes of Health (NIH); Figures 11.3B and 11.3C, CDC; Figure 11.5, *A Photographic Atlas for the Microbiology Laboratory*, 4th ed., by Leboffe and Pierce, © 2011 Morton Publishing

Chapter 12 Figure 12.3, *A Photographic Atlas for the Botany Laboratory*, 7th ed., by Rushforth, Robbins, Crawley, and Van De Graaff, © 2016 Morton Publishing; Figure 12.10C, *A Photographic Atlas for the Microbiology Laboratory*, 4th ed., by Leboffe and Pierce, © 2011 Morton Publishing; Figure 12.11A, CDC; Figures 12.12B, 12.14, and 12.15, *A Photographic Atlas for the Microbiology Laboratory*, 3rd ed., by Leboffe and Pierce, © 2005 Morton Publishing; Figure 12.16B, Thomas Kaczmarczyk

Chapter 13 Did You Know, page 257, photographer unknown, public domain

Chapter 14 Figures 14.9, 14.18, and 14.19, Champion Paper Co.; Figure 14.22C, Craig Lorenz/Science Source

Chapter 15 Chapter opener, Will Kelley; Figures 15.2, 15.7, and 15.22, *A Photographic Atlas for the Botany Laboratory*, 7th ed., by Rushforth, Robbins, Crawley, and Van De Graaff, © 2016 Morton Publishing

Chapter 17 Figure 17.13, H.J. Larsen, www.bugwood.org; Figure 17.14, U.S. Government; Figure 17.16, Andy Crawford and Tim Ridley/Dorling Kindersley/Science Source; Figure 17.21, Donald Groth, www.bugwood.org

Chapter 18 Figures 18.2–18.10, 18.12–18.14, and 18.16, William B. Winborn, Ph.D.

Chapter 19 Figure 19.2D, Mark Wilson; Figure 19.8, *A Photographic Atlas for the Microbiology Laboratory*, 4th ed., by Leboffe and Pierce, © 2011 Morton Publishing

Chapter 21 Figure 21.5, NURC/UNCW and NOAA/FGBNMS; Figure 21.11, Tom McHugh/Science Source; Figure 21.15C, Linda Snook; Figure 21.39A, Henry Firus; Figure 21.41, *Van De Graaff's Photographic Atlas for the Zoology Laboratory*, 7th ed., by Adams and Crawley, © 2013 Morton Publishing

Chapter 23 Figure 23.26, Good-Lite Co.

Chapter 24 Chapter opener, Zephyr/Science Source

Chapter 25 Chapter opener, Biophoto Associates/colorist Mary Martin/Science Source; Figures 25.9 and 25.10, U.S. Coast Guard

Safety in the Laboratory

The laboratory should provide students and instructors alike an environment conducive to accomplishing specific scientific tasks. It is imperative that everyone involved in the laboratory recognize the importance of safety. In addition, everyone using the laboratory must be aware of potential safety hazards, such as faulty electrical outlets, frayed wires, broken glassware and slides, chemical spills, and potentially dangerous organisms. Students must follow proper safety practices in the laboratory and immediately report any safety hazards or accidents to the instructor.

Basic Rules for the Laboratory

1 Follow your laboratory instructor's directions, and look for yellow warning signs throughout this manual.

2 Be familiar with the location of safety equipment (first-aid kit, eyewash, gas shutoff, fire blanket, fire extinguisher), emergency telephone numbers, and exits.

3 Be familiar with the activity of the day and potential safety issues. Read labels carefully.

4 Treat all laboratory equipment, such as microscopes, with care, and store the equipment as instructed.

5 Treat all living things with respect. Avoid causing unnecessary stress or discomfort to living animals.

6 Do not open specimen jars unless instructed.

7 Avoid horseplay in the laboratory.

8 Do not eat, drink, or smoke in the laboratory.

9 Always keep your work area clean and uncluttered. Thoroughly clean your laboratory station before and after each activity.

10 Always wash your hands with soap and water before and after the laboratory experience. Keep your hands away from your face.

11 Properly dispose of broken glass, slides, and disposable laboratory equipment.

12 Wear closed-toe shoes, eye protection, gloves, and laboratory coats when instructed.

13 When dissecting specimens, always wash them thoroughly before you begin the dissection. At the completion of the laboratory activity, properly dispose of the specimen as instructed.

14 Use caution when employing sharp instruments, such as scalpels and dissecting pins. Immediately report any cuts or punctures to your instructor.

15 Place a stopper in any chemical bottle when it is not in use. Follow your instructor's directions when carrying bottles and pouring chemicals. Report any spills immediately to your instructor. Do not taste any chemicals.

16 Keep flammable chemicals away from open flames, and take precautions when handling hot items. Roll up your sleeves when working around open flames.

17 If you have long hair, tie it back.

18 During outdoor activities, be aware of poisonous plants, venomous animals, and potentially dangerous environments, such as cliffs and water. Work in teams.

I have read and understand the basic safety rules for the laboratory.

Name _____

Class _____ Date _____

Contents

The Starting Point
Understanding the Scientific Method

To myself I seem to have been only like a boy playing on the seashore, diverting myself in now and then finding a smoother pebble or prettier shell than ordinary, whilst the great ocean of truth lay all undiscovered before me.

— Isaac Newton (1642–1727)

Just wondering . . .

Consider the following questions prior to coming to lab, and record your answers on a separate piece of paper.

1 What is meant by the statement that we have embraced technology but perhaps not the responsibility of using technology?

2 What traits constitute a healthy scientific attitude?

3 What are several misconceptions about science and scientists?

4 Why are many students anxious about science and pursuing science careers?

5 What can be done to improve the state of American science education?

Objectives

At the completion of this chapter, the student will be able to:

1. Define science and discuss the characteristics of science.

2. Discuss the steps of the scientific method.

3. Discuss the basic biology of termites and goldfish.

4. Make scientific observations and identify variables that may impact an experiment.

5. Construct a hypothesis and design and conduct an experiment to test the hypothesis.

6. Collect and interpret the data collected for an experiment.

7. Discuss the influence of the independent variable on the dependent variable.

Chapter Photo
The scientific method can be used to study many phenomena, including capillary action.

The wonders of science surround us every day, from the sun rising to each beat of our hearts. A descriptive definition of science, however, is elusive. Classically, science is defined as an organized body of knowledge that attempts to explain natural phenomena. This general definition does not recognize that science functions as a dynamic process, encompassing exploration, experimentation, and discovery (Fig. 1.1). Perhaps science is best described by several fundamental characteristics:

◆ Science is based upon observations that incorporate our senses, or instruments that extend our senses, to interpret natural phenomena. Science does not allow any room for mysticism, superstition, or thought contrary to observations.

◆ Science is a search for regularities. These regularities may include easily observed phenomena or patterns in nature.

◆ After observations have been recorded about regularities in nature, scientists must process information. Through qualitative and quantitative techniques such as computer analysis, scientists are better able to understand natural processes.

◆ Science is a self-correcting process in which previously existing concepts can be expanded, modified, or replaced if necessary. In some instances, changes to existing ways of thought are not easily accepted. Ideas that redefine our place in the universe may conflict with centuries of existing dogma. Scientific study is an ongoing, active process, and the scientific body of knowledge is growing exponentially each year.

Although we recognize that science is a tremendous endeavor, it can be divided into three major fields of study: physical science, earth science, and life science.

1. **Physical science** attempts to describe the laws that govern the universe and the composition of the universe. Physics and chemistry are two of the fundamental disciplines of physical science.

FIGURE **1.1**
Scientist at work in a lab.

2. **Earth science** attempts to describe our place in the universe and includes many disciplines, such as astronomy, geology, oceanography, and meteorology.

3. **Life science**, or biology, attempts to describe living things and consists of many disciplines from anatomy to zoology.

The Scientific Method

The **scientific method** is used to describe the way science works. It is a guide to solving problems and initiating investigations. Not all scientific discoveries, however, rigidly follow the scientific method. Many discoveries involve serendipity and even a little luck. Generally, the scientific method consists of the following steps (Fig. 1.2):

1. A phenomenon sparks the interest of the potential investigator, who makes observations, asks questions, and reviews the literature about the topic. For example, an investigator observes that water plants along a ditch polluted with motor oil are dying. The investigator asks questions about the phenomenon and completes a review of the literature.

2. Once the investigator has an understanding of the problem, a prediction is made, and a hypothesis (or hypotheses) is (are) constructed. The investigator predicts relationships between the independent and dependent variables. The function of the **hypothesis** is to provide direction for gathering data. The hypothesis must be testable, practical, and falsifiable. Keep in mind that a rejected hypothesis is not considered a failure. Several formats are used in constructing hypotheses. Initially, hypotheses can be written in an "if-then" form. A **null hypothesis** states that there is no relationship between the independent and dependent variables. In this example, the investigator constructs a hypothesis addressing oil pollution and plant growth such as: "Motor oil has no effect on plant survival."

3. After the hypothesis is stated, the investigator performs experiments and gathers data about the problem. In doing so, the investigator identifies and controls variables when developing an experiment so that a single variable

Did you know . . .

Some Representative Disciplines of Biology

cytology: the study of cells
histology: the study of tissues
anatomy: the study of structure
physiology: the study of function
botany: the study of plants
zoology: the study of animals
biogeography: the study of the range and distribution of organs through space and geological time
microbiology: the study of microorganisms
mycology: the study of fungi
phycology: the study of algae
entomology: the study of insects
ichthyology: the study of fishes
herpetology: the study of amphibians and reptiles.
ornithology: the study of birds
taxonomy: the identification and naming of organisms
genetics: the study of heredity
ecology: the study of the interrelationship between organisms and their environment
evolutionary biology: the study of how organisms change over time
hematology: the study of blood
paleontology: the study of fossils
ethology: the study of animal behavior

is manipulated, and a single variable responds to manipulation. A **variable** is any factor that can cause changes within a system. In setting up an experiment, three types of variables are recognized.

a. The **control variable** is held constant and is used as a baseline for comparison. This variable remains unchanged throughout the experiment. In many experiments this may be referred to as the control group.

b. The **independent variable** is the variable being manipulated or tested in the experiment.

c. The **dependent variable** is also known as the responding variable.

The investigator in this example then conducts experiments that address the effect of oil on plant growth.

4. The investigator then interprets the data collected from the experiment. Interpreting data is a composite skill consisting of communicating, predicting, and inferring. In interpreting data, statistical methods as well as charts

and graphs should be used to make the results understandable. Based upon the data, the hypothesis is rejected or accepted, and a conclusion is drawn. If the results indicate the motor oil reduces plant survival, then the hypothesis (alternative hypothesis) and the null hypothesis can be addressed. If the hypothesis states that motor oil reduces plant survival, the hypothesis is accepted, and if the motor oil does not reduce plant survival, the hypothesis is rejected. If the null hypothesis states that motor oil does not reduce plant survival, the null hypothesis is rejected, and if motor oil does reduce plant survival, the null hypothesis is accepted.

An experiment does not "prove" a hypothesis. It can only not disprove it. Occasionally, the use of the scientific method can yield a hypothesis about a certain topic or a collection of hypotheses that can be incorporated in the development of a scientific theory.

The term *theory* is often misused. In science, the statement, "Oh, it's just a theory," has no place because it suggests a theory is nothing more than a guess. Scientific theories, however, stand on their own accord and represent the current, well-supported explanation about some aspect of the natural world. A theory has been repeatedly confirmed by the processes of observation and experimentation and is agreed upon by the majority of experts in the field. Examples of well-known theories in the biological sciences are cell theory, evolutionary theory, and the germ theory of disease.

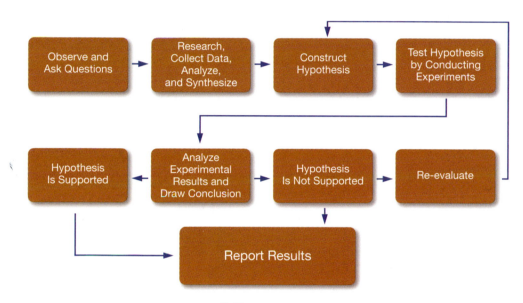

FIGURE **1.2** Scientific method.

Practicing the Scientific Method with Termites

The term *termite* is derived from the Latin word *termes*, which literally means woodworm. Nearly 2,000 species of termites constitute the insect order Isoptera (Greek: *isos* = the same; *ptera* = wing). In their reproductive form, the isopterans possess equal sized paired wings. In the environment, termites serve as decomposers, helping to break down and recycle dead wood. They become pests when they consume cellulose-based materials, such as wood and paper. In addition, termites are responsible for the destruction of valuable forest resources. To digest these cellulose-based sources, termites exist in a symbiotic relationship (an intimate relationship between two or more species), either with bacteria or ciliated protists.

Termite colonies can be composed of several dozen members or millions of members, depending upon the species and age of the colony. With the exception of the immature nymphs, a termite colony consists of three basic functional groups called castes: workers, soldiers, and reproductives.

The vast majority of members in a termite colony are wingless, sterile, and blind workers. They are milky white in color and possess hard-chewing mouth parts. Workers look after the eggs and nymphs, feed the soldiers (who defend the colony, mostly from ants) and reproductive forms, build and maintain the colony, and forage for food. The workers are responsible for the telltale signs of termite damage.

The reproductives, also called the royal caste, consist of the king and queen. Their only function is to reproduce. Members of the royal caste are dark brown in color and have functional eyes. The queen appears striped because the segments of her abdomen are distended. Swarming termites are called alates, or swarmers. They are winged reproductives that eventually establish new nests (Fig. 1.3).

Because termites are highly social insects, communication is essential. Termites use chemicals known as pheromones for communication. Specific pheromones are used in mating, producing an alarm, keeping nymphs from forming reproductive castes, and establishing a trail. Termites use the trail pheromone when they are attempting to lure other termites to follow them to a food source or other region of interest.

It has been discovered that the solvents in some Papermate and Bic ballpoint pens are similar to the trail pheromones used by termites. In this activity, students will use the scientific method and process skills of science to investigate the relationship between various inks and termite trailing behavior. The laboratory instructor and students will treat the termites in a humane manner at all times.

FIGURE **1.3** Termites are in the order Isoptera: (**A**) worker termites; (**B**) soldier termites; (**C**) reproductive termites.

Termite Movement

Materials

- ❑ Worker termites
- ❑ Petri dish lined with a moist paper towel to house the termites
- ❑ Plain white paper (4 sheets per team)
- ❑ Artist's small watercolor paintbrush
- ❑ Paper clip
- ❑ Black, blue, and red Papermate or Bic ballpoint pens
- ❑ Ruler

1 Divide into teams as directed by the instructor.

2 Procure the termites and supplies for the activity. Ensure that the termites are housed in a covered petri dish lined with a moist paper towel. When they are not being observed, cover the termites with a moist paper towel.

3 Place the typing paper on the desktop, and carefully use the brush to move and orient the termites because they are fragile.

   a. What variables may influence termite movement?

   b. Describe the movement of the termites. Does their movement exhibit any pattern?

4 Using one pen at a time, draw a 2 cm straight line on the typing paper. Using the artist's brush or paper clip, gently guide the termites toward the line. Record and describe the response of the termites to each line. Does the behavior of the termites differ in response to the lines drawn with different colors of ink?

5 Using fresh paper and the same pens, draw circles approximately 2 cm in diameter. Using the artist's brush or paper clip, gently guide the termites toward the center of the circle. Record and describe the response of the termites to each circle. Does the behavior of the termites differ when they encounter a different ink color?

6 Using fresh paper and the pen color that the termites preferred, draw several geometrical patterns with sharp turns and a figure 8. Using the artist's brush or paper clip, gently guide the termite toward the patterns, testing one design at a time. What is the response of the termites to each geometrical pattern and the figure 8?

7 Using fresh paper and the pen color the termites preferred, draw several broken straight-line patterns. Using the artist's brush or paper clip, gently guide the termite toward the line. Do the termites respond differently to broken lines? Is there any relationship between termite movement and the gaps between the lines?

8 Based upon your observations using black ink, design an investigation that tests other ink colors or types of pens.

   a. Identify the variable you are going to test.

b. Construct a hypothesis for your investigation, and state it below.

_____

_____

_____

_____

_____

_____

c. Briefly describe your procedure.

_____

_____

_____

_____

_____

_____

_____

_____

_____

d. Discuss your findings. Did you accept or reject your hypothesis?

_____

_____

_____

_____

_____

_____

_____

e. Describe several other variations of this basic activity that would be good topics for an investigation.

_____

_____

_____

_____

_____

_____

9 Place the termites back in the container, and return them and the supplies to their distribution area.

✔ Check Your Understanding

1.1 Describe characteristics and the role of different members of the termite society.

_____

_____

_____

_____

1.2 Describe how the scientific method was applied in this experiment.

_____

_____

_____

_____

EXERCISE 1.2

Applying the Scientific Method with Goldfish

This exercise is designed to have the students design and conduct an experiment that tests the effect of temperature on the respiration rate of goldfish (*Carassius auratus*) (Fig. 1.4). The student will identify variables; construct hypotheses; make observations; collect, record, and interpret data; and draw conclusions based upon the data. The laboratory instructor and students will treat the goldfish in a humane manner at all times and follow your school's specific animal safety protocols.

In aerobic respiration, an organism takes in oxygen from its environment and releases carbon dioxide as a waste product. Organisms have specialized structures to carry out respiration. Many aquatic animals use gills for respiration. In fish, the gills can be found beneath a protective covering called the operculum. The gills are made of gill filaments that serve to increase the surface area. This allows maximum exposure to the oxygen-laden water environment.

When a fish "breathes," its operculum closes, and its mouth opens. To allow water to pass over the gill filaments, the mouth closes, and the pharynx contracts. In turn, oxygen diffuses into the capillary circulatory network and is distributed throughout the fish's body. Carbon dioxide diffuses

FIGURE **1.4** Goldfish, *Carassius auratus*.

from the capillary network and enters the environment. The process of opening and closing the mouth or opening and closing the operculum constitutes one breath for a fish.

Several variables affect the respiration rate of fish. In this experiment, student teams will discover the effects of temperature on goldfish respiration and report their results in a scientific manner. Students should make every effort to ensure the survival of the experimental goldfish.

Procedure 1

Goldfish Respiration Rate

Materials

- ❏ Goldfish
- ❏ Aquarium net
- ❏ Thermometer
- ❏ Crushed ice
- ❏ Aquarium water
- ❏ 250 mL beaker
- ❏ Stirring rod
- ❏ Timing device

Read all of the instructions before beginning the experiment! In preparation, divide into working teams of four students. In this activity, one student should serve as timer, one student as recorder, one student as counter, and one student as worker. At the beginning of class, the laboratory instructor will discuss how symptoms of stress manifest in goldfish.

1 Describe variables that may influence goldfish respiration. Discuss and identify the control, independent, and dependent variables found in this experiment.

Variables: _____

Control: _____

Independent Variable: _____

Dependent Variable: _____

2 Construct a simple hypothesis pertaining to goldfish respiration and temperature before you begin.

_____
_____
_____
_____

3 Add approximately 150 mL of aquarium water to the 250 mL beaker. Place the thermometer in the beaker, and take the temperature. Practice adding small amounts of crushed ice to the water until you can consistently lower the temperature of the water approximately 2°C.

4 Empty the beaker, and refill it with 150 mL of aquarium water. This will provide the starting temperature and allow the goldfish enough water to swim.

5 Carefully capture one goldfish, and place it gently into the beaker of room temperature water (approximately 25°C). Measure and record the temperature of the water in degrees Celsius. After the goldfish has adjusted to the new environment for 3 minutes, count its breaths. Develop a consistent procedure for counting the breaths of the fish for 1 full minute using a timing device, and record the data in Table 1.1.

6 Add enough ice to lower the temperature of the water approximately 2°C, stir gently with the glass rod so as to not disturb the fish, and wait 1 minute for the fish to adjust. Count and record in Table 1.1 the number of breaths the fish takes within a 1-minute period at each temperature.

7 Each time the temperature is lowered approximately 2°C, record in Table 1.1 the number of times the fish breathes in 1 minute. If the goldfish shows signs of stress (such as altered behavior), stop the experiment, and record the temperature and the number of breaths at this point.

8 Continue lowering the temperature of the water approximately 2°C at a time, measuring the respiration rate at every 2°C interval. Allow the fish time to adjust (about 3 minutes each time) until the water reaches a temperature of 4°C.

9 At the conclusion of the experiment, *gradually* replace the cold water with aquarium water and increase the temperature of the water by increments of 2°C–4°C until the fish has recovered and the aquarium water temperature is attained. Return the goldfish to a recovery container.

10 Enter your group's data on the chart provided on the board. Complete Table 1.1 using the data from each group. Record your group's results, then graph your group's data versus the class average from the data on the board (Fig. 1.5).

11 For the next class meeting, complete a laboratory report in the format your laboratory instructor's directs.

TABLE **1.1** Your Group's Data versus the Class Average

Dependent Variable: Breaths Per Minute															
Team 1															
Team 2															
Team 3															
Team 4															
Team 5															
Team 6															
Team 7															
Team 8															
Total															
Average															
	30	28	26	24	22	20	18	16	14	12	10	8	6	4	2

Independent Variable: Temperature Degrees Celsius

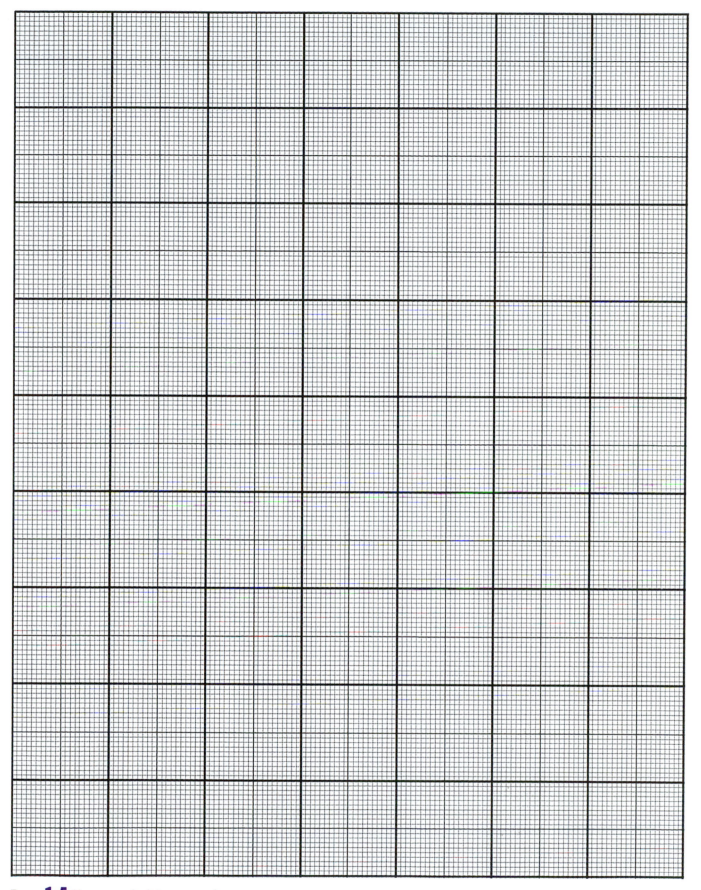

FIGURE **1.5** Your group's data versus class average.

2.1 Restate your hypothesis. Was your hypothesis rejected or accepted? Why?

2.2 Discuss how your group's results compared with the results of the entire class.

2.3 Discuss how the theme of the goldfish experiment can apply to the natural world and the care of aquatic animals.

2.4 List some possible sources of error in this experiment.

MYTHBUSTING

Tales of Toads

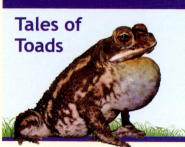

Debunk each of the following misconceptions by providing a scientific explanation. Write your answers on a separate sheet of paper.

1 Toads and frogs are the same animal.

2 In some cultures toads are good luck and in others they are bad luck.

3 Toad and frog sounds are a sure sign of upcoming rain.

4 Toads and frogs were a symbol of fertility in some ancient cultures.

1 Describe the characteristics of science.

2 What did Sir Isaac Newton mean when he made the statement: "If I have seen a little further, it is by standing on the shoulders of giants"?

3 Give several examples of things that could be done to improve scientific attitudes and education in our nation.

4 Describe several characteristics of a well-designed experiment.

5 In an experiment designed to test the effects of outboard motor oil on the growth of algae, identify the control, the independent variable, and the dependent variables. Construct a simple hypothesis.

6 The term *theory* is often used in everyday life, yet the scientific method involves the formulation of hypotheses. Describe how the term *theory* is misused in everyday life.

7 Why is there no room for superstition and mysticism in science?

8 Why is a fundamental knowledge of biology necessary for all people?

For Good Measure
Understanding Scientific Notation and the Metric System

2

The Universe is a grand book of philosophy. The book lies continually open to man's gaze, yet none can hope to comprehend it who has not first mastered the language and characters in which it has been written. This language is mathematics.

— Galileo (1564–1642)

Just wondering . . .

Consider the following questions prior to coming to lab, and record your answers on a separate piece of paper.

1 How were some of the early units of measurement derived?
2 What are the advantages of "going metric"?
3 Why is the United States so resistant to "going metric"?
4 Where can I see signs of the metric system in everyday life?
5 What is an AU in metric units and why is it important in space travel?
6 How many liters of methane can an average cow produce daily? How about carbon dioxide?

W hat is the price of oil per barrel? How many milliliters are in a tablespoon? What is your respiratory rate? How many grams of fat are in that cookie? What does this job pay per hour? Numbers are an integral part of our everyday lives, from checking gas mileage to determining a batting average. Seldom does a day pass in which we do not use mathematics or measure something.

It is said that math is the backbone of science. To succeed in scientific endeavors, we have to acquire sound mathematics and measurement skills (Fig. 2.1). Occasionally, scientists use extremely small or large numbers, such as when measuring the width of the cell membrane or the distance to another galaxy. For example, the distance to the bright star Alpha Centauri is approximately 38,000,000,000,000,000 kilometers, and the mass of a mitochondrion in a nerve cell of a 21-day-old rat is approximately 0.00000000000308 grams. Using scientific notation, these intimidating and awkward numbers become easier to manage. In addition, universal units of measure are important in science. The metric system provides scientists a logical, precise, and easy-to-use universal system.

FIGURE **2.1** Precise measurements are essential in science.

Using Scientific Notation

In scientific notation, numbers are composed of three components: the coefficient, the base, and the exponent. Thus, the number 93,000,000 can be expressed as 9.3×10^7. In this number, the coefficient is 9.3, the base is 10, and the exponent is 7. Specific rules have been developed to express a number in scientific notation properly:

1. The base always has to be 10, the coefficient has to be greater than or equal to 1 but less than 10.

2. The exponent has to reflect the number of places the decimal has to be moved to change the number to its standard notation.

3. Numbers greater than 1 should be written with positive exponents.

4. Numbers less than 1 should be written with negative exponents.

Example numbers:
> 9,000 can be expressed as 9×10^3
> 229,000,000 can be expressed as 2.29×10^8
> 7.63×10^5 can be expressed as 763,000
> 0.0008 can be expressed as 8×10^{-4}
> 0.000000000452 can be expressed as 4.52×10^{-10}

Multiplying and Dividing with Scientific Notation

To multiply numbers written in scientific notation, multiply the coefficients, and add the exponents. Always convert the answer to properly written scientific notation. To divide numbers written in scientific notation, divide the coefficients, and subtract the exponent of the divisor from the exponent of the dividend. Convert the resulting answer to properly written scientific notation.

Example calculations:
> $(3.0 \times 10^{12}) \times (6.0 \times 10^3) = 1.8 \times 10^{16}$
> $(8.1 \times 10^7) \times (3.5 \times 10^{-3}) = 2.8 \times 10^5$
> $(6.4 \times 10^{-5}) \times (2.4 \times 10^{-3}) = 1.5 \times 10^{-7}$
> $(8.0 \times 10^6) \div (4.0 \times 10^2) = 2.0 \times 10^4$
> $(5.25 \times 10^8) \div (2.75 \times 10^{-3}) = 1.91 \times 10^{11}$
> $(7.64 \times 10^{-7}) \div (3.22 \times 10^{-4}) = 2.37 \times 10^{-3}$

Adding and Subtracting with Scientific Notation

When adding and subtracting numbers written in scientific notation, all of the numbers should be converted to the same exponent value prior to performing the calculations. In many cases, this requires changing the decimal place of the coefficient as well. Add or subtract the coefficients, and leave the base and exponent the same. Convert the resulting answer to properly written scientific notation.

Example calculations:
> $4.0 \times 10^7 + 2.0 \times 10^8 = 0.40 \times 10^8 + 2.0 \times 10^8 = 2.4 \times 10^8$
> $7.25 \times 10^8 + 9.60 \times 10^7 = 7.25 \times 10^8 + 0.960 \times 10^8 = 8.21 \times 10^8$
> $5.4 \times 10^4 + 3.1 \times 10^{-3} = 5.4 \times 10^4$ (the second value is too small to affect the first)
> $8.0 \times 10^9 - 4.0 \times 10^9 = 4.0 \times 10^9$
> $6.4 \times 10^3 - 2.2 \times 10^{-3} = 6.4 \times 10^3$ (the second value is too small to affect the first)
> $9.0 \times 10^{-5} - 6.10 \times 10^{-4} = 0.90 \times 10^{-4} - 6.10 \times 10^{-4} = -5.2 \times 10^{-4}$

Procedure 1

Scientific Notation

Materials
None required

1 Convert 1234.95000000 to scientific notation.

2 Convert 0.00000004567 to scientific notation.

3 Convert 2.94×10^5 to standard notation.

4 Convert 5.43×10^{-6} to standard notation.

5 Multiply $(3.5 \times 10^5) \times (1.1 \times 10^6)$

6 Multiply $(4.6 \times 10^{-4}) \times (2.5 \times 10^{-4})$

7 Multiply $(5.9 \times 10^8) \times (3.7 \times 10^{-5})$

8 Divide $4 \times 10^{10} \div 1.9 \times 10^5$

9 Divide $6.8 \times 10^{-4} \div 3.4 \times 10^{-2}$

10 Divide $8.8 \times 10^{12} \div 4.2 \times 10^{-2}$

11 Add $5.2 \times 10^9 + 3.2 \times 10^5$

12 Subtract $7.3 \times 10^5 - 2.5 \times 10^2$

13 An average human has 125 trillion cells; convert this number to scientific notation.

14 *Bacillus anthracis* has a length of 4.1 µm. Convert this value to scientific notation.

15 A picometer is 0.000000000001 m; convert this to scientific notation.

16 Convert Avogadro's number (6.02×10^{23}) to standard notation.

17 Approximately 4.3×10^9 kg of matter is converted to energy by the sun each second; convert this number to standard notation.

18 The approximate number of stars in the Milky Way is 2×10^{11}; convert this value to standard notation.

19 If an average person produces 2 million red blood cells per second, how many red blood cells will be produced in a 24-hour period? Express the value in both standard and scientific notation.

20 If the sun is 1.5×10^8 km from Earth, how long does it take light to strike Earth when traveling at 3.0×10^5 km/s?

Hints & Tips

1 In the United States, the *meter* and *liter* spellings are commonly used. In other nations, the spellings *metre* and *litre* are used more frequently.

2 Do not follow unit symbols by a period except when they are the last word in a sentence. For example, 75 m. is written incorrectly.

3 Unit symbols are case sensitive. For example, the abbreviation for meter is m and liter is L.

4 Unit symbols are singular not plural forms. Do not write 62 ms. It is correct to use the plural form in writing the metric units. For example, 27 grams is written correctly.

5 A space must separate digits from unit symbols. For example, 4.3 m is correct, not 4.3m. In measuring temperature, do not use the space. For example, use 37°C.

6 In writing a quotient of two units, do not use a p to represent per. For example 88 km/h is correct, not 88 kph.

7 Always use a zero before the decimal point when the number is less than one. For example, use 0.61 instead of .61.

8 Never mix unit symbols such as 7.2 m 14 cm.

✔ Check Your Understanding

1.1 Explain why the use of scientific notation is important.

1.2 What does a negative exponent signify in scientific notation?

Learning the Metric System and Conversions

The metric system had its origin in France in 1790. During the same year, Thomas Jefferson (1743–1826) proposed a decimal system of measurement for the United States, but the United States continues to use the English system based upon units such as inches, pounds, and gallons. The modernized version of the metric system is called *Le Système International d'Unités*, or the SI system.

Today, in science and industry, the metric system serves as the universal system of measurement (Table 2.1). The metric system is based on units of 10. Conversions between metric units can be performed by simply shifting the decimal place. The basic units of measurement in the metric system are the **meter** (length), **gram** (mass), and **liter** (volume) (Table 2.2). In addition, temperature in the metric system is measured in **degrees Celsius** (Table 2.3). In the metric system, Greek or Latin prefixes are placed before the base unit to denote powers of 10.

TABLE **2.2** *Le Système International d'Unités* Base Units

Quantity	Unit	Symbol
Length	meter	m
Mass	kilogram	kg
Time	second	s
Electric current	ampere	A
Thermodynamic temperature	kelvin	K
Amount of substance	mole	mol
Luminous intensity	candela	cd

TABLE **2.1** Commonly Used Metric Prefixes

Prefix	Symbol	Factor	Equivalent	Name
pico	p	10^{-12}	0.000000000001	trillionth
nano	n	10^{-9}	0.000000001	billionth
micro	μ	10^{-6}	0.000001	millionth
milli	m	10^{-3}	0.001	thousandth
centi	c	10^{-2}	0.01	hundredth
deci	d	10^{-1}	0.1	tenth
base unit		10^{0}	1	one
deka	da	10^{1}	10	ten
hecto	h	10^{2}	100	hundred
kilo	k	10^{3}	1,000	thousand
mega	M	10^{6}	1,000,000	million
giga	G	10^{9}	1,000,000,000	billion
tera	T	10^{12}	1,000,000,000,000	trillion

TABLE **2.3** Commonly Used Metric Units

Measurement	Unit	Symbol
Length	millimeter	mm
	centimeter	cm
	meter	m
	kilometer	km
Mass	milligram	mg
	gram	g
	kilogram	kg
Volume	milliliter	mL
	cubic centimeter	cm^3
	liter	L
	cubic meter	m^3
Temperature	degrees Celsius	°C

For more information on the metric system and the SI system, refer to the U.S. Metric Association: http://www.us-metric.org/.

Procedure 1

Conversions

Materials
None required

Practice metric-to-metric conversions.

Example: The wingspan of an eagle is 2 m. Convert this to mm.

Answer: 2 m × 1,000 mm/1 m = 2,000 mm

1 3.9 m × _____ / _____ = _____ cm

2 19.7 kg × _____ / _____ = _____ g

3 22.4 L × _____ / _____ = _____ mL

4 89.04 g × _____ / _____ = _____ kg

5 769 nm × _____ / _____ = _____ km

6 Adult humans average 5,000 mL of blood; this equates to _____ L.

7 If the average human heart has a mass of about 300 g, what is its mass in mg? _____

8 Some individuals with diabetes insipidus can produce as much as 15 L of urine a day. This value equates to _____ mL a day.

9 A ciliate can be 10 mm in length. What is its length in m? _____

10 The Mariana Trench has a maximum known depth of 10,994 m. What is this depth in mm? _____

Because the English system is still commonly used in the United States, we have to perform conversions between the metric system and the English system. Several conversion factors are included in Hints & Tips. We suggest you learn just one conversion factor per unit. To practice English-to-metric and metric-to-English conversions, multiply conversion factors and cancel out units.

Example: You are going to run a 5K race. Convert this to mi.

Answer: 5 km × 1 mi./1.6 km = 3.1 mi.

11 90 mi. × _____ / _____ = _____ km

12 38.6 gal. × _____ / _____ = _____ L

13 37°C × _____ / _____ = _____ °F

14 1,650 lb. × _____ / _____ = _____ kg

15 2.65 mL × _____ / _____ = _____ tsp.

16 During the Carboniferous, some dragonflies had a wingspan of 70 cm. Convert this value to in.: _____

17 If an average woman's brain has a mass of 1,450 g, what is its equivalent in lb.? _____

18 If an average cow produces 60 L of saliva daily, how many gal. does this equal? _____

19 Some redwood trees have a diameter of 11 m. What is their diameter in ft.? _____

20 −60°F equals _____ °C.

✔ Check Your Understanding

2.1 What are the advantages of using the metric system?

_____

_____

_____

2.2 What are the three basic units of measure for the metric system?

_____

_____

_____

EXERCISE 2.3

Measuring Length, Mass, Volume, and Temperature

Procedure 1

Length

Materials
- ❑ Meterstick
- ❑ Metric ruler
- ❑ Penny
- ❑ Objects of instructor's choice (3)

In the metric system, the basic unit of linear measurement is the meter, defined as the distance traveled in 1/299,792,458 of a second by electromagnetic waves in a vacuum. A meter is 39.37 inches long, a little longer than a yard. In the laboratory, nanometers (1×10^{-9} m) and micrometers (1×10^{-6} m) are used to measure microscopic objects. Millimeters (1×10^{-3} m) and centimeters (1×10^{-2} m) are commonly used to measure the size of macroscopic organisms, and kilometers (1×10^{3} m) are used to measure longer distances (Fig. 2.2). To practice measuring the length of objects:

1 Obtain a meterstick and a metric ruler from your instructor.

2 Examine the meterstick and the metric ruler. How many cm are found in each instrument?

_____

How many mm are found in each instrument?

_____

Approximately how many cm are there in 1 yd.?

_____

Approximately how many cm are there in 1 ft.?

_____

Approximately how many cm are there in 1 in.?

_____

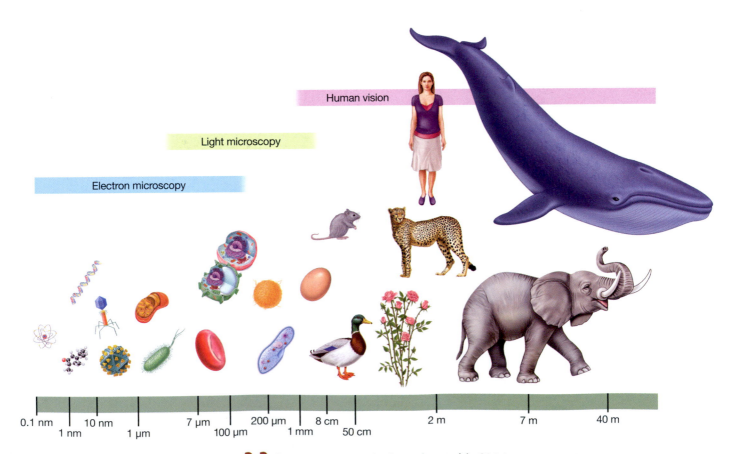

Human vision

Light microscopy

Electron microscopy

| 0.1 nm | 10 nm | 7 μm | 200 μm | 8 cm | 2 m | 7 m | 40 m |
| 1 nm | 1 μm | 100 μm | 1 mm | 50 cm | | | |

FIGURE **2.2** Comparative lengths from the world of biology.

3 Measure the following using the appropriate metric unit of length:

Length and width of the laboratory table _____

Length and width of this page _____

Diameter of a penny _____

4 The laboratory instructor will provide three objects to be measured. Record their measurements below:

1. _____

2. _____

3. _____

5 Return the materials to the designated location.

Procedure 2

Mass

Materials

- ❏ Balance
- ❏ Marble
- ❏ Quarter
- ❏ Penny
- ❏ Paper clip
- ❏ Cell phone
- ❏ Ring
- ❏ 100 mL cylinder
- ❏ Tap water

The terms *mass* and *weight* are sometimes used interchangeably but they are not the same. That is, although a person's weight is different depending on gravity, his mass remains the same. **Mass** is defined as the measure of the quantity of matter. The basic unit of mass in the metric system is the gram. In the SI system, the basic unit of mass is the kilogram. There are 1,000 grams in a kilogram (Fig. 2.3). The standard for the kilogram has been a platinum-iridium cylinder, which is 39 millimeters tall and 39 millimeters in diameter and is housed in Sèvres, France. However, the new standard for the kilogram will be introduced in May 2019 and will be based on the Planck constant.

Most laboratories use an electronic balance, and in some instances a triple-beam balance, to measure the mass of an object (Fig. 2.4). A variety of electronic balances exist. Electronic balances usually can determine mass in hundredths or thousandths of a gram. The triple-beam balance and electronic balance each use a thin piece of paper or weighing pan (boat) to hold the sample.

The weighing pan must be *dry and clean* before using it in an experiment. Before working with the sample, both the triple-beam balance and the electronic balance must be set to compensate for the weight of the pan or paper. This process is called **zeroing**, or **taring**, the balance. To zero the balance, place the paper or pan on the balance, and set the weight to 0. The laboratory instructor will provide directions for zeroing the specific balance used in this laboratory.

1 Using the laboratory instructor's directions, zero your balance with either the paper or the weighing pan. This should be done each time a specimen is measured.

2 Place a marble in the center of the paper or the weighing pan.

Electronic balances are delicate instruments, so use extra care when working with them.

A

B

1.63 × 10⁻²⁴ g   2.8 × 10⁻² g   68 kg   240 kg   6,000 kg   1 × 10⁵ kg   6 × 10²⁴ kg
     1.5 × 10⁻³ g   5 kg

1.63×10^{-24} g 2.8×10^{-2} g   68 kg   240 kg   6,000 kg 1×10^{5} kg 6×10^{24} kg 1.5×10^{-3} g   5 kg

FIGURE **2.3** Comparative masses from the world of biology.

FIGURE **2.4** (**A**) Triple-beam and (**B**) electronic balances are commonly used in the biology laboratory.

3 If an electronic balance is used, follow the instructor's directions for using this device and handle with extra care. If a triple-beam balance is used, slide the masses all the way to the left. Move one mass at a time toward the right, starting with the largest unit, until the balance arm lines up with the zero mark. The mass of the object is the sum of the masses on the beams of the balance.

What is the mass of the marble? _____

4 Determine the mass of the following:

A quarter _____

A penny _____

A small paper clip _____

A cell phone (not provided) _____

A ring (not provided) _____

Your choice _____

Your choice _____

Your choice _____

Empty 100 mL graduated cylinder _____

100 mL graduated cylinder filled with 100 mL of water _____

100 mL of water alone _____

100 mL graduated cylinder filled with 50 mL of water _____

50 mL of water alone _____

100 mL graduated cylinder filled with 30 mL of water _____

30 mL of water alone _____

5 Thoroughly clean your glassware and laboratory station, and return the materials to the designated location.

Procedure 3

Volume

Materials
- ❏ 100 mL cylinder
- ❏ Tap water
- ❏ 250 mL beaker
- ❏ 100 mL beaker
- ❏ Stirring rod
- ❏ 100 mL flask
- ❏ 10 mL pump or bulb pipette
- ❏ Dropper pipette
- ❏ Red food coloring

Volume is a measure of the space occupied by an object (Fig. 2.5). The liter is the basic unit of volume in the metric system. One liter is equal to 1,000 cubic centimeters, or 1.06 quarts, and a typical 2-liter soft drink bottle is equal to 0.5283 gallons. Milliliters (1×10^{-3} L) and microliters (1×10^{-6} L) are used to measure small volumes. Often, volume is measured in cubic centimeters (cm^3); 1 milliliter is equal to 1 cubic centimeter. In the laboratory, several devices are used to measure volume, including flasks, beakers, graduated cylinders, and pipettes (Fig. 2.6).

1 From the laboratory instructor, obtain one 250 mL beaker, one 100 mL beaker, one stirring rod, one dropper pipette, one 100 mL flask or beaker, one 100 mL graduated cylinder, and one 10 mL pump or bulb pipette.

2 Fill the 250 mL beaker with tap water to the designated line, and add 5 drops of red food coloring. Stir the solution gently.

3 Most beakers and flasks have volume markings. This marking is a close approximation and not as accurate as other measuring devices. Beakers and flasks should be used when measuring approximate volumes. Carefully pour 100 mL of the red solution into a 100 mL flask or beaker. Add or remove the solution with a dropper pipette to obtain 100 mL.

4 Graduated cylinders range in volume capacity. In this activity, a 100 mL graduated cylinder is to be used. This graduated cylinder

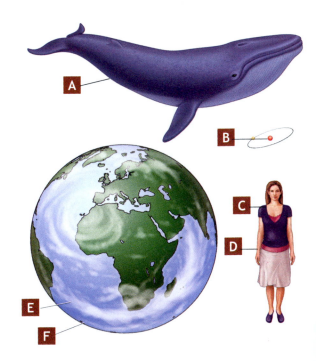

FIGURE **2.5** Comparative volumes from the world of biology: (**A**) volume of blood in a blue whale: 7.098×10^3 L; (**B**) volume of a proton: 1×10^{-12} L; (**C**) inspiration and expiration of an adult human: 0.5 L of air; (**D**) cardiac output of an adult human heart at rest: 5.25 L/minute; (**E**) water in Earth's oceans: 1.3×10^{21} L; (**F**) planet Earth: 1×10^{24} L.

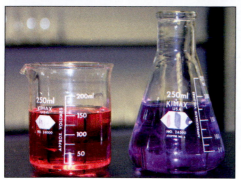

WARNING

Never use your mouth to extract liquid with a pipette.

FIGURE **2.6** Several different devices are used to measure volume.

is marked at 1 mL increments, and the total volume is labeled every 10 mL. Carefully pour the solution from the 100 mL beaker into a 100 mL graduated cylinder. To read a glass graduated cylinder, hold the cylinder at eye level and read the bottom of the curve; this is known as the **meniscus**. The meniscus results from the surface tension of the solution and adhesion of the solution to the sides of the glass. No meniscus is formed in plastic graduated cylinders (Fig. 2.7).

a. How many mL of solution did the 100 mL beaker contain? _____ mL

b. How close did you come to pouring 100 mL in the beaker or flask? _____

5 Carefully remove 25 mL of the solution from the graduated cylinder, and place it in the 250 mL beaker. To ensure accuracy, add or remove small amounts of solution with a dropper pipette.

6 Pipettes are used to distribute and extract liquid. Several sizes of pipettes are available, each measuring different volumes. The maximum and minimum volumes are denoted on the side of the pipette along with the measurement increments. The 0 mark serves as the initial point of

measure. One common type of pipette is a bulb pipette. It uses a bulb to draw liquid into the pipette. Pump or pi-pump pipettes have a thumbwheel and plunger used for extracting and distributing liquids. Dropper pipettes are used for extracting and distributing small amounts of liquid. Dropper pipettes provide only approximate values. State the range and increments of the 10 mL pipette.

Range _____

Increments _____

7 If it is not assembled already, place the bulb or pump on the pipette as directed by your laboratory instructor. Practice extracting various amounts of liquid from the 250 mL beaker. Several attempts with a bulb pipette may be needed to obtain the desired amount of liquid. In pump pipettes, the thumbwheel is used to obtain and distribute liquids. The plunger is used to flush the pipette (Fig. 2.8). Moving the thumbwheel clockwise extracts the liquid, and moving the thumbwheel counterclockwise distributes the liquid. Always remember to use the plunger to expel the residual liquid.

8 Using the 10 mL pipette, extract the following amounts of liquid from the graduated cylinder: 10 mL, 7 mL, 1 mL, 4.5 mL, 3.7 mL, and 2.2 mL. After you perform each extraction, discard the entire amount of the liquid back into the 250 mL beaker.

9 Extract 10 mL of liquid from the graduated cylinder. Release the following amounts of liquid back into the beaker: 2 mL, 3.5 mL, 2.6 mL.

a. How many total mL did you place into the beaker? _____

b. How many total mL are left in the pipette? _____

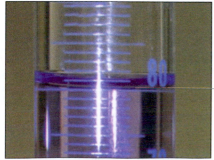

FIGURE **2.7** For accuracy when measuring the volume of a liquid in a glass graduated cylinder, the bottom of the meniscus is viewed.

— Meniscus

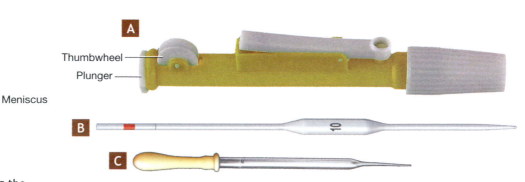

Thumbwheel
Plunger

FIGURE **2.8** Various types of pipettes: (**A**) pump pipette; (**B**) bulb pipette; (**C**) Pasteur pipette.

c. How do your values compare with the instructed amount? Account for possible discrepancies between the two values.

10 Dispose of the liquid as indicated by your laboratory instructor, and return the materials to the designated location. Thoroughly clean your glassware, and clean and disinfect your laboratory station.

Procedure 4

Temperature

Materials
- ❏ Celsius thermometer
- ❏ Tap water
- ❏ Heated water
- ❏ Ice

Temperature measures the average kinetic energy of molecules (Fig. 2.9). In the metric system, temperature is measured in degrees Celsius, named in honor of Swiss astronomer Anders Celsius (1701–1744). In his scale, the freezing point of water is 0°C, and the boiling point of water is 100°C. Room temperature is considered to be between 20 and 25°C, while the average human body temperature is 37°C.

Thermometers are used to measure temperature. A variety of thermometers and probes are available in science (Fig. 2.10).

1 Obtain a Celsius thermometer from your laboratory instructor. What is the range of your thermometer?

2 Conduct the following measurements:

a. Record the room temperature. _____

b. After holding the bulb of the thermometer tightly in your hand for 3 minutes, record the temperature.

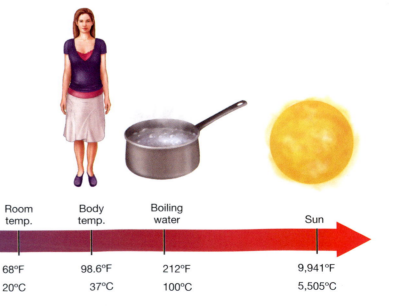

FIGURE **2.10** Thermometers: (**A**) Celsius; (**B**) digital.

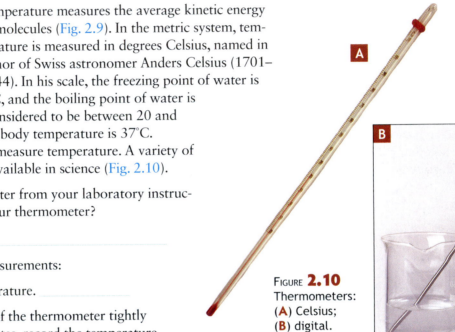

Dry ice	Convergence of °F and °C	Ice	Room temp.	Body temp.	Boiling water	Sun
−109°F	−40°F	32°F	68°F	98.6°F	212°F	9,941°F
−78.3°C	−40°C	0°C	20°C	37°C	100°C	5,505°C

FIGURE **2.9** Comparative temperatures from the world of biology.

c. What is the temperature of the tap water at your station? _____

d. What is the temperature of the hot water at your station? _____

e. Place 50 g of ice in 250 mL of water and record the following:

Initial reading _____ After 1 min. _____

After 2 min. _____ After 3 min. _____

f. What is the temperature inside the laboratory refrigerator? _____

g. What is the temperature inside the laboratory freezer? _____

h. Your choice: _____

i. Your choice: _____

j. Your choice: _____

3 Return the materials to the designated location. Thoroughly clean your glassware, and clean and disinfect your laboratory station.

✔ Check Your Understanding

3.1 What is the meniscus?

_____

_____

_____

_____

3.2 Why is taring a balance important?

_____

_____

3.3 Liquid nitrogen is nitrogen cold enough to exist in the liquid form. It has a boiling point of –320.4°F. Convert this value to degrees Celsius.

_____

_____

For Good Measure
Understanding Scientific Notation and the Metric System

MYTHBUSTING

Measuring Up

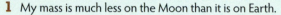

Debunk each of the following misconceptions by providing a scientific explanation. Write your answers on a separate sheet of paper.

1 My mass is much less on the Moon than it is on Earth.

2 I will never use metric values when I make Grandma's cookie; it is too confusing.

3 There are no metric units small enough to use in microscopy.

4 "Going metric" will require repackaging many products.

5 The United States, Liberia, and Burma are the only countries that don't use the metric system.

1 What is the advantage of using scientific notation?

_____

_____

_____

2 Why is a standard unit of measurement necessary in science and industry?

_____

_____

_____

_____

3 Match the following prefixes with the correct name:

_____ mega

_____ kilo

_____ nano

_____ deci

_____ micro

_____ giga

_____ centi

_____ pico

A. trillionth

B. millionth

C. million

D. billion

E. billionth

F. tenth

G. thousand

H. hundredth

4 Indicate the standard units for measuring length, mass, volume, and temperature in the metric system.

length = _____   volume = _____

mass = _____   temperature = _____

5 Conversion between metric units is performed by shifting _____ .

6 Convert the following to standard notation:

5.28×10^{-7} = _____ 3.43×10^{5} = _____

7.16×10^{8} = _____ 9.452×10^{-6} = _____

7 Convert the following to scientific notation:

0.0000061 = _____   29,420,000,000 = _____

0.000256 = _____   47,000,000 = _____

8 Perform the following calculations:

$(6.02 \times 10^{23}) \times (7.1 \times 10^{3})$ = _____ $6.4 \times 10^{6}/4.2 \times 10^{-3}$ = _____

$(8.46 \times 10^{-4}) \times (2.23 \times 10^{3})$ = _____ $7.2 \times 10^{8} + 3.3 \times 10^{7}$ = _____

$3.6 \times 10^{7}/2.2 \times 10^{4}$ = _____ $9.4 \times 10^{6} - 4.6 \times 10^{5}$ = _____

9 Perform the following conversions:

27.03 cm $\times$ _____ = _____ nm

3.3×10^{4} g $\times$ _____ = _____ kg

83 mph $\times$ _____ = _____ km/h

3.33×10^{-3} mL $\times$ _____ = _____ gal.

75 μm $\times$ _____ = _____ mm

4°C $\times$ _____ = _____ °F

12 mi. $\times$ _____ = _____ mm

2.9 g $\times$ _____ = _____ lb.

11.7 mg $\times$ _____ = _____ kg

3.62×10^{-3} L $\times$ _____ = _____ mL

36 gal. $\times$ _____ = _____ L

100 yd. $\times$ _____ = _____ m

77°F $\times$ _____ = _____ °C

15 nm $\times$ _____ = _____ m

−16°C $\times$ _____ = _____ °F

29 μg $\times$ _____ = _____ oz.

Back to Basics
Understanding Basic Chemistry

3

The four most common chemically active elements in the universe—hydrogen, oxygen, carbon, and nitrogen—are the four most common elements of life on Earth. We are not simply in the universe. The universe is in us.

— Neil deGrasse Tyson (1958–present)

Consider the following questions prior to coming to lab, and record your answers on a separate piece of paper.

1 Why is oxygen the most abundant element in Earth's atmosphere if hydrogen is the most abundant gas in the universe?
2 Why can hot pepper containing capsaicin be used to deter small mammals from a feeder but not birds?
3 Why is an ultraviolet light packaged in certain pet-urine ridding products?
4 Why are ripening apples placed at a certain distance from other fruits in the market?
5 Why is the surface of Mars red?

Objectives
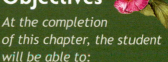

At the completion of this chapter, the student will be able to:

1. Discuss the role of chemistry in the biological sciences.
2. Differentiate among ionic, covalent, and hydrogen bonds.
3. Define and describe the properties of an acid and a base.
4. Interpret a pH scale.
5. Recall the pH of various substances, such as gastric juice, urine, tears, and blood.
6. Compare and contrast organic compounds and hydrocarbons.
7. Define and describe the characteristics of carbohydrates.
8. Discuss the composition and functions of monosaccharides, disaccharides, and polysaccharides.
9. Define and describe the characteristics of lipids and waxes.
10. Define and describe the characteristics of amino acids and proteins.

Chapter Photo
Universal indicator pH scale.

To develop a meaningful understanding of biology, a fundamental knowledge of chemistry is necessary. Chemistry is the study of the composition, structure, properties, and interaction of matter. Matter is anything that occupies space and has mass. Matter, whether it is this page, a meteorite, or a gallstone, is composed of elements. An **element** is a substance that cannot be broken down into other substances by ordinary chemical means. Although more than 118 elements have been described, only 92 elements occur naturally; the others are human made.

Of the naturally occurring elements, approximately 25 are common to living systems. A typical human is composed of 65% oxygen (O), 18.5% carbon (C), 9.5% hydrogen (H), and 3.2% nitrogen (N). Other important elements in humans and other living systems include phosphorus (P), sulfur (S), calcium (Ca), potassium (K), sodium (Na), chlorine (Cl), magnesium (Mg), iron (Fe), and silicon (Si).

Elements are composed of **atoms**. An atom is the smallest part of an element that retains the properties of that element. Atoms are composed of a variety of subatomic particles, the best known of which are protons, neutrons, and electrons. Positively charged protons and uncharged neutrons can be found in the nucleus, and negatively charged electrons exist in orbitals surrounding the nucleus.

The **atomic number** of an element reflects the number of protons in the nucleus. For example, copper (Cu) has an atomic number of 29. The atomic number appears above the symbol for the element in the periodic table (see inside front cover). The **atomic mass** of an element indicates the number of protons plus the number of neutrons in the nucleus. The atomic mass of copper is 63.546. The atomic mass appears below the symbol for the element in the periodic table. The atomic mass reflects that the number of neutrons can vary in the nucleus. Atoms that have the same number of protons and a varied number of neutrons are called **isotopes**. For example, the element tin (Sn) has 10 isotopes, and uranium has 29. Today radioactive isotopes are commonly used in medical imaging (iodine 131), tracing (barium 140), and dating some fossils and artifacts (carbon 14).

A **molecule** results from the chemical union of two or more atoms. Some molecules are simple, such as a molecule of oxygen (O_2) and a molecule of water (H_2O). Others are quite large, such as a molecule of chlorophyll ($C_{55}H_{72}O_5N_4Mg$). **Compounds** are molecules composed of different elements. Not all molecules are compounds; some molecules such as ozone (O_3) and hydrogen gas (H_2) consist of just one type of atom. Water and chlorophyll are compounds. Other examples of compounds are glucose ($C_6H_{12}O_6$), hydrogen peroxide (H_2O_2), baking soda ($NaHCO_3$), and ethyl alcohol (C_2H_5OH).

Atoms combine with other atoms by means of chemical bonds. Ionic compounds such as salt (NaCl) are held together by an attraction between positively and negatively charged ions. Loss or gain of electrons forms ions; the attraction of oppositely charged ions forms an **ionic bond**. When placed in water, ionic compounds dissociate (dissolve).

Covalent bonds result from sharing of electrons. Covalent bonds are strong and are common in living systems.

- Nonpolar covalent bonds involve an *equal sharing* of electrons. Nonpolar covalent bonds between carbon and hydrogen form a stable framework for building larger molecules such as octane (C_8H_{10}).
- Polar covalent bonds involve an *unequal sharing* of electrons (for example, in water molecules).

Nonpolar covalent substances and polar covalent substances do not mix. That is why oil, a nonpolar covalent substance, and water, a polar covalent substance, do not mix. This phenomenon is illustrated in a bottle of Italian dressing because vinegar is mostly water.

The most important polar molecule to life on Earth is water (H_2O). **Hydrogen bonds** are weak bonds between the positively charged region of a hydrogen atom of a polar covalent molecule and the negatively charged region of oxygen or nitrogen of another polar covalent molecule. Hydrogen bonds give shape and three-dimensional structure to complex molecules, such as proteins and DNA. Hydrogen bonds also give water several special properties.

■ Acid-Base Chemistry

A **solution** is a liquid composed of a uniform mixture of two or more substances. In a solution, the dissolving medium is the **solvent** and the dissolved substance is the **solute**. One of the unique properties of water is that it is an excellent solvent. An aqueous solution uses water as the solvent. Some inorganic molecules are held together by ionic bonds. In an aqueous solution, these substances undergo dissociation and produce positive and negative ions.

For example, sodium chloride (NaCl) dissociates into sodium ions (Na^+) and chloride ions (Cl^-). Substances that release ions in an aqueous solution are called electrolytes. Other biologically important electrolytes include potassium chloride (KCl), calcium chloride ($CaCl_2$), and sodium bicarbonate ($NaHCO_3$).

Although many people think acids and bases are chemicals best left in the laboratory, many common substances are classified as acids or bases. These substances can be important and, in some instances, potentially dangerous to living systems.

Acids

Both inorganic and organic acids are common in nature. An **acid** is a substance that yields (donates) a hydrogen ion in solution. Acids share a number of structural characteristics and properties.

- Acids contribute one or more hydrogen atoms to a solution when they dissociate in water.
- Acids have a sour taste.
- Acids may be corrosive or poisonous.
- Acids react with certain metals to liberate hydrogen gas.
- Acids neutralize bases.
- Acids affect the color of certain indicators.

WARNING
Never taste unknown chemicals!

Some common inorganic acids are:
- Sulfuric acid (H_2SO_4);
- Nitric acid (HNO_3);
- Phosphoric acid (H_3PO_4); and
- Hydrochloric acid (HCl).

Some common organic acids are:
- Citric acid (lemon juice);
- Acetic acid (vinegar);
- Carbonic acid (carbonated water);
- Malic acid (apple juice);
- Formic acid (bee stings); and
- Lactic acid (sour milk).

Bases

Bases are commonly known as alkalines. These chemicals share several structural characteristics and properties.

- Bases decrease the hydrogen ion concentration of their aqueous solution or release hydroxide ions (OH^-) in solution.

◗ Bases have a bitter taste.
◗ Bases feel slippery.
◗ Bases may be corrosive or poisonous.
◗ Bases neutralize acids.
◗ Bases affect the color of certain indicators.

Some common bases are:
◗ Sodium hydroxide (NaOH) (lye);
◗ Calcium hydroxide ($Ca(OH)_2$) (slaked lime); and
◗ Magnesium hydroxide ($Mg(OH)_2$) (antacid).

Examples of industrial bases are:
◗ Potassium hydroxide (KOH) (caustic potash);

◗ Barium hydroxide ($Ba(OH)_2$) (baryta); and
◗ Strontium hydroxide ($Sr(OH)_2$) (stabilizer in plastics).

Chemists use a variety of indicators to determine whether a substance is an acid or a base. Acids turn blue litmus indicators red, are colorless in phenolphthalein, and turn methyl orange indicator red. Bases turn red litmus indicators blue, turn phenolphthalein pink, and turn methyl orange indicator yellow.

pH Scale

The potential of hydrogen (pH) scale is the most commonly used method to measure the acidic or basic nature of a substance. The pH of a solution is the negative logarithm of the hydrogen ion (H^+) concentration, in moles per liter ($pH = log[H^+]$). Because the scale is based on logarithms, a substance with a pH of 2, such as lemon juice, has 10 times more H^+ ions than a substance such as grapefruit juice, with a pH of 3, and 10,000 times more H^+ ions than human urine, with a pH of 6. In a solution, as the H^+ increases the OH^- decreases, and as the H^+ decreases the OH^- increases.

The pH scale has a range from 0–14. A solution with a pH less than 7 is considered acidic, a substance with a pH of 7 is considered neutral, and a substance with a pH greater than 7 is considered basic. Ethyl alcohol and some substances do not dissociate in water and do not have a pH. The pH value of these substances reflects the water from which they were made or other ingredients in the product (Fig. 3.1).

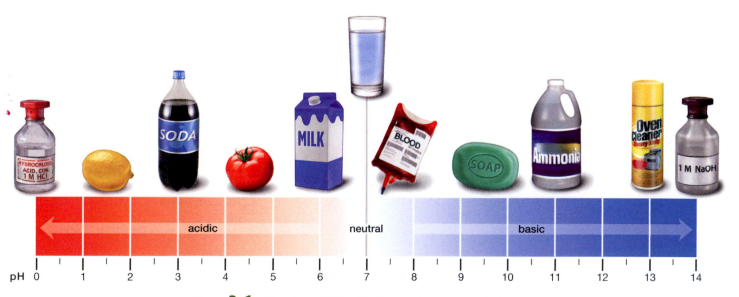

FIGURE **3.1** pH scale and the pH of several common substances.

Measuring pH

Carbon dioxide (CO_2) is a gaseous by-product released during expiration. When combined with water, it forms carbonic acid (H_2CO_3):

$$CO_2 + H_2O \longrightarrow H_2CO_3$$

Procedure 1

Using Phenolphthalein

Materials

- ❏ Distilled water
- ❏ 500 mL beaker or Erlenmeyer flask
- ❏ Phenolphthalein
- ❏ Graduated cylinder
- ❏ Scale
- ❏ Baking soda
- ❏ Clean straw
- ❏ Stirring rod
- ❏ Dropper

The formation of carbonic acid can be observed by blowing CO_2 through a straw into a basic solution containing the indicator phenolphthalein (Fig. 3.2). The indicator turns bright pink in a basic solution, and in an acidic solution the phenolphthalein is colorless.

1 Pour 250 mL of distilled water into a 500 mL beaker or Erlenmeyer flask. Add 0.5 g of baking soda ($NaHCO_3$) to the water, and stir it with a stirring rod.

2 Add 20 drops of phenolphthalein to the solution, and stir. What color is the solution?

3 Place a clean, unused straw into the solution, and begin blowing steadily into the straw. Observe the color changes in the solution. This may take a few minutes. Describe the color changes in the solution.

> When blowing into the straw, be sure to cover the end of the straw with your hand and do not ingest the solution.

4 Discuss the results of the demonstration and what happened as CO_2 from your breath was blown into the basic solution.

FIGURE **3.2** Carbon dioxide exhaled forms carbonic acid in the water. In turn, the carbonic acid neutralizes the basic solution.

Did you know . . . ❓

Why Are Fresh Eggs Harder to Peel?

Fresh eggs have a higher carbon dioxide content than older eggs. The carbon dioxide dissolves in the eggs' natural moisture and forms carbonic acid (H_2CO_3). As the egg ages, the carbon dioxide diffuses out of the egg, and the egg becomes less acidic. This condition weakens the inner membrane, preventing it from sticking to the white and making it easier to peel. *Be careful. You do not want an egg to be too old!*

Procedure 2

Using pH Paper, pH Meters, and Purple Cabbage Extract

Materials

- ❏ pH paper
- ❏ pH meter
- ❏ Purple cabbage extract
- ❏ Substances of a known pH (7), provided by instructor
- ❏ Variety of common harmless substances. (Liquids such as milk and clear soda can be easily pipetted. Solid substances need to be dissolved in water. It may be necessary to crush some substances such as aspirin to dissolve in water.)
- ❏ Transfer pipettes
- ❏ Test tubes
- ❏ Test-tube rack
- ❏ Reaction wells
- ❏ Toothpicks
- ❏ Wax pencil or Sharpie
- ❏ Colored pencils
- ❏ Beakers
- ❏ Paper towels

The pH of a substance can be measured using a variety of methods. Using pH paper (Fig. 3.3) is an inexpensive and easy method to measure the pH of a substance. The pH paper has been treated with an indicator sensitive to certain pH values. A color chart provided with the pH paper relates the color of the pH paper to a specific pH. Additionally, many labs have a pH meter (Fig. 3.4) that measures the pH.

One of the most interesting properties of acids and bases is their ability to change the color of some plant materials. Purple cabbage, hydrangea, and elderberry extracts respond in an amazing manner to an acidic or basic solution. In this activity, you will also use purple (red) cabbage to develop an acid or a base scale and determine the approximate pH value of various substances.

1. Use pH paper to determine the pH of 2 to 4 mL of each substance for investigation. Record the pH in the first column of Table 3.1.

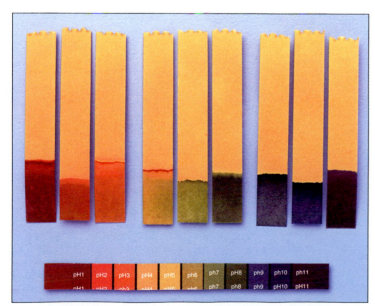

FIGURE **3.3** pH paper and indicator strip.

2. If a pH meter is available, follow your instructor's directions for its use in this activity. Record the pH in the second column in Table 3.1.

3. To develop the acid or base scale using purple cabbage, divide into working groups of at least two students per group.

4. Procure a beaker of prepared purple cabbage extract, eight test tubes, a reaction well plate, pipettes, toothpicks, substances to test, and several paper towels.

FIGURE **3.4** pH meter.

5. To establish a standard for comparison, place seven test tubes in a rack, and label them using a wax pencil or Sharpie: pH 2, pH 4, pH 6, pH 7, pH 8, pH 10, and pH 12.

6. Pipette 5 drops of each substance provided by the instructor (aspirin, baking soda, milk, clear soda, etc.) into the test tubes and label them according to each substance used.

7. Pipette 8 drops of purple cabbage juice extract into each test tube, and gently swirl the test tube, mixing the substances.

8. Record the color of each test tube in the final column of Table 3.2. Let the substance sit in the test tube labeled pH 12 for a few minutes. The color will shift because of the instability of the pigments at a high pH. Record the initial and final colors at pH 12 in Table 3.2. Once completed, Table 3.2 will serve as the standard for comparison.

9. Place 5 drops of each of the 11 substances or the dry substance to be tested into an individual well on the reaction well plate, and record its position. If no well plates are available, test tubes can be used. Label the well or test tube with a wax pencil or Sharpie.

10. Place 5 drops of the purple cabbage extract into each reaction well, and stir it with a toothpick.

TABLE **3.1** pH of Common Substances

Substance	pH Paper Value	pH Meter Value	Cabbage Extract Color	pH Corresponding to Cabbage Extract Color	Actual pH
1					
2					
3					
4					
5					
6					
7					
8					
9					
10					
11					

TABLE **3.2** Color of Standard Solutions in Purple Cabbage Juice Extract

pH	Substance	Color
2		
4		
6		
7		
8		
10		
12		

11 Record the color changes in Table 3.1. Using Table 3.2 as a reference, determine the approximate pH of the substance.

12 Look up the pH values for the substances tested in the experiment. How did the cabbage juice pH values compare with the actual values?

_____

_____

13 Procure an unknown substance from the instructor, and follow the instructions in steps 9–11. Place the unknown in the empty 12th well.

14 Determine the approximate pH of your unknown from the scale you have developed.

  a. What was the color of the unknown?

  _____

  b. What is the estimated pH of the unknown?

  _____

  c. If permitted, use pH paper or a pH meter to determine the pH of your unknown.

  _____

  d. How did the two values compare?

  _____

  _____

  _____

✔ Check Your Understanding

1.1 Carbon dioxide is bubbled through a pink basic solution and a color change occurs. What color is seen and what compound is formed?

_____

_____

1.2 Which of the two methods to measure the pH of a sample is more accurate: pH paper or the pH meter?

_____

_____

1.3 The commonly used indicator phenolphthalein will turn _____ in a basic solution and in an acidic solution is

_____

Neutralizing Acids

The pH of the gastric juices in the stomach is much more acidic than the esophageal environment. Classically, "heartburn" results from an irritation of the lower esophagus by stomach acid, causing discomfort, chest pain, and difficulty swallowing. A number of food products can initiate heartburn, such as citrus fruits, acidic vegetables, spices, caffeinated beverages, and fatty foods. Commonly, over-the-counter antacids are used to neutralize "acid stomach." These products should absorb excess hydrogen ions from stomach acid and relieve the sufferer.

Procedure 1

Using Antacids

Materials

- ❏ Distilled water
- ❏ Mortar and pestle
- ❏ 250 mL beaker
- ❏ Pipette
- ❏ Test tubes
- ❏ 0.1 M hydrochloric acid solution (provided by the instructor)
- ❏ Over-the-counter antacids (instructor's choice)
- ❏ Phenolphthalein indicator

A number of antacid makers claim their product is the best. The following exercise is designed to test which antacid neutralizes acid best. Keep in mind this activity does not promote any product over another.

1 Procure the equipment and supplies for the lab table.

2 Record the name of the antacid in Table 3.3.

3 If you have an antacid in the solid form, use a mortar and pestle to pulverize it into a fine powder.

4 Place the pulverized antacid into a beaker, add 100 mL of distilled water, and stir vigorously.

5 After the powder is dissolved in the solution, put 10 mL of the solution into a test tube. Gently swirl the solution.

6 Add 5 mL of the indicator phenolphthalein using a pipette, and gently swirl the test tube. Note the color of the solution.

7 Carefully add the prepared 0.1 M hydrochloric acid (HCl) solution 1 drop at a time using a pipette. Swirl the solution gently. Continue adding the HCl into the test tube 1 drop at a time and counting each drop until the solution turns clear. When the solution turns clear, it has become neutralized. Record the number of drops used in Table 3.3.

8 Repeat steps 2 through 7 for the remaining antacids, and then answer the following questions.

a. How many drops of 0.1 M HCl were used to neutralize the antacids recorded in Table 3.3?

b. Which antacid is potentially the strongest (i.e., which neutralized more acid)?

TABLE **3.3** Drops of 0.1 M HCl Required to Neutralize an Antacid

Antacid	Drops of 0.1 M HCl

✓ Check Your Understanding

2.1 What causes heartburn and how do over-the-counter antacids work to help people suffering with heartburn?

Organic Molecules

Organisms are composed primarily of organic compounds. Originally the term *organic* referred to the belief that such compounds could be made only by living things. Today, organic chemists study the compounds of carbon.

Carbon is a unique element. Carbon is tetravalent, meaning that it has the ability to interact with other atoms on four independent sites. This means that carbon can combine with itself and other elements to form an almost limitless number of compounds. More than 2 million organic compounds have been catalogued.

The most fundamental group of organic molecules is the hydrocarbons, constructed of hydrogen and carbon atoms. Hydrocarbons serve as the scaffold upon which more complex molecules are made. By removing hydrogen and adding functional groups and other elements, the properties of the base molecule is changed. Organic molecules commonly found in living things include carbohydrates, lipids, and proteins.

Carbohydrates are organic compounds that contain carbon, hydrogen, and oxygen in a 1:2:1 ratio. The number of carbons in carbohydrates varies from three to more than one thousand. This category includes sugars, starch, and cellulose.

Lipids are diverse organic compounds that include fats, waxes, phospholipids, and steroids. These compounds are insoluble in water and soluble in nonpolar compounds, such as ether. Lipids consist mostly of carbon and hydrogen atoms with few oxygen atoms. Lipids also may contain small amounts of other atoms, such as phosphorus and sulfur.

Proteins are the most numerous and complex molecules in living organisms. Proteins provide support, movement, storage, defense, and regulation. In addition, hormones and enzymes are proteins. Common proteins include keratin (in fingernails and hair; Fig. 3.5), hemoglobin (in blood), ovalbumin (in egg whites), actin and myosin (in muscle), insulin and glucagon (in the pancreas), lysozyme (in tears), and collagen (in connective tissues).

FIGURE **3.5** Common examples of keratin: (**A**) feathers; (**B**) spider web; (**C**) human hair.

Testing for Carbohydrates

Monosaccharides

The simplest forms of carbohydrates (Fig. 3.6) are known as single sugars, simple sugars, or **monosaccharides** (mono = one). These molecules are composed of three to seven carbon atoms and their appropriate hydrogen and oxygen atoms. Primarily, monosaccharides serve as fuels for organisms and building blocks for larger molecules.

Examples of simple carbohydrates are ribose ($C_5H_{10}O_5$), deoxyribose ($C_5H_{10}O_4$), and glucose ($C_6H_{12}O_6$). Ribose and deoxyribose are five-carbon sugars called pentoses (pent = five), important components of the nucleic acids RNA and DNA. Glucose is a six-carbon sugar known as a hexose (hex = six). Other common prefixes are given in Table 3.4. Glucose serves as an important source of cellular fuel. Fructose and galactose are also hexoses, but they have a different molecular architecture than glucose and are known as **isomers** of glucose. Fructose is derived from honey and sugar cane. Galactose is less sweet than glucose and serves as a building block in dairy products and mucilage (a sticky substance produced by some plants and microorganisms).

Disaccharides

Two monosaccharides combine in a process known as dehydration synthesis, or condensation, to form a sugar composed of two monomeric units, or disaccharide (di = two). Dehydration synthesis involves the removal of an OH^- from one sugar and an H^+ from another sugar. As a molecule of water (H_2O) is removed, the simple sugars link by means of a covalent bond, forming a disaccharide.

- Maltose is a common disaccharide composed of two glucose molecules. It is used in making beer, malts, and malted milk balls. When the enzyme salivary amylase breaks down a starch in your saliva, maltose is produced.

- Sucrose, or table sugar, is a common disaccharide composed of glucose and fructose. Sucrose is found in plants. Table sugar is sucrose harvested from the stems of sugarcane or the roots of sugar beets. Molasses and brown sugar are made from sucrose. In the digestive system sucrose is broken down into glucose and fructose by the enzyme sucrase.

- Lactose, or milk sugar, is formed from a dehydration synthesis reaction between glucose and galactose. Infant mammals use the lactose in mother's milk. Some individuals suffer from lactose intolerance. This is the inability to process lactose, which can lead to potentially severe gastrointestinal problems.

Disaccharides can be broken down into their simpler sugars through a process called **hydrolysis**, which is a chemical reaction in which water is used to break down a compound. It does this by breaking a covalent bond and inserting a molecule of water across the bond.

Polysaccharides

Polysaccharides are complex carbohydrates built from simple carbohydrates and linked through dehydration synthesis. Polysaccharides form large molecules that can be linear or highly branched. Some polysaccharides serve as energy storage molecules. For example, starch is a storage polysaccharide that consists of moderately branched glucose molecules in plants. Potatoes, corn, and rice are sources of starch in the human diet. Glycogen, or animal starch, is a highly branched, glucose-rich polysaccharide stored in the liver and skeletal muscle of animals.

Polysaccharides also can serve as structural molecules. The structural polysaccharide cellulose, in the cell wall of plants, is the most abundant carbohydrate on Earth. Cotton

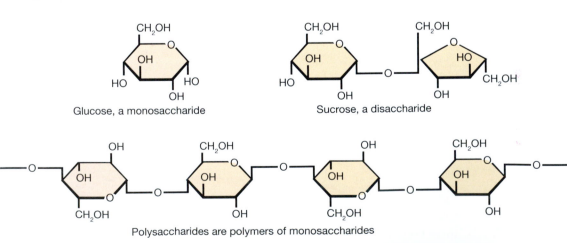

Glucose, a monosaccharide

Sucrose, a disaccharide

Polysaccharides are polymers of monosaccharides

FIGURE **3.6** Carbohydrates are classified as monosaccharides, disaccharides, and polysaccharides.

TABLE **3.4** Commonly Used Prefixes

Prefix	Number
Mono-	1
Di-	2
Tri-	3
Tetra-	4
Pent-	5
Hex-	6
Oligo-	a few or several
Poly-	many

consists of more than 90% cellulose, and some woods are about 50% cellulose. Herbivores, such as cows and rabbits, have modified digestive systems that allow them to digest cellulose. In the human diet, cellulose serves as a fiber, which is important in the proper functioning of the digestive tract.

Chitin, a modified polysaccharide, is the main component in the cell walls of some fungi and the exoskeleton of insects and other arthropods. Some chitin-based exoskeletons, such as those of crabs and crayfish (Fig. 3.7), are reinforced and hardened by calcium carbonate.

FIGURE 3.7 (A) Structural polysaccharides, such as cellulose, constitute the cell wall of plants such as aloe. (B) Chitin is found in the exoskeleton of arthropods like the crayfish.

Procedure 1

Iodine Test

Materials

- ❏ I₂KI solution
- ❏ Test-tube rack
- ❏ Test tubes (11–15)
- ❏ Wax pencil or Sharpie
- ❏ Graduated cylinder
- ❏ Dropper or transfer pipette
- ❏ Test solutions:
  - ▪ distilled water
  - ▪ clear diet soda
  - ▪ clear nondiet soda
  - ▪ onion juice
  - ▪ potato juice
  - ▪ milk
  - ▪ glucose solution
  - ▪ sucrose solution
  - ▪ paper
  - ▪ cotton
  - ▪ cornstarch

The **iodine test** using iodine–potassium iodide (I_2KI) has been developed to distinguish starch from other carbohydrates. Because starch is a coiled glucose polymer, the I_2KI solution interacts with the starch, producing a bluish-black color. The I_2KI does not react with noncoiled carbohydrates and remains a yellowish-brown color. Some carbohydrates, such as dextrin and glycogen, will produce an intermediate color reaction.

1 Procure 11 clean test tubes and a wax pencil or Sharpie. Label the test tubes 1 through 11.

2 Using the order of the test materials in Table 3.5, fill Test Tubes 1 through 8 with 7.5 mL of the solution to be tested. Place a small amount of paper and cotton in test tubes 9 and 10, respectively,

and cornstarch in Test Tube 11. Your instructor may assign additional substances.

3 Carefully add 5 drops of I_2KI to each test tube, and swirl it. Record the color changes in Table 3.5, as well as any additional observations about the reaction.

4 Discard the solutions according to the instructor's directions, and return the test tubes to the collection area.

5 Which substances contained starch?

_____

_____

6 Compare and contrast the reaction of potato juice and onion juice.

_____

_____

_____

_____

TABLE **3.5** Iodine Test for Starch

Solution	Color	Observations
1. Distilled water		
2. Diet soda		
3. Nondiet soda		
4. Onion juice		
5. Potato juice		
6. Milk		
7. Glucose solution		
8. Sucrose solution		
9. Paper		
10. Cotton		
11. Cornstarch		
12.		
13.		
14.		
15.		

✔ Check Your Understanding

3.1 What color did cornstarch and potato juice turn in the iodine test and why?

3.2 Clear diet soda and distilled water do not turn the classic bluish-black color in the presence of I_2KI when the iodine test is performed. Why?

Did you know . . .

Is It Counterfeit?

One safeguard taken when producing paper currency is to remove all traces of starch from the paper. Fortunately, counterfeiters have not mastered this technique. A common starch known as amylose turns blue in the presence of iodine. If iodine is applied to the money with a counterfeit detector pen, the money will turn blue if it is counterfeit because of the presence of starch.

Testing for Lipids

Triglycerides, more commonly referred to as fats, are the most abundant lipids in living organisms. In the body, fat serves as an energy reserve. One gram of fat stores significantly more energy than one gram of starch or one gram of glycogen. In addition, fats are important in insulation and cushioning.

A triglyceride is composed of a glycerol molecule attached to three fatty acid molecules. Fatty acid chains may vary in their length and in the way their carbon atoms join.

Saturated Fats

In **saturated fatty acids**, all of the carbon atoms are linked by single bonds (Fig. 3.8). Saturated fatty acids contain the maximum number of hydrogen atoms.

Saturated fats consist of saturated fatty acids with the maximum number of hydrogen atoms. Saturated fats include animal fats and vegetable shortenings. These substances, such as butter and lard, are solid at room temperature. Some plant products, such as coconut oil, palm kernel oil, and chocolate, have saturated fats. The American Heart Association recommends limiting the amount of saturated fats consumed as part of a diet considered heart healthy.

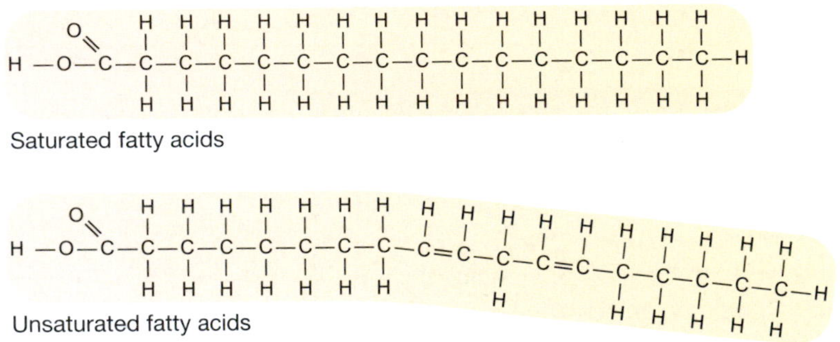

Unsaturated Fats

Unsaturated fatty acids do not have the maximum number of hydrogen atoms. They have one or more double bonds between the carbon atoms (Fig. 3.8). If an unsaturated fatty acid has one double bond, it is known as monounsaturated, as in sesame oil, corn oil, and grapeseed oil. If more than one double bond is present, the fatty acid is known as polyunsaturated, as in tung oil, linseed oil, and poppy seed oil.

Unsaturated fats consist of unsaturated fatty acids that contain at least one double bond in the fatty acid chain. Examples of food products containing unsaturated fats include many plant oils, such as olive, sunflower, canola, avocado, soybean, and nut. Fish oils are also unsaturated. Unsaturated fats tend to be liquid at room temperature.

Waxes

Waxes consist of an alcohol bonded with a long-chain fatty acid. Waxes are solid at room temperature and repel water. The cuticle found on the surface of leaves and fruits consists of waxes and another lipid, cutin. The cuticle helps conserve water and repel insects. Birds preen using waxes, fatty acids, and fats to waterproof their feathers. Cerumin, a waxy secretion better known as earwax, traps foreign substances that potentially may enter the middle ear.

Saturated fatty acids

Unsaturated fatty acids

FIGURE **3.8** Saturated and unsaturated fats.

Procedure 1

Grease-Spot Test

Materials

- ❏ Brown paper bag or brown wrapping paper
- ❏ Scissors
- ❏ Ruler
- ❏ Dropper
- ❏ Timing device
- ❏ Test substances:
  - • water
  - • vegetable oil
  - • egg white
  - • soda
  - • potato chip
  - • honey
  - • salad dressing
  - • butter
  - • rubbing alcohol
  - • white candle wax

You probably have noticed that lipids are greasy, especially after eating a bag of potato chips. The **grease-spot test** is a simple test used to identify the lipid nature of substances.

1 Cut a brown paper bag or brown wrapping paper into 2.5 cm squares. On the back of each square, write the substance to be tested.

2 Following the order in Table 3.6, either gently rub a solid substance or place a drop of liquid on the brown paper. Your instructor may assign additional substances.

3 Record your results and any additional observations in Table 3.6.

4 Let the substance and paper stand for approximately 15 minutes, and then hold the brown paper up to a light source. The presence of a translucent spot indicates a lipid. If you rub some nose grease (oil from the bridge of your nose) on the paper, what will happen?

5 Clean the lab station, and discard the paper.

TABLE **3.6** Grease-Spot Test for Lipids

Substance	Lipids Present? (Y/N)	Observations
1. Water		
2. Vegetable oil		
3. Egg white		
4. Soda		
5. Potato chip		
6. Honey		
7. Salad dressing		
8. Butter		
9. Rubbing alcohol		
10. White candle wax		
11.		
12.		
13.		
14.		
15.		

✔ Check Your Understanding

4.1 Compare the results of the grease-spot test for potato chips, white candle wax, vegetable oil, and rubbing alcohol.

3

Testing for Proteins

Amino Acids

Proteins are composed of building blocks known as amino acids. Amino acids are compounds that have an amino group ($-NH_2$) and a carboxyl group ($-COOH$). A third group, known as the R group, varies tremendously and influences the characteristic of the molecule. Twenty amino acids serve as the building blocks of proteins (Fig. 3.9).

Organisms can synthesize the majority of amino acids. Amino acids that animals cannot synthesize and that must be obtained from diet are called essential amino acids. In

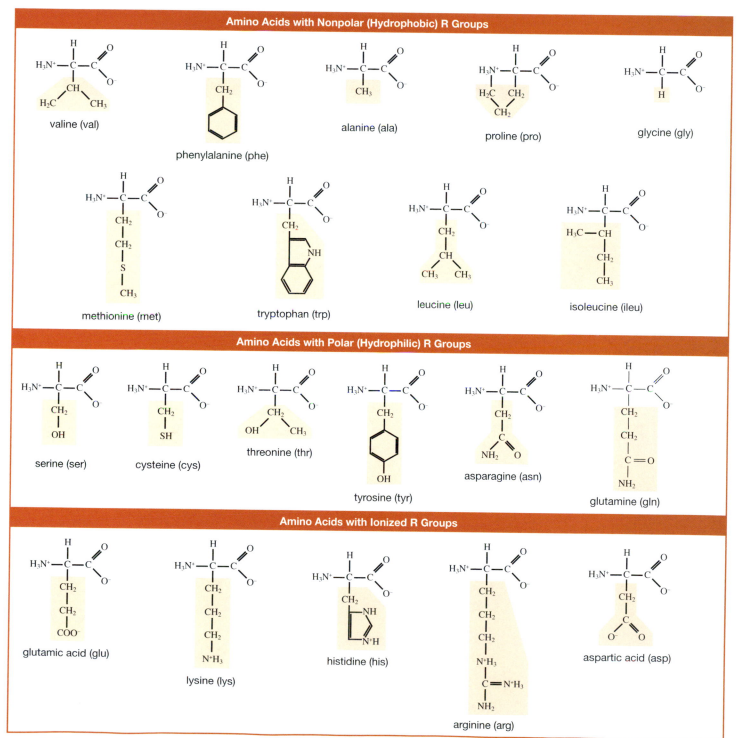

FIGURE **3.9** The 20 amino acids; the R groups are shaded.

humans, the essential amino acids are: histidine, leucine, lysine, methionine, phenylalanine, isoleucine, threonine, tryptophan, valine, and arginine (in children).

3 Polypeptides

Amino acids are linked by peptide bonds. A peptide bond is a covalent bond that forms between the amino group of one amino acid and the carboxyl group of another amino acid during a dehydration synthesis reaction. Additional amino acids can be joined to the chain, forming polypeptides. A polypeptide is a structural unit composed of more than 100 amino acids linked by peptide bonds.

A protein is a functional unit composed of one or more polypeptides with a precise three-dimensional shape that performs a specific function. The conformation (shape) of a protein refers to the final folded arrangement of the polypeptide chain of a protein. Proteins can vary in size and three-dimensional structure. Four levels of protein structural complexity are found in nature: primary structure, secondary structure, tertiary structure, and quaternary structure.

Procedure 1

Biuret Test

Materials

- ❏ Biuret reagent
- ❏ Test tubes (9)
- ❏ Test-tube rack
- ❏ Wax pencil or Sharpie
- ❏ Dropper or transfer pipette
- ❏ Timing device
- ❏ Test substances:
  - ▪ distilled water
  - ▪ sucrose solution
  - ▪ whole milk
  - ▪ bread and water solution
  - ▪ chicken broth
  - ▪ vegetable oil
  - ▪ egg white
  - ▪ egg yolk
  - ▪ plain Greek yogurt

The **Biuret test** is commonly used to detect the presence of a protein. The Biuret reagent is a blue-green colored solution containing 1% copper sulfate ($CuSO_4$) and sodium hydroxide (NaOH) or potassium hydroxide (KOH). The reagent changes color from blue-green to violet in the presence of proteins. The change in color results from the interaction between the copper ions and the peptide bonds of the protein. The more peptide bonds, the darker is the resulting color.

WARNING

Biuret reagent is extremely corrosive. Wear gloves and eye protection, and handle this chemical with great care. If you get some on your skin, wash the area with mild soap and water, and notify your laboratory instructor. Carefully follow your instructor's directions regarding the use and disposal of Biuret reagent.

1 Procure 9 test tubes, a wax pencil or Sharpie, a dropper, and a test tube rack. Label the test tubes 1 through 9.

2 Using the order of the test materials in Table 3.7, fill Test Tubes 1 through 9 with 15 mL of the substance to be tested. Your instructor may assign additional substances. Place the test tubes in the rack.

3 Add 3 drops of Biuret reagent to each test tube. Gently tap each test tube with your finger to mix the solution. Allow 2 minutes for the colors to develop, and record the observed color changes in Table 3.7.

4 Which substances were proteinaceous?

_____

_____

5 Discard the solutions according to the instructor's directions, and return the test tubes to the collection area.

TABLE **3.7** Biuret Test for Proteins

Substance	Color	Observations
1. Distilled water		
2. Sucrose solution		
3. Whole milk		
4. Bread and water solution		
5. Chicken broth		
6. Vegetable oil		
7. Egg white		
8. Egg yolk		
9. Greek yogurt		
10.		
11.		
12.		
13.		
14.		

✔ Check Your Understanding

5.1 What is a peptide bond?

5.2 Describe and explain the colors resulting from the Biuret tests for egg yolk, vegetable oil, and Greek yogurt.

5.3 The Biuret test is used to detect the presence of a _____ . When more peptide bonds are present, the resulting color will be _____ (darker/lighter).

MYTHBUSTING

It's Not Magic

Debunk each of the following misconceptions by providing a scientific explanation. Write your answers on a separate sheet of paper.

1 All acids are dangerous and will cause damage to living things.

2 In all covalent bonds, an equal sharing of the electron pair occurs.

3 Gases exist between the nucleus and the orbiting electrons.

4 Fats are the only kind of lipid.

5 Glucose is the only metabolic fuel.

6 Carbon dating is the only type of absolute dating based upon isomers.

7 Bee stings (acidic) can be neutralized by an alkaline substance such as toothpaste and a wasp sting (basic) can be neutralized by an acidic substance such as lemon juice.

1 Compare and contrast an acid and a base, and provide examples of each.

2 The pH scale has a range from 0 to _____ . A substance with a pH less than 7 is considered a/an _____ and a substance with a pH greater than 7 is a/an _____ .

3 Which of the following reactions are seen with an acid? (*Circle the correct answer.*)

a. An acid turns blue litmus paper red, is colorless in phenolphthalein, and turns methyl orange indicator yellow.

b. An acid turns blue litmus paper red, is colorless in phenolphthalein, and turns methyl orange indicator red.

c. An acid turns red litmus paper blue, turns phenolphthalein pink, and turns methyl orange indicator yellow.

d. An acid turns red litmus paper blue, is colorless in phenolphthalein, and turns methyl orange indicator yellow.

4 Recall from Chapter 1 what constitutes a well-designed experiment, and suggest why water or distilled water was tested in each procedure.

5 What is an ionic bond? Give an example.

6 What is a covalent bond? Give an example.

_____

7 Label each characteristic or example below as describing either acids (A) or bases (B):

_____ These have a bitter taste.

_____ These contribute one or more hydrogen atoms to a solution when it dissociates in water.

_____ These have a sour taste.

_____ These have a pH greater than 7.

_____ These react with certain metals, which results in liberation of hydrogen gas.

_____ These have a slippery feel.

_____ The chemical compound delivered by a bee sting is an example.

_____ These release hydroxide ions in solution.

_____ Carbonated water is an example.

_____ Sodium hydroxide is an example.

8 An old saying is "oil and water don't mix." This is the same situation for oil and vinegar. Explain why.

_____

_____

9 What is a saturated fat? Give two examples of these fats.

_____

_____

10 What is an unsaturated fat? Give two examples.

_____

11 Define monosaccharide, disaccharide, and polysaccharide and give an example of each.

_____

_____

_____

_____

12 What is a peptide bond? What units are linked by peptide bonds?

_____

13 What is an essential amino acid? Name two found in humans.

_____

The Invisible World
Understanding Microscopy and Cell Structure and Function

The motion of most of these little animalcules in the water was so swift and varied—upward, downward, and round about—that 'twas wonderful to see! I judge that some of these little creatures had a volume less than one thousandth that of the smallest ones that I have ever seen upon the rind of cheese, in wheaten flour, and the like.

— Antony van Leeuwenhoek (1632–1723)

Just wondering . . .

Consider the following questions prior to coming to lab, and record your answers on a separate piece of paper.

1 Why are there no giant cells like the blob that devours cities?
2 Why is Leeuwenhoek known as the "Father of the Microscope" even though he did not invent the microscope?
3 How do bacterial cells survive near hydrothermal vents?
4 What makes a cancer cell different from a normal cell?
5 What is a phase-contrast microscope and what is it used for in biology?

Objectives

At the completion of this chapter, the student will be able to:

1. Discuss the importance of the microscope in biology and identify and describe the function of the parts of a compound microscope.
2. Determine the total magnification of a compound microscope using different objectives.
3. Define the following and describe their interrelationships: magnification, illumination, plane of focus, and depth of field.
4. State the cell theory.
5. Define and give examples of unicellular, colonial, and multicellular organisms.
6. Compare and contrast prokaryotic and eukaryotic cells.
7. Describe the structural components of a typical prokaryotic cell and their functions.
8. Compare and contrast plant and animal cells.
9. Describe the structural components of a plant cell and an animal cell and their functions.

Chapter Photo
Light microscope examining a blood sample.

Microscopy

The microscope is one of the most important and frequently used tools in the biological sciences. It allows the user to peer into the world of the cell as well as discover the fascinating world of microscopic organisms. A typical compound microscope is capable of extending the vision of the observer more than a thousand times (Fig. 4.1). Other microscopes, such as the transmission electron microscope, can magnify objects up to 1 million times. Since its invention more than 300 years ago, the microscope has greatly improved our understanding of the cell, tissues, disease, and ecology.

The most commonly used microscope in the biology laboratory today is the light microscope. A simple light microscope can have a single lens, similar to the early microscopes. Compound microscopes use two sets of lenses to magnify an object. They are capable of a magnification range of 10–2,000× and a resolution of 300 nanometers.

An example of the compound light microscope is the bright field microscope. Light is transmitted directly through the specimen, and the specimen generally appears as a dark object against

FIGURE **4.1** The compound microscope is an essential instrument in the biology laboratory.

a light background. This microscope is used to examine various types of cells, microscopic organisms, and tissues (Fig. 4.2).

Electron microscopes use beams of electrons to magnify a specimen. They are capable of greater magnification than compound microscopes. The transmission electron microscope (TEM) uses extremely thin sections of specimens treated with heavy metal salts and is capable of magnifications of 200–1,000,000× (Fig. 4.3). The TEM has a resolution of 0.1 micrometers because of the shorter wavelength of the electron beam. This instrument is used to study the ultrastructure of cells and certain biochemicals.

The scanning electron microscope (SEM) provides three-dimensional views of objects and has a greater depth of focus. It is capable of magnifications from 10–500,000× and a resolution of 5–10 nanometers. The SEM is useful in studying the surface features of specimens (Fig. 4.4). The scanning transmission electron microscope (STEM) is a combination of the TEM and SEM. It is used in the analysis of specimens and various chemicals.

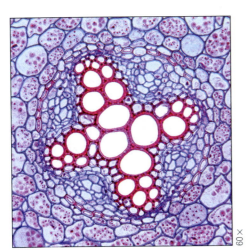
60×

FIGURE **4.2** Transverse section of a dicot root viewed through a bright light field microscope.

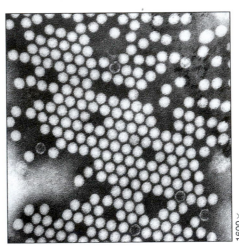
1600×

FIGURE **4.3** Virus viewed through a transmission electron microscope (TEM).

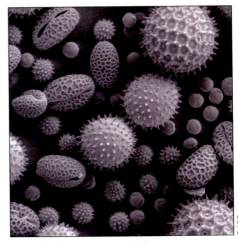
2400×

FIGURE **4.4** Pollen viewed through a scanning electron microscope (SEM).

Learning about Microscopy

Microscopes are expensive and delicate instruments, so proper care and handling procedures must be followed when working with them. Carefully procure a microscope from its storage location according to the instructor's directions, and place it on the lab table. Identify the following parts of the compound microscope (Table 4.1 and Fig. 4.5).

TABLE **4.1** Parts of a Compound Microscope

Part	Description
Ocular (eyepiece)	The uppermost lens or series of lenses through which a specimen is viewed. Most oculars have a magnification of 10×. A microscope with one ocular is called a monocular microscope, and a microscope with two oculars is called a binocular microscope. The distance between the oculars can be adjusted to match different distances between pupils. Many times, a trinocular head is added to a microscope for teaching and photography purposes. Some oculars contain a pointer or a micrometer disk used to determine the size of an object. *Never touch the ocular lens with your fingers.*
Draw tube	Connects the ocular to the body tube.
Body	Holds the nosepiece at one end and includes the draw tube.
Arm	Serves as a handle.
Nosepiece	Revolves and holds the objectives. When changing objectives, always turn the nosepiece instead of using the objectives themselves. The objectives will click into place when properly aligned.
Objectives	Lower lenses attached to the nosepiece. The magnification of each objective is stamped on the housing of the objective. The magnification may vary with different brands. *Never touch the objective lenses.*
Scanning objective	Used for viewing larger specimens or searching for a specimen; the shortest objective usually magnifies an object 4× or 5×.
Low-power objective	Used for coarse and preliminary focusing; magnifies an object approximately 10×.
High-power objective	Used for final and fine focusing, magnifies an object approximately 40×, 43×, or 45×.
Oil-immersion objective	Uses the optical properties of immersion oil to help magnify a specimen. Oil-immersion objectives are capable of magnifications of 93×, 95×, or 100×. Care should be taken with oil-immersion objectives in that the oil should not be allowed to build up on the objective or contaminate other objectives. After use, the oil-immersion objective and slide should be thoroughly wiped off and cleaned.
Stage	Platform on which slides are placed. Some microscopes have a mechanical stage to accurately control the movement of slides. Stage clips secure the slide.
Light source (illuminator)	Serves as the source of illumination for the microscope.
Iris diaphragm	Regulates light entering the microscope; usually is controlled by a mechanical lever or rotating disk.
Condenser	A lens system found beneath the stage; used to focus the light on the specimen.
Coarse-focus adjustment knob	Used to adjust the microscope on scanning and low power only.
Fine-focus adjustment knob	Used to adjust the specimen into final focus.
Base	The supportive portion of the microscope, which rests on the laboratory table.

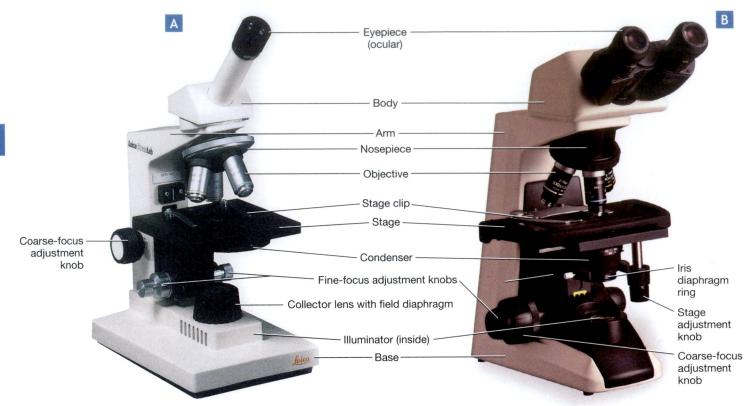

FIGURE **4.5** Light microscopes: (**A**) compound monocular microscope; (**B**) compound binocular microscope.

Labels in figure:
- Eyepiece (ocular)
- Body
- Arm
- Nosepiece
- Objective
- Stage clip
- Stage
- Coarse-focus adjustment knob
- Condenser
- Fine-focus adjustment knobs
- Collector lens with field diaphragm
- Illuminator (inside)
- Base
- Iris diaphragm ring
- Stage adjustment knob
- Coarse-focus adjustment knob

Procedure 1

The Compound Microscope

Materials
❏ Compound microscope
❏ Lens paper
❏ Prepared slide
❏ Colored pencils

1 Ensure that your work area is clean and uncluttered.

2 After procuring your microscope from its storage location, carry the microscope close to your body in an upright position with one hand placed under the base and the other hand holding the arm. Place the microscope directly in front of you on the table. After it is in place, do not drag the microscope across the table for others to view. Doing so can damage the intricate optics and mechanisms of the microscope.

3 Examine the anatomy of the microscope, and identify all the parts.

4 Carefully clean the ocular and objectives with *lens paper only*. If a smudge or scratch persists, consult the laboratory instructor.

5 Make sure that the microscope was stored with the scanning or low-power objective in place. Never begin a session with the high-power objective in place, to prevent breaking your slide or scratching the objective. If necessary, use the revolving nosepiece ring to click the low-power objective into place.

6 Inspect the electrical cord, making sure it is not frayed or damaged. Plug in the electrical cord as instructed so it will not get in your way, trip other students, or damage the microscope. Turn the switch to the "on" position.

7 A good-quality clean slide will be provided for observation. Carefully place your slide on the stage, and use the stage clips or the mechanical stage to hold it in place. Center the specimen under the objective.

8 If using a binocular microscope, adjust the distance between the oculars to match the distance between your pupils. One of the oculars on some binocular microscopes can be focused individually for precision. If using a monocular microscope, keep both eyes open to view the object, because closing one eye will result in eyestrain and a headache. If you are having difficulty keeping both eyes open, try covering the opposite eye with your hand, alternating eyes periodically.

9 Use the coarse-focus adjustment knob to focus the specimen. This knob is to be used to view specimens under scanning and low power only. Depending upon the brand of the microscope, either the nosepiece will move toward the stage or the stage will move toward the nosepiece. Practice your microscopy skills by viewing various parts of the slide. While viewing the slide, do not rest your hand on the stage.

10 Reposition the slide to attain the desired view. Use the iris diaphragm and condenser to focus and regulate the light entering the microscope. On scanning and low power, the fine-focus adjustment knob may be used to fine-tune the specimen. Sketch and record the name and magnification of your specimen in the space provided.

11 If your specimen is in the center of the field of view and in focus, you are now ready to move to the next power. Using the revolving nosepiece ring, rotate the high-power objective into position until you feel it click into place.

12 Many microscopes are **parfocal** (after the image is focused with one objective, it should be in focus with others) and require only minor adjustments in focusing. Using the fine-focus adjustment knob only, focus your specimen. You may have to reposition your specimen carefully and adjust the iris diaphragm and condenser. Sketch and record the name and magnification of your specimen in the space provided below.

13 Before removing the slide, return the microscope to the lowest objective.

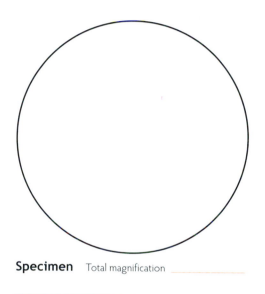

Specimen   Total magnification _____

_____

_____

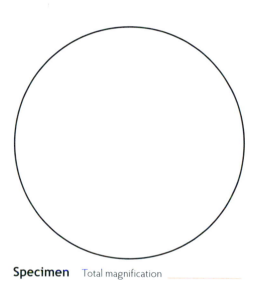

Specimen   Total magnification _____

_____

_____

✔ # Check Your Understanding

1.1 Why is it important to start with the scanning objective in place, rather than the high-power objective?

_____

_____

1.2 *True or false, and justify your answer*: The coarse-focus adjustment knob should be used with scanning objective, low-power objective, and the high-power objective.

_____

_____

_____

Developing the Skills of Microscopy

To become proficient with the microscope, you must understand fundamental principles of microscopy and acquire basic skills. The key to becoming skillful with a microscope is *practice*.

Microscopes are designed to magnify objects. The **magnification** of a specimen is the product of the power of the ocular and the power of the objective. Microscopes generally are equipped with a 10× ocular, and objectives in the ranges of a 4× scanning objective, a 10× low-power objective, a 40× high-power objective, and a 100× oil-immersion objective. Thus, the total magnification of a microscope on low power is the 10× ocular times the 10× low-power objective, or 100× magnification.

The ability to resolve objects (distinguish two closely spaced, minute objects as separate entities) is an important characteristic of microscopes and a measure of lens quality. The **resolving power** of a microscope depends upon the design and quality of the objective lenses. Quality lenses have a high resolving power, or the ability to deliver a clear image in detail. In comparing microscopes, lower values represent better resolving power. For example, a resolution of 5–10 nanometers means you can see more detail than if the resolution were 300 nanometers. The human eye is only capable of distinguishing objects 0.1 millimeters apart, and a quality magnifying glass improves the resolution to 0.01 millimeters.

When using the microscope, magnifying a specimen is not always accompanied by an increase in the clarity of the detail within the specimen. At a certain point, greater magnification of the specimen yields progressively fuzzy images. Oil-immersion objectives can increase the resolving power of a microscope.

When viewing the image under the microscope, movement of the slide is in reverse direction. For example, when you move your slide to the right, the image will move to the left. This is important when tracking motile organisms in a **wet mount**. It is imperative that you become comfortable with this when making adjustments to your slide. Using your laboratory microscope, perform the following tasks.

Procedure 1

Orientation

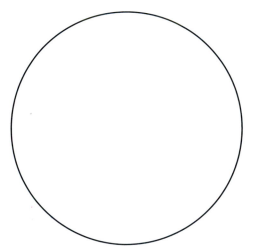

Materials
- ❏ Compound microscope
- ❏ Lens paper
- ❏ Prepared slides of the letter "e"

1 Place a prepared slide of the letter "e" on the stage of your microscope. Using low power (remember to always begin on low power), observe the slide. After you have found the letter "e" and it is in the center of your field of view, sketch in the circle what you see. Make sure under your sketch, in the blanks provided, to label the name of your slide and what the total magnification is (ocular magnification × objective magnification). What is the difference between seeing the orientation of the letter with the unaided eye and the microscope?

_____

2 Move the slide to the left and to the right. In which direction did the image move?

_____

3 After you finish, return the microscope to its lowest power and return the slide to the slide tray.

Letter "e" slide Total magnification _____

Procedure 2

Illumination

Materials

- ❏ Compound microscope
- ❏ Lens paper
- ❏ Prepared slides

When using a microscope, proper illumination is essential because improper illumination can result in poor images and eyestrain. Some microscopes are equipped with a mirror to gather and concentrate light. Generally, the mirror has a smooth side and a concave side. The smooth side should be used in most applications, and the concave side should be used when it is necessary to concentrate light on the specimen. If your microscope uses a mirror, adjust the mirror to deliver maximum light on the specimen, and use the iris diaphragm to regulate the intensity of the light.

Most microscopes are equipped with an illuminator attached to the base. The **iris diaphragm** is used to adjust the amount of light entering the objective lens by rotating a disk or moving an aperture-adjustment control. The iris diaphragm can be used to regulate the light passing through an object as well as the contrast of the object. Although closing the diaphragm results in poorer resolution, it can be useful in viewing certain specimens.

Some microscopes have a **substage condenser**. For an introductory laboratory, the condenser should be left in its uppermost position. Many microscopes have a light-intensity control; use this control as directed by the instructor.

1 Carefully place a slide (instructor's choice) on the stage, and focus using low power. What kind of specimen are you looking at?

2 Use the iris diaphragm to attain the best illumination for the specimen. Open and close the iris diaphragm, and notice changes in the contrast of the specimen. How did the appearance of the specimen change under various levels of illumination?

3 Follow the same procedure using the high-power objective. When increasing magnification, what happens to the light available? Why?

4 After you finish, return the microscope to its lowest power and return the slide to the slide tray.

Procedure 3

Focus and Depth

Materials

- ❏ Compound microscope
- ❏ Lens paper
- ❏ Prepared slides of different colored threads, tissues, and microorganisms
- ❏ Colored pencils

Microscope lenses have a restrictive **plane of focus**, a specific distance from the lens where the specimen can be sharply focused. It is possible to focus through different planes of the specimen by changing the plane of focus. The plane of focus has some depth called the **depth of field**, or the thickness of the specimen in focus at any one time. When using lens systems, including cameras, the greater the magnification, the less is the depth of field.

1 Procure a prepared slide containing three different colored threads, and place it on the stage of your microscope.

2 Locate the point on the slide where the threads intersect, and center it under the objective.

3 Using the low-power objective, slowly focus until the first thread comes into focus. The fine adjustment can be used in addition to the coarse adjustment. What is the color of the first thread?

4 Continue focusing through the threads, noting the order of the colored threads. What is the color of the second thread?

What is the color of the third thread?

5 Draw the levels of the threads in the space provided.

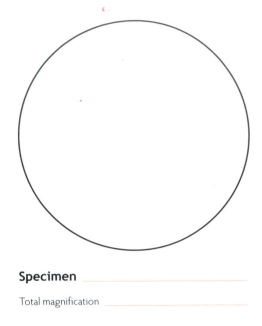

Levels of threads _____

Total magnification _____

Specimen _____

Total magnification _____

6 Switch to the high-power objective, and notice that the depth of field is shallower. If you constantly use the fine adjustment, the three-dimensional form of the threads can be determined.

7 The laboratory instructor will provide a slide of a tissue such as columnar epithelium or an organism such as a rotifer. Focus through your specimen and record your observations. Draw what you see in the space provided. When will an understanding of planes of focus and depth of field be valuable in the laboratory?

8 Discuss planes of focus and depth of field based upon your observations of both the low-power and high-power objectives.

_____

_____

_____

_____

Procedure 4

Oil Immersion

Materials
- ❑ Compound microscope
- ❑ Lens paper
- ❑ Immersion oil
- ❑ Prepared slide of human blood
- ❑ Colored pencils

Many microscopes have an oil-immersion objective for observing small organisms and the fine detail of specimens (Fig. 4.6). It appears longer than the other objectives and is clearly marked. Oil-immersion objectives are capable of magnifications of 93×, 95×, or 100×, depending on the microscope.

1 Procure a prepared slide of human blood, and place it on the stage of the microscope. The red blood cells, or erythrocytes, will appear as lightly stained, biconcave disks with a lighter-colored center. White blood cells, leukocytes, will appear large with a dark-stained nucleus.

2 Focus and view the specimen under low power. Click the high-power objective into place, focus, and view the

specimen. Sketch and record the magnification of your observations in the space provided.

3 Carefully swing the high-power objective away from the slide. Place one drop of immersion oil (confirm that you are using immersion oil) in the center of the viewing area of the slide. Do not use any other objectives with immersion oil, and be careful not to get the oil on anything. Click the oil-immersion objective into place. The oil-immersion objective will dip into the oil. Focus on the specimen using the fine-focus adjustment knob only. Sketch and record the magnification of the specimen in the space provided.

4 After the exercise is complete, clean the slide and objective thoroughly with lens paper. Lens paper with a drop of xylene is often used to clean the objective and slide. Return the lowest-power objective into place.

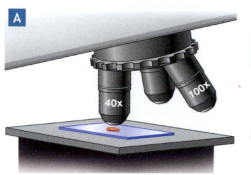  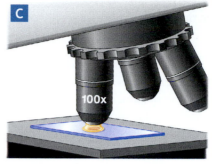

FIGURE **4.6** Using an oil-immersion objective. (**A**) Focus on the specimen using the high-power objective. (**B**) Swing the high-power objective away from the specimen, and place a drop of immersion oil on the slide, covering the specimen. (**C**) Swing the oil-immersion objective over the specimen so the tip of the objective is in contact with the oil, and adjust focus.

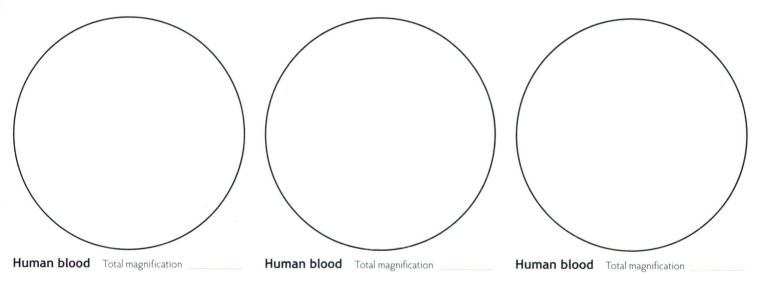

Human blood   Total magnification _____

Human blood   Total magnification _____

Human blood   Total magnification _____

✔ Check Your Understanding

2.1 If a 4× objective is used, what is the total magnification of a microscope? _____

2.2 Which objective lens will give you the greatest resolving power? _____

2.3 To move a specimen to the upper left-hand field of view, which way must one move the slide?

_____

2.4 As magnification increases, does the amount of light increase or decrease? _____

2.5 With increased magnification, what happens to the field of view?

_____

2.6 As magnification increases, the depth of field _____.

2.7 _____ is the distance between the objective lens and the specimen.

Preparing Slides

Prepared slides are commonly used in the biology laboratory. These slides are the products of tedious work. Many of the specimens on prepared slides have been stained for better contrast and observation. The slides must be carried by the edge or the end; never place your fingers over the viewing area. The label of the slide may indicate that the specimen is a whole mount (w.m.), a cross section (c.s. or x.s.), or a longitudinal section (l.s.).

When viewing slides, always note the type of mount you are viewing. In addition, do not become dependent on dye colors when learning specimens. Prepared slides are expensive, so treat them carefully, and report any broken or damaged slides to your laboratory instructor. Wet mounts are often made in the laboratory to view fresh specimens (Fig. 4.7).

Safety First

Microscope slides and many coverslips are made of glass and easily broken. Please handle them with care and report any broken glass to your instructor immediately.

FIGURE **4.7** Preparing a wet mount. (**A**) Add a drop of pond water or a solution over the center of the slide. (**B**) Place the angled coverslip next to the droplet along one edge as shown in the illustration. The side resting against the glass will act as the pivot point as you lower the coverslip over the sample and will minimize air bubbles. (**C**) Lower the coverslip into place. As you do, the drop will spread outward and suspend the sample between the slide and coverslip.

Procedure 1

Wet Mounts

Materials
- ❏ Compound microscope
- ❏ Blank microscope slides
- ❏ Coverslips
- ❏ Lens paper
- ❏ Pond or aquarium water
- ❏ *Elodea* leaf
- ❏ Protoslo, methyl cellulose, or dishwashing detergent
- ❏ Eyedropper
- ❏ Forceps
- ❏ Paper towels
- ❏ Colored pencils

1 Clean and dry a slide and coverslip thoroughly.

2 When mounting bits of tissue, place the specimen to be studied in the center of the slide. Prepare it according to your instructor's directions. For pond culture studies (Fig. 4.8), place a drop of the culture liquid in the center of the slide, and lower a coverslip. Always place a coverslip over wet mounts.

3 Holding the coverslip at a 45-degree angle (see Fig. 4.7), place its edge in the margin of the drop of liquid on the slide, and release the coverslip slowly. The procedure will minimize the odds of air bubbles forming beneath the coverslip.

4 If the coverslip floats on the liquid, it could be caused by excess water on the slide. The excess water may be siphoned off with the edge of a paper towel. Too much liquid (water or stain) should be removed. If some of the reagents get on the microscope, they may corrode it and cause serious damage. You may have to add water at the edge of the coverslip to compensate for evaporation in wet mounts, such as in pond cultures.

5 At certain times it will be necessary to support the coverslip to prevent crushing the organisms being observed. A small piece of lens tissue at each corner of the coverslip or small pieces of broken coverslips may be placed beneath the top coverslip.

6 Many times, when observing living specimens, they appear to move rapidly. To slow them down, place a drop of a prepared slowing solution, such as Protoslo, methyl cellulose, or dishwashing liquid, in the drop of water. The result will be equivalent to a person swimming in a pool of molasses.

7 Prepare and observe a wet mount of an *Elodea* leaf and pond water. Use the scanning, low-power, and high-power objectives in this activity.

8 Sketch your observations in the spaces provided on the following page.

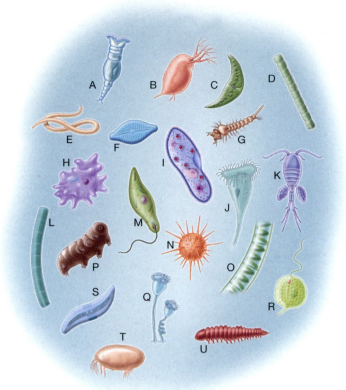

FIGURE **4.8** Life in a drop of pond water. Common pond organisms include: (**A**) rotifer; (**B**) *Daphnia*; (**C**) desmid; (**D**) cyanobacteria (*Oscillatoria*); (**E**) nematode; (**F**) diatom; (**G**) mosquito larva; (**H**) amoeba; (**I**) *Paramecium*; (**J**) *Stentor*; (**K**) copepod; (**L**) cyanobacteria; (**M**) *Euglena*; (**N**) radiolarian; (**O**) green algae (*Spirogyra*); (**P**) tardigrade; (**Q**) *Vorticella*; (**R**) *Phacus*; (**S**) diatom; (**T**) ostracod; (**U**) oligochaete.

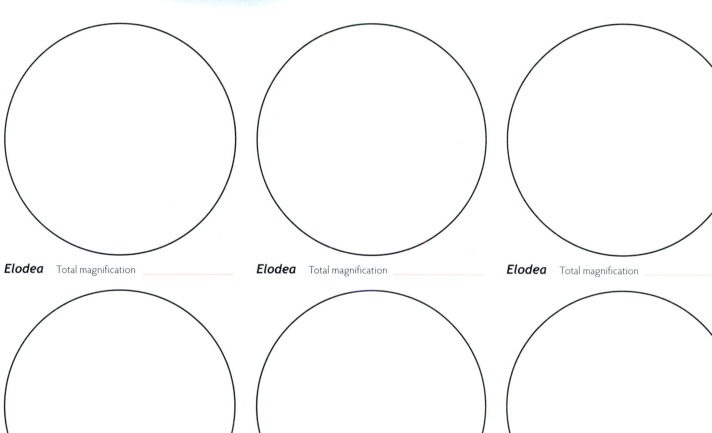

Elodea   Total magnification _____

Elodea   Total magnification _____

Elodea   Total magnification _____

Pond water   Total magnification _____

Pond water   Total magnification _____

Pond water   Total magnification _____

✔ Check Your Understanding

3.1 Why is a coverslip placed on a slide at a 45-degree angle?

4

_____

_____

_____

_____

3.2 Why are wet mounts used to view biological specimens in the laboratory?

_____

_____

_____

_____

_____

Using the Stereomicroscope (Dissecting Microscope)

A stereomicroscope (Fig. 4.9) also is known as a dissecting microscope or binocular microscope. The dissecting microscope has two oculars and is capable of magnifications of 4× to 50×. These microscopes provide a significantly greater field of view and depth of field than compound microscopes. This type of microscope is advantageous when viewing larger objects and dissecting.

The anatomy of a dissecting microscope is similar to that of a compound microscope. The two oculars can be adjusted for the individual viewer by moving the oculars or using an interpupillary adjustment. The specimen can be lighted from above (reflection) with an external or built-in illuminator, or from below (transmission) with a substage light or both.

When viewing a specimen, determine which type of light is best for your specimen. Some dissecting microscopes have fixed magnifications, and others have dial-type or zoom-magnification ability. Dissecting microscopes have a single focusing knob. Treat this microscope with the same respect afforded a compound microscope.

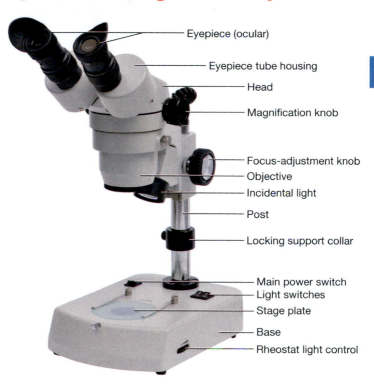

Eyepiece (ocular)
Eyepiece tube housing
Head
Magnification knob
Focus-adjustment knob
Objective
Incidental light
Post
Locking support collar
Main power switch
Light switches
Stage plate
Base
Rheostat light control

FIGURE **4.9** Dissecting microscope.

Procedure 1

Light and Magnification

Materials

- ❏ Dissecting microscope
- ❏ Ring with stone
- ❏ Coin
- ❏ Dollar bill
- ❏ Fossil
- ❏ Feather
- ❏ Leaf
- ❏ Flower
- ❏ Pond water
- ❏ Other specimens, instructor's choice
- ❏ Colored pencils

1 View the following objects using a dissecting microscope: your fingernail, the surface of a stone on a ring, a coin, a dollar bill, a fossil, a feather, a leaf, several biological specimens provided by your instructor, and some pond water in a petri dish.

2 Try using various light and magnification settings with your dissecting microscope. Sketch several of your observations in the spaces provided on the following page.

Hints & Tips

To properly store a microscope:

1 At the completion of each laboratory experience, rotate the lowest-power objective into place, and remove the slide.

2 Clean the ocular and objectives with *lens paper only*. If you have been using oil immersion, clean the slide and objectives as instructed in the procedure for using oil immersion.

3 If you have been using wet mounts, clean the stage with a cleaning tissue or a clean cloth.

4 Turn off the light and unplug the microscope. Wrap the cord around the base as directed by the instructor. If a plastic dust cover is available, cover the microscope, and return it to its storage location.

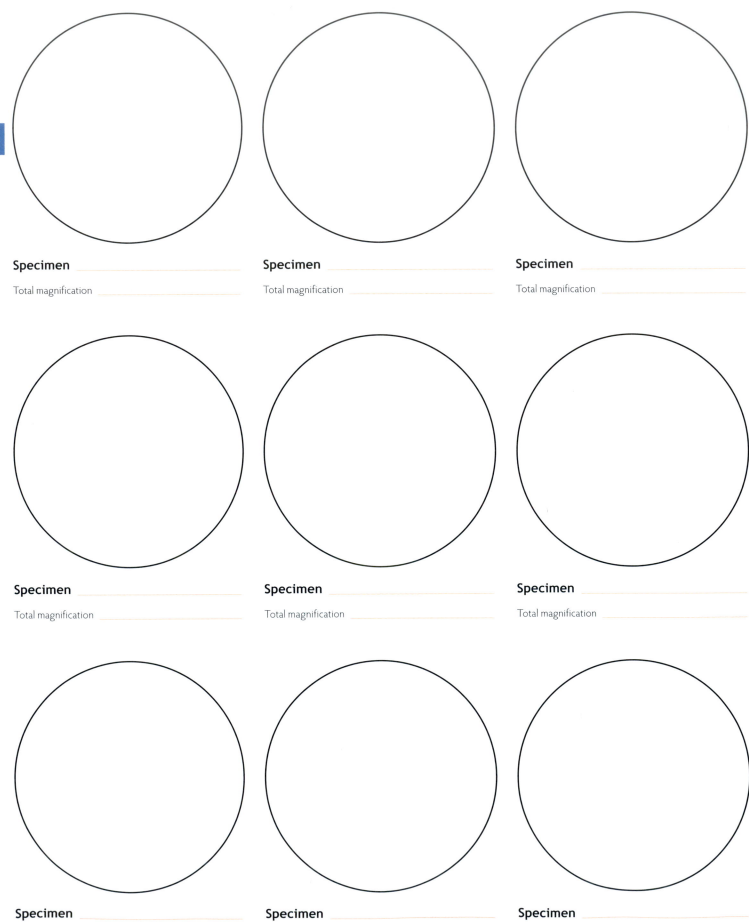

Specimen _____

Total magnification _____

Specimen _____

Total magnification _____

Specimen _____

Total magnification _____

Specimen _____

Total magnification _____

Specimen _____

Total magnification _____

Specimen _____

Total magnification _____

Specimen _____

Total magnification _____

Specimen _____

Total magnification _____

Specimen _____

Total magnification _____

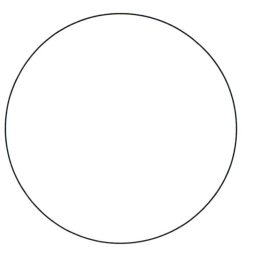

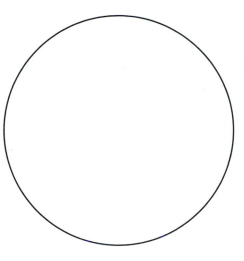

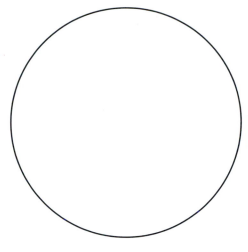

Specimen _____

Total magnification _____

Specimen _____

Total magnification _____

Specimen _____

Total magnification _____

✔ Check Your Understanding

4.1 A stereomicroscope has a _____ (larger/smaller) field of view and a _____ (larger/smaller) depth of field than compound microscopes.

4.2 What are two types of illumination used in a stereomicroscope?

_____

_____

■ Cell Structure and Function

From tiny bacteria to the great blue whale, the cell serves as the fundamental building block of all living things (Fig. 4.10). A student reading this paragraph consists of nearly 125 trillion cells working together to maintain a state of biological balance, or **homeostasis**. Even the salad and pizza you ate last night were made up of a multitude of plant and animal cells. Yes, the salad and pizza also had their share of bacterial cells as well.

Cell Theory

The **cell** is the smallest unit of biological organization that can undergo the activities associated with life, such as metabolism, response, and reproduction. British scientist Robert Hooke (1635–1703) first described the cell in 1665 while observing cork. In the late 1830s, two German scientists—botanist Matthias Schleiden (1804–1881) and zoologist Theodor Schwann (1810–1882)—provided a powerful understanding of the structure and function of plant and animal cells with their cell theory. Basically the **cell theory** states that all living things are composed of cells, and that the cell is the basic unit of structure and function of all living things.

In the 1850s, German physician Rudolf Virchow (1821–1902) added to the cell theory that cells come only from pre-existing cells. Virchow also pointed out that the cell is the fundamental link in the biological levels of organization, which include tissues, organs, systems, and, ultimately, the complete organism. Today, the biological levels of organization have been expanded to include populations, communities, ecosystems, and the biosphere.

Although cells vary in size from a bacterium 1–10 micrometers in diameter to a chicken egg larger than 3.8 centimeters in diameter, most cells are microscopic. The inclusions and organelles within the cell are much smaller and are measured in nanometers. The reason for the absence of giant cells is the ratio of the surface area to volume (Fig. 4.11). If the surface area of a cell increases, the volume does

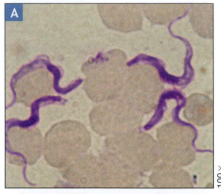

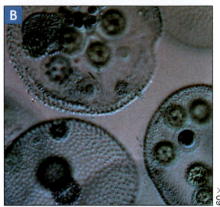

FIGURE **4.10** (**A**) *Trypanosoma* sp. is a unicellular organism; (**B**) *Volvox* sp. is a colonial organism; (**C**) *Diceros bicornis* (black rhinoceros) is a multicellular organism.

not increase in direct proportion; the volume increases proportionally faster. Thus, the surface area could not support the metabolic needs of the increased volume. Some cells, such as frog eggs, chicken eggs, and ostrich eggs, can become large because they are not metabolically active until they begin to divide. Other cells, such as nerve cells, can possess extensions of more than a meter, but the extensions are narrow and have little volume.

Unicellular and Multicellular Organisms

Bacteria and many protists, such as the green alga *Spirogyra* and the paramecium, are composed of one cell and are called **unicellular**. Despite having just one cell, these organisms carry on all of the life processes efficiently. Several species of protists exist as colonies, or loosely connected groups or aggregates of cells. An example of a **colonial organism** is the green alga *Volvox*.

Organisms composed of many cells, such as the azalea, the mushroom, and the walrus, are called **multicellular**. The biological levels of organization of these organisms exhibit a division of labor and have a variety of specialized tissues.

Two Types of Cells

Although innumerable forms of cells exist in nature, only two basic types of cells constitute life on Earth: prokaryotic cells and eukaryotic cells.

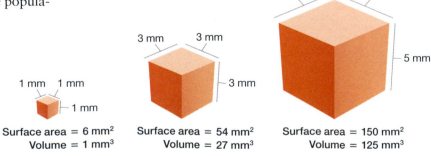

FIGURE **4.11** Ratio of surface area to volume.

1. **Prokaryotic cells** lack a membrane-bound nucleus and organelles such as mitochondria; they are also much smaller than eukaryotic cells. The cytoplasm of prokaryotes is surrounded by a plasma membrane, and the majority of prokaryotes are encased in a protective cell wall. Prokaryotic organisms are placed within the kingdoms Archaebacteria and Eubacteria.

2. **Eukaryotic cells** are more structurally complex, are surrounded by plasma and a cell membrane, and are larger than prokaryotic cells. They have a membrane-bound nucleus and a variety of organelles. Members of the kingdoms Protista, Plantae, Fungi, and Animalia possess eukaryotic cells.

The cell is the basic unit of structure and function of all living things. A cell serves as the exclusive functional unit in unicellular organisms. In multicellular organisms ranging from the giant sequoia to a minute mushroom, however, cells differentiate to perform a variety of specialized functions. Multicellular organisms involve a division of labor, with certain groups of cells becoming highly specialized to perform duties that benefit the entire organism. Groups of cells and their intercellular substances similar in structure and function are called **tissues**.

Tissues are a fundamental part of the **biological levels of organization** (Fig. 4.12), which begin with **atoms**, which make up **molecules**, which eventually form cells. Cells, in turn, form tissues. To perform specific functions, tissues are organized into **organs**. Organs may contain several representative tissues, and the arrangement of these tissues determines the organ's structure and function. In turn, several organs working together to perform a particular function form an **organ system**. The complete **organism** consists of an individual containing several systems working together.

  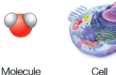   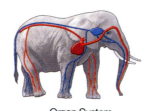 

| Atom | Molecule | Cell | Tissue | Organ | Organ System | Organism |

FIGURE **4.12** Biological levels of organization.

EXERCISE 4.5

Observing Prokaryotic Cells

The most cosmopolitan organisms on Earth today are the prokaryotes. Bacterial fossils have been dated at older than 3.5 billion years. They exist in every possible environment, even those that do not seem conducive to life. Although the prokaryotes are small in size (1 to 50 micrometers in width and diameter), they are economically, ecologically, and medically important. Two distinct groups of prokaryotic organisms are archaebacteria and eubacteria.

1. The **archaebacteria**, or ancient bacteria, can be found living in extreme environments, such as exceedingly salty habitats (extreme halophiles), exceptionally hot environments (extreme thermophiles), the anaerobic mud of swamps, and the guts of termites and many mammals (methanogens).

2. The **eubacteria**, or true bacteria, are better known to the general public. Three basic shapes of eubacteria

exist. The cocci are generally spherical in shape, such as *Neisseria meningitidis* (bacterial meningitis); the bacilli are generally rod-shaped, such as *Escherichia coli* (fecal contamination); and the spirilli are spiral in shape, such as *Borrelia burgdorferi* (Lyme disease). Although the majority of eubacteria are harmless or helpful, such as *Lactobacillus acidophilus* in yogurt, there are several medically dangerous species. Examples of these organisms are *Yersinia pestis* (black plague), *Clostridium perfringens* (gangrene), *Helicobacter pylori* (ulcers), *Vibrio cholerae* (cholera), *Staphylococcus aureus* (boils), and *Bacillus anthracis* (anthrax).

The **cyanobacteria**, once classified as the blue-green algae, are photosynthetic eubacteria. The cyanobacteria are common and can be found in a number of environments, including in the soil, on sidewalks, on the sides of buildings, on trees, and in bodies of water such as ditches.

Procedure 1

Cyanobacteria

Materials

- ❏ Compound microscope
- ❏ Prepared slides of cyanobacteria (*Gloeocapsa* and *Nostoc*)
- ❏ Blank slides and coverslips
- ❏ Living specimens of cyanobacteria (*Oscillatoria* and *Anabaena*)
- ❏ Forceps
- ❏ Paper towels
- ❏ Colored pencils

These rather large prokaryotes do not possess chloroplasts; the chlorophyll *a* is located in the thylakoid membranes (Fig. 4.13). The cyanobacteria have a number of accessory pigments that can mask the green color of chlorophyll. As a result, species of cyanobacteria appear red, yellow, brown, or blue-green.

1 Procure a microscope, prepared slides, blank slides, and coverslips. Using proper microscopy techniques, observe the prepared slides of *Gloeocapsa* and *Nostoc*.

2 Sketch and describe the gelatinous *Gloeocapsa* in the space provided.

3 Sketch and describe the filamentous and gelatinous *Nostoc* in the space provided.

4 Properly prepare a wet mount of *Oscillatoria* and *Anabaena*. Using proper microscopy techniques, observe *Oscillatoria* and *Anabaena*.

700×

FIGURE **4.13** *Microcoleus* sp. is one of the most common cyanobacteria in and on soils throughout the world. It is characterized by several filaments in a common sheath.

Gloeocapsa

Total magnification

5 Sketch and describe the filamentous *Oscillatoria* in the space provided.

6 Sketch and describe the filamentous *Anabaena* in the space provided.

7 After completion of the activity, clean up your work area, disinfect the lab surface, and return or dispose of the materials as instructed.

Nostoc   Total magnification _____

Oscillatoria   Total magnification _____

Anabaena   Total magnification _____

Procedure 2

Bacteria

Materials

- ❏ Compound microscope
- ❏ Immersion oil
- ❏ Prepared slides of mixed bacteria and yogurt smear
- ❏ Blank slides and coverslips
- ❏ Tap water
- ❏ Toothpick
- ❏ Pipette
- ❏ Plain yogurt
- ❏ Paper towels
- ❏ Colored pencils

Most bacteria are significantly smaller than the cyanobacteria. The bacteria are simple in form and anatomy and exhibit three basic shapes: **bacillus** (rod-shaped), **coccus** (spherical-shaped), and **spirillum** (spiral-shaped). An electron microscope is used to observe the anatomical detail of a typical bacterium (Fig. 4.14 and Table 4.2).

To understand the cell structure of bacteria:

1 Procure a microscope, immersion oil, prepared slides, blank slides, coverslips, and a toothpick. Using proper microscopy techniques,

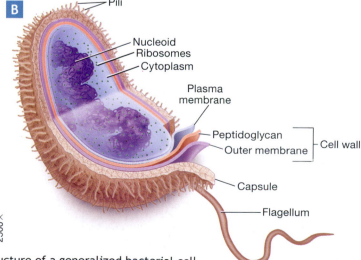

B
- Pili
- Nucleoid
- Ribosomes
- Cytoplasm
- Plasma membrane
- Peptidoglycan ⎤ Cell wall
- Outer membrane ⎦
- Capsule
- Flagellum

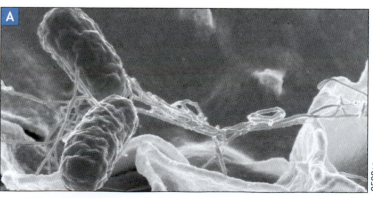

A

2500 ×

FIGURE **4.14** (**A**) SEM showing *Salmonella typhimurium*; (**B**) basic structure of a generalized bacterial cell.

TABLE **4.2** Common Anatomical Features of a Generalized Bacterium

Structure	Function
Cell wall	In eubacteria, a peptidoglycan envelope that provides protection and shape
Plasma membrane	Phospholipid bilayer that provides support and regulates the movement of substances into and out of the cell
Cytoplasm	Semifluid medium within the cell
Nucleoid	Region that houses the bacterial DNA in a single chromosome; some bacteria possess small circular fragments of DNA called plasmids
Ribosome	Site of protein synthesis
Fimbriae	Short, hairlike structures that aid in attachment
Pili	Rigid, hairlike structures important for attachment and the exchange of genetic information
Flagellum	Elongated structure used for locomotion; the number of flagella and their location are important in determining the species of bacteria
Capsule	Protective slime-like area lying outside the cell wall that helps the bacterium adhere to certain surfaces, keeps it from drying out, and protects the bacterium from phagocytosis by other organisms or cells

observe the prepared slide of mixed bacteria. To view the specimen properly, use an oil-immersion objective, if available.

2 Sketch and describe the shapes of the three types of bacteria in the space provided.

3 Procure a small amount of plain yogurt on the tip of a toothpick. Rub the yogurt onto the central portion of a blank slide. Place one drop of water on the yogurt with a pipette, and mix it with a toothpick. Gently place the coverslip on the water/yogurt mixture. Observe the bacteria in the yogurt under high power. (The majority of bacterial cells in yogurt are *Lactobacillus acidophilus*.)

4 Sketch and describe *Lactobacillus* in the space provided.

5 After completing the activity, clean up your work area, disinfect the lab surface, and return or dispose of the material as instructed.

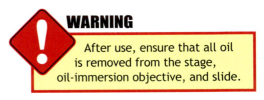

WARNING

After use, ensure that all oil is removed from the stage, oil-immersion objective, and slide.

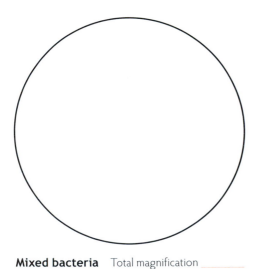

Mixed bacteria   Total magnification _____

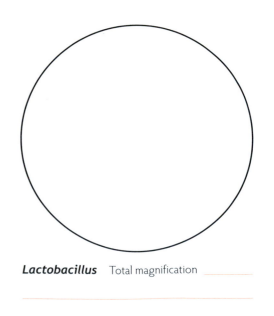

Lactobacillus   Total magnification _____

✔ Check Your Understanding

5.1 What does "lacto-" mean, and what does "bacillus" mean? Did your observation of the shape of this bacterium support its name?

5.2 *Gloeocapsa* and *Nostoc*, two of the organisms that you viewed under the microscope, are classified as cyanobacteria. How do they differ from other eubacteria?

Observing Eukaryotic Cells

Eukaryotic cells originated nearly 2 billion years ago. Eukaryotes include the protists, fungi, plants, and animals. The cells of eukaryotes possess a membrane-bound nucleus and a variety of membrane-bound organelles.

4

Procedure 1

Protists

Materials

❏ Compound microscope
❏ Prepared slides of *Volvox* sp. and *Amoeba proteus*
❏ Blank slides and coverslips
❏ Pipette
❏ Toothpicks
❏ Culture of *Spirogyra* sp. and *Paramecium* sp.
❏ Protoslo or methyl cellulose
❏ Forceps
❏ Paper towels
❏ Colored pencils

Protists include a diverse group of organisms (Fig. 4.15). In fact, the former kingdom Protista is undergoing reorganization and one day will consist of several new kingdoms. Presently, the protists can be separated into the plantlike protists (algae), fungus-like protists (slime and water molds), and animallike protists (protozoans). The protists are discussed in more depth in Chapter 12.

1 Procure a microscope, prepared slides, blank slides, coverslips, and toothpicks. Using proper microscopy techniques, observe the prepared slides of the colonial alga *Volvox* sp. and the protozoan *Amoeba proteus*.

2 Sketch and describe *Volvox* sp. and *Amoeba proteus* in the space provided.

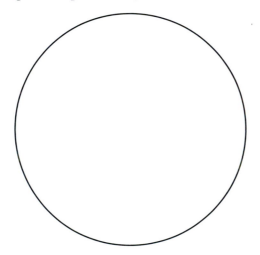

Volvox **sp.**   Total magnification _____

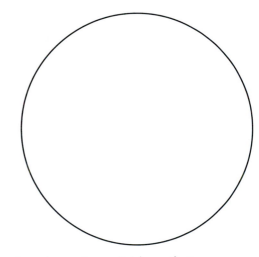

Amoeba proteus   Total magnification _____

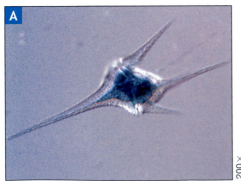

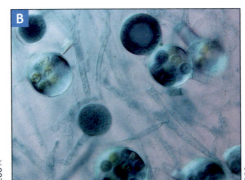

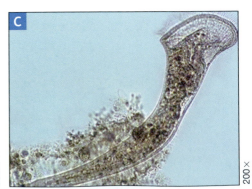

FIGURE **4.15** Examples of protists include: (**A**) *Ceratium* sp., a dinoflagellate (a plantlike protist); (**B**) *Saprolegnia* sp., a water mold (a fungus-like protist); (**C**) *Stentor* sp., a protozoan (an animallike protist).

3 Carefully prepare a wet mount of *Spirogyra* sp. and *Paramecium* sp., and observe the living protists. Protoslo may have to be added to the slide with paramecia to slow them down for observational purposes.

4 Sketch and describe *Spirogyra* sp. and *Paramecium* sp. in the space provided.

5 After completing the activity, clean up your work area, disinfect the lab surface, and return or dispose of the materials as instructed.

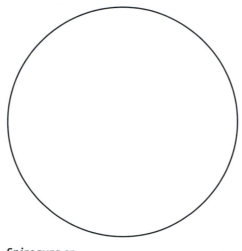

Spirogyra sp.

Total magnification _____

_____

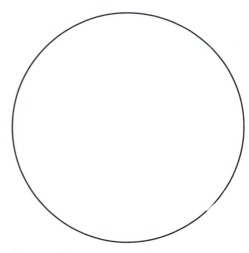

Paramecium sp.

Total magnification _____

_____

Procedure 2

Fungi

Materials

- ❏ Compound microscope
- ❏ Blank slides and coverslips
- ❏ Culture of baker's yeast (*Saccharomyces cerevisiae*)
- ❏ Culture of *Paramecium* sp.
- ❏ Pipettes
- ❏ Toothpicks
- ❏ Methylene blue
- ❏ Protoslo or methyl cellulose
- ❏ Yeast cells stained with Congo red dye
- ❏ Colored pencils
- ❏ Timing device
- ❏ Paper towel

⚠ WARNING

Avoid inhalation and skin contact with methylene blue. Immediately rinse it off the skin with mild soap and water because methylene blue will stain clothing. Follow additional guidelines dictated by your instructor and university.

Kingdom Fungi includes a diverse group of mostly multicellular heterotrophic organisms. Examples of fungi are mushrooms, truffles, morels, rusts, bread mold, ringworm (a fungal infection involving the skin), and yeast (Fig. 4.16). The fungi are discussed in more depth in Chapter 17. An example of a unicellular fungus is *Saccharomyces cerevisiae*, or baker's yeast.

1 Procure a microscope, blank slides, coverslips, and a pipette. Carefully prepare a wet mount of baker's yeast. Observe the slide. If the yeast is difficult to see, carefully add one drop of methylene blue stain to the wet mount with a pipette.

2 Describe and sketch baker's yeast (*Saccharomyces cerevisiae*) in the space provided on the following page.

FIGURE **4.16** Microscopic and macroscopic examples of kingdom Fungi: (**A**) *Aspergillus*; (**B**) *Amanita*, or death angel mushroom.

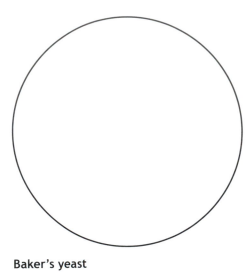

Baker's yeast

Total magnification _____

_____

_____

3 Describe the smell of the yeast culture. Why does the culture have a characteristic smell?

_____

_____

4 Place a drop of yeast cells stained with Congo red dye on a blank slide. Add a drop of paramecia from the culture. Then add a drop of Protoslo with a toothpick

to the slide. Next place the coverslip on the slide. Let the slide sit for 10 minutes to allow the paramecia to begin feeding.

5 Describe and sketch the interaction between the yeast and the paramecia in the space provided.

6 After the completion of the activity, clean up your work area, disinfect the lab surface, and return or dispose of the material as instructed.

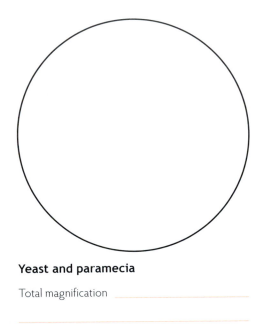

Yeast and paramecia

Total magnification _____

_____

_____

✔ Check Your Understanding

6.1 Would you classify the yeast cells you observed as prokaryotic or eukaryotic cells and why?

_____

_____

_____

6.2 Does *Saccharomyces* have a membrane-bound nuclei? Explain your answer.

_____

_____

6.3 After observing the characteristics and movement of the protist *Paramecium* sp., does it appear more plantlike, fungus-like, or animallike? Why?

_____

_____

Observing Plant Cells

Kingdom Plantae includes some of the most conspicuous organisms on Earth. The plant kingdom contains approximately 280,000 species of multicellular, photosynthetic autotrophs. Plants vary in size and complexity from the minute duckweed to the giant redwood tree.

One of the primary cell types found in plants is **epidermal cells**, which cover and protect the underlying cells and tissues in leaves and stems. Plant epidermis is composed of a closely packed layer of single cells. Epidermal cells do not have chloroplasts and do not undergo photosynthesis.

4

Procedure 1

Representative Plant Tissues

Materials

❏ Compound microscope
❏ Blank slides and coverslips
❏ Pipette
❏ Prepared slide of plant epidermal tissue
❏ Specimen of onion skin
❏ Forceps
❏ Paper towels
❏ Colored pencils

1 Describe and sketch the prepared slide of plant epidermal tissue.

2 Peel a single layer of epidermal tissue from the skin of an onion (Fig. 4.17). Prepare a wet mount of the onion skin, and sketch and describe it in the space provided.

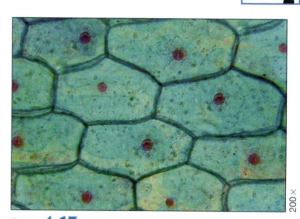

200 ×

FIGURE **4.17** Epidermal cells from onion skin.

Plant epidermal tissue

Total magnification _____

_____

_____

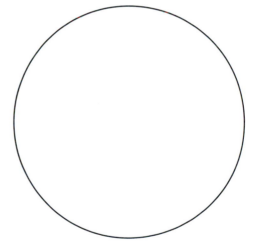

Wet mount of onion skin

Total magnification _____

_____

_____

Procedure 2
Elodea

4

Materials
- ❏ Compound microscope
- ❏ Living specimen of *Elodea*
- ❏ Blank slides and coverslips
- ❏ Transfer pipette
- ❏ Tap water
- ❏ Forceps
- ❏ Paper towels
- ❏ Colored pencils

Elodea is a common plant that lives in freshwater habitats such as ponds and lakes. It provides an excellent example for studying basic plant cell anatomy. The leaves of *Elodea* are only a few cells thick and allow light to pass through the leaf without special preparation techniques. Refer to Figure 4.18 and Table 4.3 for references to plant cell anatomy.

1 Procure a microscope, a blank slide, coverslips, and a transfer pipette. Carefully remove a single healthy leaf from the *Elodea*. Place the leaf in a drop of water on the blank slide with the top surface facing upward. (The cells on the upper surface are much larger and easier to observe.) Place a coverslip over the *Elodea*. Periodically check the leaf, making sure it does not dry out. If the leaf begins to dry, add a drop of water with a pipette.

2 Examine the leaf surface with the scanning and low-power objectives. Focus through the cell layers of the *Elodea*.

3 Describe and sketch *Elodea* in the space provided. In this specimen, only the cell wall, cytoplasm, and chloroplast will be easily observed.

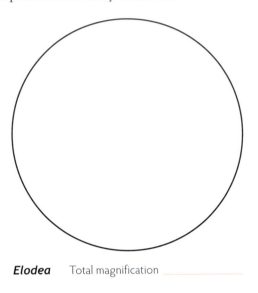

Elodea Total magnification _____

_____

_____

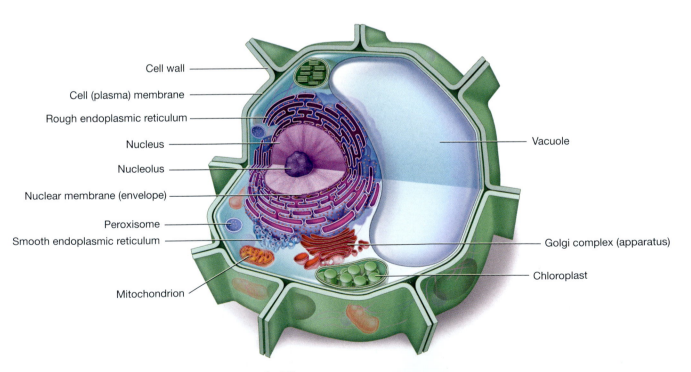

FIGURE **4.18** Typical eukaryotic plant cell.

TABLE 4.3 Common Anatomical Features of Eukaryotic Cells

Structure	Function
Cell wall	In plant cells, a cellulose envelope that provides protection and shape
Plasma membrane	Phospholipid bilayer that provides support and regulates the movement of substances into and out of the cell
Cytoplasm	Semifluid medium located between the plasma membrane and nucleus; inclusions and organelles are found in the cytoplasm
Nucleus	Control center of the cell
Nuclear envelope	Membrane surrounding the nucleus; possesses numerous nuclear pores
Nucleoplasm	Cytoplasm within the nucleus
Nucleolus	Chromatin-rich region that serves to combine proteins and RNA to make ribosomal subunits; many cells possess numerous nucleoli
Chromatin	Diffuse, threadlike strands composed of DNA and proteins
Mitochondrion	Site of aerobic cellular respiration
Endoplasmic reticulum (ER)	Network of membranes throughout the cytoplasm; synthesis of protein and nonprotein products
Rough ER	Lined with ribosomes; involved in the synthesis and assembly of a variety of proteins and production of membranes
Smooth ER	Not associated with ribosomes; main site of steroid, fatty acid, and phospholipid synthesis; site of detoxification
Golgi complex (apparatus)	Stacks of flattened membranous sacs or cisternae; receives, packages, stores, and ships protein products; produces lysosomes and other vesicles
Peroxisome	Vesicle containing enzymes that help in breaking down fatty acids and neutralizing hydrogen peroxide
Lysosome	In animal cells, vesicle containing hydrolytic digestive enzymes used in destroying cellular debris and worn-out organelles; also important in programmed cell death
Centrioles	Found in animal cells with the exception of roundworms (nematodes); appear as a pair of cylindrical structures made of microtubules; form the spindle apparatus in cell division
Ribosomes	Sites of protein synthesis
Cytoskeleton	Structures that help the cell maintain its shape, anchor organelles, and move; three kinds of cytoskeletal elements are recognized: microtubules, microfilaments, and intermediate fibers
Chloroplasts	In plant cells, sites of photosynthesis; contain grana, or "stacks," composed of chlorophyll-rich thylakoids
Central vacuole	In plant cells, large fluid-filled sac that helps maintain the shape of the cell and stores metabolites
Middle lamellae	Region between adjacent plant cells that cements the cell walls together

4 Using the high-power objective, examine a single cell of *Elodea*. Attempt to locate the structures indicated in Figure 4.19. The gray-colored nucleus may be difficult to locate. The nucleus may become more evident whether a drop of iodine is placed upon the leaf. In a good preparation, the nucleolus may be evident. Carefully notice whether the cytoplasm and chloroplasts are moving. This process is called **cytoplasmic streaming.**

5 Describe and sketch *Elodea* in the space provided.

6 After completion of the activity, clean up your work area, disinfect the lab surface, and return or dispose of the material as instructed.

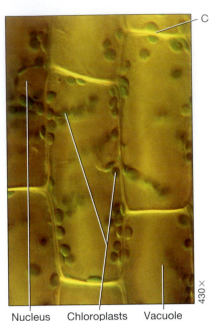

Cell wall

430×

Nucleus   Chloroplasts   Vacuole

FIGURE **4.19** *Elodea* is a common plant found in freshwater ponds and lakes.

Elodea   Total magnification _____

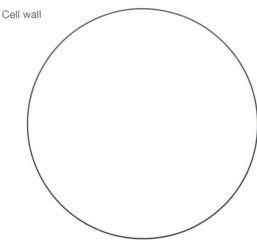

✔ Check Your Understanding

7.1 Name the structures you observed within the *Elodea* cells. For each cellular structure observed, list its function.

_____

_____

_____

7.2 An *Elodea* leaf is an excellent example to use for studying plant cells. Why?

_____

7.3 Describe the shape of the chloroplasts seen in the *Elodea* cells.

_____

7.4 Where in the cell do you see the nucleus and most of the chloroplasts in both the onion and the *Elodea* cells?

_____

7.5 Describe cytoplasmic streaming and suggest a function for this process.

_____

_____

Observing Animal Cells

Kingdom Animalia encompasses more than 1.5 million species of multicellular heterotrophs. Members of the animal kingdom vary tremendously, from simple sponges to humans. The cells lining your mouth along the inside of your cheeks are excellent examples of typical animal cells. These simple cells, known as squamous epithelial cells, are flat and thin and possess an obvious nucleus. Epithelial cells appear in regions of wear and tear and are constantly being sloughed away. In this specimen, only the cell membrane, cytoplasm, and nucleus will be easily observed. Refer to Figure 4.20 and Table 4.3 for references to animal cell anatomy.

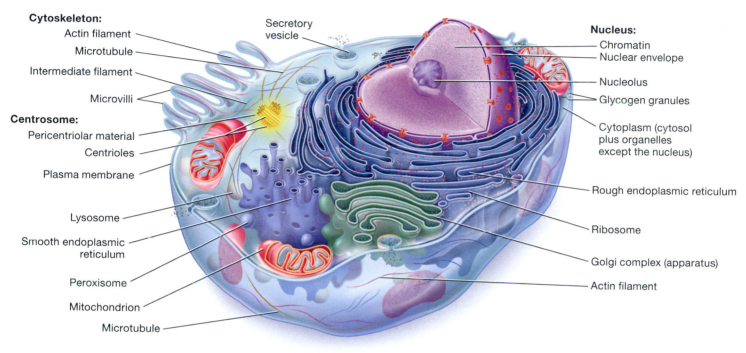

Cytoskeleton:
Actin filament
Microtubule
Intermediate filament
Microvilli

Centrosome:
Pericentriolar material
Centrioles
Plasma membrane

Lysosome
Smooth endoplasmic reticulum
Peroxisome
Mitochondrion
Microtubule

Secretory vesicle

Nucleus:
Chromatin
Nuclear envelope
Nucleolus
Glycogen granules
Cytoplasm (cytosol plus organelles except the nucleus)
Rough endoplasmic reticulum
Ribosome
Golgi complex (apparatus)
Actin filament

FIGURE **4.20** Typical eukaryotic animal cell.

Procedure 1

Human Epithelial Cells

Materials
- Gloves
- Compound microscope
- Blank slides and coverslips
- Methylene blue
- Toothpicks
- Transfer pipette
- Tap water
- Paper towels
- Colored pencils

WARNING

Avoid inhalation and skin contact with methylene blue. Immediately rinse it off the skin with mild soap and water because methylene blue will stain clothing. To avoid contact with skin, wear gloves for the duration of this procedure. Follow additional guidelines dictated by your instructor and university.

1 Procure a microscope, a blank slide, coverslips, clean toothpicks, and a pipette. Using a clean toothpick, gently scrape the inside of your cheek.

2 Place a small drop of water on a blank slide. Gently roll and swirl the end of the toothpick with the epithelial scrapings in the drop of water. Discard the used toothpick into the designated biohazard container.

3 Carefully place a drop of methylene blue in the drop of water. Place a coverslip over the specimen and make your observations.

4 Describe and sketch the epithelial tissue (Fig. 4.21) in the space provided.

5 After completion of the activity, clean up your work area, disinfect the lab surface, and return or dispose of the materials as instructed.

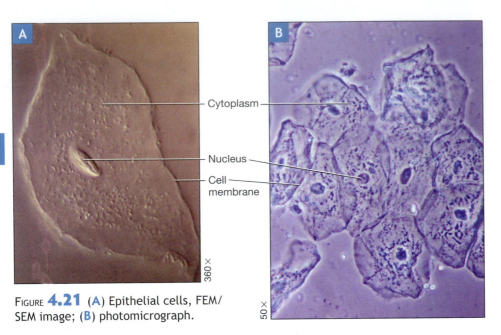

Cytoplasm

Nucleus

Cell membrane

360×

50×

FIGURE **4.21** (**A**) Epithelial cells, FEM/ SEM image; (**B**) photomicrograph.

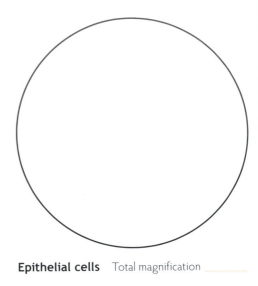

Epithelial cells   Total magnification _____

Procedure 2

Human Bone Cells

Materials

- ❏ Compound microscope
- ❏ Prepared slides of bone tissue
- ❏ Colored pencils

Humans are composed of several tissue types, including epithelial, connective, muscle, and nervous tissue. Detailed overviews of these tissues will be provided in Chapter 18. Of the connective tissues, one of the most distinguishable types is bone, or osseous tissue (Fig. 4.22). Bone consists of living cells dispersed in an organic and mineralized matrix. The most prominent portion of a transverse section of bone is the Haversian canal. These longitudinal channels contain nerves and blood vessels. Concentric rings of lamellae surround the Haversian canal and are made up of "little houses" called **lacunae**

that contain the bone cell, or osteocyte. Canaliculi are used for communications and appear as canals radiating from a lacuna.

To understand the structure of various human cells:

1 Procure a microscope and the prepared slide of bone tissue. Observe bone tissue using the low-power objective, and locate the Haversian canal, lamellae, lacunae, and canaliculi.

2 Describe, sketch, and label the bone specimen in the space provided.

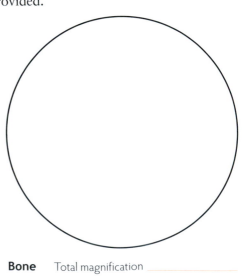

Bone   Total magnification _____

Lacunae

Canaliculi

Central (Haversian) canals

Lamellae

200×

FIGURE **4.22** Cross section of two osteons in bone tissue.

✔ Check Your Understanding _____

8.1 Name the structures you observed within the epithelial cells.

8.2 Describe the shape and size of the nucleus within the epithelial cells you observed. Did you see a nucleolus? If so, describe the nucleolus.

8.3 Bone, or osseous, tissue is what type of tissue? (*Circle the correct answer.*)
   a. Nervous.
   b. Connective.
   c. Epithelial.
   d. Muscle.

MYTHBUSTING

The Mind's Eye

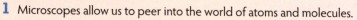

Debunk each of the following misconceptions by providing a scientific explanation. Write your answers on a separate sheet of paper.

1 Microscopes allow us to peer into the world of atoms and molecules.

2 All cells are microscopic.

3 All cells have mitochondria.

4 Viruses and bacteria can be easily seen with a compound microscope.

5 A compound microscope can reveal all of the internal structures of a typical eukaryotic cell in great detail.

1 A student wants to study the legs of an insect and the margin, or edge, of a leaf. Which microscope is the best choice?

2 Name the two types of electron microscopes. Describe what types of materials are studied using each type.

3 When you move the letter "e" slide toward you, does the letter "e" move toward you or away from you?

4 What is the total magnification of a microscope with a 10× ocular and a 100× oil-immersion objective?

5 Compare and contrast the characteristics of prokaryotic and eukaryotic cells.

6 Match the following functions with the prokaryotic cell structures:

_____ Cell wall	A.	Structure used for locomotion
_____ Pili	B.	Rigid, hairlike structures used for attachment and genetic material exchange
_____ Fimbriae	C.	Peptidoglycan envelope that gives protection and shape to a cell
_____ Nucleoid	D.	Short, hairlike structures important for attachment
_____ Flagellum	E.	Region of cell that contains bacterial DNA

7 Categorize the following organisms as unicellular (u), colonial (c), or multicellular (m).

A. _____ *Elodea*   D. _____ *Salmonella typhimurium*   F. _____ *Amoeba proteus*

B. _____ yeast   E. _____ *Volvox* sp.   G. _____ onion skin

C. _____ *Paramecium* sp.

8 Label the parts and functions of the microscope (**Fig. 4.23**).

1. _____

2. _____

3. _____

4. _____

5. _____

6. _____

7. _____

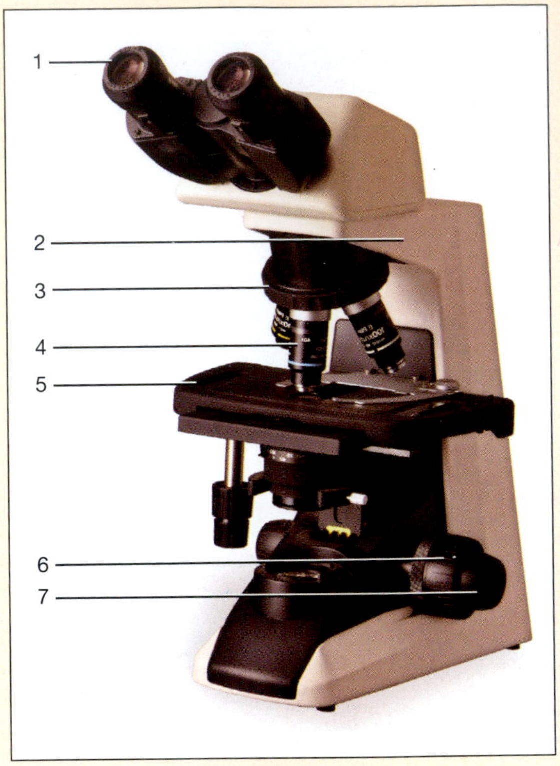

Figure **4.23** Light microscope.

9 Which of the following organelles/structures are found in plant cells, animal cells, or both plant and animals? Use the following key for your answers:

P = Plant cell only   A = Animal cell only   B = Both plant and animal cells

A. _____ cell wall   D. _____ cytoplasm   G. _____ lysosomes

B. _____ plasma membrane   E. _____ central vacuole   H. _____ centrioles

C. _____ chloroplasts   F. _____ nucleus

10 Label the anatomical features of a bacterium (**Fig. 4.24**).

1. _____

2. _____

3. _____

4. _____

5. _____

6. _____

7. _____

8. _____

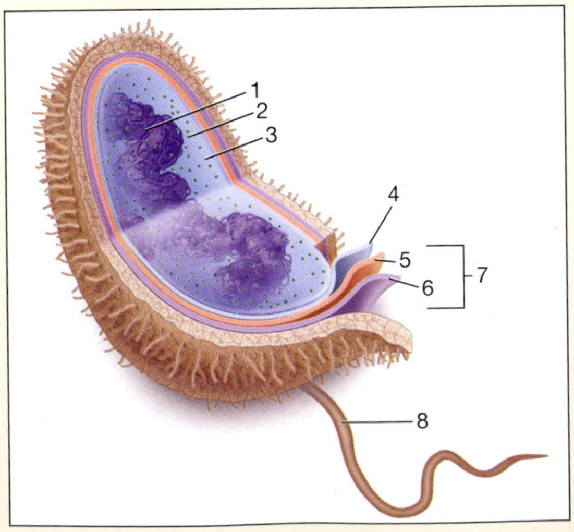

Figure **4.24** Basic structure of a generalized bacterial cell.

Metabolism I
Understanding Enzymes and Cellular Transport

5

"It seems to me that the natural world is the greatest source of excitement; the greatest source of visual beauty; the greatest source of intellectual interest. It is the greatest source of so much in life that makes life worth living."

— Sir David Attenborough (1926–Present)

Just wondering . . .

Consider the following questions prior to coming to lab, and record your answers on a separate piece of paper.

1 Is the color pattern in Siamese cats influenced by climatic conditions and age?

2 Is there a relationship between enzymes and puppy breath? Explain.

3 A medical technologist notes that in one of the samples to be tested many red blood cells are crenated (shriveled up). What is an explanation for this observation?

4 When salt water intrudes into a freshwater marsh as the result of a hurricane or the actions of humans, what happens to many of the freshwater plants?

5 How can some organisms such as blue crabs, oysters, and some fishes (even some sharks) live in varied salinities?

6 What is the relationship between hypertonicity and the genetic disorder cystic fibrosis?

Objectives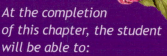

At the completion of this chapter, the student will be able to:

1. Define metabolism, enzyme, activation energy, substrate, and active site.

2. Discuss the importance of enzymes to living systems and list the characteristics of enzymes.

3. Discuss how enzymes work.

4. Describe the effects of the enzyme bromelain on gelatin.

5. Describe the effects of temperature and pH upon enzyme activity.

6. Compare, contrast, and describe several means of passive and active transport.

7. Compare and contrast endocytosis and exocytosis.

8. Define equilibrium, solution, solvent, solute, cytolysis, and plasmolysis.

9. Compare and contrast hypotonic, hypertonic, and isotonic solutions.

Chapter Photo
Molecular model of oxygen-producing oxygenase.

Within an organism, the sum total of chemical processes is called **metabolism.*** Some processes break down substances and are called **catabolic** (degradation). Other processes build new substances and are called **anabolic** (synthesis).

Enzymes

Most of the chemical reactions within living systems are controlled by specialized proteins called enzymes. An **enzyme** is a biological catalyst that accelerates a chemical reaction without itself being affected by the reaction.

Within an organism, many metabolic pathways are involved in breaking down and forming products. By lowering the activation energy required for a reaction to take place, enzymes ensure these pathways do not slow down and become congested. The **activation energy** is the original input of energy necessary to initiate a reaction. In Figure 5.1, the activation energy is described as the amount of energy necessary to push the reactants over a barrier so the reaction can begin.

Two physical means of attaining the activation energy are heat and agitation, which increase the number of collisions between reactants and speed up the reaction. These means, however, may damage living systems. Thus, in living systems, enzymes serve as catalysts to lower the amount of activation energy required. Notice in Figure 5.1 that in the presence of

*Thank you to Sarah Jean Rayner for her contributions to Chapters 5 through 9.

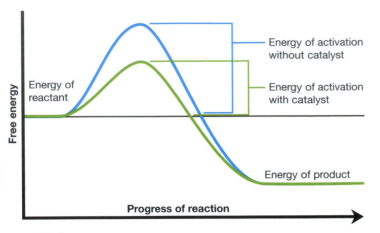

FIGURE **5.1** Less activation energy is needed in the presence of a catalyst.

plus water, which yields carbonic acid (CO_2 + H_2O = H_2CO_3), occurs slowly. Blood transports carbon dioxide from cells to the lungs, and the amount that can be transported is 300 times greater when carbonic anhydrase converts the carbon dioxide to carbonic acid. Without enzymes like carbonic anhydrase, the chemistry of life would be too slow.

Enzyme action often is assisted by other molecules (Fig. 5.2). **Cofactors** are usually nonprotein or metallic ions such as copper, zinc, and manganese, which are required for enzymatic activity. **Coenzymes** are nonprotein organic molecules that improve enzymatic action. An important coenzyme involved in energy metabolism is nicotinamide adenine dinucleotide (NADH) derived from niacin, a B vitamin. Vitamins or their derivatives commonly function as coenzymes.

an enzyme, less free energy (G) is needed to overcome the barrier, but the change in free energy (ΔG) remains the same.

Mechanism of Enzymatic Action

In an enzymatic reaction, the reactant the enzyme acts upon is the **substrate**. Enzymes are substrate-specific. The specificity of an enzyme results from the enzyme's unique, three-dimensional molecular conformation (shape). When the enzyme and the substrate join to make an **enzyme-substrate complex**, catalytic actions of the enzyme convert the complex into one or more products. The product is released from the enzyme, freeing the enzyme to react with another molecule of substrate.

A portion of the enzyme called the **active site** binds to the substrate. Usually, weak ionic or hydrogen bonds link the substrate and the active site. In some cases, an enzyme reacts with a specific substrate thousands of times per second. The **induced-fit model** explains how an active site on an enzyme changes its shape slightly to accommodate the substrate. In addition, many enzyme-catalyzed reactions are reversible; the same enzyme can catalyze a reaction in either the forward or the reverse direction.

More than a thousand different enzymes have been described. Enzymes may be specific to a certain type of cell and a certain species of organism. Enzyme names usually end in ase, such as sucrase and amylase, although others, such as bromelain, also exist. An example of a metabolically important enzyme found in vertebrate red blood cells is carbonic anhydrase. Without the presence of carbonic anhydrase, the reversible chemical reaction carbon dioxide

Factors Affecting Enzymatic Action

Enzymatic action can be affected by a number of external factors (e.g., concentration of substrate, concentration of enzymes, temperature, pH, salinity). An enzyme functions best within certain parameters called optimal conditions. All enzymes have an optimal temperature at which they react more rapidly. Most human enzymes are more efficient near 37°C.

Increasing the temperature of an enzyme-driven reaction speeds up the reaction to a certain point. If the temperature is below or above the optimum temperature, the reaction of the enzyme and the substrate is impeded. The optimum pH for most enzymes is between 6 and 8. Exceptions include the digestive enzyme pepsin, which digests proteins in the stomach at pH 2.

Any factor that may alter the unique shape of an enzyme may affect its ability to serve as a catalyst. If the shape of an enzyme is altered and loses its function, the enzyme is said to be **denatured**. For example, when egg white solidifies during cooking, it has been denatured. **Inhibitors** bind to an enzyme and decrease its activity. Usually, the end product of a given reaction inhibits the action of an enzyme. Conversely, **activators** increase enzyme activity.

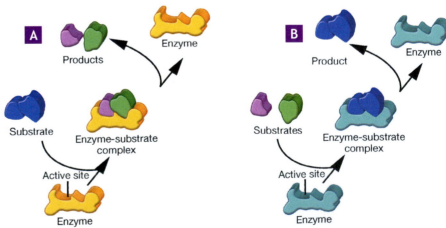

FIGURE **5.2** Enzymatic action: (**A**) catabolism, or degradation; (**B**) anabolism, or synthesis.

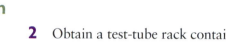

EXERCISE 5.1

Bromelain as an Enzyme

The following laboratory procedures have been designed to investigate an enzyme that occurs in pineapples (*Ananas comosus*) called **bromelain**. Pineapple plants, along with orchids, are classified as bromeliads. Although native to South America, pineapples are grown in many tropical areas, including Hawaii, Brazil, and Thailand. The enzyme bromelain can be found in the leaves, stems, and fruit of pineapples. Bromelain is a proteolytic enzyme (protease) that breaks down proteins into their amino acids by hydrolysis.

The preparation directions of many gelatins, including Jell-O, recommend against the user placing certain fruits, including fresh pineapple, into the product. Gelatin is a protein obtained from collagen (a structural protein component of fibrous connective tissues such as animal hooves). The proteins essentially trap and absorb water, allowing the gelatin to set. Bromelain and other proteolytic enzymes degrade the gelatin proteins and prevent the gelatin from setting. A number of variables, including temperature and pH, can affect how bromelain reacts with gelatin.

5

Procedure 1

The Effect of Bromelain on Jell-O Formation

Materials

- ❏ Test-tube rack
- ❏ Test-tube clamp
- ❏ Test tubes (3)
- ❏ Syringes or transfer pipettes
- ❏ Water
- ❏ Canned pineapple juice
- ❏ Fresh pineapple juice
- ❏ Eye protection
- ❏ Laboratory apron
- ❏ 3 mL warm gelatin
- ❏ Wax pencil or Sharpie
- ❏ Ice and ice bucket
- ❏ Timing device

⚠ WARNING

Eye protection is required. Wear a laboratory apron while completing this procedure.

1 Construct a simple hypothesis addressing the effect of bromelain on the formation of Jell-O. Identify the control, independent, and dependent variables addressed in this experiment.

2 Obtain a test-tube rack containing 12 test tubes. In this activity, three test tubes will be used. Using a wax pencil or sharpie, label the test tubes: 1 Water, 2 Fresh, and 3 Canned. To each of the three test tubes, add 3 mL of warm gelatin provided by the instructor.

3 With separate syringes or transfer pipettes, obtain 2 mL of water, 2 mL of fresh pineapple juice, and 2 mL of canned pineapple juice. Rapidly push the following into the test tube to mix the gelatin with the solution:

Test Tube 1: 2 mL water

Test Tube 2: 2 mL fresh pineapple juice

Test Tube 3: 2 mL canned pineapple juice

4 Place the test tubes on ice. Carefully watch Test Tube 1; its contents should solidify within 5 to 10 minutes. After it solidifies, observe Test Tubes 2 and 3. Record your observations in Table 5.1.

5 Return the supplies to the main table, clean and disinfect the lab surface, and dispose of the materials as directed by the laboratory instructor.

TABLE **5.1** Results of Bromelain Activity in Jell-O

Test Tube	Contents	Does Solidification Occur? (Y/N)	Observations
1	Water		
2	Fresh pineapple juice		
3	Canned pineapple juice		

Procedure 2

The Effect of Temperature on Bromelain Activity

Materials

- ❏ Test-tube rack
- ❏ Test-tube clamp
- ❏ Test tubes (3)
- ❏ Syringes or transfer pipettes
- ❏ Ice container
- ❏ Hot plate
- ❏ Thermometer
- ❏ Container for hot-water bath
- ❏ Water
- ❏ Fresh pineapple juice
- ❏ Ice
- ❏ Eye protection
- ❏ Gloves
- ❏ Laboratory apron
- ❏ 3 mL warm gelatin
- ❏ Sharpie or wax pencil

1 Construct a simple hypothesis addressing the effect of temperature on bromelain. Identify the negative control, independent variable, and dependent variable addressed in this experiment.

_____

_____

_____

_____

_____

2 Procure three test tubes. Using a Sharpie or wax pencil, label the test tubes 1 through 3.

 WARNING

Eye protection is required. Wear gloves and a laboratory apron while completing this procedure.

3 With separate syringes or transfer pipettes, add 2 mL of water to Test Tube 1, and place it in the test-tube rack. Add 2 mL of fresh, not canned, pineapple juice to Test Tubes 2 and 3.

4 Leave Test Tubes 1 and 2 at room temperature. Place Test Tube 3 into a 70°C water bath for 5 minutes. After 5 minutes, carefully remove Test Tube 3 from the water bath.

5 Add 3 mL of warm gelatin provided by your instructor to each test tube. Place all three test tubes in ice until the contents of Test Tube 1 begin to solidify.

6 Observe the test tubes, and record your results in Table 5.2.

7 Return the supplies to the main table, dispose of the materials as directed by the laboratory instructor, and clean and disinfect the lab surface.

TABLE **5.2** Effects of Temperature on Bromelain Activity

Test Tube	Contents	Does Solidification Occur? (Y/N)	Observations
1	Water		
2	Cold juice		
3	Hot juice		

Did you know . . .

Got Enzymes?

Enzymes determine the color pattern in Siamese cats. A heat-sensitive enzyme that helps to control melanin production is less active in the warmer (lighter-colored) regions of the body. The cooler extremities are darker in color.

Procedure 3

The Effect of pH on Bromelain Activity

Materials
- Test-tube rack
- Test-tube clamp
- Test tubes (6)
- Syringes or transfer pipettes
- Ice container
- Water
- Fresh pineapple juice
- Ice
- 0.1 M HCl
- 0.1 M NaOH
- Eye protection
- Gloves
- Laboratory apron
- 3 mL warm gelatin
- Wax pencil or Sharpie

WARNING

Eye protection is required. Wear gloves and a laboratory apron while completing this procedure.

1 Construct a simple hypothesis addressing the effect of pH on bromelain. Identify the negative control, independent variable, and dependent variable addressed in this experiment.

2 Procure six test tubes. Using a wax pencil or Sharpie, label the test tubes 1 through 6. Using a pipette, carefully add the following to the test tubes:

Test Tube 1: 2 mL water

Test Tube 2: 2 mL fresh pineapple juice

Test Tube 3: 1 mL HCl and 1 mL fresh pineapple juice

Test Tube 4: 1 mL HCl and 1 mL water

Test Tube 5: 1 mL NaOH and 1 mL fresh pineapple juice

Test Tube 6: 1 mL NaOH and 1 mL water

3 Mix the components by gently and carefully swirling the tubes and letting them sit for 3 minutes. Add 3 mL of gelatin to each test tube.

4 Place all six test tubes in ice until Test Tube 1 begins to solidify. Observe the test tubes, and record your results in Table 5.3.

5 At the completion of the activity, return the supplies to the main table, dispose of the materials as directed by the laboratory instructor, and clean and disinfect the lab surface.

TABLE **5.3** Effects of pH on Bromelain Activity

Test Tube	Contents	Did the pH Level Have an Effect? (Y/N)	Observations
1	2 mL of water		
2	2 mL of fresh pineapple juice		
3	1 mL HCl + 1 mL fresh pineapple juice		
4	1 mL HCl + 1 mL water		
5	1 mL NaOH + 1 mL fresh pineapple juice		
6	1 mL NaOH + 1 mL water		

✔ Check Your Understanding

1.1 When canned pineapple juice is added to gelatin, it solidifies. This is not the case when fresh pineapple juice is added to gelatin. Explain why.

1.2 _____ , the enzyme occurring naturally in pineapples, is considered a

enzyme, which breaks down proteins into their amino acids by hydrolysis.

1.3 Explain how extremes of temperature and pH alter the activity of an enzyme.

EXERCISE 5.2

Catalase and Reusability of Enzymes

An enzyme serves as a biological catalyst, speeding up a reaction without itself being consumed or physically altered. Another characteristic of an enzyme is that it is reusable within reactions. Hydrogen peroxide (H_2O_2) is produced as a poisonous metabolic by-product in the liver of animals. In order to break down the H_2O_2 into water and oxygen, the liver produces the enzyme **catalase**. This enzyme acts quickly and can break down millions of H_2O_2 molecules per minute.

Procedure 1

The Effect of Catalase on Hydrogen Peroxide

Materials

- Test-tube rack
- Test-tube clamp
- Test tubes (9)
- 3% hydrogen peroxide solution
- Small pieces of beef or chicken liver (9): 3 raw, 3 boiled, and 3 ground
- Sugar water (10% sucrose solution)
- Tap water
- Eye protection
- 25 mL graduated cylinder (2)
- Wax pencil or Sharpie

WARNING

Hydrogen peroxide can irritate the skin.

1 Procure the materials needed for the activity. Using a wax pencil or Sharpie, label the test tubes 1 through 9.

2 Using a graduated cylinder, pour 10 mL of hydrogen peroxide into Test Tubes 1, 4, and 7. Using a clean graduated cylinder, pour 10 mL of tap water into Test Tubes 2, 5, and 8. Pour about 10 mL of sugar water into Test Tubes 3, 6, and 9.

3 Drop a small piece of raw liver into Test Tubes 1, 2, and 3. Record your observations in Table 5.4.

4 Drop a small piece of boiled liver into Test Tubes 4, 5, and 6. Record your observations in Table 5.4.

5 Drop a small piece of ground liver into Test Tubes 7, 8, and 9. Record your observations in Table 5.4.

6 Dispose of the materials in Test Tubes 4 through 9 as directed by your instructor. Leaving the liver in Test Tubes 1, 2, and 3, pour off the liquid as directed by the laboratory instructor.

7 Add an additional 5 mL hydrogen peroxide to Test Tube 1, 5 mL more tap water to Test Tube 2, and 5 mL more sugar water to Test Tube 3. Record your observations in Table 5.4.

8 Repeat step 7 three more times. Record your observations.

9 Thoroughly clean and disinfect your laboratory station as instructed.

TABLE **5.4** Observations of Various Solutions on Liver

Test Tube	Contents	Observations
1	Hydrogen peroxide with raw liver	
2	Tap water with raw liver	
3	Sugar water with raw liver	
4	Hydrogen peroxide with boiled liver	
5	Tap water with boiled liver	
6	Sugar water with boiled liver	
7	Hydrogen peroxide with ground liver	
8	Tap water with ground liver	
9	Sugar water with ground liver	
1	One addition of hydrogen peroxide	
2	One addition of tap water	
3	One addition of sugar water	
1	Three additions of hydrogen peroxide	
2	Three additions of tap water	
3	Three additions of sugar water	

✔ Check Your Understanding

2.1 Because enzymes are proteins, they can be destroyed by heat. If the beef or chicken liver was boiled for 5 minutes before placing it in the test tubes, what may have been the result?

2.2 In the liver of animals, the toxic compound hydrogen peroxide (H_2O_2) is produced. In order to break down this compound, the liver produces the enzyme _____ .

2.3 Enzymes have the ability to be reusable within biological reactions. Why is this important?

Cellular Transport

One of the major functions of the plasma membrane is to regulate the movement of substances into and out of the cell. This process is essential in maintaining the homeostatic state of the cell. The plasma membrane is composed primarily of a phospholipid bilayer and specialized proteins. The unique structure of the plasma membrane allows it to be **selectively permeable**. The permeability of a plasma membrane to a given molecule is dependent on the molecule's characteristics—size, charge, lipid solubility (Fig. 5.3)—and its concentration, as well as external factors, such as temperature and pressure. Living systems have two primary mechanisms for moving substances in and out of the cell: passive and active transport.

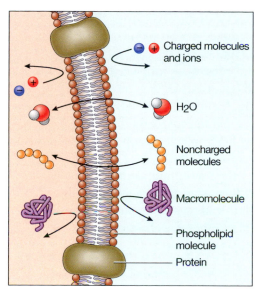

FIGURE **5.3** The plasma membrane serves as the active interface between the cell and its environment.

Passive Transport

Passive transport of essential substances (such as oxygen and water) through the plasma membrane does not require the use of cellular energy in the form of ATP.

Diffusion

The most fundamental means of passive transport is **diffusion**, the random movement of molecules from regions of greater concentration to regions of lesser concentration (Fig. 5.4). This random movement also is known as **Brownian motion**. A state of **equilibrium** is attained when an equal distribution of molecules exists throughout the system. The movement of water across the plasma membrane in living systems is called **osmosis**.

Other mechanisms of passive transport include facilitated diffusion and filtration.

1. In **facilitated diffusion**, carrier proteins along the cell membrane are required to carry specific molecules, such as glucose, across the membrane into the cell.

FIGURE **5.4** Diffusion of food coloring in water from regions of greater concentration to regions of lesser concentration.

2. **Filtration** involves hydrostatic pressure (water pressure) forcing molecules through a cell membrane. Filtration is an essential mechanism that takes place in the kidneys in the formation of urine.

Osmosis

Recall that osmosis is the diffusion of water across a selectively permeable membrane, and diffusion always occurs from regions of greater concentration to regions of lesser concentration. A typical **solution** consists of two components—the **solvent** as the dissolving medium and the **solute** as the substance dissolved in the solvent. In a saltwater solution, the water serves as a solvent and the salt as the solute. **Tonicity** refers to the concentration of solute in the solvent.

1. In a **hypotonic** solution, there is a lower concentration of solute relative to the inside of the cell. If a cell, such as a red blood cell or a potato cell, is placed in a hypotonic solution, water will rush into the cell in an attempt to reach a state of equilibrium. An ideal hypotonic solution is distilled water because it is devoid of solutes. As the cell begins to fill with solvent, the cell will swell and perhaps burst. The bursting of cells in a hypotonic solution is called **cytolysis**, or osmotic lysis (Fig. 5.5A).

2. In a **hypertonic** solution, there is a higher concentration of solute relative to the inside of the cell. If a cell, such as a red blood cell or potato cell, is placed in a hypertonic solution, water will be drawn out of the cell into the outside solution in an attempt to reach a state of equilibrium. This is called **crenation** in red blood cells and **plasmolysis** in plant cells (Fig. 5.5B). In plants, the swelling of cells placed in a hypotonic solution results in **turgor pressure**. The framework cell wall protects the cell from bursting. Turgor pressure keeps the plant erect. If the turgor pressure is lost in a plant, the plant will wilt. Just think of the plants in your yard on a hot summer day.

3. A solution that contains the same concentration of solutes as the inside of a cell is **isotonic** (Fig. 5.5C). If the concentration of solutes in plasma changes, water will move

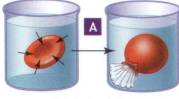
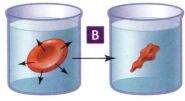

Hypotonic solution

FIGURE **5.5** Osmosis and animal cells. (**A**) When the outer solution is hypotonic in comparison to the cell, the solution will move into the cell and the cell will lyse. (**B**) When the solution is hypertonic in comparison to the cell, the solution will move out of the cell and the cell will shrink, or become crenate. (**C**) In an isotonic solution, homeostasis is achieved.

Hypertonic solution

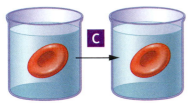

Isotonic solution

into or out of the red blood cell—from where there is more water to where there is less water—until equilibrium is established. Red blood cells are normally in an isotonic state with the plasma. Many marine invertebrates are osmoconformers and can maintain an internal salinity that is equal to the surrounding water. Some fishes have evolved a number of mechanisms in order to osmoregulate in a variety of environments.

Active Transport

The second mechanism for moving substances into or out of a cell occurs when the substance is to be moved from an area where the substance is present in a relatively low concentration to an area where the substance is in a higher concentration—that is, the substance is to be moved against its **concentration gradient**. This can be compared to moving an object uphill, which requires that the cell use energy, in the form of ATP, to move the substance across the plasma membrane. This process, called **active transport**, is made possible by a variety of "pumps" in the plasma membrane; the type of pump and its location are correlated to the structures

and function of the cell in which it occurs. Some examples include:

1. Proton pumps, responsible for moving hydrogen ions into the lumen of the stomach;

2. Calcium pumps, responsible for normal function of neurons and muscle cells; and

3. Sodium-potassium pumps, found in a wide variety of cells and integral to normal cellular metabolism.

Endocytosis and Exocytosis

Macromolecules, such as polypeptides and polysaccharides, are too large to be moved through a cell membrane by any of the processes described so far. Instead, they must be transported into the cell by **endocytosis** and out of the cell by **exocytosis**. Both endocytosis and exocytosis require the formation of a vesicle made of plasma membrane for transporting substances in and out of the cell.

Transport of solid substances into a cell involves a form of endocytosis called phagocytosis (Fig. 5.6), which translates as cellular eating. Examples of phagocytosis include the ingestion of yeast by a paramecium and the engulfing of foreign substances by macrophages, a type of white blood cell that protects us from pathogens.

Relatively large volumes of fluid, with their solutes, can be taken up by a cell through a form of endocytosis called pinocytosis, or cellular drinking. For example, specialized cells in the roots of plants ingest a variety of dissolved nutrients via pinocytosis.

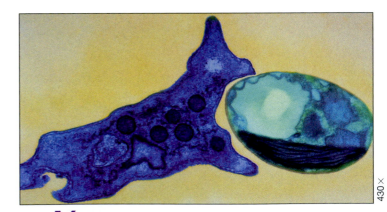

FIGURE **5.6** Amoeba capturing its food through phagocytosis.

Procedure 1

Simple Diffusion Exercises

Materials
- ❏ 100 mL beakers (3)
- ❏ Ice water
- ❏ Room-temperature water
- ❏ Hot water
- ❏ Green food coloring
- ❏ Timing device

In Part 1 of this activity, the instructor will spray perfume or scented air freshener in a front corner of the classroom. Timing devices will be used to determine how long it takes for the molecules to diffuse throughout the room in an attempt to attain a state of equilibrium. In Part 2, diffusion of a colored chemical (green food coloring) throughout water at different temperatures is observed and the results recorded.

Part 1

1 Students should be equally dispersed throughout the classroom. Record your distance from the instructor.

_____

2 The instructor will spray a small amount of a scent in the front corner of the room, and this will serve as time zero. The molecules will diffuse throughout the room in an attempt to attain a state of equilibrium. Each student will record the time and raise their hand when they detect the odor.

_____

3 Discuss your results. How long did it take for the scent to reach the farthest point of the room? What variables can affect the rate of diffusion of the scent?

_____

_____

_____

_____

Part 2

1 Procure three beakers and label them 1, 2, and 3.

2 In Beaker 1, put 200 mL of ice water. In Beaker 2, put 200 mL of room-temperature water. In Beaker 3 put 200 mL of hot water.

3 Predict which beaker will diffuse food coloring the fastest and why.

_____

_____

_____

4 In each beaker, put 5 drops of green food coloring. Observe as the colored chemicals diffuse in the water. Record your observations in Table 5.5.

5 Which beaker of water diffuses the food coloring the fastest? Which beaker diffuses the food coloring the slowest?

_____

_____

_____

TABLE **5.5** Diffusion of Food Coloring in Water

Beaker	Contents	Speed of Diffusion
A	Ice water	
B	Room-temperature water	
C	Hot water	

Diffusion across a Membrane

Materials

- ❏ 100 mL beaker
- ❏ 25 mL graduated cylinder
- ❏ 1% cornstarch solution
- ❏ Iodine solution (4% iodine in 200 mL of water)
- ❏ Tap water
- ❏ Dialysis tubing
- ❏ String
- ❏ Timing device
- ❏ Ruler
- ❏ Paper towels
- ❏ Clip

In this activity, dialysis tubing will serve as the selectively permeable membrane. After students make a bag with the dialysis tubing and fill it with colorless cornstarch solution, the bag will be immersed in a beaker containing a caramel-colored iodine solution (Fig. 5.7). Movement of the iodine molecules across the membrane can be detected by a change in the cornstarch solution to a purplish-brown color.

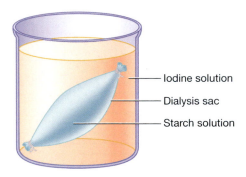

FIGURE **5.7** Color changes in the bag and solution.

- Iodine solution
- Dialysis sac
- Starch solution

1 Procure glassware and needed accessory materials.

2 Measure and cut a 12 cm length of dialysis tubing. Place the tubing in water until it becomes soft and pliable. While it is still under the water, gently rub the tubing with your fingers to open it at the ends. After the tubing begins to open, use your finger to create a larger opening.

3 Form a bag with the dialysis tubing by closing one end with string or tying one end into a knot. Fill the dialysis tubing bag with 15 mL of the cornstarch solution. Make sure the bag remains flaccid. With another length of string, securely tie off the top end of the bag or drape the other end of the dialysis tubing over the beaker and clip it to the beaker with a clip.

4 Immerse the bag containing the cornstarch solution into a beaker containing 100 mL of iodine solution.

   a. What is the color of the iodine solution?

   b. What is the color of the cornstarch solution?

   c. What do you predict will happen in this activity?

5 Leave the apparatus undisturbed for 15 minutes. Every 3 minutes record the colors in Table 5.6. Remove the bag from the solution, and place it on a paper towel. Observe the color changes in the dialysis bag.

   a. What color changes did you observe in the bag and in the solution?

   b. Explain the color changes in the solution and in the bag.

TABLE **5.6** Laboratory Data for Membrane Diffusion

Time (minutes)	Color of Bag	Color of Beaker
0		
3		
6		
9		
12		
15		

✓ Check Your Understanding

3.1 Define diffusion.

3.2 What factors influence the rate of diffusion? Do they cause an increase or a decrease?

3.3 How do you know that iodine molecules diffused through the dialysis tubing in Procedure 2?

Did you know . . .

Why Do Sea Turtles Cry?

Have you ever watched a nature show and seen sea turtles crying as they lay their eggs on some lonesome beach? Why do turtles cry? Is it because they are anticipating the future of their young, or perhaps they are sentimental, or maybe it hurts? No, it's not any of these reasons. Marine turtles have glands in the regions of their eyes that help them remove excess salt consumed from the hypertonic solution in which they live. That is why these turtles appear to cry!

Crying sea turtle.

Osmosis

Procedure 1

Observing Osmosis

5

Materials
- ❏ Freshly peeled potato
- ❏ Compound microscope
- ❏ Blank microscope slides (3)
- ❏ Coverslips (3)
- ❏ Wax pencil or Sharpie
- ❏ Tap water
- ❏ Distilled water
- ❏ 10% sodium chloride (salt) solution
- ❏ Paper towels
- ❏ Forceps
- ❏ Scalpel
- ❏ Colored pencils

1 Procure the needed equipment, and bring it to your lab station. With a wax pencil or Sharpie, label the corner of one microscope slide C for the control, another microscope slide O for the hypotonic solution, and the third slide E for the hypertonic solution.

2 With a scalpel, remove a small cube (25 cm × 25 cm × 25 cm) from a freshly peeled potato. Using tap water, prepare a wet mount (Slide C) of the potato, and observe the slide with the microscope at 40×. Record your observations, and sketch the control slide in the circle labeled "Control."

3 Using distilled water, prepare a wet mount (Slide O) of the potato, and observe the slide with the microscope at 40×. Record your observations, and sketch the hypotonic solution slide in the circle labeled "Hypotonic solution."

4 Using a 10% sodium chloride solution, prepare a wet mount (Slide E) of the potato, and observe the slide with the microscope at 40×. Record your observations, and sketch the hypertonic solution slide in the circle labeled "Hypertonic solution."

5 Compare and contrast what happens to the potato cube under all three conditions.

_____

_____

_____

_____

_____

6 Clean your lab station thoroughly as directed by the instructor.

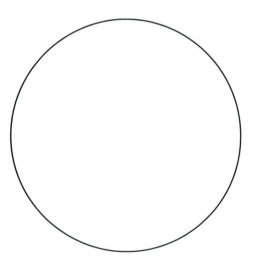

Control (C)

Total magnification _____

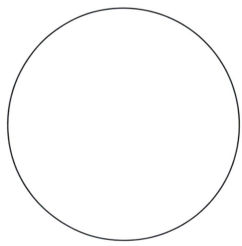

Hypotonic (O) solution

Total magnification _____

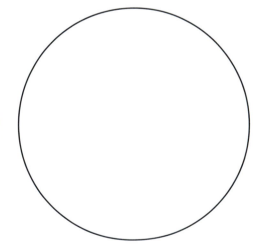

Hypertonic (E) solution

Total magnification _____

Procedure 2

Hypotonicity and Hypertonicity in Plant Cells

Materials

- ❏ *Elodea* leaves
- ❏ Compound microscope
- ❏ Blank microscope slides (3)
- ❏ Coverslips (3)
- ❏ Wax pencil or Sharpie
- ❏ Tap water
- ❏ Distilled water
- ❏ 10% NaCl solution
- ❏ Paper towels
- ❏ Forceps or scalpel
- ❏ Dropper

Elodea, or water weed, is a common freshwater plant native to North America. The plant features dark green leaves about 1.2 cm long existing in whorls of three around a green stalk. In this activity, a leaf of *Elodea* will be placed in hypotonic and hypertonic solutions, and observations will be made and recorded.

1 Procure the needed equipment, and bring it to your lab station.

2 With a wax pencil or Sharpie, label in the corner of three microscope slides as follows:

Slide 1: C for the control

Slide 2: O for the hypotonic solution

Slide 3: E for the hypertonic solution

3 With a scalpel or forceps, remove a healthy leaf from a stalk of *Elodea*. Prepare a wet mount (Slide C) of the *Elodea* leaf. To do this, add a leaf plus 1 to 2 drops of tap water. Drop a coverslip at a 45° angle and allow it to wick.

4 Observe the slide with the microscope at 40×. Record your observations, and sketch the control slide in the space provided.

5 Prepare a wet mount (Slide O) of the *Elodea* leaf by adding a leaf plus 1 to 2 drops of the distilled water solution. Drop a coverslip, allow it to wick, and then observe the slide with the microscope at 40×. Record your observations, and sketch the hypotonic solution slide in the space provided.

6 Prepare a wet mount (Slide E) of the *Elodea* leaf by adding a leaf plus 1 to 2 drops of the 10% NaCl solution. Drop a coverslip, allow it to wick, and then observe the slide with the microscope at 40×. Record your observations, and sketch the hypertonic solution slide in the space provided.

7 Compare and contrast the shape of the cells within the *Elodea* leaf under all three conditions.

_____

_____

_____

_____

8 Clean and disinfect your lab station thoroughly as directed by the instructor.

Control (C) (tap water)

Total magnification _____

_____

_____

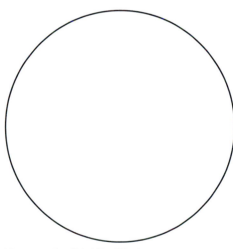

Hypotonic (O) (distilled water)

Total magnification _____

_____

_____

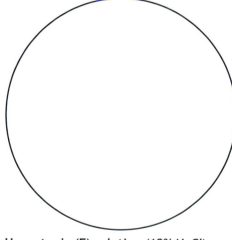

Hypertonic (E) solution (10% NaCl)

Total magnification _____

_____

_____

Procedure 3

Hypotonicity and Hypertonicity in Animal Cells

Materials

- ❏ Compound microscope
- ❏ Microscope slides and coverslips
- ❏ Whole sheep blood
- ❏ Test tubes (3)
- ❏ Test-tube rack
- ❏ Wax pencil or Sharpie
- ❏ 0.9% NaCl
- ❏ 10% NaCl
- ❏ Distilled water
- ❏ Paper towels
- ❏ Pipette
- ❏ Eyedropper
- ❏ Colored pencils

In this activity, sheep red blood cells will be placed in hypotonic and hypertonic solutions and observations made and recorded.

1 Procure the needed equipment, and bring it to your lab station.

2 With a wax pencil or Sharpie, label the three test tubes as follows:

Test Tube 1: 0.9% NaCl

Test Tube 2: 10% NaCl

Test Tube 3: Distilled water

3 Using a pipette, add 5 mL of 0.9% NaCl solution to Test Tube 1. Add 5 mL of 10% NaCl solution to Test Tube 2. Add 5 mL of distilled water to Test Tube 3.

4 Using the eyedropper, add 5 drops of whole sheep blood to each test tube. Let the solutions sit undisturbed for 1 minute. Hold each test tube in front of this printed page to determine which one is most clear.

5 Prepare a wet mount from each test tube. Using the compound microscope, observe each slide, and sketch and label your observations in the space provided.

a. Which solution was the clearest? Why?

b. Which solution was less clear? Why?

c. Which solution more closely resembles the tonicity of plasma? Why?

6 Clean and disinfect your lab station, and return the equipment.

 **WARNING**

Human red blood cells are not to be used. Sheep blood will be used for safety reasons.

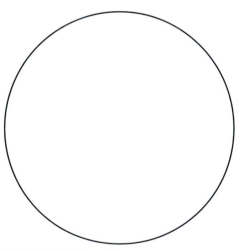

0.9% NaCl (isotonic)

Total magnification _____

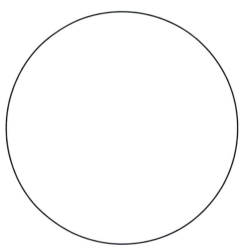

10% NaCl (hypertonic)

Total magnification _____

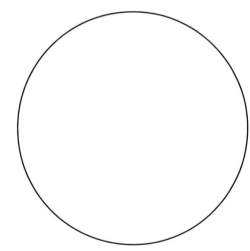

Distilled water (hypotonic)

Total magnification _____

✔ Check Your Understanding

4.1 During hurricanes, salt water is blown into freshwater marshes. Why do many of the freshwater plants and fish die?

4.2 At the cellular level, what would you expect to happen if you placed a piece of raw potato into tap water, distilled water, and a solution of 10% sodium chloride?

4.3 The aquatic plant *Elodea* was placed in distilled water and a 10% sodium chloride solution. Which caused the cells to swell? Which keeps the cells from bursting?

MYTHBUSTING

Covering All the Bases

Debunk each of the following misconceptions by providing a scientific explanation. Write your answers on a separate sheet of paper.

1 The low pH of the stomach destroys stomach enzymes.

2 Enzymes are the only biological catalysts.

3 Meat tenderizers contain acids that help break down meat.

4 Dolphin urine lacks salt.

5 Because many sharks are marine creatures, they cannot visit freshwater environments.

1 What is an enzyme? Identify three features of an enzyme.

2 Explain the induced-fit model of enzyme action.

3 Enzymes that break down substances are known as _____ enzymes.

4 What is activation energy? Is it increased or decreased by a catalyst?

5 Enzymatic action can be affected by which of the following? (*Circle the correct answer.*)

a. Enzyme and substrate concentration.

b. Temperature.

c. pH.

d. Salinity change.

e. All of the above.

6 An enzyme that has lost its function because of a change in its shape is said to be _____ . (*Circle the correct answer.*)

a. activated

b. inactivated

c. denatured

d. inhibited

e. saturated

7 What was the optimum temperature for bromelain activities?

_____

8 Referring to **Figure 5.8**, is the enzymatic action catabolism (degradation) or anabolism (synthesis)?

_____

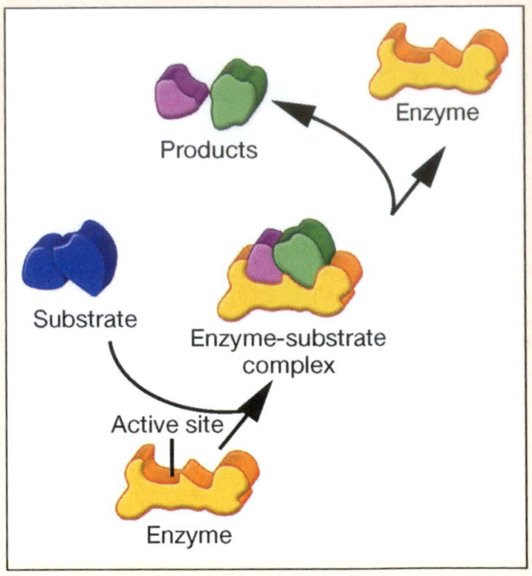

FIGURE **5.8** Enzymatic action.

9 Match the following types of molecules associated with each enzyme activity:

_____ Molecule that increases enzyme activity

_____ Nonorganic molecule, such as copper and zinc, which aids the action of an enzyme

_____ Molecule that binds to an enzyme, decreasing its activity

_____ Nonprotein organic molecule that improves enzymatic action

A. Inhibitor

B. Coenzyme

C. Cofactor

D. Activator

10 Match the following terms with the correct description.

_____ Solute

_____ Hypotonic

_____ Tonicity

_____ Hypertonic

_____ Isotonic

_____ Solvent

A. Lower concentration of solute relative to the inside of the cell

B. Concentration of solute in the solvent

C. Same concentration of solute relative to the inside of the cell

D. Dissolving medium

E. Higher concentration of solute relative to the inside of the cell

F. Substance dissolved in the solvent

11 Ingestion of foreign substances by macrophages and yeast cells by an amoeba is known as _____ .

12 Permeability of a plasma membrane to a given molecule is dependent on the particular molecule's characteristics of _____ . (*Circle the correct answer.*)

a. size

b. charge

c. lipid solubility

d. both a and c

e. all of the above

13 A red blood cell is placed in a hypotonic solution. What is the fate of the cell? (*Circle correct answer.*)

a. Initially the cell will fill with water but does not lyse.

b. The cell will crenate.

c. The cell will remain the same as homeostasis is achieved.

d. The cell will lyse.

Metabolism II
Understanding Photosynthesis and Cellular Respiration

6

In eating the plants, we combine the carbohydrates with oxygen dissolved in our blood because of our penchant for breathing air, and so extract the energy that makes us go. In the process we exhale carbon dioxide, which the plants then recycle to make more carbohydrates.

— Carl Sagan (1934–1996)

Just wondering . . .

Consider the following questions prior to coming to lab, and record your answers on a separate piece of paper.

1 Why should I "thank a plant"?
2 Why does the grass turn yellow under a brick after only a short period of time?
3 How much lumber and paper is produced annually?
4 What is the relationship between coal and photosynthesis?
5 What is the relationship between ATP and rigor mortis?

Objectives

At the completion of this chapter, the student will be able to:

1. Describe the role of photosynthetic organisms.
2. Describe the relationship between the visible spectrum and electromagnetic spectrum.
3. Discuss the location and products of the light-dependent and light-independent reactions.
4. Write and describe the overall reaction of photosynthesis.
5. Discuss the role of pigments in photosynthesis.
6. Describe the internal anatomy of a leaf.
7. Describe the role of the stomata in a leaf.

All living organisms share the ability to transform energy from one form to another. Life is a conquest of energy that all begins with the sun. The radiant energy of sunlight is converted into chemical energy (primarily glucose) through photosynthesis. In turn, the glucose is used to produce **adenosine triphosphate (ATP)**, the energy currency of living systems, through the process of **cellular respiration**. Plants and animals have a unique evolutionary relationship based upon each using the other's products. As a result of photosynthesis, some bacteria, algae, and plants produce oxygen as a waste product. In turn, oxygen is a substrate for **aerobic cellular respiration**. The plants ultimately use the CO_2 produced in cellular respiration to build carbohydrates.

■ Photosynthesis

It is time plants and other photosynthetic organisms receive a much-deserved "thank you." Without them we would not exist. On the early Earth, primitive photosynthetic bacteria and algae produced copious amounts of oxygen, changing the atmosphere forever and making it conducive to the development of more complex life forms. It is hard to believe, but the algae that inhabit primarily aquatic environments produce nearly 80% of the oxygen in the atmosphere today and serve as the basis of many food chains.

Plants such as grasses, shrubs, and trees dominate terrestrial landscapes, providing habitat, food, and commercial products. Plants are eloquent green machines that harness sunlight and produce carbohydrates and oxygen necessary for themselves as well as other living things. This remarkable process is termed **photosynthesis**. Keep in mind that photosynthesis is not the opposite of respiration, which will be discussed later in this chapter.

Chapter Photo
Close-up of a cottonwood leaf, *Populus deltoides*.

Absorption of Sunlight by Pigments

Sunlight powers photosynthesis. Using a prism, English physicist Isaac Newton (1642–1727) demonstrated that white light consists of a variety of colors ranging from red at one end of the **visible spectrum** to violet at the other end. In the mid-1800s, James Clerk Maxwell (1831–1879) illustrated that the visible spectrum was a minute portion of a continuous spectrum, or **electromagnetic spectrum**, which includes radio waves, visible light, X-rays, and cosmic rays (Fig. 6.1). Radiations of the spectrum travel in waves measured in nanometers (1 nm = 10^{-9} m). Radiations with longer wavelengths (radio waves) have less energy, and those with shorter wavelengths (X-rays) have more energy.

For an organism to use light energy, it has to absorb that energy. In living systems, **pigments** absorb light energy. Some pigments, such as melanin, absorb all wavelengths of light, and they appear black. At the other end of the spectrum, many pigments absorb only certain wavelengths of light and reflect the other wavelengths. For example green leaves contain the pigment chlorophyll, which absorbs red and blue light and reflects the green portion of the spectrum. Chlorophyll is the most important pigment in photosynthesis.

Several types of chlorophyll exist in nature. **Chlorophyll** *a* is the main photosynthetic pigment in some cyanobacteria and in plants. Other pigments important in plants but not involved directly in photosynthesis are called **accessory pigments**. Chlorophyll *b*, xanthophyll, and carotene are examples of accessory pigments that broaden the spectrum of visible light that can be absorbed and used for photosynthesis.

Chloroplasts reside within plant cells and serve as the organelles of photosynthesis. A chloroplast consists of a double membrane (inner and outer) that surrounds a semi-fluid matrix called the **stroma**. A third membrane system forms a series of flattened sacs or disks called **thylakoids**. In some chloroplasts, the thylakoids become stacked, forming a **granum** (plural = grana). Pigment molecules embedded in the membranes of the thylakoids initiate photosynthesis. Sugars are synthesized in the stroma.

Chemical Reaction of Photosynthesis

The general reaction for photosynthesis is:

$$6\ CO_2 + 6\ H_2O + \text{Light energy} \longrightarrow C_6H_{12}O_6 + 6\ O_2$$

This reaction is the result of a series of chemical reactions controlled and carried out by specific enzymes. The reactions of photosynthesis are divided into two distinct metabolic pathways (Fig. 6.2):

1. In the light reactions, or **light-dependent reactions**, chlorophyll *a* absorbs light energy, which leads to the production of ATP, oxygen gas, and the electron

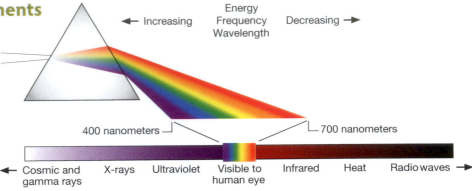

FIGURE 6.1 Electromagnetic spectrum.

carrier (nicotinamide adenine dinucleotide phosphate, or NADPH).

2. In the **light-independent reactions**, or Calvin cycle, carbon dioxide gas is incorporated into organic molecules (carbohydrates). The light-independent reactions take place in the stroma of the chloroplasts. They are responsible for fixing carbohydrates (including glucose). Photosynthetic organisms produce an estimated 200 billion metric tons of carbohydrate annually.

The oxygen gas produced during photosynthesis is used by plants and other organisms for performing aerobic cellular metabolism. The carbohydrates produced during photosynthesis are converted into many plant products, such as fiber, wood, and other structural materials. In addition, the simple sugars the plant produces can be converted into disaccharides and polysaccharides (such as starch) for energy storage. Sugars also are used in the synthesis of amino acids to form proteins and other cellular components. Life on Earth depends on the ability of photosynthetic organisms to convert the radiant energy of the sun into ATP, oxygen, and other nutrients.

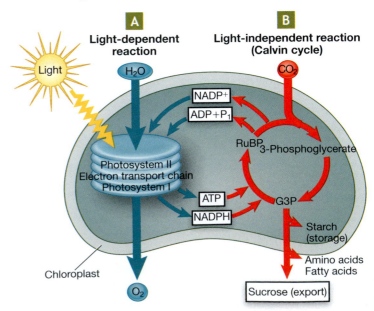

FIGURE 6.2 (**A**) Light-dependent and (**B**) light-independent (Calvin cycle) reactions.

Leaf Structure

Leaves are the most conspicuous part of a plant. They vary tremendously in shape and size, and some large trees have more than 100,000 leaves. One of the major functions of a leaf is that it serves as a photosynthesis factory. Leaves generally consist of a blade and a petiole (Fig. 6.3). The petiole attaches the flattened blade to the stem.

Procedure 1

Macroanatomy of a Leaf

Materials

❏ Dicot leaf, such as tomato, sunflower, or ivy
❏ Dissecting microscope or hand lens
❏ Colored pencils

In this activity, you will examine a typical dicot leaf.

1 Obtain a leaf from your instructor, and observe it using the dissecting microscope or hand lens.

2 Sketch the leaf, and label the blade and petiole in the space provided.

— Margin
— Blade
— Midrib
— Petiole

FIGURE **6.3** Generalized leaf structure.

6

External leaf anatomy

Procedure 2

Microanatomy of a Leaf

Materials

❏ Compound microscope
❏ Prepared slide of a dicot leaf, such as tomato, sunflower, or ivy
❏ Colored pencils

The internal anatomy of a typical leaf is complex (Fig. 6.4). A waxy **cuticle** covers the upper side of the leaf, and an **epidermis** completes the upper and lower layers of a typical leaf. Scattered primarily throughout the lower epidermis are **stomata** (sing. = stoma), tiny openings regulated by **guard cells**. The stomata allow the carbon dioxide from the atmosphere to enter the leaf and oxygen to exit.

The center of the leaf consists of the **mesophyll**, composed of **palisade parenchyma** and **spongy parenchyma**. The palisade parenchyma is columnar in shape and usually appears beneath the upper epidermis. The spongy parenchyma is loosely packed and surrounded by numerous air spaces. Most of the photosynthetic activity occurs in the cells of the palisade parenchyma.

In this activity, you will observe the internal structures of a leaf.

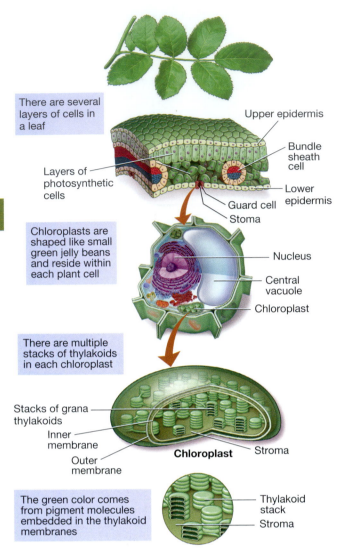

There are several layers of cells in a leaf

Upper epidermis

Bundle sheath cell

Layers of photosynthetic cells

Lower epidermis

Guard cell

Stoma

Chloroplasts are shaped like small green jelly beans and reside within each plant cell

Nucleus

Central vacuole

Chloroplast

There are multiple stacks of thylakoids in each chloroplast

Stacks of grana thylakoids

Inner membrane

Chloroplast   Stroma

Outer membrane

The green color comes from pigment molecules embedded in the thylakoid membranes

Thylakoid stack

Stroma

Figure **6.4** Leaf hierarchy.

1 Obtain a prepared slide of a leaf from your instructor.

2 Using a compound microscope, observe the leaf. Sketch a cross section of the leaf, and label the following structures: cuticle, epidermis, mesophyll, palisade parenchyma, spongy parenchyma, stomata, and guard cells.

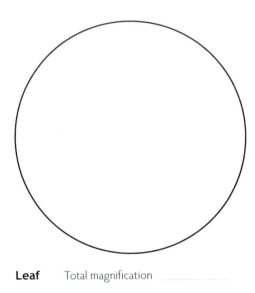

Leaf   Total magnification _____

Procedure 3

Observing Stomata and Guard Cells

Materials

- ❏ Compound microscope
- ❏ Slide
- ❏ Clear fingernail polish
- ❏ Clear tape, such as packing tape
- ❏ Scissors
- ❏ Colored pencils
- ❏ Fresh leaf (geranium, *Tradescantia*, or any fresh leaf) exposed to sunlight
- ❏ Fresh leaf (geranium, *Tradescantia*, or any fresh leaf) in darkness for several hours

In this activity, you will observe stomata and guard cells from a leaf.

1 Procure the materials needed for this activity. Paint a patch of clear fingernail polish about 1 cm² in size on the underside of the leaf exposed to sunlight. Let the fingernail polish dry completely, about 15 minutes.

2 Secure a piece of clear tape over the dried fingernail polish.

3 Carefully and gently begin peeling the tape and the fingernail polish beneath it from the leaf until the tape and fingernail polish are free of the leaf.

4 Place the tape (impression) on a clean microscope slide, and use scissors to trim away the excess tape. Place the slide on the microscope.

5 Scan the slide under low power, and find stomata and guard cells. Observe the specimen with high power as well. Sketch and describe your specimen in the circles provided.

6 Follow steps 1–3 using a leaf specimen that has been in the dark.

7 Return the equipment, and dispose of the leaves as instructed.

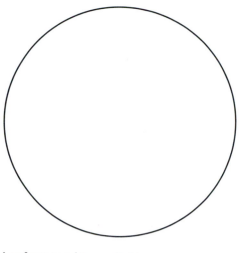

Leaf exposed to sunlight

Total magnification _____

_____

_____

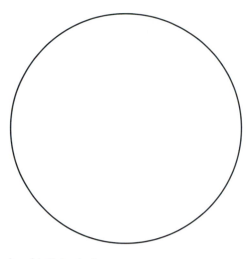

Leaf left in darkness

Total magnification _____

_____

_____

✔ Check Your Understanding

1.1 Referring to Figure 6.1, which has more energy: green light or purple light?

_____

1.2 What is the function of guard cells?

_____

_____

1.3 Where are the stomata and guard cells typically found on the leaf?

_____

1.4 What differences and similarities did you observe in the stomata and guard cells in the different leaves in Procedure 3?

_____

_____

_____

_____

_____

_____

Plant Pigments

The following laboratory activity has been designed to study plant pigments. Plant pigments function in the absorption of light. The principal pigments found in the thylakoids of plants are chlorophylls *a* and *b*, both of which absorb red and blue light and reflect green light. In addition to chlorophyll, accessory pigments such as xanthophylls and carotenes absorb light and transfer energy to chlorophyll *a*. Xanthophylls reflect yellow light, and carotenes reflect orange light. Chlorophyll can mask the xanthophylls and carotenes in the leaves, but in autumn, when the chlorophyll begins to break down, the other less abundant pigments can be seen.

Procedure 1

Paper Chromatography

Materials
- Safety goggles and gloves
- Scissors
- Chromatography paper strip
- Wax pencil or Sharpie
- Capillary tube
- Control plant pigment extract or spinach leaves
- Test tube
- Cork stopper
- Graduated cylinder
- Chromatography solvent (alternate isopropyl alcohol)
- Metric ruler
- Timing device
- Hook or fashioned paper clip
- Paper towels
- Test-tube rack
- Mortar and pestle

Paper chromatography is a method used to separate the individual plant pigments in a mixture. In this technique, the pigment mixture is applied to paper, considered a "stationary phase," and the size of each pigment and its relative solubility in a solvent, called the "mobile phase," is exploited to elute the pigments from where they are applied to the paper and move them up the paper with the solvent.

The separated components, or chromatogram, will appear as bands of color parallel to the place on the paper where the pigment mixture was first applied (Fig. 6.5). The relative rate of migration (the R_f value) for each pigment can be determined from the chromatogram. R_f values in this procedure establish a control, or baseline, for examining extracts of leaves for which plant pigments have not been identified.

$$R_f = \frac{\text{Distance moved by the pigment}}{\text{Distance moved by solvent}}$$

1 Obtain a strip of chromatography paper, and cut it so it fits into a test tube and barely touches the bottom of the tube. It is suggested to cut the part of the paper immersed in the liquid into a "v" shape. Securely attach the top of the strip to a hook, push pin, or fashioned paper clip at the bottom of a cork stopper (Fig. 6.6). Test for fit. Remove the cork and the strip from the test tube.

2 Using a wax pencil or Sharpie, draw a faint line across the strip about 2 cm from the bottom tip of the strip. Place the cork and strip into the top of the test tube, and with a wax pencil or Sharpie mark the test tube 1 cm below the bottom of the cork.

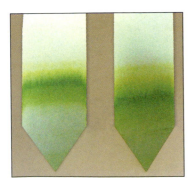

FIGURE **6.5** Simple paper chromatogram showing the different distances plant pigments have traveled.

⚠️ **WARNING**

When handling the chromatography strip, touch the top of the paper only. Work in the fume hood with gloves on when working with solvents.

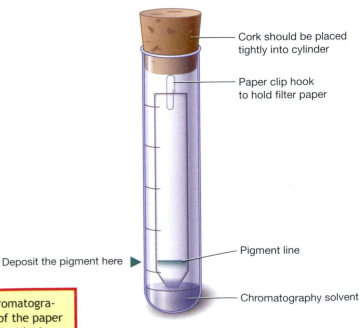

Cork should be placed tightly into cylinder

Paper clip hook to hold filter paper

Deposit the pigment here ▶

Pigment line

Chromatography solvent

FIGURE **6.6** Apparatus of paper chromatography.

3 Remove the strip of chromatography paper from the test tube and place on a paper towel. Dip a capillary tube into the plant pigment extract provided by the instructor. The tube will fill on its own. Apply the extract to the pencil line on the paper. Blow the strip dry, and repeat this application process three or four more times.

4 Using a graduated cylinder in the fume hood, carefully measure 5 mL of chromatography solvent, and carefully pour it into the test tube. Place the chromatography strip from the steps above in the test tube, and position it so the tip of the strip just touches the solvent. Be careful not to let the plant pigment extract touch the solvent. Keeping the test tube capped, place the test tube in a test-tube rack.

5 Record your observations in Table 6.1 as the solvent rises up on the paper.

6 When the solvent has moved up to the wax pencil line drawn on the test tube, remove the cork with the attached strip of paper. Set aside the paper to dry. Identify the pigment bands. The chlorophyll *a* will be blue-green,

the chlorophyll *b* will be olive-green, the xanthophylls will be yellow, and the beta-carotene will be a bright orange-yellow band.

7 Measure the distance of the solvent from its origin to the highest point it traveled on the paper. Then measure the distance the different pigments travel from the origin (pencil mark) to the center of each pigment band. Record your measurements in Table 6.1. Calculate the R_f (rate of migration) for each pigment.

8 Dispose of chromatography solvent as instructed.

(Optional activity) The laboratory instructor will provide several leaf specimens (at least three) collected from a nearby source, or with permission, you can collect leaves of interest. During the fall, try to collect different-colored leaves; during the remainder of the year, try to collect leaves of different colors. If you collect your leaves, be sure to identify them properly. Using a mortar and pestle, crush the leaves individually and collect a sample of extract. Follow the procedure just outlined.

TABLE **6.1** Control Plant Pigments

Color of Band	Pigment	Migration (mm)	R_f Value

✔ Check Your Understanding

2.1 Name two accessory pigments found in plants and what light they reflect. Why do we not see these colors in leaves during most of the year?

2.2 What is the basic theory of paper chromatography?

2.3 Which of the pigments isolated from your leaf migrated the farthest from the point of origin? Why?

2.4 Sketch and label the results of your chromatogram.

Photosynthesis in *Elodea*

Procedure 1

Uptake of Carbon Dioxide

Materials

- ❏ Large, leafy stem of *Elodea*
- ❏ Test tube
- ❏ Test tube covered with foil as a negative control
- ❏ Test-tube rack
- ❏ Medicine dropper
- ❏ 1% solution of phenol red (pH indicator)
- ❏ Stopper
- ❏ Straw
- ❏ Tap water
- ❏ Light source

This activity illustrates the plant's uptake of carbon dioxide from the environment during the light-independent reactions, or Calvin cycle, of photosynthesis (Fig. 6.7).

⚠ WARNING

Do not ingest the phenol red.

1 Fill two-thirds of a test tube with water. Place the *Elodea* in the tube. Add 4 or 5 drops of the phenol red to the test tube.

2 Insert a straw into the test tube, and blow gently to release carbon dioxide. The water will become an orange-yellow color as carbonic acid is formed and the water becomes more acidic. Immediately place the stopper on the test tube.

3 Place the test tube in a well-lit area for 10–20 minutes. Record your observations below.

Light-dependent reactions

SUN

↓

Light energy

↓

Absorbed by chlorophyll *a* and accessory pigments

↓

Chlorophyll *a* serves as an energy carrier and becomes energized. Supplies energy to . . .

↓

Split H_2O Photophosphorylation P + ADP = ATP

Oxygen   2H trapped by NADP to form 2 NADPH   Energy for light-independent reaction

By-product   2 NADPH to light-independent reactions

Light-independent reactions

CO_2

↓

CO_2 combines with RuBp (a 5-carbon sugar) to form an unstable 6-carbon molecule. This is carried out by an enzyme called Rubisco.

↓

6-carbon molecule breaks down into two 3-carbon molecules called glyceraldehyde 3-phosphate (G3P).

↓

Regeneration of RuBp   G3P is important in forming glucose, other sugars, cellulose, and starches. It is also important in amino acid and fatty acid synthesis.

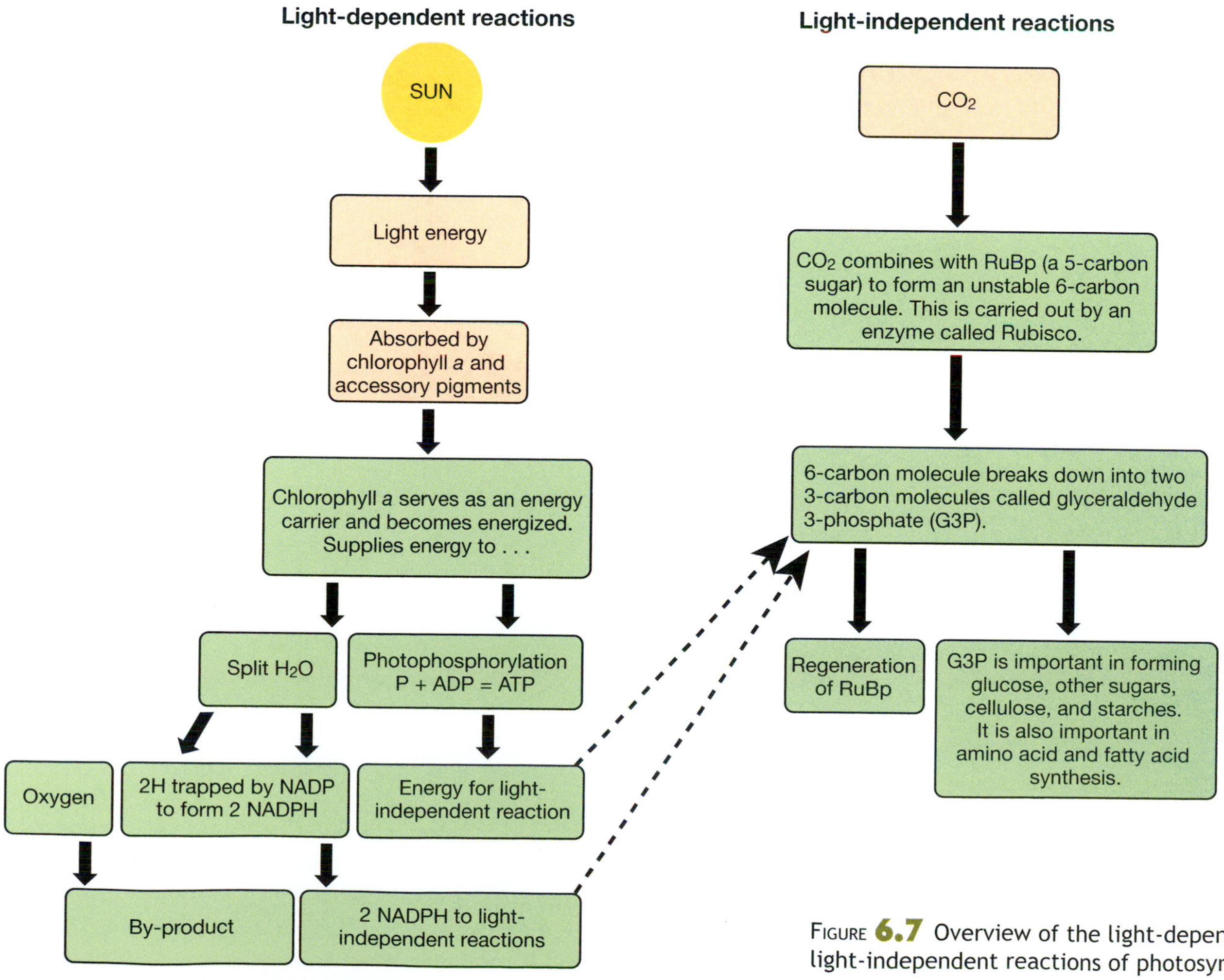

FIGURE 6.7 Overview of the light-dependent and light-independent reactions of photosynthesis.

Procedure 2

Oxygen Production

Materials

- ❏ Large, leafy stem of *Elodea*
- ❏ Test tube and fitted stopper
- ❏ Test-tube holder
- ❏ 500 mL beaker
- ❏ Glass funnel
- ❏ Tap water
- ❏ Safety glasses and gloves
- ❏ Matches

This activity illustrates the production of oxygen by a plant during the light-dependent reactions.

1 Procure a healthy piece of *Elodea* from the laboratory instructor. Place the *Elodea* in a 500 mL beaker containing 350 mL of water. Place the glass funnel into the beaker, and completely cover the *Elodea* so any oxygen the *Elodea* produces will pass through the funnel. Ensure that the stem of the funnel is under the water (Fig. 6.8).

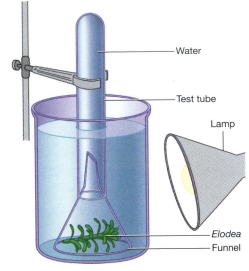

FIGURE **6.8** Apparatus of oxygen production.

> ### WARNING
> Safety glasses must be worn. Review the locations of the fire extinguisher, first-aid kit, and other safety equipment.

2 Completely fill a test tube with water. Place your thumb over the open end of the test tube. Keeping your thumb over the open end of the test tube, invert the test tube and place it over the stem of the funnel (Fig. 6.8), dipping it down into the water.

3 Place the apparatus in direct sunlight, or expose it to lights provided by the laboratory instructor.

4 When the test tube is approximately 3/4 filled with oxygen, and while it is still in the beaker, carefully immerse your hand in the beaker and place a stopper over the end of the test tube. Carefully remove the stoppered test tube from the beaker and place it in a test-tube holder.

5 Put on your safety glasses and gloves. Holding the test tube in the holder remove the stopper and strike a match, moving it to the mouth of the test tube. Record your observations below.

_____

_____

_____

> ### WARNING
> Your university may have safety protocols in place that prevent the use of matches. If so, skip step 5.

Did you know . . .

Variations in Photosynthesis

Most plants (perhaps 90%) undergo typical photosynthesis as described and prefer the more moderate heat of spring. These plants, such as rice, zinnias, and oaks, are known as C_3 plants. Alternative pathways have evolved in other plants as adaptations to availability of CO_2, water, and light intensity.

Some plants, especially those that thrive in the intense heat of summer, use another pathway and are known as C_4 plants. Examples of C_4 plants are sugarcane, corn, and crabgrass. C_4 plants fix carbon into a four-carbon compound that serves as a reserve of carbon for the Calvin cycle, and they can keep their stomata closed during most of the hot part of the day. These adaptations prevent photorespiration—fixing oxygen gas into molecules that cannot sustain the Calvin cycle. In contrast, photorespiration can occur in C_3 plants that keep their stomata closed to prevent water loss: oxygen gas builds up, and carbon dioxide cannot get in. This prevents plant growth.

Desert plants, such as cacti, undergo the Crassulacean acid metabolism (CAM) pathway. Other plants that use the CAM pathway include Spanish moss, pineapple, quillwort, and *Welwitschia*. CAM plants also keep their stomata closed during the day to prevent water loss.

✔ Check Your Understanding

3.1 For the carbon uptake procedure (Procedure 1), describe the color change that occurred in the test tube.

_____

_____

3.2 Plants like *Elodea*, zinnias, oaks, etc., undergo typical photosynthesis involving the light reactions and the light-independent reactions. They are known as _____ plants. Desert plants use the _____ pathway, which enables them to keep stomata closed during the day to conserve water.

3.3 In Procedure 1, CO_2 is blown into a test tube containing the aquatic plant *Elodea*. What is the indicator used in this procedure? Does this procedure illustrate events in the light-dependent or light-independent reactions?

_____

_____

3.4 In the oxygen production procedure (Procedure 2), how do you know oxygen was produced? Where did it originate?

_____

_____

_____

_____

■ Cellular Respiration

In nature, two major types of cellular respiration have evolved:

1. Some bacteria and fungi undergo **anaerobic cellular respiration**, which occurs in the absence of oxygen. The ancestors of anaerobic organisms first appeared on Earth approximately 3.5 billion years ago. Today, many species of anaerobic organisms still abound. Anaerobic bacteria can be found in certain soils, sediments in bodies of water, and the guts of some animals. Several species of anaerobic bacteria are responsible for diseases, such as gangrene, tetanus, and botulism. Animal cells normally perform aerobic respiration; however, during vigorous activity (such as exercise), lactic acid fermentation occurs in muscle cells if oxygen is in short supply.

2. **Aerobic cellular respiration** takes place in the presence of oxygen. The evolution of aerobic cellular respiration began approximately 2.7 billion years ago, and the vast majority of organisms on Earth today are aerobic. The overall reaction for aerobic cellular respiration is the reverse of photosynthesis, although the enzymes used in these pathways are quite different:

$$C_6H_{12}O_6 + 6\,O_2 \longrightarrow 6\,CO_2 + 6\,H_2O + ATP\ (energy)$$

Glycolysis

Anaerobic and aerobic respiration including fermentation begins with a molecule of glucose produced through photosynthesis. With few exceptions, most living things catabolize glucose to release the potential energy stored in covalent bonds between carbons. The first sequence of reactions involved in the release of potential energy from glucose is referred to as **glycolysis** (Fig. 6.9). This complex chemical pathway occurs anaerobically in the cytoplasm of a cell. Because glycolysis is common to life on Earth, it arose early in the evolution of life.

The major end net products of glycolysis are two molecules of ATP, two molecules of nicotinamide adenine dinucleotide (NADH, a coenzyme), and two molecules of pyruvate. If oxygen is available, pyruvate molecules are shuttled into mitochondria, and ATP formation occurs via aerobic cellular respiration. In the absence of oxygen, pyruvate molecules are metabolized in the cytoplasm via anaerobic fermentation reactions.

Anaerobic Fermentation Reactions

The two primary fermentation pathways are **alcoholic fermentation** and **lactic acid fermentation**. In both pathways, pyruvate is reduced by NADH. In alcoholic fermentation, ethyl alcohol (C_2H_5OH) and carbon dioxide (CO_2) are formed; in lactic acid fermentation, lactic acid ($C_3H_5O_3$) is formed. Both pathways yield only two molecules of ATP.

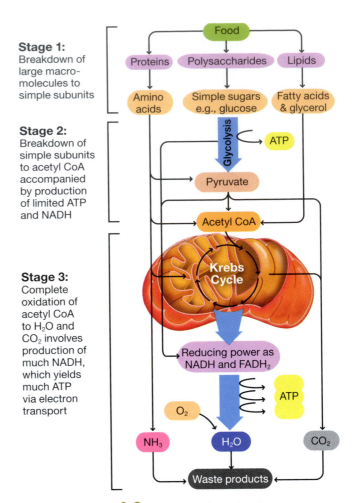

Stage 1: Breakdown of large macromolecules to simple subunits

Stage 2: Breakdown of simple subunits to acetyl CoA accompanied by production of limited ATP and NADH

Stage 3: Complete oxidation of acetyl CoA to H_2O and CO_2 involves production of much NADH, which yields much ATP via electron transport

FIGURE **6.9** Overview of respiration.

Combined with the two ATPs produced from glycolysis, the net yield of ATP during each reaction is only four molecules of ATP. Thus, anaerobic organisms have no "energy to spare." That explains why anaerobic life forms cannot even "putt" around like a paramecium. Other types of commercially important microbial fermentation processes yield acetone and methanol.

Aerobic Cellular Respiration

One of the most significant events in the history of life was the evolution of aerobic cellular respiration. In aerobic cellular respiration, the two molecules of pyruvate that result from glycolysis are converted further to two molecules of **acetyl coenzyme A** and two molecules of CO_2 in a mitochondrion. This is the CO_2 you exhale. NAD⁺ facilitates the removal of electrons from pyruvate and forms two molecules of NADH. The acetyl coenzyme (CoA) carries the two-carbon (acetyl) group.

The acetyl group enters the **citric acid cycle**, or **Krebs cycle**, named after biochemist Hans Krebs (1900–1981). The citric acid cycle is a complex series of metabolic reactions occurring in the matrix of a mitochondrion. The end products of the citric acid cycle, per original glucose molecule, are two molecules of ATP, four molecules of CO_2, six molecules of NADH, and two molecules of flavin adenine dinucleotide ($FADH_2$).

The final step of aerobic respiration occurs in the inner mitochondrial membrane of eukaryotes and in the plasma membrane of prokaryotes. This process is known as the **electron transport chain**. The product of the electron transport chain is water and a proton gradient across the inner mitochondrial membrane that drives the synthesis of ATP via chemiosmosis.

The combined production of water via electron transport and ATP via chemiosmosis is referred to as oxidative phosphorylation. In this process, oxygen is required because it serves as the final electron acceptor by combining with two hydrogen atoms to form water. Approximately 32 ATP molecules form by the completion of oxidative phosphorylation. In the electron transport chain, oxygen serves as the final electron acceptor by combining with two hydrogen atoms to form water.

Did you know...

A Tidbit from Biohistory

In the mid-1800s, a French winery that produced alcohol from the fermentation of sugar beets ran into a big problem: Its prized wine tasted like vinegar. The owner, Monsieur Bigo, contacted the renowned Louis Pasteur (1822-1895) to come up with a solution to the problem. At that time no one, including Pasteur, really understood how sugar ferments into alcohol.

Pasteur sampled a healthy vat of fermenting sugar beets and, to his surprise, noted healthy yeasts under the microscope. Next he examined a vat that contained poor-tasting wine. In this vat, instead of finding yeasts, he observed rod-shaped bacteria. These bacteria were collected and grown in a nutritive media. Pasteur determined that instead of producing alcohol as a by-product, the bacteria produced lactic acid. Thus, with this discovery, Pasteur was able to save the French wine industry by ensuring that yeasts, not bacteria, are introduced to wine-making vats.

Observing Alcoholic Fermentation

Yeasts are unicellular organisms placed in kingdom Fungi that produce energy (ATP) anaerobically in a two-stage pathway called the alcoholic fermentation reaction (Fig. 6.10). In the first stage of the pathway, yeast chemically breaks down glucose into pyruvate in a series of metabolic reactions called glycolysis. A molecule of CO_2 then is removed from the pyruvic acid. This leaves a two-carbon compound.

In the second stage of the pathway, two hydrogen atoms from NADH and H^+ are added to the two-carbon compound to form ethyl alcohol. The NADH is oxidized to re-form NAD^+, an essential molecule that allows the glycolysis pathway to continue.

Alcoholic fermentation is essential in making wine, beer, and bread. In making bread, the CO_2 produced causes the bread dough to rise (Fig. 6.11). The ethyl alcohol evaporates during baking. Have you ever eaten a slice of pizza or a piece of bread that smells a little of beer? The beer odor is a bit of residual alcohol that has not completely evaporated!

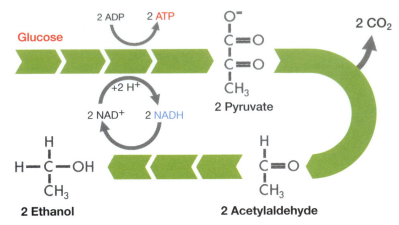

FIGURE **6.10** Overview of alcoholic fermentation.

FIGURE **6.11** Bread making depends on the fermentation of sugar by yeast. (**A**) The CO_2 produced in the reaction forms bubbles in the bread dough, causing it to rise at the beginning of the process; (**B**) 4 hours later.

Procedure 1

Demonstrating Yeast Fermentation

Materials

- Compound microscope
- Microscope slides
- Coverslips
- Package of baker's yeast (1)
- Granulated sugar (1 tablespoon)
- Tap water
- 250 mL beaker
- Empty glass bottle (e.g., soda bottle)
- Stirring rod
- Glass dropper
- Balloons
- Hot plate
- Thermometer
- Rubber band
- Measuring tape
- Goggles
- Gloves
- Laboratory apron
- Paper towels

The following activity has been designed to demonstrate alcoholic fermentation in yeasts (see Fig. 6.11).

1 Procure the materials to be used in this activity, and bring them to your lab station.

2 Add 200 mL of tap water to the beaker, and, using a hot plate, heat the water to approximately 35°C.

3 Carefully add the yeast and the sugar to the water, and stir. Wash your hands after handling the yeast.

4 Carefully pour the mixture into the glass bottle to approximately the halfway mark. Reserve the remaining mixture in the beaker for use in step 9.

5 Blow up a balloon several times, letting the air out each time, to make the balloon more flexible.

6 Cover the lip of the bottle with the balloon, and secure the base of the balloon to the bottle with a rubber band.

7 Observe and record what happens to the balloon during the next 30 minutes. Measure the diameter of the balloon every 10 minutes, and record your results below. You may also want to take photos with your camera or phone.

Initial diameter of balloon	
Diameter of balloon at 10 minutes	
Diameter of balloon at 20 minutes	
Diameter of balloon at 30 minutes	

8 At the end of 30 minutes, remove the balloon from the glass bottle, waft your hand over the bottle, and smell the contents. Describe the smell.

_____

9 Using a glass dropper, take a drop of the reserved mixture from the beaker.

10 Make a wet mount of the mixture.

11 Observe the mixture using the microscope on low and high power.

12 Sketch your observations in the space provided.

13 Return your materials to the designated area, and discard the liquid as instructed.

14 Clean your laboratory station, glassware, and hands.

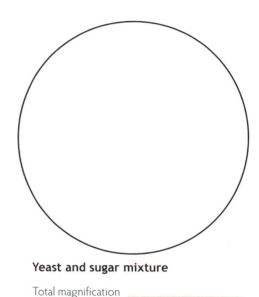

Yeast and sugar mixture

Total magnification _____

✔ Check Your Understanding

4.1 What gas caused the balloon to expand in Procedure 1?

_____

4.2 What process caused the gas production in Procedure 1?

_____

_____

4.3 Describe the odor of the mixture. What caused the odor?

_____

_____

_____

Aerobic Respiration

Many people do not realize that both plants and animals carry out aerobic respiration. In aerobic respiration in plants, as in animals, oxygen functions as the final electron acceptor for the electron transport chain. As oxygen combines with two hydrogen atoms at the end of the electron transport chain, it forms water. In aerobic respiration, oxygen is used to release energy from glucose. Carbon dioxide is formed in the process.

Procedure 1

Oxygen Consumption

Materials
- ❏ Germinating green peas
- ❏ Nongerminating green peas
- ❏ Glass beads
- ❏ Absorbent cotton
- ❏ KOH pellets
- ❏ Test tubes (2)
- ❏ Rubber stoppers (two-hole) to fit test tubes (2)
- ❏ 500 mL beakers (2)
- ❏ Test-tube clamps (2)
- ❏ Ring stands (2)
- ❏ Tap water
- ❏ Rubber tubing
- ❏ Graduated pipettes (2)
- ❏ Tubing clamps (2)
- ❏ Black cloth
- ❏ Brodie's manometer fluid
- ❏ Pasteur pipette
- ❏ Wax pencil or Sharpie
- ❏ Timing device
- ❏ Forceps
- ❏ Gloves
- ❏ Safety goggles

In this experiment, oxygen consumption in germinating and nongerminating green peas will be compared. The peas contain a plant embryo that will germinate when conditions are favorable. During germination, the carbohydrates stored in the peas serve as fuel for growth of the embryonic plant. Your laboratory instructor will initiate the germination process by soaking the peas in water in a dark environment for 3 days. The nongerminated peas will be heat-killed by roasting the peas in a 200°F oven for 20 minutes.

To measure the amount of oxygen consumed by the green peas during aerobic respiration, potassium hydroxide (KOH) will be used to remove the carbon dioxide produced in the reaction. This reaction is:

$$CO_2 + 2KOH \longrightarrow K_2CO_3 + H_2O$$

⚠ WARNING

Do not handle the KOH with your bare fingers because it is a caustic base. Wear gloves, and take precautions in the handling and disposal of KOH.

1. Procure the supplies needed for this experiment from your instructor.

2. To one test tube add 30 germinating peas, and to the second test tube add 30 nongerminating peas (Fig. 6.12).

3. Place a small wad of absorbent cotton on top of the peas in each test tube.

4. Using forceps, carefully place six pellets of KOH on the cotton in the test tube with the germinating peas (Fig. 6.12).

5. In the test tube containing the nongerminating peas, add 10 glass beads on top of the cotton (Fig. 6. 12).

6. Carefully insert the graduated pipette and rubber tubing into the rubber stoppers, as shown in Figure 6.13.

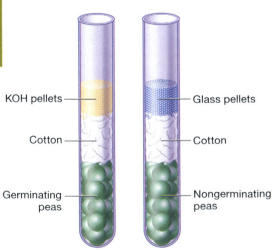

FIGURE 6.12 Test tubes with germinating and nongerminating peas.

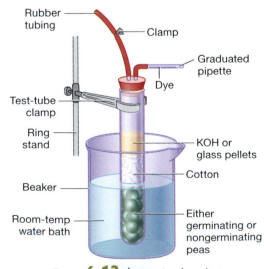

FIGURE 6.13 Apparatus housing room-temperature water bath.

7 Insert the rubber stopper apparatus into the test tubes containing the peas, cover each test tube with a black cloth or aluminum foil to block the light, and attach test-tube clamps to each ring stand as shown in Figure 6.13.

8 Prepare two room-temperature water baths by adding approximately 350 mL of tap water to each beaker.

9 Attach the test tubes with the apparatus to each test-tube clamp. Gently submerge the test tubes into the water bath to the level just below the cotton (Fig. 6.13).

10 Using a Pasteur pipette, carefully add a drop of Brodie's manometer fluid to the open end of the graduated pipette. Brodie's manometer fluid is an aqueous solution used to detect gas production or uptake.

11 Let the apparatus sit for 3 minutes for equilibration.

12 Pinch-clamp the rubber tubing, and mark the location (Initial) of the manometer fluid droplet in the graduated pipette with the wax pencil or Sharpie. Record the location on the graduated pipette as manometer fluid 0 in Table 6.2 and Table 6.3.

13 At 10-minute intervals during the next 30 minutes, mark the position of the manometer fluid movement on the graduated pipette.

14 After marking the location of the manometer fluid movement, measure and record the distance the manometer fluid has traveled, and record it in the appropriate table.

How did temperature influence the rate of oxygen consumption?

15 Remove the pinch-clamp from the rubber tubing, and pipette the manometer fluid back into the Pasteur pipette. Disassemble your apparatus. Clean and return your equipment as directed by your laboratory instructor. Clean your laboratory stations.

TABLE **6.2** Oxygen Consumption in Green Peas Water Bath (room temperature)

Initial		10 Minutes		20 Minutes		30 Minutes	
G	NG	G	NG	G	NG	G	NG
Oxygen consumed in mL							

TABLE **6.3** Oxygen Consumption in Green Peas Water Bath (35°C temperature)

Initial		10 Minutes		20 Minutes		30 Minutes	
G	NG	G	NG	G	NG	G	NG
Oxygen consumed in mL							

Procedure 2

Carbon Dioxide Production

Materials

- ❏ 250 mL glass beakers (8)
- ❏ Squares of aluminum foil (8, approximately 16 cm² each)
- ❏ Dechlorinated water
- ❏ Bromothymol blue (BTB) solution
- ❏ Small water snails or goldfish (4)
- ❏ Leafy stems of *Elodea* (4)
- ❏ Glass dropper
- ❏ Wax pencil or Sharpie
- ❏ Artificial light source
- ❏ Dark closet or box
- ❏ Safety goggles
- ❏ Gloves

In cellular respiration, plants and animals use oxygen to release energy from carbohydrates. In the process, carbon dioxide is formed as a waste product. We have observed that in the process of photosynthesis the plants use carbon dioxide to produce oxygen and carbohydrates. The interrelationship between plants and animals is crucial to life.

In this activity, the chemical indicator bromothymol blue (BTB) will be used to detect the production of carbon dioxide in closed systems. The color of the BTB will change from blue to greenish-yellow in the presence of an acid. As carbon dioxide is released during respiration, the solution will become more acidic and will shift in color. To appreciate this concept fully, this activity will use eight closed systems.

- ◆ Two of the eight systems will serve as negative controls, containing only dechlorinated water and BTB.

- ◆ Two systems will contain the dechlorinated water, BTB, and a small water snail or goldfish.

- ◆ Two systems will have the dechlorinated water, BTB, and a small leafy stem of *Elodea*.

- ◆ The last two systems will contain the dechlorinated water, BTB, a water snail or goldfish, and a leafy stem of *Elodea*.

The role of light in the process also will be investigated by placing one of each system in a darkened environment (Fig. 6.14).

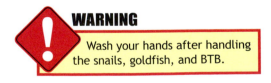

WARNING

Wash your hands after handling the snails, goldfish, and BTB.

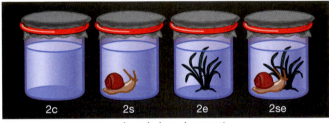

In strong artificial light

In a dark environment

FIGURE **6.14** Beakers placed in light and dark environments.

1 Procure eight 250 mL glass beakers from your instructor. Label the eight beakers 1c, 2c; 1s, 2s; 1e, 2e; 1se, and 2se.

2 Add 150 mL of dechlorinated water to each beaker. Using a glass dropper, add approximately 8 drops of BTB to each beaker.

3 Cover the top of the control beakers (1c and 2c) with aluminum foil.

4 Carefully place one small water snail or goldfish in the beakers labeled 1s and 2s. Cover the top of each beaker with aluminum foil.

5 To beakers 1e and 2e add a leafy stem of *Elodea*, and cover the top of each beaker with a piece of aluminum foil.

6 Carefully place one small water snail or goldfish and one leafy stem of *Elodea* into the beakers labeled 1se and 2se. Cover the top of each beaker with a piece of aluminum foil.

7 Place beakers 1c, 1s, 1e, and 1se under a strong artificial light. Put beakers 2c, 2s, 2e, and 2se in a dark closet or a box. After 24 hours, observe each closed system, and record your observations in Table 6.4.

8 Dispose of your materials as directed by the lab instructor. Clean up your lab station.

TABLE **6.4** Carbon Dioxide Production Results

System	Color of Water	Acidic? (Y/N)	Organism	System	Color of Water	Acidic? (Y/N)	Organism
1c				2c			
1s				2s			
1e				2e			
1se				2se			

✔ Check Your Understanding

5.1 In Procedure 1, what material was used to remove CO_2 produced in the reaction? Why was this material not added to the nongerminating peas?

_____

_____

5.2 Complete the equation for aerobic cellular respiration.

$$C_6H_{12}O_6 + 6\ O_2 \longrightarrow \underline{\hspace{2cm}} + 6\ H_2O$$

5.3 Using the equation in question 5.2, what was the source of the carbohydrate utilized in Procedure 1?

_____

5.4 In aerobic respiration, _____ is used to release energy from glucose.

Metabolism II
Understanding Photosynthesis and Cellular Respiration

MYTHBUSTING

Shedding Light

Debunk each of the following misconceptions by providing a scientific explanation. Write your answers on a separate sheet of paper.

1 Plants just undergo photosynthesis and do not undergo respiration.

2 Plants only undergo respiration at night.

3 The dark reaction, or light-independent reaction, only occurs at night.

4 Plants get their "food" from the soil.

5 Respiration only involves breathing.

---

1 Write the general equation for photosynthesis.

2 In the Calvin cycle, which of the following occurs? (*Circle the correct answer.*)

a. CO_2 is incorporated into carbohydrates.

b. Chlorophyll *a* absorbs light energy resulting in production of ATP.

c. Reactions take place in the stroma of the chloroplasts.

d. Both a and c.

e. All of the above.

3 In the electromagnetic spectrum, the longer the wavelength is, the less the energy it has. (*Circle the correct answer.*)

True / False

4 Sketch and label the basic internal anatomy of a cross section of a leaf.

Internal leaf anatomy

Total magnification _____

Type of leaf _____

_____

5 Describe the role of the stomata in leaves.

_____

_____

_____

6 Describe the light-dependent and light-independent reactions of photosynthesis.

_____

_____

_____

7 The final step of aerobic respiration, which is known as _____ , occurs in the _____ of eukaryotic cells and the _____ of prokaryotes. (_Circle the correct answer._)

   a. electron transport chain, plasma membrane, mitochondria

   b. alcohol fermentation, mitochondria, plasma membrane

   c. electron transport chain, mitochondria, plasma membrane

   d. Krebs cycle, plasma membrane, mitochondria

   e. Krebs cycle, mitochondria, plasma membrane

8 The interrelationship between photosynthesis and cellular respiration is shown by the following: (_Circle the correct answer._)

   a. Photosynthesis (which occurs in algae, plants, and some bacteria) produces the waste product oxygen.

   b. The overall reaction for aerobic cellular respiration is the reverse of photosynthesis:
$$C_6H_{12}O_6 + 6\,O_2 \longrightarrow 6\,CO_2 + 6\,H_2O.$$

   c. Plants use CO_2 produced in cellular respiration to produce carbohydrates.

   d. Both a and c.

   e. All of the above.

9 Glycolysis occurs in the mitochondria of a typical eukaryotic cell producing 4 ATP, 2 NADH, and 2 molecules of pyruvate. (_Circle the correct answer._)

True / False

10 How many ATP molecules are produced in the Krebs cycle? Contrast this with the amount of ATP formed during glycolysis and the electron transport chain.

_____

_____

_____

_____

_____

11 Which of the following are true statements regarding cellular respiration? (_Circle the correct answer._)

   a. Lactic acid fermentation is an aerobic process.

   b. The final electron acceptor in the electron transport chain is oxygen.

   c. Two ATP are produced in the citric acid cycle.

   d. Both b and c.

   e. All of the above.

12 Why was the development of aerobic respiration one of the most important events in the evolution of life on Earth?

_____

_____

6

Splitting Up
Understanding Cell Division and Meiosis

Where a cell arises, there a cell must have previously existed (omnis cellula e cellula), just as an animal can spring only from an animal, and a plant only from a plant.

— Rudolf Virchow (1821–1902)

Just wondering . . .

Consider the following questions prior to coming to lab, and record your answers on a separate piece of paper.

1 What are the major types of cancer and their incidence?
2 What are the major causes of cancer?
3 What are some new treatments for cancer?
4 What is trisomy 21 and some of its characteristics?
5 What is the significance of crossing over?
6 Is there any relationship between mother's age and the incidence of trisomy 21?

Objectives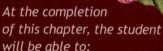

At the completion of this chapter, the student will be able to:

1. Discuss the functions of cell division.
2. Describe the cell cycle.
3. Describe, draw, and label the stages of interphase and mitosis.
4. Identify the stages of mitosis in prepared slides of a whitefish blastula and an onion root tip.
5. Discuss control of the cell cycle.
6. Define *meiosis* and its evolutionary significance.
7. Identify which cells undergo the process of meiosis.
8. Explain the relevance of meiosis to sexual reproduction.
9. Describe how the chromosome number is reduced from the diploid number ($2n$) to the haploid number (n) in meiosis I.
10. Compare and contrast cell division and meiosis.

Many times the terms **mitosis** and **meiosis** are confused because they sound alike. Ultimately, both processes are essential for the proliferation of life on Earth. However, each process is distinctly different. Mitosis is a process that occurs during **cell division**, which results in a parent cell giving rise to two identical **daughter cells**. Meiosis, on the other hand results in the formation of **gametes**, or sex cells, through a series of reduction divisions involving the chromosomes. Keep in mind that life cycles can be complex and that in some lower plants and fungi meiosis may produce asexual spores.

■ Cell Division

Living organisms as well as the cells that compose tissues are capable of reproduction and growth. **Cell division** is the mechanism by which new cells are produced, whether for growth, repair, replacement, or forming a new organism. In 1875, German botanist Eduard Strasberger (1844–1912) first described cell division in plants, but the processes of cell division were not described in detail until 1876. In that year, German zoologist Walther Flemming (1843–1905), described the development of salamander eggs. Flemming coined the terms **chromatin** (from the Greek word for colored or dyed) and **mitosis** (from the Greek word for thread) and also established the framework for understanding the stages of cell division.

Cell division is the biological process by which cellular and nuclear materials of **somatic cells** (body cells; nonsex cells) are divided between two daughter cells formed from an original parent cell. The resulting daughter cells are structurally and functionally similar to each other and to the parent cell. In prokaryotic organisms (Bacteria and Archaea) the distribution of exact replicas of genetic material is comparatively simple. In eukaryotic organisms, such as animals, the process of cell division is much more complex. The complexity is the result of the presence of a larger cell, a nucleus, and more DNA (larger genome) found in individual linear chromosomes. Thus, any study of cell division in eukaryotes will include a discussion of the **cell cycle** and its two components, **interphase** and mitosis.

Chapter Photo
Light micrograph of a pig cell undergoing mitosis.

The Cell Cycle

Dividing cells pass through a regular sequence of cell growth and division known as the cell cycle (Fig. 7.1). **Checkpoints** in the cell cycle ensure that cell division is occurring properly. The first checkpoint, which occurs at the end of G_1 (recovery and normal function), monitors the size of the cell and whether the DNA has been damaged. If a problem is detected, either repair or **apoptosis** (programmed cell death) occurs. At the second checkpoint, which occurs at the end of G_2 (preparation for mitosis), the cell will proceed to mitosis only if the DNA is undamaged and if DNA replication has occurred without error. The final checkpoint at the transition from metaphase to anaphase monitors the spindle assembly and controls the onset of anaphase.

Cell types, hormones, and growth factors as well as external conditions influence the time required to complete the cell cycle. The cell cycle is divided into two major stages: interphase (when the cell is not actively dividing) and mitosis (when the cell is actively dividing). At any given time, the majority of cells in an organism are in interphase.

Stages of the Cell Cycle

Interphase

Interphase (Fig. 7.3A) typically accounts for 90% of the time that elapses during each cell cycle. Classically called the *resting stage*, interphase is actually a busy part of the cell cycle. During interphase and before a cell begins the process of mitosis, it must undergo DNA replication, synthesize important proteins, produce enough organelles to supply both daughter cells, and assemble the structures used during cell division.

Interphase is divided into three phases: gap 1 (G_1), synthesis (S), and gap 2 (G_2). In recent years, a period known as G_0 has been described. G_0, or the quiescent phase, occurs when a nondividing cell exits G_1 and represents the time during which a cell is metabolically active but not proliferative. Actively dividing cells and cancer cells either skip G_0 or pass through the stage quickly. Some cells, such as nerve cells, may never exit G_0 once formed.

The G_1 phase, occurring after mitosis, serves as the cell's primary growth phase. During this time, the cell recovers from the previous division, increases in size, and synthesizes proteins, lipids, and carbohydrates. Also, the number of organelles and inclusions increases in number including the centrosome that contain two centrioles.

In cells that contain **centrioles**, the two centrioles begin to form during G_1 (note that the cells of flowering plants, fungi, and roundworms do not contain centrioles). The G_1 phase occupies the major portion of the life span of a typical cell. Slow-growing cells, such as some liver cells, can remain in G_1 for more than a year. Fast-growing

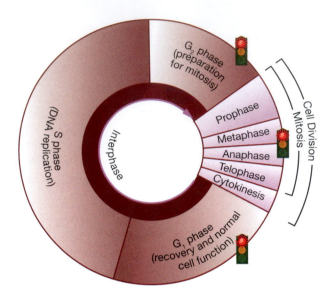

FIGURE **7.1** Cell cycle.

cells, such as epithelial cells and those of bone marrow, remain in G_1 for 16–24 hours.

In the S phase of interphase, which follows the G_1 phase, each chromosome replicates to produce two daughter copies, or **sister chromatids**. The two copies remain attached at a point of constriction called the **centromere**. During this time, these structures are not visible under the light microscope because they have not yet condensed. Among the numerous proteins manufactured during this phase are histones and other proteins that coordinate the various events taking place within the nucleus and cytoplasm. Duplication of the centrioles is completed, and they organize and begin to migrate to the opposite poles of the cell. In addition, the microtubules that will become part of the spindle apparatus are synthesized.

The G_2 phase of interphase, which occurs after the S phase, involves further replication of membranes, microtubules, mitochondria, and other organelles. In addition, the newly replicated sister chromatids diffusely distributed throughout the nucleus begin to coil and become more compact (Fig. 7.2). The start of chromosome condensation

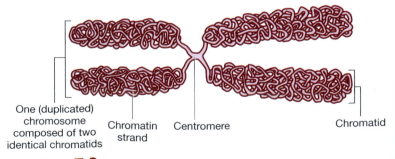

One (duplicated) chromosome composed of two identical chromatids | Chromatin strand | Centromere | Chromatid

FIGURE **7.2** Sister chromatids in prophase at the end of G_2.

at the completion of G_2 signals the beginning of mitosis. By the end of G_2, the volume of the cell has nearly doubled.

Mitosis

Cell division includes both the division of the genetic material, mitosis, or **karyokinesis**, and the division of the cytoplasm, **cytokinesis**. The overall goal of karyokinesis is to distribute identical sets of chromosomes from the parent cell to each of the daughter cells. Keep in mind that cell division is a dynamic and continuous process, but for ease in explanation and study it has been divided into several stages—prophase, metaphase, anaphase, and telophase (Fig. 7.3).

Prophase

Prophase (Fig. 7.3B) is the first, longest, and most active stage of mitosis, characteristically accounting for 70% of the time a cell spends in the mitotic process. During this stage, the chromosomes progressively become more visible as they condense and thicken. Upon close examination in late prophase, the sister chromatids and their centromeres can be seen easily. Prophase also is characterized by migration of the centriole pairs toward opposite poles and disintegration of the nuclear envelope and nucleolus.

Short microtubules known as **asters** appear and begin to radiate from the centrioles. Asters are believed to stiffen the point of microtubular attachment during the retraction of the spindle. Polar microtubules between the centrioles begin to form the **spindle fibers** that eventually will expand from one pole to another, meeting at the equatorial plane. Polar microtubules function to keep centrosomes at opposite ends of the cell.

The term **prometaphase** has become increasingly popular to describe events of late prophase. It usually is distinguished by the attachment of sister chromatids to the spindle fibers via a proteinaceous hook, or **kinetochore**. In addition, the nuclear envelope completely breaks down during prometaphase.

Metaphase

Metaphase (Fig. 7.3C) is the second stage of mitosis, when the centromere joining each pair of sister chromatids is attached to the spindle. Eventually, the pairs of sister chromatids appear to align midway between the centrioles along what is called the **equatorial plane**, or **metaphase plate**.

Anaphase

Anaphase (Fig. 7.3D) is the third stage of mitosis. Although brief, this stage is characterized by the separation of the sister chromatids from their centromere and the movement of the chromatids as they are pulled back along the spindle to opposite poles. Errors during anaphase could result in an unequal distribution of genetic material with devastating consequences.

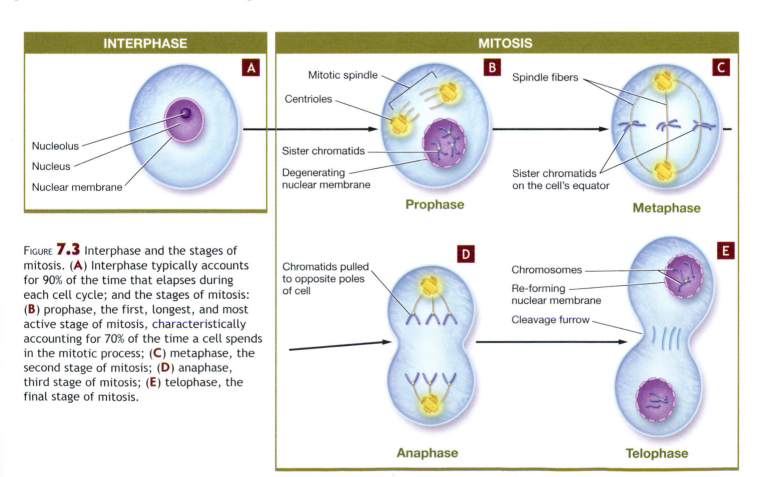

FIGURE **7.3** Interphase and the stages of mitosis. (**A**) Interphase typically accounts for 90% of the time that elapses during each cell cycle; and the stages of mitosis: (**B**) prophase, the first, longest, and most active stage of mitosis, characteristically accounting for 70% of the time a cell spends in the mitotic process; (**C**) metaphase, the second stage of mitosis; (**D**) anaphase, third stage of mitosis; (**E**) telophase, the final stage of mitosis.

INTERPHASE

A
Nucleolus
Nucleus
Nuclear membrane

MITOSIS

B
Mitotic spindle
Centrioles
Sister chromatids
Degenerating nuclear membrane
Prophase

C
Spindle fibers
Sister chromatids on the cell's equator
Metaphase

D
Chromatids pulled to opposite poles of cell
Anaphase

E
Chromosomes
Re-forming nuclear membrane
Cleavage furrow
Telophase

Telophase

In the final stage of mitosis, **telophase** (Fig. 7.3E), the cell resembles a dumbbell with a set of chromosomes at each end. During telophase, the spindle is disassembled, the nuclear envelope and nucleolus are re-formed, and cellular inclusions and organelles are organized. Cytokinesis, the division of cytoplasm upon completion of nuclear division, begins between the nuclei during late telophase.

During cytokinesis in animal cells, a cleavage furrow develops, and the cell appears to be pinched in two. Eventually the cleavage furrow is completed, and two daughter cells are formed. Upon completion of telophase and cytokinesis, the daughter cells enter the G_1 phase, and the cell cycle repeats. In plant cells, instead of undergoing cytokinesis, a **cell plate** is formed, dividing the mother cell into two daughter cells.

Procedure 1

Animal Cell Division

7

Materials
❑ Compound light microscope
❑ Prepared slide of whitefish blastula
❑ Colored pencils

In this procedure, you will observe the cell cycle in prepared slides of a whitefish blastula (Fig. 7.4). The **blastula** occurs in the early stage in the embryonic development of an animal and appears as a ball of cells, each cell in one of the stages of interphase or mitosis. The blastula can be thought of as a basketball with each "pebble" on the basketball representing a cell.

1 Procure a prepared slide of a whitefish blastula. Obtain a compound light microscope, and follow the safety and handling procedures for a light microscope.

2 Observe the whitefish blastula first on low power. In this field you will be able to see many slices of depth of the blastula. Focusing on low power, you should see individual cells.

3 Switch to high power, and, in the space provided, sketch and describe in detail what is happening within the cell in each of the stages of interphase and mitosis.

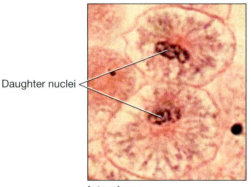

Interphase
Two daughter cells result from cytokinesis.

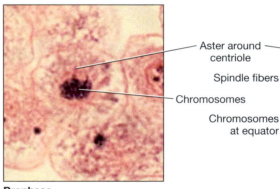

Prophase
Chromatin begins to condense to form chromosomes. Nuclear envelope is intact.

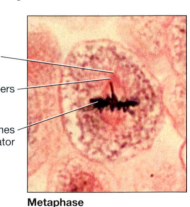

Metaphase
Duplicated chromosomes are each made up of two chromatids, at equatorial plane.

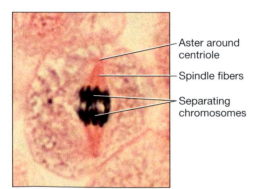

Early anaphase
Sister chromatids are beginning to separate into daughter chromosomes and daughter chromosomes are nearing poles.

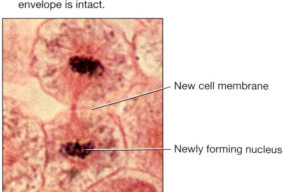

Telophase
Daughter chromosomes are at poles and cleavage furrow is forming.

FIGURE **7.4** Stages of animal cell (whitefish blastula; all 500×) mitosis followed by cytokinesis.

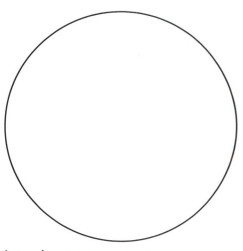

Interphase Total magnification _____

_____

_____

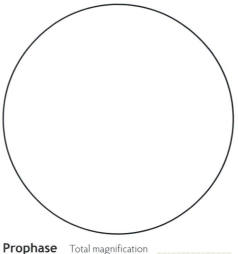

Prophase Total magnification _____

_____

_____

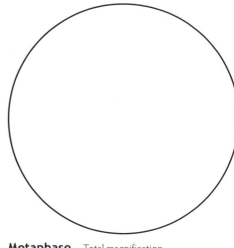

Metaphase Total magnification _____

_____

_____

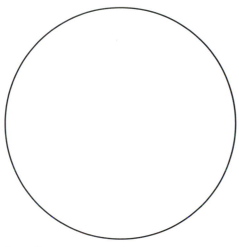

Anaphase Total magnification _____

_____

_____

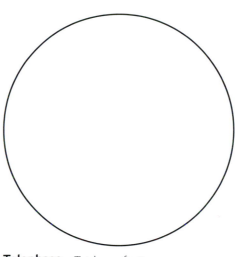

Telophase Total magnification _____

_____

_____

Procedure 2

Plant Cell Division

Materials

- ❏ Compound light microscope
- ❏ Prepared slide of green onion root tip mitosis
- ❏ Colored pencils

In this procedure, you will observe the cell cycle in prepared slides of green onion (scallions) root tip mitosis. In plants, cell growth occurs in the **meristematic** regions, primarily located at the tips of stems and roots, so the root tips often are used to study the cell cycle (Fig. 7.5).

1 Procure a prepared slide of onion root tip mitosis. Obtain a compound light microscope, and follow the safety and handling procedures for a light microscope.

2 Observe the onion root tip mitosis first on low power. In this field, you should see individual cells. You will be able to see many stages of interphase and mitosis.

3 Switch to high power, and sketch and describe in detail each of the stages of interphase and mitosis in the space provided.

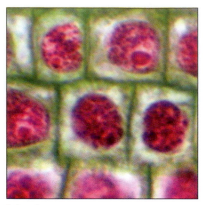

Interphase
Two daughter cells result from cytokinesis.

Early prophase
Chromatin begins to condense to form chromosomes.

Late prophase
Nuclear envelope is intact, and chromatin condenses into chromosomes.

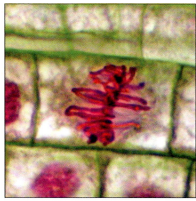

Early metaphase
Duplicated chromosomes are each made up of two chromatids, at equatorial plane.

Late metaphase
Duplicated chromosomes are each made up of two chromatids, at equatorial plane.

Early anaphase
Sister chromatids are beginning to separate into daughter chromosomes.

Late anaphase
Daughter chromosomes are nearing poles.

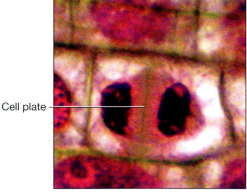

Cell plate

Telophase
Daughter chromosomes are at poles, and cell plate is forming.

FIGURE **7.5** Stages of mitosis in plant root tip (all 430×) followed by cell plate formation and cytokinesis.

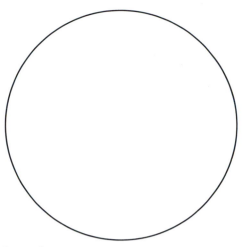

Interphase   Total magnification _____

_____

Prophase   Total magnification _____

_____

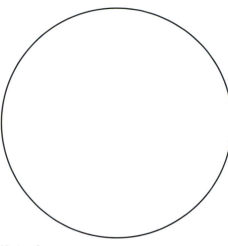

Metaphase   Total magnification _____

_____

Anaphase   Total magnification _____

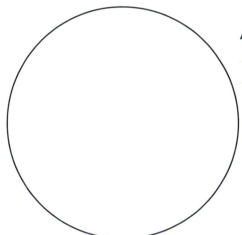

Telophase   Total magnification _____

Procedure 3

Squash Mount

Materials

- ❑ Compound light microscope
- ❑ Microscope slide
- ❑ Coverslip
- ❑ Forceps
- ❑ Paper towels
- ❑ Vial of pretreated *Allium* root tips
- ❑ Methyl green–pyronin Y stain in a dropper bottle
- ❑ Wood macerating stick
- ❑ Pencil with eraser
- ❑ Timing device
- ❑ Safety goggles
- ❑ Gloves
- ❑ Fixative solution
- ❑ Colored pencils

In this procedure, you will prepare the tip of an onion root, *Allium* sp. To observe the stages of the cell cycle, you will prepare what is called a squash mount. In preparing a squash mount, tissue taken from the plant's meristem region (the region of cell division) is removed and treated with a fixative to stop the cells from dividing. The onion root tissue then is placed on a microscope slide containing methyl green–pyronin Y stain.

This stain contains two different stains to distinguish between the DNA and the RNA. The methyl green stains the DNA blue, and the

⚠ WARNING

Be careful when handling the stain because it will stain your clothing and your skin. Gloves and safety goggles should be used. When handling coverslips, do not apply too much pressure because the coverslip and slide could break. If you break the coverslip or the slide, notify your instructor.

pyronin Y stains the RNA pink. The onion root tip then is squashed, making a single layer of tissue. A coverslip is placed over the squashed tissue, and you then will view the squash mount under the microscope.

1 Obtain materials from your instructor. Take a blank microscope slide, and place it on top of several folded paper towels. Carefully add two drops of the methyl green–pyronin Y stain to the center of the microscope slide.

2 Using forceps, remove a prepared onion root tip from the vial, and carefully transfer it to the stain on the microscope slide. With the wooden macerating stick in your hand, gently but firmly smash the onion root tip into the stain, using a straight up-and-down motion.

3 Let the squashed root tip stain for 15 minutes. Periodically add more stain if needed to keep the squashed root tip from drying out. After the 15-minute period, carefully place a coverslip on top of the stained squashed onion root tip. Place a folded paper towel over the coverslip, and using the eraser part of the pencil, press firmly straight down (do not twist) on the folded paper towel.

4 Observe the slide first using low power. After you have the cells in focus, use caution when moving to a higher

power. This is a thick mount. Scan the slide for cells that are actively going through mitosis. Sketch and record your observations in the space provided.

5 After making your observations, return your materials to the proper place as directed by your lab instructor, and clean your lab station.

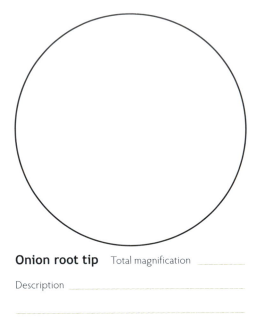

Onion root tip   Total magnification _____

Description _____

_____

✔Check Your Understanding

1.1 The prepared slides you observed in Procedure 2 (Plant Mitosis) and the slides you prepared in Procedure 3 (Squash Mount) were made with onion root tips. Why is this tissue typically used to study stages of the cell cycle?

_____

_____

1.2 List the most common stages that you observed in the whitefish blastula slides and the prepared onion root tip slides.

_____

_____

1.3 How does cytokinesis vary in animal and plant cells?

_____

_____

1.4 What purpose does methyl green-pyronin Y stain serve?

_____

■ Meiosis

In addition to the evolution of aerobic respiration, one of the most significant events in the history of life was the evolution of sexual reproduction. Sexual reproduction is responsible for the great diversity of life on Earth, because offspring are no longer "chips off the old cell" but, rather, produced by sexual reproduction. This creates unique individuals with characteristics inherited from past generations along with new, hopefully beneficial, traits.

The underlying mechanism behind sexual reproduction is **meiosis.** Unlike mitosis, which involves a division of the somatic, or body, cells, meiosis involves reduction division resulting in the formation of sex cells, or **gametes.** This is achieved in two distinct nuclear divisions, **meiosis I** and **meiosis II,** during which genetic recombination generates genetic variation.

With a few exceptions to the normal condition, species possess a genetically determined number of chromosomes. Eukaryotic chromosomes occur in pairs called homologs, or homologous pairs. An individual receives one of each pair from each parent. Homologous chromosomes contain the same genes in the same order along the chromosome; the DNA sequences are not necessarily identical. The **diploid number** ($2n$) represents the total number of chromosomes in the nucleus of a cell. In humans, the diploid number is 46. Example diploid numbers from the living world include goldfish (94), Adder's tongue fern (1,440), horse (64), chimp (48), cat (38), corn (20), and fruit fly (8). Clearly, the size or complexity of an organism is not determined by the number of chromosomes.

The number of pairs of chromosomes is the **haploid number** (n), one-half of the diploid number. In humans, the haploid number is 23, and in a fruit fly it is 4. Meiosis involves the reduction of the number of chromosomes in the diploid state ($2n$) to the number of chromosomes in the haploid state (n). These haploid cells serve as gametes, or sex cells, such as the sperm and ovum (egg cell). Meiosis also ensures the stability in the number of chromosomes passed from one generation to the next.

Sexual reproduction is based upon **fertilization,** the union of haploid male and female gametes to form a diploid **zygote,** or fertilized egg. The zygote possesses homologous chromosomes from each parent. The process of meiosis varies slightly in the kingdoms Plantae, Fungi, and Animalia.

Meiosis consists of two distinct cell divisions designated meiosis I and meiosis II. Interphase, including the S phase that creates sister chromatids, is completed prior to meiosis I.

Meiosis I

The first stage of meiosis I is prophase I. In this stage, homologous chromosomes pair up in a process called **synapsis** and form **bivalents** or **tetrads**. A synaptonemal complex is formed between homologous chromosomes to ensure proper pairing. The physical exchange of genetic material between homologous pairs known as **crossing over** results in the formation of new combinations of genetic material, making that sex cell unique (Fig. 7.6). The arms of the sister chromatids are held together at regions known as chiasmata (sing. = chiasma).

In metaphase I, the homologous pairs line up along the metaphase plate (Fig. 7.7). This independent alignment of chromosomes of maternal and paternal origin allows for the genetic diversity seen in families. In contrast to metaphase during mitosis, the sister chromatids are lined up in a double row rather than a single row.

Anaphase I follows metaphase I. In this stage, homologous chromosomes separate and migrate toward opposite poles.

Telophase I, the final stage of meiosis I, is characterized by the sister chromatids reaching the opposite poles and eventually forming two new haploid cells. Although the two cells produced at the end of meiosis I are considered haploid by convention, they possess two sister chromatids in a chromosome.

Did you know . . .

Preformation!

The idea of "where do we come from" confounded early scientists. Perhaps Pythagoras (circa 530 BC) was the first to state that the father contributed his essence to the offspring and the mother contributed the material substrate. The individuals that supported this idea were known as spermists and included a number of well-known scientists through the centuries (Galen, Leeuwenhoek, and von Baer). In 1664, Nicolaas Hartsoeker actually described a little human (homunculus) curled up inside the sperm ready to enter the egg. On the other hand, ovists such as William Harvey conjectured that the egg contained the homunculus. The ovists contended that women carried eggs containing boy and girl children, and that the gender of the offspring was determined before conception. The battle between the spermists and the ovists and whether we were preformed raged for centuries. So you think that meiosis is complicated!

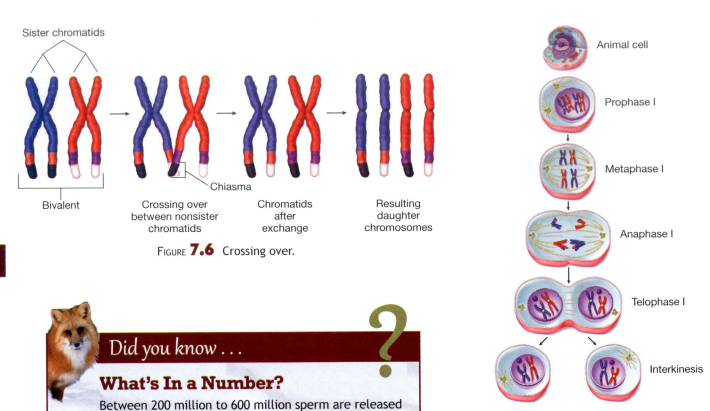

Sister chromatids

Bivalent

Crossing over
between nonsister
chromatids

Chiasma

Chromatids
after
exchange

Resulting
daughter
chromosomes

FIGURE **7.6** Crossing over.

Animal cell

Prophase I

Metaphase I

Anaphase I

Telophase I

Interkinesis

FIGURE **7.7** Overview of meiosis I
in animal cells.

Did you know . . .

What's In a Number?

Between 200 million to 600 million sperm are released
during each ejaculation in humans.

Meiosis II

Meiosis II is similar to mitosis and results in division of the sister chromatids from meiosis I. Prophase II is characterized by
attachment of the sister chromatids to the spindle. During metaphase II, the sister chromatids line up along the metaphase
plate. In anaphase II, the sister chromatids are separated and migrate toward the centrioles. Telophase II results in the formation of four unique haploid daughter cells with one chromatid per chromosome (Fig. 7.8).

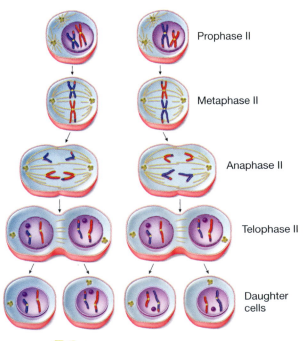

Prophase II

Metaphase II

Anaphase II

Telophase II

Daughter
cells

FIGURE **7.8** Overview of meiosis II in animal cells.

Did you know . . .

EGFs and Dog Saliva

The Greek God of Medicine, Asclepius, was best
known for prescribing rest and proper diet for his
patients. Great temples of healing were built in his
honor. In these temples, holy snakes and dogs were
allowed to slither and walk among the sick and injured.
Supposedly, the snakes would keep away evil spirits and
the dogs would promote the healing by licking the
wounds.

Today, it is known that dog saliva contains a great
amount of epidermal growth factors (EGFs), which promote cell division and speed healing. The next time your
dog detects a scratch on your arm, notice the concern!
EGFs are an integral part of modern medicine, used in
cornea transplants and burn treatment.

Meiosis in Animals

The majority of animals are **dioecious**, or have separate sexes. Haploid cells produced in meiosis II become gametes (egg or sperm). The process of forming gametes is termed **gametogenesis**. In females, **oogenesis** occurs in the ovaries. In males, **spermatogenesis**, the production of sperm, occurs in the seminiferous tubules in the testes.

In humans, the ovaries are female endocrine glands. Here mature egg cells, or ova, are formed. Oogenesis is the process that results in egg formation. **Primordial germ cells** (diploid cells) develop during the first month of embryonic development. These cells divide by mitosis and undergo differentiation to form diploid **oogonia**, equivalent to the male spermatogonia. Oogonia undergo further differentiation to form diploid **primary oocytes**, again equivalent to the primary spermatocyte in males.

The primary oocytes form before birth in human females. It is estimated that each ovary contains nearly a million primary oocytes at birth. Once formed, the primary oocyte begins meiosis I. This process occurs prenatally. Prophase I is an extended phase, not completed until the female reaches puberty. After she reaches puberty, each month one primary oocyte held in suspension continues its meiotic division. A haploid **secondary oocyte** is formed along with a first **polar body.** This small haploid polar body is a result of unequal division of the cytoplasm. The first polar body from meiosis I

also may divide to form two additional haploid polar bodies. The polar body or polar bodies eventually degenerate. Their biological importance is to provide the egg cell massive amounts of cytoplasm and cellular organelles (Fig. 7.9).

When a mature egg, or ovum, is released from the ovary, it begins its journey through the fallopian tube, awaiting fertilization. If fertilization does not occur, the ovum is lost in the next menstrual cycle. If fertilization occurs, meiosis II is completed, a diploid zygote is formed, and a haploid polar body is released. The zygote then can divide by mitosis and develop into an **embryo**.

In human males, diploid primordial germ cells divide by mitosis to form diploid **spermatogonia**. Some of the spermatogonia divide by mitosis, forming diploid **primary spermatocytes**. Others continue to divide, forming many spermatogonia.

Spermatogenesis (Fig. 7.10) begins at puberty and continues throughout the male's lifetime. The primary spermatocyte will enter meiosis I. At the end of meiosis I, two haploid **secondary spermatocytes** are formed. The two secondary spermatocytes undergo meiosis II, eventually forming four **spermatids**. After undergoing metamorphosis, or **spermiogenesis**, the spermatids mature into haploid sperm cells, each with one flagellum. The whole process of spermatogenesis, including spermiogenesis, takes about 2 months.

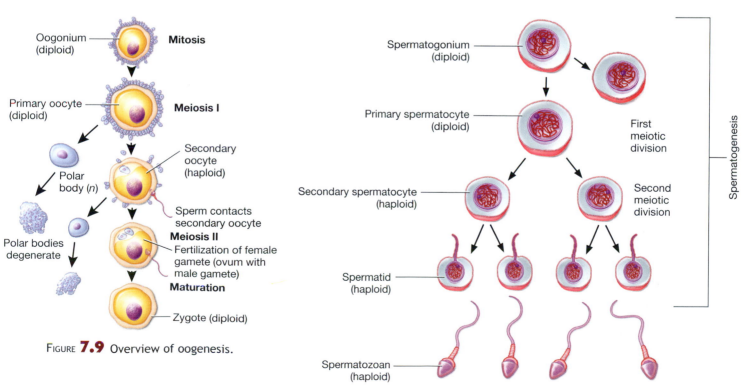

FIGURE **7.9** Overview of oogenesis.

FIGURE **7.10** Overview of spermatogenesis.

Meiosis in *Ascaris* Ovaries

Materials

- ❏ Compound light microscope
- ❏ Colored pencils
- ❏ Prepared slides of *Ascaris* ovaries

In this activity, you will observe the stages of meiosis using a series of prepared slides of *Ascaris*, a species of roundworm, or nematode. *Ascaris* is a parasitic worm that can infect primarily the digestive system of animals, including humans. The female roundworm can exceed lengths of 30 centimeters. Female *Ascaris* have large, elongated ovaries that contain as many as 27 million ova (Fig. 7.11). Within the nucleus of their cells, four chromosomes can be easily observed.

1 Obtain a microscope and the series of slides of meiosis in the *Ascaris* ovary.

2 Place your first slide on the stage of the microscope, and focus first on low power before moving the objective to a higher power. Locate, sketch, and label the stages of meiosis you observe in the space provided.

3 Repeat step 2 for the remaining slides.

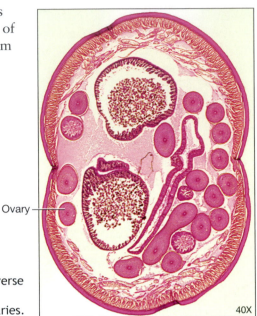

Ovary

FIGURE **7.11** Transverse section of female *Ascaris* showing ovaries.

40X

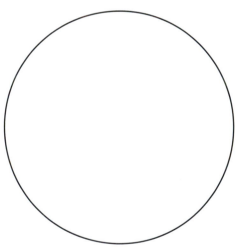

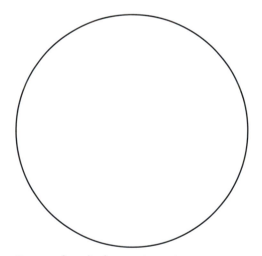

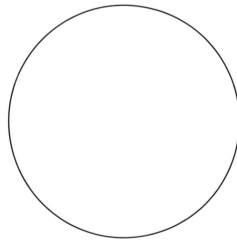

Stage of meiosis Total magnification _____

Stage of meiosis Total magnification _____

Stage of meiosis Total magnification _____

Procedure 2

Meiosis in Sperm Cells

Materials

❏ Compound light microscope

❏ Prepared slides of grasshopper, frog, bull, or other animal testis

❏ Colored pencils

In this activity you will observe gametogenesis, using prepared slides from the testes of an animal. The formation of sperm (spermatogenesis) occurs in the seminiferous tubules of the testes (Fig. 7.12).

1 Obtain a microscope and the slide of a testis.

2 Place the slide on the stage of the microscope, and focus first on low power before moving the objective to a higher power. Locate, sketch, and label the stages of meiosis in the space provided.

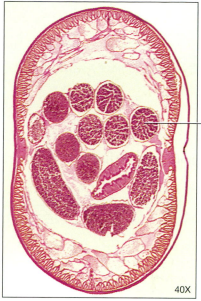

FIGURE **7.12** Transverse section of male *Ascaris* showing testes.

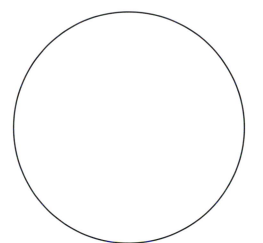

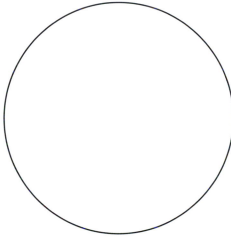

Stage of meiosis _Total magnification_ _____

Stage of meiosis _Total magnification_ _____

Did you know . . .

Every Chromosome Counts

The majority of humans exhibit euploidy, having the normal number of 23 pairs of chromosomes for a total of 46 chromosomes. Unfortunately, 1 in 250 live births exhibit aneuploidy, resulting in an individual with an incorrect number of chromosomes. The majority of aneuploid individuals spontaneously abort. Aneuploidy can result from meiotic errors (nondisjunction), and the affected individual will have too many or too few chromosomes.

Some aneuploid conditions, such as trisomy 21, have been related to a mother's age at the time of pregnancy. As a result, it is recommended that mothers 35 years old and older should be screened for fetal chromosomal abnormalities. Aneuploid conditions that affect the somatic chromosomes (pairs 1–22) are more lethal than those that affect the sex chromosomes (pair 23). Examples of aneuploid states include trisomy 21 (Down syndrome), trisomy 18 (Edward syndrome), trisomy 13 (Patau syndrome), Klinefelter syndrome (XXY), and Turner syndrome (XO). Trisomy 21 also occurs in chimpanzees and causes similar syndromes.

✔ Check Your Understanding

2.1 How many chromosomes does the nucleus of an *Ascaris* ovum have? How many pairs of chromosomes are seen in the cells in meiosis I?

_____

_____

2.2 Why is the female *Ascaris* worm a good choice for meiosis studies?

_____

_____

2.3 What are polar bodies? How are they formed?

_____

_____

2.4 The process of forming sex cells is known as _____. In females, oogenesis occurs in the

_____ and in males, spermatogenesis takes place in _____.

2.5 In most cases humans exhibit euploidy, having the normal number of chromosomes. Aneuploidy may occur in some instances, which results from _____ and causes _____. (*Circle the correct answer to fill in the blanks.*)
  a. mitotic errors, individuals affected to have too many or too few chromosomes
  b. mitotic errors, individuals affected to have too many chromosomes.
  c. meiotic errors, individuals affected to have too many chromosomes
  d. meiotic errors, individuals affected to have too many or too few chromosomes
  e. meiotic errors, individuals affected to have too few chromosomes

2.6 Identify each term as having either the diploid number or haploid number of chromosomes, using the following key:

D = Diploid (2n)   H = Haploid (n)

_____ spermatogonia

_____ secondary spermatocytes

_____ spermatids

_____ primary spermatocytes

_____ spermatozoa

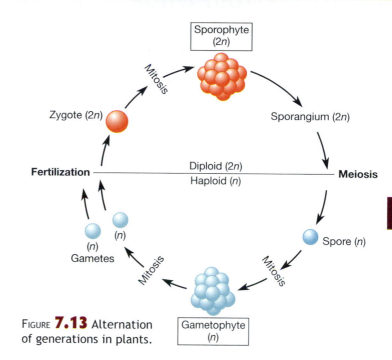

Meiosis in Plants

Many plants have an "alternation of generations" and undergo sexual reproduction. Therefore, meiosis is integral in the formation of gametes in plants. The life cycle of plants is different from that of animals in that plants undergo an alternation of generations in which they alternate between the haploid and diploid forms. Meiosis occurs in the ovary of female plants. The male gametes (pollen grains) are produced in the anthers.

The alternation of generations has two distinct phases: the sporophyte generation (*2n*) and the gametophyte generation (*n*), with both being multicellular. The diploid sporophyte generation produces haploid spores that develop into a new organism. A pine tree is an example of a sporophyte. Visible true mosses are an example of the gametophyte generation. The gametophyte generation produces haploid gametes that undergo fertilization to form a diploid zygote (Fig. 7.13).

FIGURE **7.13** Alternation of generations in plants.

Procedure 1

Meiosis in *Lilium* Anther and Ovary

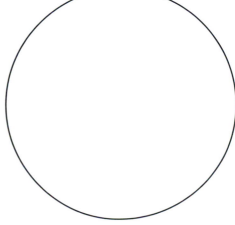

Materials

❏ Compound light microscope
❏ Prepared slides of meiosis in the *Lilium* anther and *Lilium* ovary
❏ Colored pencils

In this lab activity, you will observe and describe stages of meiosis in prepared slides of the *Lilium* (lily) anther and ovary.

1 Obtain a microscope and the series of slides of meiosis in the *Lilium* anther and the *Lilium* ovary.

2 Place your first slide on the stage of the microscope, and focus first on low power before moving the objective to a higher power. Repeat for the remaining slides. Locate, sketch, and label the stages of meiosis in the space provided.

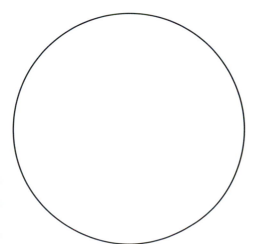

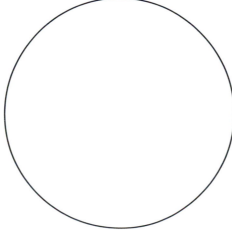

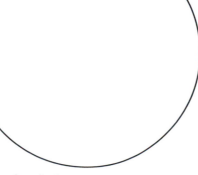

Stage of meiosis Total magnification _____

Stage of meiosis Total magnification _____

Stage of meiosis Total magnification _____

✔Check Your Understanding

3.1 Name the process that is the basis for increased variety and diversity in the plant kingdom.

_____

3.2 Male gametes, also called _____, are produced in the _____ of a flower.

3.3 List one difference between plant and animal meiosis.

_____

7

MYTHBUSTING

Don't Let Misconceptions Multiply!

Debunk each of the following misconceptions by providing a scientific explanation. Write your answers on a separate sheet of paper.

1 During interphase, the cell is resting and not active.

2 The number of chromosomes is halved during mitosis.

3 In some organisms, single cells develop into multicellular adults through meiosis.

4 The size of an organism is determined by the number of chromosomes.

5 Trisomy 21 is inherited.

6 Twins result when polar bodies are fertilized.

1 Explain what happens during each phase of interphase: G_1, S, and G_2.

_____

_____

_____

2 Identify the following structures as animal mitosis and/or plant mitosis using the following key:

AM = Animal mitosis   PM = Plant mitosis

_____ centrioles   _____ cell plate   _____ anaphase   _____ interphase

_____ cleavage furrow   _____ metaphase   _____ telophase

3 Which of the following statement(s) is incorrect? (*Circle the correct answer.*)

a. All animal cells contain centrioles.

b. In mitosis, the division of genetic material is known as karyokinesis.

c. Interphase accounts for approximately 90% of the time that elapses during each cell cycle.

d. The G_2 phase occurs following mitosis and serves as a cell's primary growth phase.

e. Both a and d are incorrect.

4 Animal cells go through the process of cytokinesis, or division of cytoplasm, upon the completion of nuclear division.

Plant cell division, unlike animal cell division, involves formation of _____, which divides the mother

cell into _____.

5 What cells undergo meiosis? Does meiosis produce haploid or diploid cells?

_____

_____

6 Draw and label one complete cell cycle.

7

7 In plants, _____. (*Circle the correct answer.*)
   a. the life cycle is different from that of animals in that plants alternate between diploid and haploid forms
   b. meiosis occurs in the ovary of female plants and male gametes are produced in the anthers of flowers
   c. there are two distinct phases: *n* sporophyte generation and 2*n* gametophyte generation
   d. both a and b
   e. both a and c

8 What are polar bodies? What are they used for?

_____

_____

_____

9 Describe three differences between mitosis and meiosis.

_____

_____

_____

10 The cellular process meiosis ensures that _____. (*Circle the correct answer.*)
   a. the same number of chromosomes are passed to the next generation
   b. genetically unique daughter cells are produced
   c. genetic variation will be found in the species
   d. both a and c
   d. all of the above

It's All in the Genes
Understanding Basic Genetics and DNA

8

Yes, evolution by descent from a common ancestor is clearly true. If there was any lingering doubt about the evidence from the fossil record, the study of DNA provides the strongest possible proof of our relatedness to all other living things.

— Francis Collins (1950–Present)

Just wondering . . .

Consider the following questions prior to coming to lab, and record your answers on a separate piece of paper.

1. In RNA, why does uracil combine with adenine instead of thymine?
2. How can mitochondrial DNA be used in forensics?
3. What is the Human Genome Project?
4. What is the significance of the wobble hypothesis?
5. Is there a heterozygous advantage to genetic disorders such as sickle cell disease and cystic fibrosis?

Objectives

At the completion of this chapter, the student will be able to:

1. Define the following terms: *chromosome, gene, allele, homozygous, heterozygous, genotype, phenotype, dominant trait, recessive trait.*
2. Discuss expected results and observed results based on the inheritance of one trait.
3. Discuss basic principles of population genetics and the Hardy-Weinberg model.
4. Construct a simple pedigree.
5. Compare and contrast DNA and RNA.
6. Describe the makeup of chromosomes.
7. Outline the processes of DNA replication and protein synthesis.
8. Compare and contrast transcription and translation.
9. Describe mitochondrial and chloroplast DNA.
10. Isolate DNA using the spooling technique.

Chapter Photo
Map treefrog, *Hypsiboas geographicus,* with different-colored eyes.

Genetics

"She has her father's eyes." "Do you think I'll be bald like Dad?" "Can two blue-eyed parents have a brown-eyed child?" "If my black-and-white cat mates with my neighbor's yellow cat, will we have any calico kittens?" "How can I increase the chances of my next litter of puppies being champions?" "What are the odds of two carrier parents having a child with cystic fibrosis?" The questions are endless when people begin discussing inheritance. People are naturally curious about how traits are inherited from one generation to the next. That's *genetics*!

As we approach the dawn of the age of genetics with its vast potential of understanding the human genome and engineering the genes themselves, it is hard to believe the science of genetics had its humble origins in an obscure monastery garden in Austria in the mid-1800s. Our fundamental knowledge of genetics is a result primarily of the experiments conducted by an Austrian monk, Gregor Mendel (1822–1844), known as the father of modern genetics.

His investigation into the inheritance patterns of certain characteristics of pea plants serves as the origin of modern genetics. However, the significance of Mendel's work was not appreciated until the early 1900s. In fact, his work would have slipped into obscurity if not for the independent work of three botanists: Hugo de Vries (1848–1935), Carl Erich Correns (1864–1933), and Erich von Tschermak (1871–1962). These men rediscovered Mendel's work in 1900 and helped expand the awareness of Mendel's laws of inheritance to the scientific world.

Chromosomes, Genes, and Alleles

Each **chromosome** consists of thousands of structural and functional units called **genes** (units of heredity), or segments of DNA that code for protein. Today, we are beginning to understand how genes work and where they are located on the chromosomes. As examples, one of the genes for Alzheimer's disease was found on chromosome 21, and the gene for cystic fibrosis has been identified on chromosome 7. In a typical human, between 20,000 and 25,000 genes are responsible for producing the traits that make you human as well as the characteristics that make you unique. A trait may be controlled by one or more genes.

Because genes occur in pairs, one on each homologous chromosome, the two DNA sequences of one gene pair may be identical or different in sequence. The possible form a gene may take is called an **allele.** If an individual possesses two identical alleles, they are **homozygous,** such as (AA) or (aa). If an individual possesses two different alleles, they are **heterozygous,** such as (Aa). An individual's genetic makeup, or **genotype,** in turn influences his or her physical characteristics, the **phenotype.**

In many cases, the expression of one allele may mask the expression of the other allele. The allele that is expressed and observed is called the **dominant allele,** and the allele not expressed and therefore not observed is called the **recessive allele.** The dominant trait is represented by an uppercase letter, such as (A), for cystic fibrosis, and the recessive trait is represented by a lowercase letter, such as (a).

Mendel's Laws

Mendel's laws are fundamental to a basic understanding of genetics. The seven traits that Mendel investigated in pea plants were either dominant (A) or recessive (a). His laws are based upon his initial findings that one allele is dominant to another allele. His **law of segregation** states that during the formation of gametes, the alleles responsible for each trait separate so that each gamete contains only one allele for that trait. Thus, female and male parents contribute equally in the formation of an embryo. His **law of independent assortment** states that every trait is inherited independently of every other trait carried on a different chromosome. As a result of the law of independent assortment, all of the possible combinations of the alleles can occur in gametes.

Punnett Squares

If the genotypes of both parents are known, it is possible to calculate the expected genotypes and phenotypes of their offspring. The expected results can be calculated either by mathematical methods or by using a **Punnett square.** This handy table is named in honor of British geneticist Reginald C. Punnett (1875–1967). Keep in mind that expected results are specific figures and are not the result of chance. However, in nature the **expected results** may not agree with the observed results. **Observed results** are those that appear in the offspring because of random combinations of the genes.

A **monohybrid cross** represents a mating between organisms. This involves one specific trait, such as flower color

in some plants and number of digits in some mammals. In complete dominance, the expression of the dominant allele masks the expression of the recessive allele. For example, the trait for polydactyly in humans (having extra digits; Fig. 8.1) (P) is dominant to that of having the normal number of digits (p). If a homozygous dominant polydactyl individual (PP) mates with a heterozygous individual (Pp), what will be the potential genotypes and phenotypes of the offspring? A Punnett square can be constructed to solve the question.

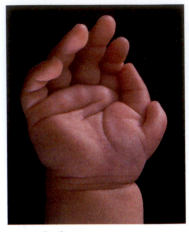

FIGURE **8.1** Polydactyly is a common congenital deformity of the hand, although it also occurs in the foot.

Looking at the solved Punnett square (Table 8.1), the potential genotypes are 50% homozygous dominant polydactyl and 50% heterozygous polydactyl. The resulting phenotype is 100% polydactyl. In the blank Punnett square (Table 8.2), solve the following problem. If a heterozygous polydactyl individual (Pp) mates with a heterozygous polydactyl individual (Pp), what will be the potential genotypes and phenotypes of the offspring? Solve the problem, and write the potential genotypes and phenotypes in the space provided.

TABLE **8.1** Solved Punnett Square for Polydactyly

	P	P
P	PP	PP
p	Pp	Pp

TABLE **8.2** Punnett Square

	P	p
P		
p		

In the natural world, not all traits are either dominant or recessive. Some traits exhibit **incomplete dominance** (Fig. 8.2). In these cases the heterozygous condition has an intermediate phenotype. For example, in snapdragon flowers, the heterozygous condition results in a pink color, and the homozygous conditions result in either red (RR) or white (rr) flowers (Table 8.3).

Another condition known as **codominance** also exists. In this case, both alleles in the heterozygous condition are expressed equally. A classic example of codominance is the ABO blood groups. In a person with type AB blood, both alleles are expressed.

8

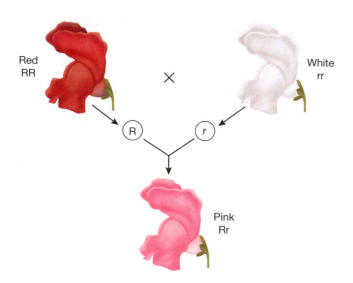

Red
RR

×

White
rr

R   r

Pink
Rr

FIGURE **8.2** Incomplete dominance in snapdragons.

TABLE **8.3** Solved Punnett
Square for Incomplete Dominance

	R	R
R	RR	RR
r	Rr	Rr

For simplicity, this chapter has been designed to consider traits controlled by only one gene (one pair of alleles). Keep in mind that most human traits are controlled by more than one gene. In these exercises, students' knowledge of the fundamental mechanisms of genetics will be reinforced by developing a simple model of inheritance and participating in a human genetic traits survey.

8

Understanding Heredity

Punnett squares make predictions about expected results based upon laws of probability. In nature, however, the expected results may not agree with the observed results. Observed results appear in the offspring as a result of random combinations of the genes. This exercise has been developed to help students understand the concepts of expected results and observed results. Students will work in teams of two. Assume that in pennies, heads (H) are dominant to tails (h).

Procedure 1

Determining Genotypes and Phenotypes Using Coin Tosses

Materials
❏ Paper
❏ Pennies

1 Complete Table 8.4 based upon two heterozygous parents. Record the genotypes and phenotypes of the results.

TABLE **8.4** Punnett Square

	H	h
H		
h		

2 Determine the expected genotypes and phenotypes for the Punnett square.

Genotypes: _____

Phenotypes: _____

3 Using the expected genotypes and phenotypes, predict the probability of genotype and phenotype combinations for 100 coin tosses. Record this information in the expected genotype column in Table 8.5 and the expected phenotype column in Table 8.6.

4 Place two pennies in your hand, and then toss them onto the tabletop. Tally the letter combinations (H = heads; h = tails) below, and record your group's results in the observed genotype and phenotype columns in Tables 8.5 and 8.6.

5 Compile class data and complete Tables 8.5 and 8.6.

TABLE **8.5** Expected and Observed Genotypes

	Expected Genotype	Observed Genotype
Group Data (100 tosses)		
HH		
Hh, hH		
hh		
Class Data		
HH		
Hh, hH		
hh		

TABLE **8.6** Expected and Observed Phenotypes

	Expected Phenotype	Observed Phenotype
Group Data (100 tosses)		
Heads		
Tails		
Class Data		
Heads		
Tails		

✓ Check Your Understanding

1.1 What did the pennies represent in the exercise? Were they an accurate representation? Why or why not? Why were two coins used?

_____

_____

_____

_____

1.2 What is a Punnett square?

_____

_____

1.3 Observed results appear in the offspring as a result of _____ combination of genes.

1.4 How did the observed genotypes and phenotypes from your group compare to the rest of the class?

_____

_____

_____

_____

1.5 If more coin tosses are completed, you would expect the observed results to approach the expected results. (*Circle the correct answer.*)

True / False

Population Genetics

Evolution does not occur at the level of individuals but rather at the level of populations. It involves changes in the frequency (proportion) of a given allele in that population. For example, in a population of pea plants that originally has only alleles for yellow peas, a new allele might appear in the population (the result of a mutation in the DNA sequence that determines pea color) that causes peas to be green. Initially, the frequency of the green allele would be low, but over time the frequency might increase so eventually 10%, 50%, 75%, or even more of the alleles for pea color might be green. This change in allele frequency over time represents evolution in that population.

Allele frequencies are expressed as a fraction of 1.0. Thus, if 80% of the alleles are yellow and 20% of the alleles are green, the frequencies would be 0.8 and 0.2, respectively. When the frequency of alleles for a gene are constant over time (no change in allele frequency), that gene is in **genetic equilibrium**. According to the **Hardy-Weinberg model**, genetic equilibrium will be maintained in a population if the following conditions are met:

1. There must be no mutation.

2. There must be no movement of individuals into or out of the population.

3. Mating between individuals must be completely random.

4. The population must be sufficiently large so the laws of probability apply.

5. There is no selection; no allele is favored over another allele. (*Example:* Green and yellow peas are equally likely to survive and reproduce.)

In a population, the frequency of the two alleles for a given gene adds up to 1.0. If the two alleles are designated p and q, then p + q = 1.0. If the conditions of the Hardy-Weinberg model are met in the population, then the frequencies of the genotypes pp, pq, and qq will not change over time, and their proportions can be estimated by the Hardy-Weinberg equation:

$$p^2 + 2pq + q^2 = 1.0$$

Procedure 1

Determining Allele Frequencies over Generations

Materials
❏ Cup or beaker
❏ Red and white plastic pieces; any contrasting material will suffice (50 of each color)
❏ Graph paper

In this procedure, we will be looking at allele frequencies over several generations in a population in which these conditions are generally met. Then we will study the effect of violating two of the above conditions on the allele frequencies in the population.

1 Procure a cup or beaker and 50 red (allele R) and 50 white (allele r) plastic pieces from the instructor. This represents an allele frequency of 0.50 for R and 0.50 for r.

2 Shake the cup to mix the pieces, and without looking, choose 2 pieces to represent the genotype of a single offspring from that population. Note and record the genotype (RR, Rr, rr) in the first cell of Generation 1 in Table 8.7.

3 Replace the pieces in the cup, shake to mix, and pull out 2 more pieces. Continue this procedure until you have genotypes for a total of 50 individuals (this constitutes one generation).

4 Record your genotype totals in Table 8.8. Note that each RR individual represents two copies of the R allele, each rr individual represents two copies of the r allele, and each Rr individual represents one copy of R and one copy of r.

5 Based on your tally, calculate the allele frequencies for this next generation of the population and record these calculations in Table 8.9.

Example
See the example row in Tables 8.8 and 8.9 to follow the math below:

RR 12 individuals = 24 copies of R

Rr 28 individuals = 28 copies of R + 28 copies of r

rr 10 individuals = 20 copies of r

Total 50 individuals: _____

Total 100 alleles: _____

The total number of R alleles is 24 + 28 = 52

Allele frequency = 52 ÷ 100, or 0.52

The total number of r alleles is 20 + 28 = 48

Allele frequency = 48 ÷ 100, or 0.48

In this example, for the next generation to reflect this new allele frequency, you would add 2 red pieces to your cup and take out 2 white pieces to give a total of 52 red and 48 white. For your own experiment, you will be adding and subtracting whatever number of pieces is necessary to give you the appropriate allele frequencies.

6 Repeat the entire experiment (50 individuals) starting with your new allele frequencies to find out the allele frequency of the next generation. Continue this for a total of three generations, using your newly calculated allele frequencies to begin each subsequent generation.

7 Graph the frequency of the red allele (R) from the beginning (0.5) through the three generations on Figure 8.3 on the next page.

TABLE **8.7** Data for Determining Allele Frequencies over Generations

Generation 1				Generation 2				Generation 3			
1		26		1		26		1		26	
2		27		2		27		2		27	
3		28		3		28		3		28	
4		29		4		29		4		29	
5		30		5		30		5		30	
6		31		6		31		6		31	
7		32		7		32		7		32	
8		33		8		33		8		33	
9		34		9		34		9		34	
10		35		10		35		10		35	
11		36		11		36		11		36	
12		37		12		37		12		37	
13		38		13		38		13		38	
14		39		14		39		14		39	
15		40		15		40		15		40	
16		41		16		41		16		41	
17		42		17		42		17		42	
18		43		18		43		18		43	
19		44		19		44		19		44	
20		45		20		45		20		45	
21		46		21		46		21		46	
22		47		22		47		22		47	
23		48		23		48		23		48	
24		49		24		49		24		49	
25		50		25		50		25		50	

TABLE **8.8** Compilation of Data for Determining Allele Frequencies over Generations

Round	Total Individuals	RR Individuals	R Alleles	Rr Individuals	R Alleles	r Alleles	rr Individuals	r Alleles
Sample	50	12	24	28	28	28	10	20
1	50							
2	50							
3	50							

TABLE **8.9** Calculation of Allele Frequencies over Generations

Round	Total Individuals	Total Number of Alleles	Total Number of R Alleles	Allele Frequency of R	Total Number of r Alleles	Allele Frequency of r
Sample	50	100	52	0.52	48	0.48
1	50	100				
2	50	100				
3	50	100				

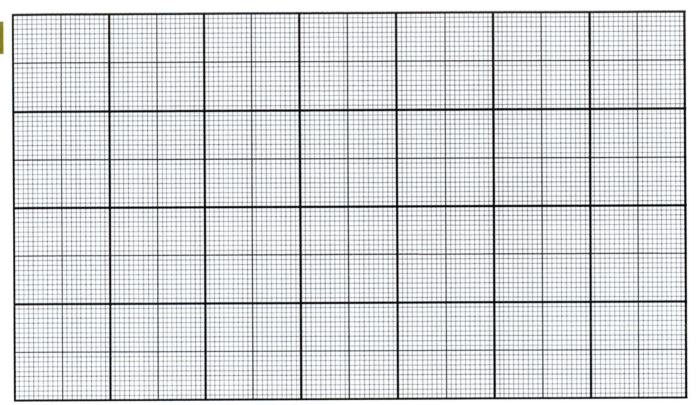

FIGURE **8.3** Frequency of red allele through three generations.

Procedure 2

Effects of Selection on Allele Frequencies

This procedure will be performed the same way as Exercise 8.2, Procedure 1, except you will now assume that homozygous white (rr) is lethal. Individuals who have this lethal genotype will not survive and reproduce. In other words, r is being selected against, and R is being selected for.

1 Beginning with allele frequencies of R = 0.5 and r = 0.5, select 50 individuals as before. This time, however, do not count the r alleles that occur in rr individuals. Record your data in Table 8.10.

Example
See the example row in Tables 8.11 and 8.12 to follow the math below:

RR 12 individuals = 24 copies of R

Rr 28 individuals = 28 copies of R + 28 copies of r

rr 10 individuals = LETHAL

Total 50 individuals: _____

Total 80 alleles: _____

Total R alleles: 24 + 28 = 52 Allele frequency of R = 52 ÷ 80, or 0.65

Total r alleles: 0 + 28 = 28 Allele frequency of r = 28 ÷ 80, or 0.35

In this example, for the next generation, you would have to add 15 red pieces (for a total of 65) and take away 15 white pieces (for a total of 35) to give the proper allele frequencies.

2 Continue the procedure for a total of three generations.

3 Compile your totals and calculate the allele frequencies in Tables 8.11 and 8.12.

4 Graph the frequency of the R allele over time on Figure 8.3.

TABLE **8.10** Data for the Effects of Selection on Allele Frequencies

Generation 1				Generation 2				Generation 3			
1		26		1		26		1		26	
2		27		2		27		2		27	
3		28		3		28		3		28	
4		29		4		29		4		29	
5		30		5		30		5		30	
6		31		6		31		6		31	
7		32		7		32		7		32	
8		33		8		33		8		33	
9		34		9		34		9		34	
10		35		10		35		10		35	
11		36		11		36		11		36	
12		37		12		37		12		37	
13		38		13		38		13		38	
14		39		14		39		14		39	
15		40		15		40		15		40	
16		41		16		41		16		41	
17		42		17		42		17		42	
18		43		18		43		18		43	
19		44		19		44		19		44	
20		45		20		45		20		45	
21		46		21		46		21		46	
22		47		22		47		22		47	
23		48		23		48		23		48	
24		49		24		49		24		49	
25		50		25		50		25		50	

TABLE **8.11** Compilation of Data for Effects of Selection on Allele Frequencies

Round	Total Individuals	RR Individuals	R Alleles	Rr Individuals	R Alleles	r Alleles	rr Individuals	r Alleles
Sample	50	12	24	28	28	28	LETHAL	
1	50						LETHAL	
2							LETHAL	
3							LETHAL	

TABLE **8.12** Calculations of Allele Frequencies after Selection

Round	Total Individuals	Total Number of Alleles	Total Number of R alleles	Allele Frequency of R	Total Number of r alleles	Allele Frequency of r
Sample	50	80	52	0.65	28	0.35
1	50					
2						
3						

Procedure 3

Effects of Small Population on Allele Frequencies

This time you will be looking at allele frequencies in 5 individuals rather than 50. Simply pick out genotypes as before, but this time for only 5 individuals.

1 Begin the experiment with 5 red pieces and 5 white pieces.

Example
See the example row in Tables 8.13 and 8.14 to follow the math below:

RR 2 individuals = 4 copies of R

Rr 2 individuals = 2 copies of R and 2 copies of r

rr 1 individual = 2 copies of r

Total 5 individuals: _____

Total 10 alleles: _____

Frequency of R = 4 + 2 = 6   6 ÷ 10 = 0.6

Frequency of r = 2 + 2 = 4   4 ÷ 10 = 0.4

2 Repeat this experiment six times (always beginning with 5 red and 5 white).

3 Record your data in Table 8.13. Compile your totals in Table 8.14, and calculate the allele frequencies for each group in Table 8.15.

TABLE **8.13** Data for Effects of Small Population on Allele Frequencies

	Round 1		Round 2		Round 3		Round 4		Round 5		Round 6
1		1		1		1		1		1	
2		2		2		2		2		2	
3		3		3		3		3		3	
4		4		4		4		4		4	
5		5		5		5		5		5	

TABLE **8.14** Compilation of Data for Effects of Small Population on Allele Frequencies

Round	Total Individuals	RR Individuals	R Alleles	Rr Individuals	R Alleles	r Alleles	rr Individuals	r Alleles
Sample	5	2	4	2	2	2	1	2
1	5							
2	5							
3	5							
4	5							
5	5							
6	5							

TABLE **8.15** Calculations of Allele Frequencies in a Small Population

Round	Total Individuals	Total Number of Alleles	Total Number of R Alleles	Allele Frequency of R	Total Number of r Alleles	Allele Frequency of r
Sample	5	10	6	0.60	4	0.40
1	5	10				
2	5	10				
3	5	10				
4	5	10				
5	5	10				
6	5	10				

✔ Check Your Understanding

2.1 Complete the Hardy-Weinberg Equation: $p^2 +$ _____ $+ q^2 = 1$

2.2 List the conditions that will maintain genetic equilibrium according to the Hardy-Weinberg model.

2.3 What is the impact of small population size upon population genetics?

2.4 What is the advantage of using larger populations in genetic and other types of studies?

2.5 Compare the change in the allele frequencies of R over three generations both without selection (Procedure 1) and with selection (Procedure 2). Which shows more change? Is evolution more or less likely to occur when selection is acting on a particular allele? Why?

2.6 In Procedure 2, will selection against the rr genotype ever lead to complete removal of the r allele from the population? Explain your answer.

2.7 Evolution _____ . (*Circle the correct answer*)
- a. involves no change in the frequency of a given allele in a population
- b. occurs at the level of populations
- c. occurs at the level of the individual
- d. involves changes in the frequency of a given allele in a population
- e. both b and d

8

The Corn Maize

The world is dependent upon corn (*Zea mays*) as a source of food, industrial products, and fuel. Over the last 5,000 years, corn has been developed through artificial selection and has changed tremendously from its ancestor teosinte. Among the many varieties of corn are sweet corn, popcorn, and Indian corn (one of the most colorful varieties).

In this lab activity, students will investigate the inheritance patterns of color and sweetness in Indian corn (Fig. 8.4). The kernel color (purple or yellow) in Indian corn is controlled by the alleles P (purple) and p (yellow), with purple (P) being dominant; and carbohydrate content (starchy or sweet) is controlled by the alleles S (starchy) and s (sweet), with starchy (S) being dominant. Sweet kernels can be distinguished from starchy kernels because sweet kernels are wrinkled when they dry, and starchy kernels are smooth.

The marbling observed in some of the kernels is a result of **transposons**, or "jumping genes," first described by American geneticist Barbara McClintock (1902–1992). She is best known as a cytogeneticist who dedicated her life to studying corn (*Zea mays*). She and Harriet Creighton (1909–2004) first described the process of crossing over and the recombination of genetic traits during meiosis. Through the 1940s and 1950s, her description of how genes could move within and between chromosomes was not accepted. However, her ideas were confirmed and resurfaced in the late 1960s, and she eventually was awarded the Nobel Prize in Physiology or Medicine in 1983. Barbara McClintock was the first American woman to win an unshared Nobel Prize.

FIGURE **8.4** Indian corn.

8

Procedure 1

Corn Inheritance

Materials
- ❏ Ear of Indian corn
- ❏ Plastic wrap
- ❏ Sharpie or marking pen

1 Obtain an ear of Indian corn, plastic wrap, and a marking pen from your instructor. Tightly wrap the corn with the plastic wrap. Wrapping the corn ensures that it can be reused by other groups.

2 Carefully count all the corn kernels (row by row), marking each kernel with your pen to ensure that the kernels are not counted twice. As you count the kernels, tally their phenotype in Table 8.16.

3 After you have completed counting and tallying the kernels, determine the totals for your group using Table 8.17, and report your findings to the laboratory instructor.

4 The instructor will display each group's results for analysis (document them in Table 8.17). To determine ratio, divide each phenotypic total by the total of all kernels counted.

a. Total each phenotypic column.

b. Calculate the total of all seeds counted.

c. Divide the total by 16 (the number of rows in an ear of corn).

5 Discuss your findings. What were the genotypes of the parental corn crossed in your ear of corn? What are the possible genotypes for a purple/starchy kernel, purple/sweet kernel, yellow/starchy kernel, and yellow/sweet kernel?

_____

_____

_____

_____

TABLE **8.16** Individual Tally

Purple and Starchy (Smooth)	Purple and Sweet (Wrinkled)	Yellow and Starchy (Smooth)	Yellow and Sweet (Wrinkled)	Total Kernels

TABLE **8.17** Group Tally

Group Number	Purple and Starchy (Smooth)	Purple and Sweet (Wrinkled)	Yellow and Starchy (Smooth)	Yellow and Sweet (Wrinkled)	Total Kernels Per Group
1					
2					
3					
4					
5					
6					
7					
8					
9					
10					
Phenotypic totals					✕
Total kernels for class	✕	✕	✕	✕	
Ratio (phenotypic total kernels for class)					✕

✔ Check Your Understanding

3.1 What causes the marbling effect in Indian corn? (*Circle the correct answer.*)

   a. Partial masking of the dominant gene (P).

   b. Complete dominance.

   c. Transposons ("jumping genes").

   d. Random mutations.

   e. Partial masking of the recessive gene (p).

3.2 How would you tell the difference between a sweet and a starchy kernel of Indian corn?

Fitting In

This activity asks students to participate in a survey of some common human characteristics (Fig. 8.5). In addition, the activity introduces students to the concept of genetic variability within a population.

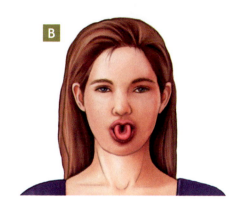
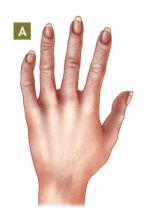

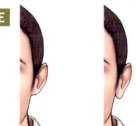

FIGURE **8.5** Some common genetic characteristics: (**A**) bent little finger; (**B**) tongue rolling; (**C**) widow's peak; (**D**) dimpled chin; (**E**) free earlobe; (**F**) hitchhiker's thumb.

8

Procedure 1

Genetic Traits

Materials

None required

1 Working with your lab partner, determine your phenotypes for the traits discussed in this procedure. The traits in this procedure are based upon two alleles and follow basic Mendelian dominance patterns. Record your phenotype and possible genotype in Table 8.18.

2 The instructor will collect the class data and place it on a chart on the board. Record these data in Table 8.18. Complete all the table columns regarding the information you collected.

Look for the following genetic traits:

1. **Bent little finger:** Examine your little finger on each hand. If the last joint of your little finger bends toward your ring finger, you have the dominant gene "C" for bent little finger. A straight little finger gene "c" is recessive. The condition of bent little finger is called **clinodactyly**.

2. **Tongue rolling:** Extend your tongue, and attempt to roll it into a U-shape. The gene for tongue rolling "R" is dominant to the gene for non-tongue rolling "r."

3. **Widow's peak:** Think of Count Dracula or Eddie Munster. These individuals have a V-shaped hairline in the middle of the forehead. This characteristic comes from the dominant allele "W" for widow's peak. The recessive allele "w" results in a straight forehead hairline.

4. **Free earlobe:** If a portion of the earlobe remains unattached below the point of attachment to the head, you have the dominant gene "E" for a free earlobe. The recessive gene "e" results in attached earlobes.

5. **Finger hair:** If you possess hair on the middle segment of your fingers, you have the dominant allele "H" for mid-digital hair. If you do not have hair, you are recessive for the "h" allele.

6. **Dimpled cheeks:** The presence of dimples in one or both cheeks, "D," is dominant to dimple absence "d."

7. **Eyebrow raising:** Are you kin to Mr. Spock on *Star Trek*? Can you raise your eyebrows? If so, you have the dominant allele "Y." If not, you are recessive "y."

8. **Ear wiggling:** The gene for having the ability to wiggle your ears, "W," is dominant to non-wiggling "w."

9. **Long toe:** Check out your second toe. If it is longer than your big toe, it is a result of a dominant allele "L"; a shorter second toe is recessive "l."

10. **Curly hair:** Curly hair "A" is dominant over straight hair "a."

11. **Freckles:** The presence of freckles results from the dominant allele "Z"; the allele "z" is recessive.

12. **PTC tasting:** Approximately 70% of the U.S. population has the ability to taste PTC (phenylthiocarbamide) paper. If you are chewing gum, remove it and wash out your mouth with water. Place a piece of PTC paper on your tongue. Record your results. PTC tasting comes from the dominant allele "P"; not tasting PTC indicates the recessive "p" allele.

T_ABLE **8.18** A Survey of Some Common Human Characteristics

Trait	Your Phenotype	Possible Genotypes	Class N	Class Phenotypes	Class Percent Phenotypes
Bent little finger					
Tongue rolling					
Widow's peak					
Free earlobe					
Finger hair					
Dimpled cheeks					
Eyebrow raising					
Ear wiggling					
Long toe					
Curly hair					
Freckles					
PTC tasting					

8

✔Check Your Understanding

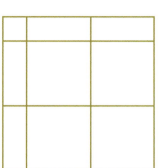

4.1 If a man who is heterozygous for PTC tasting marries a woman who is homozygous recessive, predict the potential genotypes and phenotypes for the offspring for that trait.

   a. Genotype of male _____

   b. Genotype of female _____

   c. Genotypes of offspring _____

   d. Phenotypes of offspring _____

4.2 If a man who is heterozygous for hitchhiker's thumb marries a woman who is also heterozygous, predict the potential genotypes and phenotypes for the offspring for that trait.

   a. Genotype of male _____

   b. Genotype of female _____

   c. Genotypes of offspring _____

   d. Phenotypes of offspring _____

4.3 If a man who is homozygous recessive for free earlobe marries a woman who is homozygous dominant, predict the potential genotypes and phenotypes for the offspring for that trait.

   a. Genotype of male _____

   b. Genotype of female _____

   c. Genotypes of offspring _____

   d. Phenotypes of offspring _____

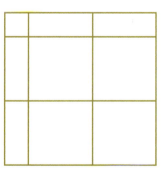

Did you know . . .

Unraveling a Mystery

Why are approximately 10% of today's Europeans resistant to HIV? After a tedious scientific endeavor, a genetic explanation is now unraveling. HIV-resistant individuals have two copies of a mutation (D32) of the CCR-5 gene that disrupts the path through which the HIV virus enters white blood cells. People who inherit just one copy of the mutation might become infected, but the disease progresses at a slower rate.

In the 1990s scientists suggested the same mutation resulted in a survival advantage against the bubonic plague, or Black Death, that ravaged Europe in the 1300s. Today, some scientists believe the mutation might have offered resistance to smallpox or some form of hemorrhagic fever. Despite disagreement, the prevalence of the CCR-5 D32 mutation in some people of European descent provides evidence of evolution in action and offers hope to one day finding better treatment or prevention for AIDS.

Blood Groups

ABO Blood Types

All of us have one of the four blood types: A, B, AB, or O. The ABO blood types are identified on the basis of the presence or absence of specific glycoproteins on the surface of red blood cells. Because the glycoproteins are complex and can stimulate an immune response, they are called antigens. The glycoproteins result from the expression of specific combinations of multiple alleles: type A ($I^A I^A$, $I^A i$), type B ($I^B I^B$, $I^B i$), type AB ($I^A I^B$), and type O (ii) blood.

The A, B, AB, and O blood types provide a great example of **codominance**. The alleles I^A and I^B are codominant to each other and cause both antigens A and B to be expressed. They are both dominant over the alleles for type O blood (ii) (Table 8.19).

Blood typing is crucial in determining the safety of blood transfusions. In 1901, Karl Landsteiner (1868–1943) developed the system for naming the blood groups. This is significant because a person with type A blood possesses anti-B proteins called **antibodies** in his or her plasma. Similarly, an individual with type B blood has anti-A antibodies.

Agglutination, or clumping of the red blood cells, can occur if anti-A antibodies from an individual with type B blood mix with red blood cells of a type A person, or if anti-B antibodies from a type A person mix with red blood cells from a person with type B blood. **Transfusion reactions** resulting from inadvertently mixing these blood types can, in severe cases, lead to death. Individuals with type AB blood do not possess antibodies and can receive blood from all blood groups (**universal recipients**). Those with type O blood have no cell surface antigens and are called **universal donors**. They have anti-A and anti-B antibodies in the plasma, but under emergency conditions, the red blood cells without plasma can be transfused safely.

Blood typing (in conjunction with DNA testing) can be important in cases in which paternity is in question. For example, if the mother has type AB blood ($I^A I^B$), and the father has type O blood (ii), the child will have either $I^A i$ or $I^B i$ because type A and type B are dominant over type O.

Blood typing can be used only in determining whether a man is a *potential* father. In order to determine true paternity, DNA testing provides more definitive results. Construct a Punnett square to determine the genotypes and phenotypes of the potential offspring of the above scenario (Table 8.20).

a. Genotype of male _____

b. Genotype of female _____

c. Genotypes of offspring _____

d. Phenotypes of offspring _____

TABLE **8.20** Paternity

Rh Factors

Other antigens are found on the surface of red blood cells. In 1937, Karl Landsteiner (1868–1943) and Alexander Wiener (1907–1976) were studying blood from Rhesus monkeys and found other antigens on the surface of their red blood cells. They called these antigens Rhesus monkey factors (Rh factors). In humans, the Rh factor is the RhD antigen found on the surface of some individuals' red blood cells. Either an individual has the antigen (D-positive) or does not have the antigen marker (D-negative). Pregnant women are checked to see whether they are Rh$^+$ or Rh$^-$. This is important, especially if the mother is Rh$^-$, because a second pregnancy could cause **erythroblastosis fetalis**, a hemolytic disease of the newborn that could be life-threatening. Today, this problem can be prevented by giving the mother RhoGAM (antibodies that are derived from human plasma).

TABLE **8.19** Synopsis of the ABO Blood Group

Blood Type (Phenotype)	Genotype	Red Blood Cell Antigen	Antibody Present in Blood Plasma
A	$I^A I^A$, $I^A i$	A	anti-B
B	$I^B I^B$, $I^B i$	B	anti-A
AB	$I^A I^B$	AB	neither
O	ii	neither A nor B	anti-A and anti-B

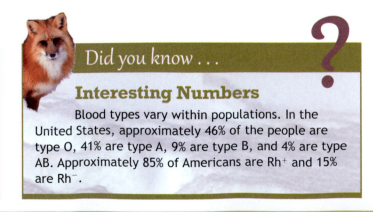

Did you know . . .

Interesting Numbers

Blood types vary within populations. In the United States, approximately 46% of the people are type O, 41% are type A, 9% are type B, and 4% are type AB. Approximately 85% of Americans are Rh$^+$ and 15% are Rh$^-$.

Determining Blood Type and Rh Factor

Materials

- ❑ Vials or kit with samples of synthetic blood
- ❑ Vials or kits for typing blood (synthetic anti-A serum, anti-B serum, and anti-D [Rh] serum)
- ❑ Safety glasses
- ❑ Gloves
- ❑ Blood-typing slides with wells
- ❑ Paper towels
- ❑ Toothpicks
- ❑ Timing device

In this procedure, you will determine blood type and Rh factor using synthetic blood.

1 Procure the materials from the instructor. Put on your gloves and safety glasses. On a paper towel, arrange the well slide for determining blood type; the toothpicks; the vials containing synthetic anti-A serum, anti-B serum, and anti-Rh serum; and three samples of synthetic blood as distributed by the instructor.

2 Carefully add a single drop of synthetic anti-A serum to the slide well labeled "A." Replace the cap on the vial.

3 Carefully add a single drop of synthetic anti-B serum to the slide well labeled "B." Replace the cap on the vial.

4 Carefully add a single drop of synthetic anti-D (Rh) serum to the slide well labeled "D" or "Rh." Replace the cap on the vial.

5 Using the synthetic blood vial, place a drop of the first synthetic blood sample in each well of the blood-typing slide. Replace the cap on the dropper vial. Make sure to replace the cap on one vial before opening the next vial to prevent cross-contamination.

6 Using a different toothpick for each well (anti-A, anti-B, and anti-D [Rh]), gently mix the synthetic blood and antiserum drops for 30 seconds. Discard the toothpicks after a single use to avoid contamination of the samples.

7 Examine the thin films of liquid mixture left behind in the wells. You might want to use a lamp for best viewing. If the film remains uniform in appearance, and there is no clumping, then there is no agglutination. Thus, the reaction is negative. If the sample in the well appears granular and clumpy, agglutination has occurred. If agglutination occurs, the reaction is positive, and it indicates blood type. Place your results (clumping or no clumping) in Table 8.21.

8 Repeat steps 2 through 5 for the next two synthetic blood vials.

9 Follow the instructor's rules for proper cleanup.

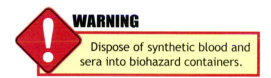

WARNING
Dispose of synthetic blood and sera into biohazard containers.

TABLE **8.21** Blood Type and Rh Factor Results

Blood Sample Vial	Anti-A Serum Reaction	Anti-B Serum Reaction	Anti-D (Rh) Serum Reaction
1			
2			
3			

✓ Check Your Understanding

5.1 An expectant mother has type B blood, and her genotype is $I^B i$; the father has type A blood, and his genotype is $I^A i$. What are the potential genotypes and phenotypes of the blood types of their offspring?

_____

_____

5.2 If a father has type A blood (genotype $I^A I^A$) and the mother has type O blood (genotype ii), what are the potential genotypes and phenotypes of the blood types of the offspring?

_____

_____

8 **5.3** How can blood type be used to establish the identity of an individual in a paternity suit? Can it be used as the only evidence of paternity?

_____

_____

5.4 What is the potential danger from receiving the wrong type of blood during a transfusion? Why?

_____

_____

_____

5.5 A patient who is waiting on a kidney transplant list gets notification that there may possibly be a kidney from another transplant center near their home. The patient has type A blood. What blood type(s) do the transplant coordinators need to consider for this patient? (_Circle the correct answer._)

a. A only.

b. AB only.

c. O only.

d. A and O.

e. B.

5.6 If a mother is type O blood, and the father is type B, what are the potential blood types of their offspring?

_____

_____

_____

5.7 Recently Mary underwent extensive surgery and needs a transfusion. She has type A blood. Her friend Paul has type AB blood, and another friend, Thomas, has type O blood. Which of her friends' blood can be safely transfused? Why?

_____

_____

_____

Developing Pedigrees

To geneticists, a **pedigree** is a valuable tool resembling a family tree used to display family relationships and to track traits through a family. In medical genetics, pedigrees are helpful in understanding how disorders appear in families.

Pedigrees are particularly valuable in understanding the inheritance of **monogenic** (single-gene) traits such as albinism, cystic fibrosis, and hemophilia. A pedigree can be used to visually represent Mendelian inheritance of a trait in an extended family. A typical pedigree consists of universally accepted symbols connected by either horizontal or vertical lines. Roman numerals represent generations, and Arabic numerals represent individuals. Filled shapes represent individuals who express a trait, and half-shaded shapes represent carriers. A few common symbols appear in Figure 8.6.

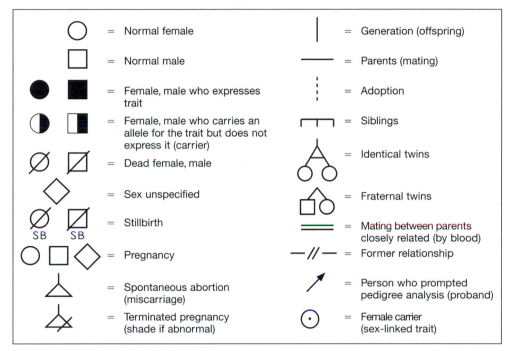

FIGURE **8.6** Pedigree symbols.

Procedure 1

Using and Constructing Pedigrees

Materials

None required

1 Using a pedigree, **autosomal recessive** traits, such as cystic fibrosis, are easy to follow through the generations. Explain the inheritance of cystic fibrosis in two children of the third generation in Figure 8.7.

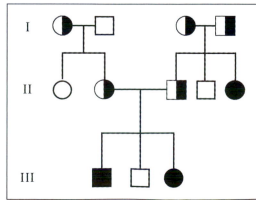

FIGURE **8.7** Cystic fibrosis pedigree.

2 The inheritance of **autosomal dominant** traits also can be explored through pedigree analysis. Polydactylism, having extra digits, results from a dominant gene (P). Using Figure 8.8, explain the appearance of polydactyly in children of generation 3.

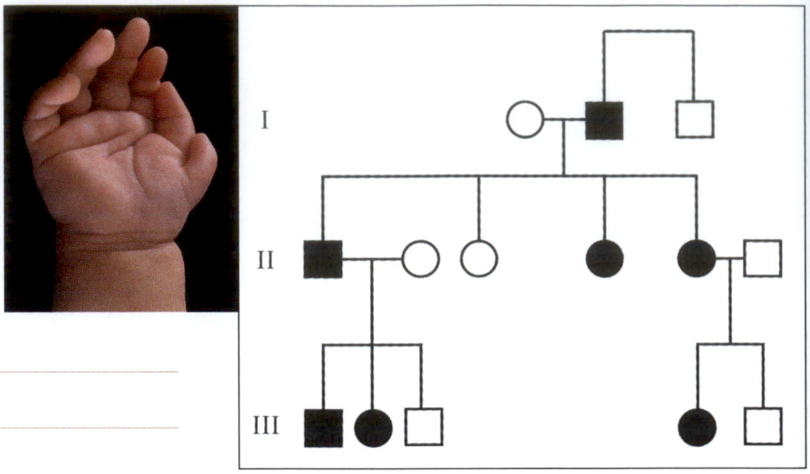

FIGURE **8.8** Polydactylism pedigree.

X-linked traits are carried exclusively on the X chromosome. Because a male possesses only one X chromosome, if he receives an X chromosome that carries an X-linked trait, he will express that trait. For a female to express an X-linked trait, she will have to inherit two copies of the gene, one on each X chromosome. Several conditions, such as hemophilia A and red-green color blindness, are X-linked recessive traits.

3 Using the following information, construct a pedigree for color blindness in three generations of a family in the sketch box provided, and answer the questions. In the first generation, neither parent was color blind. The second generation had four children. In their birth order, one was a normal female, one was a male that expressed the trait, another was a female that carried the trait but it was not expressed, and one was a normal male.

a. The first born daughter, who was normal, had one normal daughter and one son who expressed the trait. How did this happen, and what is the genotype of the husband/father in this family?

b. The second born male, who expressed the trait, had a son and daughter both of whom expressed the trait. How did this occur?

c. The third born, a female who did not express the trait, had three daughters who did not express the trait and four sons who did not express the trait. Explain the inheritance pattern.

_____

_____

_____

_____

d. The fourth born, a male who did not express the trait, had three sons who did not express the trait. How did this happen?

_____

_____

_____

_____

4 Achondroplasia, the most common type of human dwarfism, affects 1 in 15,000 to 1 in 40,000 individuals. Many cases of achondroplasia are the result of a spontaneous mutation. After the mutation occurs, however, it becomes an autosomal dominant trait. Two individuals with achondroplasia had a son of average height and two other sons and a daughter with achondroplasia. Draw a pedigree of this family, and explain how the parents could have a child of "average" height.

_____

_____

_____

_____

_____

_____

_____

5 The following pedigree illustrates the inheritance of Tay-Sachs disease in four generations of a family (Fig. 8.9). Interpret the pedigree and determine whether the trait is dominant or recessive. What does the double line symbol between individuals 2 and 3 in generation III indicate?

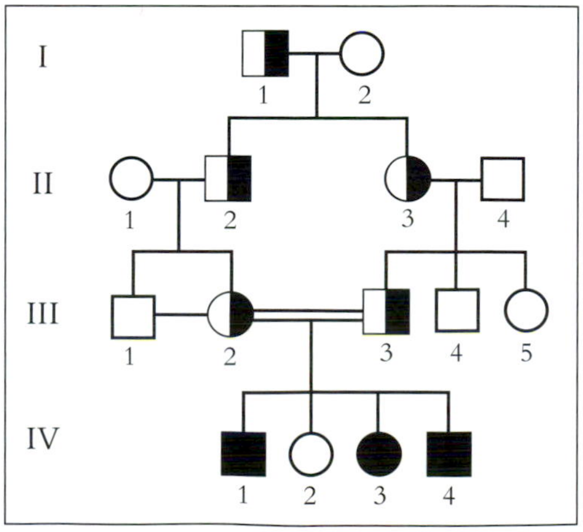

_____

_____

_____

_____

_____

FIGURE **8.9** Tay-Sachs disease pedigree.

✔ Check Your Understanding

6.1 Give the name of the tool that is used by medical geneticists and genetic counselors. Why is this invaluable in their work?

_____

_____

_____

6.2 Distinguish between the symbols for identical twins who have identical DNA (maternal twins) and fraternal twins who do not have identical DNA.

_____

_____

6.3 Consider the X-linked trait red-green color blindness. If a man who is color blind marries a woman who is a carrier for color blindness, predict the genotype and phenotype for their offspring.

   a. Genotype of male _____

   b. Genotype of female _____

   c. Genotypes of offspring _____

   d. Phenotypes of offspring _____

Did you know . . . ?

Strange!

Some traits that were believed to be monogenic, have been discovered to be polygenic and/or can be learned traits. Examples include dimpled chins, hitchhiker's thumb (the ability to bend your thumb back at a 60-degree angle), and thumb crossing. To see what thumb crossing is, swing your hands freely, and suddenly clasp your hands, interlocking your fingers. Is your left thumb uppermost, or your right? For the heck of it, force yourself to place your hands together the opposite way. Strange!

DNA

It is hard to believe the humble beginnings of our knowledge of DNA can be traced back to an obscure physician studying the chemical composition of pus-soaked rags and the sperm of salmon. During his studies in the 1870s, Swiss physician Johann Miescher (1844–1895) described a mysterious substance he coined **nuclein**. Eventually, nuclein would be known as **deoxyribonucleic acid**, or **DNA**. Scientists studying nuclein described this as a **nucleic acid**. Albert Kossel (1853–1927) labeled two distinct types of nucleic acids:

1. Thymus nucleic acid (DNA); and
2. Yeast nucleic acid (RNA).

Early scientists thought perhaps DNA was involved in heredity, and RNA was an energy source for cellular metabolism. Nucleic acids are known to be responsible for the transmission of hereditary information as well as the production of proteins by cells. The two kinds of nucleic acids are deoxyribonucleic acid (DNA) and **ribonucleic acid (RNA)**. DNA possesses instructions for making proteins and RNA. RNA serves primarily as a template of instructions used to build proteins by linking amino acids.

Composition of a Nucleic Acid

Chemically, a nucleic acid is a molecule that consists of repeating units called **nucleotides** (Fig. 8.10). Each nucleotide contains a pentose sugar (deoxyribose in DNA or ribose in RNA), a phosphate group, and a nitrogenous base. Nucleotides are composed of five different nitrogenous bases. The bases are further divided into either double-ringed purines (adenine [A] and guanine [G]) or single-ringed pyrimidines (thymine [T] and cytosine [C] in DNA, and uracil [U] and cytosine [C] in RNA).

A DNA molecule is a double helix with as many as 3 billion nucleotides. Hydrogen bonds between pyrimidines and purines hold the double helix together in **complementary base pairs**. As a result of the contributions of Erwin Chargaff (1905–2002) and others, scientists determined that in DNA the purine adenine (A) is paired with the pyrimidine thymine (T) by hydrogen bonds, and the purine guanine (G) is paired with the pyrimidine cytosine (C) by hydrogen bonds. In RNA, thymine is replaced by uracil (U) to create A-U complementary base pairs.

As a result of the X-ray diffraction studies of Rosalind Franklin (1920–1958) and Maurice Wilkins (1916–2004), scientists suspected that the molecular configuration of DNA was helical. Using the work of Chargaff, Franklin, Wilkins, and others, James Watson (1928–present) and Francis Crick (1916–2004) developed the double-helix model of DNA (Fig. 8.11). Watson and Crick described DNA as a double-stranded helical structure with a backbone of deoxyribose sugar and phosphate and rungs of **complementary base pairs (A-T, C-G)** held together by hydrogen bonds. As a result of bonding, the two strands of the DNA molecule run in opposite directions and accordingly are called **antiparallel**.

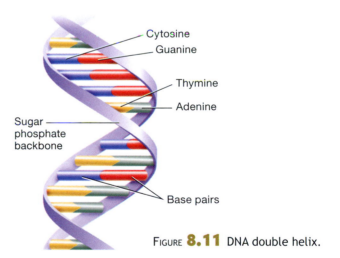

FIGURE **8.11** DNA double helix.

FIGURE 8.10 Structure of nucleotides in DNA.

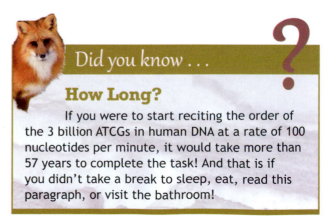

Did you know . . .

How Long?

If you were to start reciting the order of the 3 billion ATCGs in human DNA at a rate of 100 nucleotides per minute, it would take more than 57 years to complete the task! And that is if you didn't take a break to sleep, eat, read this paragraph, or visit the bathroom!

8

In eukaryotic cells, a **chromosome** consists of a continuous molecule of DNA and several types of associated proteins. Humans have 46 chromosomes per somatic cell. Genes are DNA sequences that code for the synthesis of a polypeptide or RNA molecule. All of the genes in all of an organism's chromosomes are collectively referred to as its **genome**. The human genome consists of 21,000 to 25,000 genes.

In a chromosome, DNA is tightly coiled around proteins termed *histones* that resemble a bead-like structure forming a **nucleosome** (Fig. 8.12). The nucleosome is composed of units of eight histone proteins capped by another histone known as a linker. The framework of DNA is maintained by specialized scaffold proteins. Collectively, chromosomes make up chromatin, which consists of DNA, histones, and nonhistone proteins.

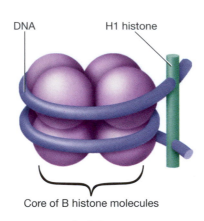

FIGURE **8.12** Nucleosome.

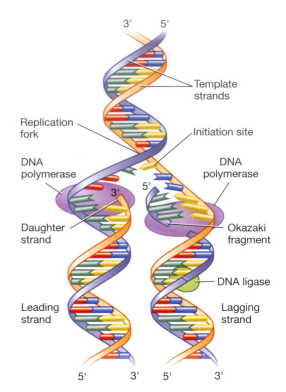

FIGURE **8.13** Overview of DNA replication.

Semiconservative Replication of DNA

Each human chromosome replicates at several hundred points along its length. In eukaryotes, from 500 to 5,000 base pairs are assembled per minute at up to 50,000 origins of replication.

1. Replication begins at a point called the origin of replication, or the initiation site, on the parental strands of DNA (Fig. 8.13). Here, an enzyme called helicase facilitates unwinding. Another enzyme, gyrase, prevents the strands of DNA from tangling. This unwinding results in a Y-shaped replication fork; gyrase forms a nick in the DNA repaired later by an enzyme called ligase. DNA demonstrates **semiconservative replication**. One strand serves as a direct template for the new strand, and the other strand is pieced together.

2. One strand of DNA is called the leading strand (continuous; 3' to 5'), continuing replication in the same direction as the movement of the replication fork. The other strand, the lagging strand (discontinuous; 5' to 3'), continues in the opposite direction. Then, at the start of each DNA segment to be replicated, an enzyme called RNA primase builds a short piece of RNA called an RNA primer.

3. The RNA primer attracts an enzyme called DNA polymerase that attracts the proper nucleotides to the template. The new strand grows as hydrogen bonds are formed. On the lagging strand, Okazaki fragments are formed and are joined by ligase (Fig. 8.14).

4. Proofreading enzymes are responsible for ensuring the fidelity of replication. DNA replication is accurate; only 1 base pair in 10,000 is matched incorrectly.

5. Repair enzymes also ensure fidelity. Excision and post-replication enzymes are common in the repair process. Unfortunately, repair systems can be damaged by prolonged exposure to sunlight; resulting mutations (errors in DNA replication) can bring about skin cancer. Examples of two genetic disorders that involve repair enzymes are xeroderma pigmentosum and ataxia telangiectasia.

During gene expression, DNA is copied onto RNA, which plays a key role in building proteins. Five fundamental differences between DNA and RNA include the following:

1. DNA has deoxyribose sugar, and RNA has ribose sugar. Ribose has a hydroxyl instead of a hydrogen attached to the 2' carbon (Fig. 8.15).

2. DNA has base pairs consisting of A-T and C-G, whereas RNA has A-U and C-G.

3. DNA is double-stranded, whereas RNA is usually single-stranded.

4. DNA is usually longer than RNA.

5. DNA is more stable than RNA because of the stable hydrogen at the 2' position in DNA and the reaction –OH group at the 2' position in RNA.

The three primary types of RNA include the following:

1. Ribosomal RNA (rRNA): forms ribosomes when combined with ribosomal proteins. rRNA is made inside of the nucleolus, whereas the other types of RNA are made in the nucleus. rRNA is 100–30,000 nucleotides long

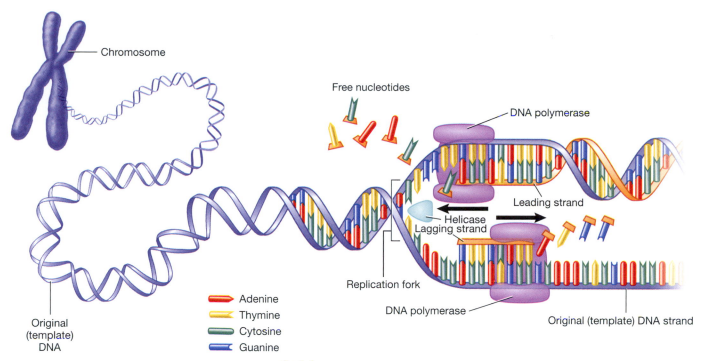

Free nucleotides

DNA polymerase

Leading strand

Helicase
Lagging strand

Replication fork

DNA polymerase

Original (template) DNA strand

Chromosome

Original (template) DNA

Adenine
Thymine
Cytosine
Guanine

FIGURE **8.14** Process of DNA replication.

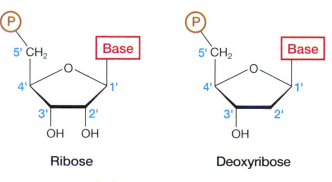

FIGURE **8.15** Comparison of DNA and RNA.

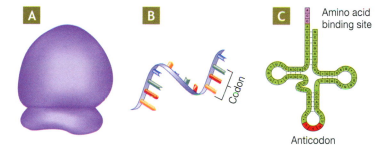

Amino acid binding site

Codon

Anticodon

FIGURE **8.16** (**A**) Ribosomal RNA; (**B**) messenger RNA; (**C**) transfer RNA.

and has two subunits: the small subunit binds the mRNA to the ribosome, and the large subunit attaches to the tRNA and helps bind it to the protein (Fig. 8.16A).

2. Messenger RNA (mRNA): a single strand of nucleotides whose bases are complementary to those of the template DNA to which the RNA was transcribed. Most mRNA consists of 500–1,000 bases. Working in groups of three, or triplets, the mRNA forms codons that specify amino acids in protein synthesis (Fig. 8.16B).

3. Transfer RNA (tRNA): connectors linking an mRNA codon to a specific amino acid. These consist of 75–80 nucleotide base pairs. A loop of tRNA has three bases complementary to the codon called anticodons. The end opposite to the anticodon covalently bonds to a specific amino acid. tRNA always attaches to specific amino acids (Fig. 8.16C).

Protein Synthesis

Protein synthesis is the process of building proteins from amino acids. This process is directed and coordinated by DNA and consists of two major steps: transcription and translation. An overview of this process is shown in Figure 8.17.

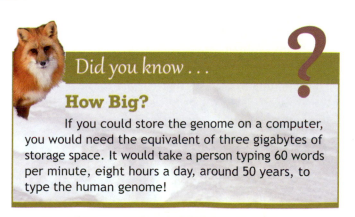

Did you know . . .

How Big?

If you could store the genome on a computer, you would need the equivalent of three gigabytes of storage space. It would take a person typing 60 words per minute, eight hours a day, around 50 years, to type the human genome!

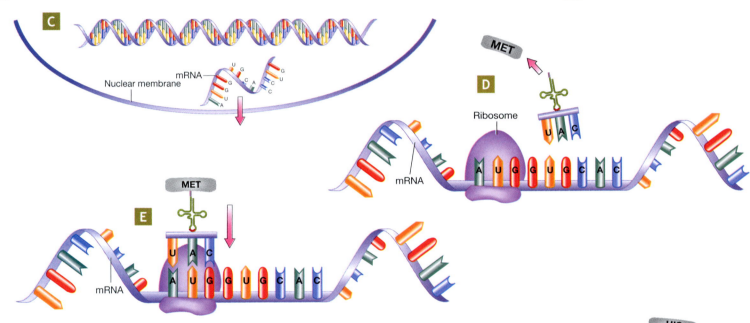

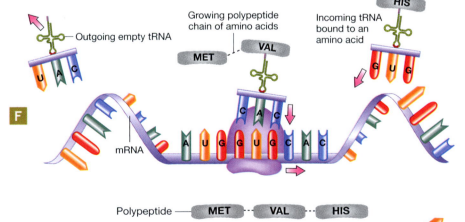

FIGURE **8.17** Overview of protein synthesis. (**A**) Portion of DNA making up a gene; (**B**) complementary message transcribed to mRNA; (**C**) mRNA moves into the cytoplasm via the nuclear membrane; (**D**) mRNA attaches to a ribosome that "reads" the message and tRNA picks up an amino acid; (**E**) tRNA with an amino acid bonds with the mRNA, bringing its amino acid into place; (**F**) a peptide bond forms between each amino acid that is brought into place by a tRNA; (**G**) a stop codon signals that the construction of the protein is complete.

Transcription

The process by which chemical information encoded in DNA is copied into RNA is called **transcription**. Generally, only one strand of the double helix of DNA is transcribed and is called the template strand. Most genes are composed of a coding region transcribed into RNA and a regulatory region that oversees transcription in the coding portion. The promoter is a specific part of the regulatory region that serves as the starting point of transcription.

RNA polymerase builds the complementary strand of RNA. Beginning at the promoter, RNA polymerase unwinds the DNA, temporarily breaking the hydrogen bonds. RNA polymerase then assembles the RNA strand based on complementary base pairing. Transcription ceases at a transcription termination sequence on the DNA. After RNA for a specific region is made, the hydrogen bonds between the two strands of the DNA double helix quickly wind together again, displacing the new RNA, messenger RNA (**mRNA**).

The next step, translation, cannot happen without post-transcriptional modifications. The beginning of the RNA sequence is called a leader, and the end part is called the trailer. The DNA sequence contains coding portions called exons and noncoding portions called introns. Before translation, introns are removed by protein complexes called spliceosomes. Introns range from 65 to 100,000 bases, and exons range from 100 to 300 bases. Many genes are riddled with introns. At one time introns were considered "genetic junk" by some scientists. (Less than 2% of the total DNA carries instructions to make proteins.) However, today introns are known to have regulatory and translation functions and aid in the coordination of development. In the future many more functions of introns will be unraveled.

Once processed, mRNA exits the nucleus via the nuclear pores and enters the cytoplasm where it attaches to a ribosome. Each triplet of mRNA is a codon, which specifies one of 20 standard amino acids. There are 64 different codons. For example, the codon GAG specifies glutamic acid. One codon, AUG, encodes for methionine and serves as a start signal for building a protein. UAA, UAG, and UGA represent stop codons, ending the formation of a protein. These codons are responsible for the genetic code that, for the most part, is universal (Fig. 8.18).

The genetic code is redundant; more than one codon can encode for a specific amino acid. For example, glycine is coded by the codons GGU, GGC, GGA, and GGG. The first two nucleotides are the same in each codon, and only the nucleotide in the third (wobble) position varies. This phenomenon, described by Francis Crick, is known as the **wobble hypothesis**.

First Base	Second Base				Third Base
	U	**C**	**A**	**G**	
U	UUU phenylalanine	UCU serine	UAU tyrosine	UGU cysteine	U
	UUC phenylalanine	UCC serine	UAC tyrosine	UGC cysteine	C
	UUA leucine	UCA serine	**UAA stop**	**UGA stop**	A
	UUG leucine	UCG serine	**UAG stop**	UGG tryptophan	G
C	CUU leucine	CCU proline	CAU histidine	CGU arginine	U
	CUC leucine	CCC proline	CAC histidine	CGC arginine	C
	CUA leucine	CCA proline	CAA glutamine	CGA arginine	A
	CUG leucine	CCG proline	CAG glutamine	CGG arginine	G
A	AUU isoleucine	ACU threonine	AAU asparagine	AGU serine	U
	AUC isoleucine	ACC threonine	AAC asparagine	AGC serine	C
	AUA isoleucine	ACA threonine	AAA lysine	AGA arginine	A
	AUG (start) methionine	ACG threonine	AAG lysine	AGG arginine	G
G	GUU valine	GCU alanine	GAU aspartic acid	GGU glycine	U
	GUC valine	GCC alanine	GAC aspartic acid	GGC glycine	C
	GUA valine	GCA alanine	GAA glutamic acid	GGA glycine	A
	GUG valine	GCG alanine	GAG glutamic acid	GGG glycine	G

FIGURE **8.18** Genetic code.

Translation

Translation is the process that represents the expression of the genetic code in the form of protein. It involves the conversion of information in the form of nucleotide codons to the chemically different form of polypeptides. tRNA molecules recognize both nucleotides and amino acids and thus function as translators. The process of translation occurs in three stages:

1. **Initiation** of translation begins when mRNA associates with a small ribosomal subunit. The AUG codon is recognized as the start site by a tRNA molecule with the anticodon to AUG. The amino acid methionine is attached to the region of the tRNA opposite the anticodon.

2. During **elongation**, each successive codon is read by tRNA molecules that bring the corresponding amino acid. Amino acids are linked by peptide bonds formed by catalytic components of the ribosome.

8

3. **Termination** occurs when a stop codon is encountered. Protein folding and the final touches to the proteins occur after termination. The process of protein synthesis is accurate, but **mutations** can happen. Protein synthesis is economical because cells can produce large amounts of a protein from just one or two copies of a gene. For example, a plasma cell in the human immune system can produce more than 2,000 identical antibodies (which are proteins).

Mitochondrial and Chloroplast DNA

Mitochondria are common organelles in the majority of eukaryotic cells (Fig. 8.19). Scientists have discovered these organelles have their own unique mitochondrial DNA (**mtDNA**). This is illustrated by the **endosymbiontic theory,** which holds that primordial cells established a symbiotic relationship with early purple bacteria. Eventually, the bacterial cell changed and became an important part of the cell's machinery, providing the cell energy.

The **mitochondrial genome** is significantly smaller than the nuclear genome. In many mammalian species, including humans, the mitochondrial genome consists of 37 genes. With the exception of a few cases of the paternal transmission of mtDNA, the majority of mtDNA is transmitted maternally.

With our knowledge of mitochondrial DNA, we are beginning to have a deeper understanding of heredity, medicine, forensics, and evolution. These applications will be covered in further depth in Chapter 9.

Chloroplasts in plant cells and some protists also have unique chloroplast DNA (**clDNA**). These organelles also were derived, according to endosymbiontic theory, as the result of early cells engulfing cyanobacteria (Fig. 8.20). The clDNA genome is also very small. Recent studies have verified that chloroplasts are inherited maternally rather than from pollen. This discovery will have great import for creating genetically modified plants.

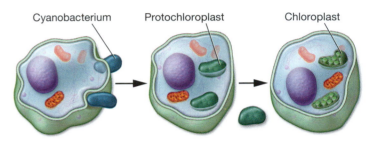

FIGURE **8.20** Chloroplasts were derived from early cyanobacteria engulfed by early cells.

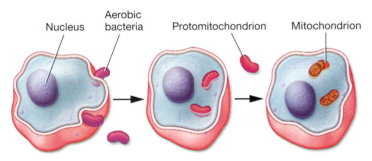

FIGURE **8.19** Mitochondria were derived from early purple bacteria engulfed by early cells.

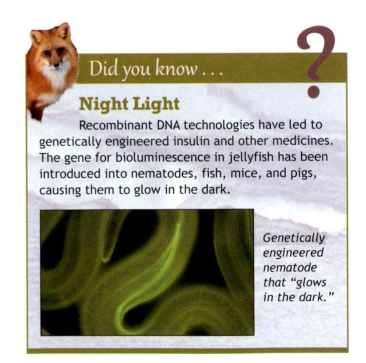

Did you know . . .

Night Light

Recombinant DNA technologies have led to genetically engineered insulin and other medicines. The gene for bioluminescence in jellyfish has been introduced into nematodes, fish, mice, and pigs, causing them to glow in the dark.

Genetically engineered nematode that "glows in the dark."

EXERCISE 8.7

Isolating DNA

The cultivated strawberry, *Fragaria ananassa*, is an outstanding candidate for DNA extraction and spooling. Although the majority of organisms have diploid cells with two sets of 7 chromosomes (14 total), the cultivated strawberry is octoploid (has eight sets of 7 chromosomes for a total of 56). Thus, it yields an abundance of DNA.

Strawberries also contain enzymes, such as pectinase and cellulase, which aid in breaking down cell walls. Other fruits, such as bananas, raspberries, and kiwis, can be used in this activity but do not yield as much DNA. Your instructor may have you keep this DNA to use in Chapter 9.

Procedure 1

Spooling DNA from Strawberries

Materials

- ❏ Water
- ❏ Mild dishwashing soap or nonconditioning shampoo
- ❏ Noniodized salt (NaCl)
- ❏ Strawberries (preferably fresh, but frozen will suffice)
- ❏ Ziploc freezer bag
- ❏ Cold 91–100% isopropyl alcohol or 95% ethanol stored in an ice bath
- ❏ Beakers for mixing (2)
- ❏ 30–50 mL glass test tube
- ❏ 25 mL graduated cylinder
- ❏ 50 mL graduated cylinder
- ❏ Pipette
- ❏ Wooden sticks or glass stirring rods
- ❏ Timing device
- ❏ Cheesecloth
- ❏ Funnel
- ❏ Rubber band
- ❏ Test-tube rack

1 Procure materials from the lab instructor.

2 In a 200 mL beaker, prepare an extraction solution by mixing 1.5 mL of mild dishwashing detergent or shampoo with 120 mL of water and a pinch (0.5 mL) of NaCl. Slowly and thoroughly mix the extraction solution with a stirring rod, avoiding the formation of soap bubbles. The soap in the extraction solution helps dissolve the cell membrane, and the salt serves to remove some proteins bound to the DNA. Soap also keeps the proteins from precipitating in the alcohol used in a later step.

3 Remove and discard the sepals (green tops) of two fresh strawberries. If frozen strawberries are used, let them thaw to room temperature. Place the strawberries into a Ziploc freezer bag. Seal the bag, and with your fingers, smash the strawberries thoroughly for at least 3 minutes.

4 Reopen the bag, and pour approximately 15 mL of the extraction solution from step 2 into the bag. Reseal the bag, and smash the strawberry extraction solution for 2 minutes, trying to avoid making soap bubbles.

5 Procure a clean 50 mL graduated cylinder. Insert a funnel into the cylinder. Place a piece of cheesecloth over the funnel with a rubber band, and pour the strawberry extract into the cylinder. Pour about 10 mL of the filtered strawberry extract into a test tube. Alternatively, your instructor may tell you to filter the strawberry extract directly into the test tube.

6 Tilt the test tube at a 45-degree angle. Pour 15 mL of the ice-cold isopropyl or ethanol slowly down the side of the test tube. Do not shake or mix the alcohol with the contents of the test tube. The alcohol will form a layer on top of the solutions. Alcohol will cause the DNA to precipitate out of solution.

7 Place the test tube in the test-tube rack, and allow it to stand for approximately 1 minute. You should observe a white cloudy or stringy mass of DNA on top of the strawberry extract solution.

8 Using a clean wooden stick or glass stirring rod, stir the DNA, spool it (wind it onto the glass stirring rod), and observe it (Fig. 8.21).

9 Follow the lab instructor's directions for disposal of all waste materials.

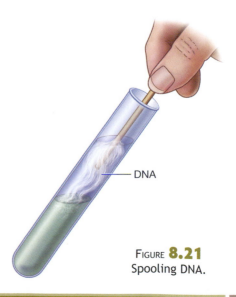

FIGURE **8.21**
Spooling DNA.

✔ Check Your Understanding

7.1 Many fruits can be used to isolate DNA, but strawberries yield more DNA. Give two reasons why this is true.

_____

_____

_____

7.2 At the end of the procedure, how is the cloudy, stringy mass of DNA consolidated so it can be observed? Describe the process.

_____

_____

7.3 The addition of alcohol to the strawberry extract causes _____ to precipitate out of the solution.

MYTHBUSTING

The Gene Pool

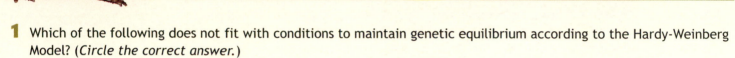

Debunk each of the following misconceptions by providing a scientific explanation. Write your answers on a separate sheet of paper.

1 All mutations are harmful.

2 Dominant traits are most likely to occur in a given population.

3 Every aspect of an organism is the product of genes alone.

4 During replication, the leading and lagging strands are identical.

5 In transcription, both strands of DNA are transcribed.

---

1 Which of the following does not fit with conditions to maintain genetic equilibrium according to the Hardy-Weinberg Model? *(Circle the correct answer.)*

a. There must be no mutation.

b. Mating between individuals must be completely random.

c. There must be no movement of individuals into or out of the population.

d. No selection; no allele is favored over another allele.

e. The population must be small so the laws of probability apply.

2 Why is mating between blood relatives (consanguinity) discouraged from a genetic perspective?

_____

_____

_____

3 What is the probability of two parents who carry the gene for albinism (an autosomal recessive disorder) having a child without albinism?

_____

_____

_____

4 In Manx cats, the alleles "TT" yield a cat with a normal long tail, the alleles "Tt" yield a cat with a short or absent tail, and the alleles "tt" are lethal to the embryo. Predict the genotypes and phenotypes of potential kittens resulting from the mating of two short-tailed cats.

a. Genotype of male _____

b. Genotype of female _____

c. Genotypes of offspring _____

d. Phenotypes of offspring _____

5 Phenylketonuria (PKU) is one of the most common recessive genetic disorders in humans. Infants with PKU are missing an enzyme called phenylalanine hydroxylase, needed to break down an amino acid called phenylalanine. If a strict diet that minimizes the amount of phenylalanine in the diet is not followed, a severe nervous disorder may result. If two heterozygous parents (Pp) have a child, what is the possible genotype and phenotype for this individual?

a. Genotype of male _____

b. Genotype of female _____

c. Genotypes of offspring _____

d. Phenotypes of offspring _____

6 Which of the following occur in the DNA replication process? (*Circle the correct answer.*)
a. At the start of each DNA segment to be replicated, RNA polymerase builds an RNA primer.
b. Termination of the process occurs when a stop codon is encountered.
c. RNA primer attracts an enzyme known as DNA polymerase.
d. Proofreading enzymes are responsible for ensuring accuracy of this process.
e. Answers a, b, and c are all correct.

7 Compare and contrast the structure of DNA and RNA.

_____

_____

_____

8 What occurs during the process of translation?

_____

_____

9 What is the significance of the start and stop codons in protein synthesis?

_____

_____

_____

Touching the Future
Understanding Biotechnology and Forensics

9

The first century of the new Millennium will belong . . . also to biotechnology which will bring unprecedented advances in human and animal health, agriculture and food production, manufacturing and sustainable environmental management.

— Ben Ngubane (1941–Present)

Just wondering . . .

Consider the following questions prior to coming to lab, and record your answers on a separate piece of paper.

1 What is the relationship between bioethics and biotechnology?

2 What are five genetically modified organisms?

3 Can biotechnology find its way to biological warfare?

4 Would gel electrophoresis be useful in examining the differences in DNA of my classmates? Why or why not?

5 Are genetically modified crops safe?

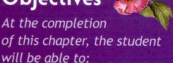

Objectives

At the completion of this chapter, the student will be able to:

1. Define biotechnology and describe several biological disciplines that contribute to biotechnology.

2. Describe how biotechnology shapes our lives.

3. Discuss how biotechnology contributes to medicine, agriculture, industry, and forensics.

4. Discuss why some people are skeptical about biotechnology.

5. Discuss how several historical cultures used biotechnology.

6. Describe the function of gel electrophoresis.

When the term *biotechnology* is mentioned, thoughts of improved agricultural products, cloning, vaccines, stem cell research, cutting edge forensics, biofuels, and cures for devastating genetic diseases come to mind. Although considered a modern endeavor, biotechnology in a literal sense has been around for thousands of years.

For centuries smallpox was a scourge of humankind, "plaguing" many generations with great suffering and death. In ancient China, in order to provide protection from smallpox, physicians collected the scabs of smallpox pustules from individuals with a slight case of the disease and prepared a dry powder of the crust. The new, powdered medicine would then be snorted into the nostril by the patient. To work properly, male patients would snort the powder into the left nostril and females into the right.

In ancient India, patients would buy crusted pustules of smallpox from a physician and allow the doctor to rub the crust into a scratch on the skin. This strategy was met with limited success. In the 1790s a British country doctor, Edward Jenner (1749–1823), developed the smallpox vaccine that later virtually eradicated the disease. Like all new technologies, his idea came with much criticism and generated fear and rumors (Fig. 9.1). Some people thought that since Jenner was using his cowpox vaccine, that recipients of the vaccine would turn into cows! In modern biotechnology applications, similar fears still pervade in some circles of society.

Biotechnology

Biotechnology refers to the controlled and deliberate manipulation of biological systems such as cells, cellular components, cellular products, and organisms for the production and manufacture of biological products. Over the past 25 years, the science of biotechnology has made tremendous strides in many fields such as agriculture, medicine, industry, and forensics. In

Chapter Photo
Electrophoresis of RNA.

FIGURE **9.1** Biotechnological advances have spawned much misunderstanding in the past and continue to flourish in the present. In this cartoon from 1802 by James Gillray, Edward Jenner is administering the smallpox vaccine to a misunderstanding public.

addition, the number of biotechnology companies has grown tremendously in recent years, creating a multitude of new jobs for qualified candidates. The foundations of biotechnology are based upon the biological sciences of molecular biology, cell biology, genetics, cell culture, embryology, microbiology, agriculture, botany, zoology, and human biology.

Any new technology will generate a multitude of relevant as well as irrelevant questions. Such questions addressing the safety of genetically engineered food are common. Also, many people question the accuracy of DNA testing. Whatever the questions may be, we are standing at the brink of the dawn of biotechnology and must face the future with skepticism, vigilance, and confidence.

Medical Applications

In medicine, biotechnology has vast potential to improve human health. In recent years, enzyme replacement therapy, stem cell therapy, vaccine development, cloning, CRISPR-Cas9, and gene therapy have been used with success in the medical world. Several new drugs, including genetically engineered insulin, have been released to the public. Xeno-transplantation of cells from various animal species and transplanting them into humans has been successfully used to replace damaged tissues such as heart valves and even skin.

In the future, advances in biotechnology will yield a better understanding and diagnosis of, and potential cures and preventatives for, such dread diseases as HIV and Ebola. In addition, in the near future, biotechnology will open the door for better diagnosis and cures for genetic disorders such as cystic fibrosis. Fields such as pre-implantation genetic diagnosis will develop, allowing for testing of embryos that

were fertilized through in vitro techniques before they are implanted into the mother. This technique has been used successfully to ensure embryos are free of Tay-Sachs disease.

Agricultural and Industrial Applications

Many aspects of biotechnology are applicable to agriculture. Biotechnology has been used in agriculture to develop disease-resistant, pest-resistant, and frost-resistant plants. Through the development of agricultural biotechnology, increased and enhanced crop production will result (Fig. 9.2). Improved nutritional value, better taste, improved transportability, and increased shelf life will benefit consumers. Some of the new, genetically engineered foods have been labeled "frankenfoods" in many circles and are considered controversial. However, many genetically modified foods are an important part of our diets. Common examples of genetically modified foods include: corn, soy, alfalfa, canola, sugar beets, and milk. Genetically modified Granny Smith and Golden Delicious apples may be on the market soon that resist browning.

The industrial applications of biotechnology are limited only by the imagination. If developed to its full potential, industrial biotechnology will have a tremendous impact upon the world. It will offer businesses a way to reduce costs and create new markets, improve products and production, and protect the environment. Biotechnology techniques will yield biocatalytic tools that can improve things such as contact lens cleaners and meat tenderizers. In addition, greener and

Did you know . . .

A Glow in the Waters

Been to an aquarium shop lately? The green, yellow, red, blue, pink, orange, and purple fluorescent tropical fishes are absolutely stunning. So how has biotechnology aided in the production of these fascinating fishes? Fluorescent fishes were originally developed to detect environmental pollutants but their fascinating colors captured the eye of consumers. The first fluorescent fishes were introduced to aquarium shops in the mid-2000s.

Through biotechnology, transgenic fluorescent fishes were developed by microinjecting a gene for bioluminescence from jellyfish, coral, or sea anemones into fish eggs. The genes in turn became part of the genetic material of the fish, causing the fish to fluoresce. Since the gene has become part of the genotype of the fish, it can be passed to the next generation through traditional breeding. Subsequently, fish sold in stores today are produced in this manner.

Many people think that these glowing fish have a "dark side" and more regulation on their production and introduction should be implemented. What do you think?

FIGURE **9.2** The melon on the right contains DNA from another source to help it appear more uniform and orange.

FIGURE **9.4** Forensic scientist taking a sample from a bloodstained shirt.

more efficient biofuels can be produced through industrial biotechnology (Fig. 9.3). The environment and climate will benefit greatly from these technologies. Nanotechnologies will provide medicine and industry with minute devices that can improve the delivery of medicine as well as sensory devices that can monitor environmental conditions.

Forensics

In one application of forensic biotechnology, investigators can use DNA to create a genetic fingerprint of a potential suspect involved in a crime (Fig. 9.4). This information can be used to find and prosecute the criminal. Today, DNA databanks of convicted criminals are a valuable resource for investigators. This same technology can also be used to identify the remains of lost individuals or victims of an accident or incident. This technology was particularly useful following the 9/11 terrorist attacks. Along with contemporary sleuthing (fingerprinting, blood types, aging maggots, and wound analysis), today's Sherlock Holmes has a plethora of tools to identify and catch criminals.

FIGURE **9.3** Biofuel electricity generator powered by gas produced by the anaerobic fermentation of microalgae.

Did you know . . .

Full of Gas

Escherichia coli (*E. coli*) is a gram-negative, rod-shaped bacterium that usually resides in the lower intestines of mammals. Most strains of *E. coli* are harmless but pathogenic strains exist causing food poisoning. *E. coli* is commonly used as an environmental indicator organism for fecal contamination and in microbiology laboratories as an example prokaryotic organism. Since 1982, human insulin has been produced in the laboratory by growing insulin proteins within *E. coli*.

In 2014 researchers discovered a pathway to produce propane by using *E. coli*. In *E. coli*, fatty acids are converted into the cell membrane through metabolic process driven by enzymes. In modified *E. coli*, the pathway to produce cell membranes is altered to produce propane. This is achieved by using three different enzymes.

Propane gas serves as the bulk component of liquid petroleum gas. This gas is extensively used worldwide, for heating and cooking. Think about your BBQ grill. Liquid petroleum gas can also serve as a gasoline alternative for motor vehicles. Bioengineered propane has the potential to be less expensive and be less destructive to the planet.

It is hoped that in the future, scientists may be able to insert this bioengineered propane-producing mechanism into different species of photosynthetic bacteria. In turn, the system would then be capable of directly converting solar energy into propane that is ready to use in a number of ways without complex and expensive processing and refining.

Gel Electrophoresis

As you learned in Chapter 8, within the double helix of DNA reside the molecular instructions that control the development and function of all living things. Since the 1970s one of the most powerful tools used to study DNA has been agarose gel electrophoresis. This technique uses agarose, a polypeptide polymer derived from seaweed, to separate DNA (fragment length) or proteins (size and/or charge). The process of gel electrophoresis is considered the standard laboratory procedure for separating DNA for study, visualization, and purification.

In gel electrophoresis, an electrical field is used to move the negatively charged DNA molecule (because of the phosphate group) toward a positive electrode (anode) through a matrix of agarose gel. The porous gel matrix allows the shorter fragments of DNA to migrate more quickly than the larger fragments. Over time, the shorter strands will move farther through the gel matrix than the larger strands. Strands of the same length will travel the same distance. As a result, the length of a DNA segment can be determined by running the DNA on agarose gel alongside a DNA segment of known length.

Keep in mind that each chromosome in a eukaryotic cell (that is 46 in you) consists of a very long, single molecule of DNA. It is virtually impossible to isolate intact DNA from a chromosome because of the fragility of the DNA molecule. Staining the groups of DNA makes them visible to the human eye.

9

Applications of Gel Electrophoresis

DNA Isolation

Materials

- ❏ Lab coat, eye protection, and gloves
- ❏ Thin slice of frozen beef liver
- ❏ 0.2 mL buffered salt solution
- ❏ Microcentrifuge
- ❏ Microcentrifuge tube
- ❏ Protease solution
- ❏ 20% SDS detergent solution
- ❏ Heat sink or block
- ❏ Buffered salt (4 M sodium chloride solution)
- ❏ Ice bath
- ❏ Ethanol
- ❏ Toothpicks
- ❏ Dropper
- ❏ Test-tube rack
- ❏ 0.1 mL deionized water
- ❏ Timing device

In this activity DNA will be isolated from beef liver tissue.

This activity can be modified to study strawberries or other specimens. The DNA isolated in the activity will then be separated by agarose gel electrophoresis and analyzed.

 **WARNING**

Wear a lab coat, safety goggles, and gloves for this activity.

1. Work in groups of 3 or 4, and place the microcentrifuge tubes in the microcentrifuge as a group.

2. Place a small, thin slice of beef liver in a microcentrifuge tube that contains 0.2 mL of a buffered salt solution. Slowly and carefully invert the tube several times in order to suspend the tissue. Label the top of the tube with your initials. The salt in the extraction solution causes the proteins and carbohydrates to precipitate out, letting the DNA remain in solution.

3. To the microcentrifuge tube add 1 drop of protease solution and 1 drop of 20% SDS detergent solution. Slowly and gently mix the mixture by inverting the tube several times. It is imperative to do this gently to keep DNA fragmentation to a minimum. The soap destroys the cell and nuclear membranes, allowing the DNA to escape.

4. Place the microcentrifuge tube upright into a heat sink or block set at 55°C, in order to further break down the tissue.

5. Ensure that the microcentrifuge is physically balanced with the microcentrifuge tubes from the class, and then turn on the microcentrifuge for 20 seconds to further break down the mixture.

6. Fill this microcentrifuge tube with ice-cold 95% ethanol taken from the ice bath that contains ethanol and gently invert the tube several times to ensure adequate mixing of the contents. The ethanol serves to precipitate the DNA. Ice-cold ethanol increases the amount of DNA that will precipitate.

7. In this tube, a cloudy white mass should begin to appear through precipitation. This is the extracted DNA.

8. Using a toothpick, gently swirl the DNA and gently pick up the stringy DNA for visual examination.

9. Stand the toothpick with its associated DNA in a small test-tube rack and let it dry for approximately 15 minutes to allow the ethanol to evaporate.

10. Pre-label a new microcentrifuge tube with your initials and add 0.1 mL of deionized water. In order to resuspend the DNA, stir the toothpick into the new tube.

11. Clean your laboratory station as instructed.

12. Write your observations in the space provided.

Procedure 2

Agarose Gel Electrophoresis

Materials

- ❏ Lab coat, eye protection, and gloves
- ❏ DNA solution from Procedure I
- ❏ Micropipetter
- ❏ Gel loading buffer
- ❏ Agarose gel in electrophoresis apparatus
- ❏ Ultraviolet lightbox

1 In this activity, continue wearing your lab coat, eye protection, and gloves.

2 The instructor will provide a premade agarose gel slab. A fluorescent dye (molecular tag) of ethidium bromide was added to the gel after it cooled. When exposed to ultraviolet light, the ethidium bromide will fluoresce an orange color. Generally, the fluorescence will intensify when it binds to DNA. Ethidium bromide is considered a mutagen and should be handled with extreme care. At this point, the gel block should appear opaque.

3 Pay close attention. The laboratory instructor will demonstrate how to hold and use an automatic pipetter. You may want to practice using this instrument prior to proceeding to the next step (Fig. 9.5).

4 Using the automatic micropipetter, add 20 microliters of gel loading buffer to the tube of DNA from Procedure 1. Gently shake the solution.

5 One well of the agarose gel electrophoresis apparatus will be loaded with 10 microliters of a known DNA marker.

6 Each group will add 10 microliters of their DNA/loading buffer solution into a well of the agarose gel in the electrophoresis apparatus (Fig. 9.6). Remember which well belongs to your group.

7 Once all of the groups' DNA/loading buffer solution is added to their wells, cover the apparatus. Verify that the red and black electrodes match up. The wells should be set up toward the cathode (negative electrode) and the DNA will migrate toward the anode (positive electrode).

8 Turn on the switch to the gel electrophoresis apparatus.

9 Observe the gel. When the bands are well separated, turn off the apparatus (Fig. 9.7).

10 Remove the gel tray from the apparatus

11 Place the gel under an ultraviolet light source for observation. Record your observations below.

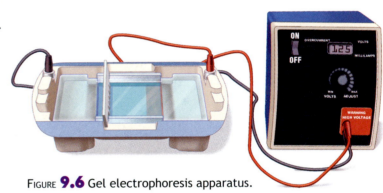

FIGURE **9.6** Gel electrophoresis apparatus.

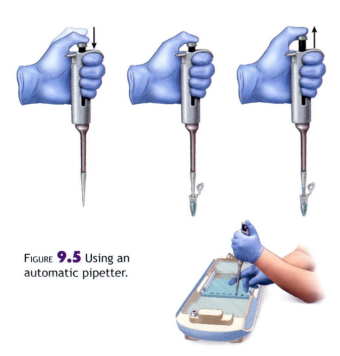

FIGURE **9.5** Using an automatic pipetter.

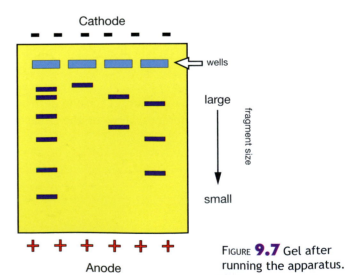

FIGURE **9.7** Gel after running the apparatus.

12 Clean up the work area as instructed.

13 Write your gel electrophoresis observations. Include a sketch of your results.

Procedure 3

Who Done It?

Materials

- ❑ Lab coat, eye protection, and gloves
- ❑ DNA solutions (4)
- ❑ Micropipetter
- ❑ Gel loading buffer
- ❑ Agarose gel in electrophoresis apparatus
- ❑ Ultraviolet lightbox

Six months ago Jeff and Maria's home was burglarized and the only clue left behind was a pool of blood near a broken window. The thief managed to steal an iPad, a digital camera, a small stereo system, DVDs, a few hockey sweaters, and some jewelry.

Detectives were able to retrieve a DNA sample from the blood left at the scene. However, it did not match any of the DNA on file in their database. Thus, the detectives began the meticulous task of trying to identify the potential suspect. After testing the DNA of Jeff and Maria and interviewing relatives, friends, and neighbors, the case was at a standstill.

As a result of the investigation, four suspects were identified. All four were interviewed and a sample of their DNA was taken and sent to the crime lab.

▸ Suspect 1, Number 1809. A neighbor's video camera recorded a stranger looking in the ornamental plants in the yards of several of the homes near Jeff and Maria's house on the day of the burglary. During his interview, the suspect stated that he was looking for his daughter's lost kitten, which had wandered away from home. However, he lives 3 miles from the victim's house. The suspect has a healed scar on his forearm that he claims was cut on the job. At the time, he was working as a waiter in a local cafe. None of the stolen

items were in the suspect's possession. However, he did have eight DVDs that coincided with those stolen from Jeff and Maria's home. The suspect had been arrested three years ago for shoplifting at a local electronics store.

▸ Suspect 2, Number 1836. The suspect was stopped for speeding on a nearby freeway while returning from a hockey game, and given a ticket. The policeman, being a hockey fan, remembered that one of the stolen items was a special edition Philadelphia Flyers sweater. The suspect was wearing the unusual sweater. When the detectives began their investigation, they discovered that the suspect was a well-respected president of a local company. He claimed he received the sweater from a friend in Louisiana as part of a trade for season tickets to New Orleans Saints football games. His friend could not be found. The suspect had no visible scars and no criminal record. Other than an extended lunch hour, he was at work the day of the burglary.

▸ Suspect 3, Number 1859. On a hunch, an investigator decided to interview and collect DNA from suspect 1836's chief assistant based upon the fact that she had just received a substantial raise. Upon examination she had a scar on her left finger. However, she claims that she cut her finger while working in her new garden. She has an exemplary record and claims she was making deliveries the day of the incident. Her driving log confirms that she was out that day as verified by suspect 1836.

◆ Suspect 4, Number 1882. This suspect was arrested for armed robbery at a nearby convenience store five months after the burglary of Jeff and Maria's home. He claims that he was out of town the day of the break in but has no witnesses to verify his story. He sports a large healed scar on his right hand. None of the stolen items were in his possession; however, he had an iPad with the serial number removed. The suspect has a record of battery, burglary, shoplifting, and possession of drugs.

In this activity, you are to devise an agarose gel electrophoresis activity to determine if any of the suspects can be convicted of the burglary based upon DNA evidence. The instructor will provide you with DNA from the scene and from the four suspects. Go get 'em, Sherlock (Fig. 9.8)!

1 Continue wearing your lab coat, eye protection, and gloves.

2 The instructor will provide DNA solutions representing the DNA from the crime and the four suspects.

3 Devise a means using agarose gel electrophoresis to analyze the DNA.

4 Determine if your DNA analysis can convict one of the suspects.

5 In the space provided, discuss your findings and sketch your agarose gel.

FIGURE **9.8** Sherlock Holmes.

6 Clean up the work area as instructed.

✔ Check Your Understanding

1.1 In gel electrophoresis, _____ is used to move negatively charged DNA molecules

toward a _____ electrode or _____ through a matrix of agarose gel. The

porous nature of the agarose allows _____ DNA fragments to migrate more quickly than

_____ fragments.

1.2 In the DNA isolation process, what substance in the extraction solution caused the proteins and carbohydrates to precipitate out, letting the DNA remain in solution?

_____

1.3 In the DNA isolation procedure, ice-cold 95% ethanol was added to the tube that originally contained a thin piece of beef liver. Why was the ethanol added? Why is ice-cold ethanol used instead of room temperature?

_____

_____

1.4 The length of an unknown DNA segment is able to be determined in gel electrophoresis by which method? (*Circle the correct answer.*)

  a. Using ethidium bromide to stain the DNA segment.
  b. Running the unknown DNA with a segment of DNA from a similar species.
  c. Running the DNA with a DNA segment of known length.
  d. Staining the segment with methylene blue, followed by ethidium bromide.
  e. None of the above.

1.5 What happens when ethidium bromide is exposed to ultraviolet light?

_____

_____

1.6 Students and others using ethidium bromide stain are warned to handle it carefully. Why?

_____

_____

Did you know . . .

The Cutting Edge

Clustered Regularly Interspaced Short Palindromic Repeats (CRISPR-Cas9) is a revolutionary new genome editing technology that allows geneticists to make minutely precise modifications to DNA within a genome. It has great potential in medicine, genetic engineering, agriculture, industry, and scientific research. In the medical world, it can be used to develop powerful individualized gene-based medicines for patients with serious diseases such as cancer, malaria, AIDS, sickle cell anemia, and cystic fibrosis. CRISPR technologies are now being used or contemplated to expedite crop and livestock breeding programs, engineer new antimicrobials, and control disease-carrying insects such as mosquitos. Like any new technology, many individuals are highly critical of CRISPR and afraid of its potentially diabolical applications. It is our duty as a scientifically literate society to see that such technologies are used to benefit the planet and humankind.

9

MYTHBUSTING

Two of a Kind

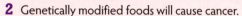

Debunk each of the following misconceptions by providing a scientific explanation. Write your answers on a separate sheet of paper.

1 One day, scientists will be able to clone Hitler and other villains.

2 Genetically modified foods will cause cancer.

3 Nanoparticles can be detrimental to our health!

4 Embryonic stem cell lines come from aborted fetuses.

5 All forensic scientists do is field work just like in the movies. What a job!

1 Define biotechnology.

_____

_____

_____

2 Describe three forensic applications of biotechnology.

_____

_____

_____

3 Biotechnology has vast potential to improve human health. Describe two examples of medical applications of biotechnology in use today.

_____

_____

_____

4 Which of the following statements does not fit with your understanding of biotechnology? (*Circle all that apply.*)

a. Biotechnology has applications in the fields of agriculture, medicine, forensics, and industry.

b. Biotechnology involves the controlled manipulation of biological systems.

c. The foundations are based on the biological sciences: molecular biology, cell culture, agriculture, human biology, cell biology, and genetics.

d. People accept and wholly support all of the new applications produced by biotechnology.

5 We think of biotechnology as a modern achievement, but in reality it has been around for thousands of years. Give two examples that illustrate this history.

_____

6 Cells and tissues from some animal species, such as pig valves, have been transplanted into humans to replace damaged heart valves and to replace seriously burned skin. The transplantation of cells from animals to humans is called _____.

7 Using the process of gel electrophoresis, DNA is separated based on which criteria? (*Circle the correct answer.*)
a. Porous gel allows smaller fragments to migrate more quickly.
b. Presence of negatively charged molecules.
c. Porous gel allows longer fragments to migrate more slowly.
d. Both a and b.
e. Both b and c.

8 How can gel electrophoresis be applied in forensics?

_____

_____

_____

_____

9 Biotechnology encompasses which of the following? (*Circle the correct answer.*)
a. Genetically engineered insulin.
b. Transgenic fluorescent fish.
c. Biofuels.
d. Both a and b.
e. All of the above.

10 Why do people have concerns about the future of biotechnology?

_____

_____

_____

_____

_____

_____

_____

Mystery of Mysteries
Understanding Evolution and Taxonomy

There is grandeur in this view of life, with its several powers, having been originally breathed by the Creator into a few forms or into one; and that, whilst this planet has gone cycling on according to the fixed law of gravity, from so simple a beginning endless forms most beautiful and most wonderful have been, and are being evolved.

— Charles Darwin (1809–1882)

Just wondering . . .

Consider the following questions prior to coming to lab, and record your answers on a separate piece of paper.

1 In the world of evolution, would you rather be a specialist or generalist species? Why?

2 What is a living fossil? What are several examples of living fossils?

3 Why are the Galapagos Islands so unique?

4 What evidence is there to place birds with the dinosaurs?

5 Do any transitional fossils exist?

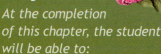

Objectives

At the completion of this chapter, the student will be able to:

1. Define evolution, and provide proofs of evolution.

2. Define and give examples of microevolution and macroevolution.

3. Explain beak variations in Darwin's finches.

4. Compare and contrast artificial selection and natural selection.

5. Cite factors that are favorable to fossilization and several means of fossilization.

6. Describe the role of taxonomy in modern biology and distinguish between taxonomy, systematics, and cladistics.

7. Describe the composition, derivation, and purpose of scientific names.

8. Discuss and outline the basic taxa used in taxonomy.

9. Use and construct a simple biological key and a cladogram.

Chapter Photo
Galapagos tortoise, *Chelonoidis nigra*.

◼ Evolution

Some questions have always intrigued enlightened minds: How and when did life begin? How did species form? Do we know our ancestors? The methods and tools of modern science are leading us to a better understanding of such mysteries. In 1973, a loyal defender of Darwinism, Theodosius Dobzhansky (1900–1975), stated, "Nothing in biology makes sense except in the light of evolution." Today, with the continued controversy surrounding evolution, creationism, and intelligent design, this statement is a rallying point for modern biologists.

Unfortunately, many people do not understand or do not wish to understand what evolution is all about. Their concern seems to be only with the origin of humans, but we are one species among vast numbers of extinct and extant forms. Sometimes, in our anthropocentric view, we tend to forget the rest of life on Earth. Evolution is all around us. It explains why the HIV virus is so diabolical, how methicillin-resistant *Staphylococcus aureus* (MRSA) is terrorizing our world, and even how cockroaches have developed resistance to our best pesticides. Evolutionary medicine has provided a greater understanding of genetic disorders such as sickle cell disease and cystic fibrosis, as well as other diseases. Through this approach to medicine, new understanding about disease will lead to new treatments. Again, "Nothing in biology makes sense except in the light of evolution."

Definition

Classically, **evolution** has been defined as a change over time in the genetic makeup of a population. Modern biologists have modified this definition in stating that evolution involves changes in allele frequencies over time. **Microevolution** involves changes in a population (with a species) over time, such as the ability of some organisms to adapt to global warming, or the bacteria that causes gonorrhea evolving resistance to penicillin. These changes can be viewed on a small scale and, geologically speaking, occur rather rapidly. **Macroevolution** occurs above the species level and involves vast amounts of time. Macroevolution addresses such questions as how dinosaurs gave rise to birds and how flowering plants originated.

Many people view evolution as a rather **gradualistic** process, as in the evolution of dolphins and whales from land ancestors. However, in 1972, Steven Jay Gould (1941–2002) and Neils Eldridge (1943–present) proposed another evolutionary process known as **punctuated equilibrium**. They proposed that some species are generally stable and change very little for long expanses of time. This period of species stability is "punctuated" by a rapid burst (from a geological standpoint) or a jump that results in a new species that does not leave many fossils in the record. The fossil records of some mollusc and bryozoan species reflect the pattern of punctuated equilibrium.

Darwin's Contribution

Among the many significant contributors in the history of evolutionary thought, Charles Robert Darwin (1809–1882), trained in theology and natural science, has been named the "father of evolution." Darwin's idea of natural selection, that "mystery of mysteries," revolutionized modern science. His voyage on the *HMS Beagle* from 1831 to 1836 was a pivotal point in the history of biology. As a young naturalist exploring the lands visited by the *Beagle*, Darwin made observations that, coupled with his observations recorded in the English countryside, would shape his ideas on **natural selection** and descent with modification.

The domestication of dogs is thought to have begun approximately 15,000 years ago from wolf ancestors. Since then, humans have shaped a number of breeds of dog through **artificial selection**. This process involved selectively breeding for desirable traits. The American Kennel Club recognizes some 150 distinct breeds, from Boston terriers to malamutes. If artificial selection can be so powerful in 15,000 years, just think of the power of **natural selection** working for 3.8 billion years! This power is reflected in the diversity of life on Earth.

In 1859, scientist T. H. Huxley (1825–1895), also known as "Darwin's bulldog," stated, "How extremely stupid for me not to have thought of that!" after reading Darwin's *On the Origin of Species*. He pointed out that Darwin's basic concepts outlined in the book were nothing more than common sense illustrated by brilliant and meaningful examples. In *On the Origin of Species*, Darwin discussed how variation exists within a species, that organisms produce more young than can be expected to survive naturally, that there is a struggle for existence in nature, and how those with favorable characteristics have a better chance to survive.

After publication of *On the Origin of Species*, support for the theory of evolution has emerged from a variety of disciplines. A better understanding of the earth's age has given scientists a frame of reference for evolution, and the fossil record has provided many important clues. Evidence from comparative anatomy and embryology has informed our knowledge of form and function. Biogeography has shed light on the distribution of species and their formation. Population genetics, classical genetics, neo-Mendelian genetics, and epigenetics have provided detailed information on inheritance and evolution, and molecular biology has produced evidence of the molecular basis of evolution.

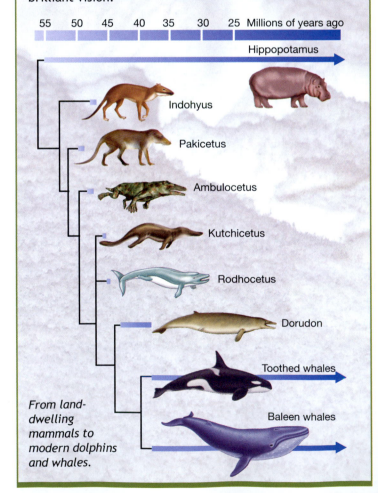

Did you know . . . ?

A Whale of a Tale!

In the first writing of *On the Origin of Species*, Charles Darwin proposed that the ancestor of modern cetaceans was a bear-like creature. Unable to substantiate his claim with fossils, Darwin dropped the statement from subsequent editions. In the late 1990s, however, Darwin was vindicated with myriad fossil discoveries of ancestral cetaceans from the shores of the ancient seas known as Tethys.

Today it is accepted that the ancestors of cetaceans are related to artiodactyls (pigs and hippos, for example). If you were to see the earliest land ancestor to a whale, you would declare—before you bolted for safety—that it resembled a wolf with hooves. The fossil record and molecular evidence continue to reinforce Darwin's brilliant vision.

55   50   45   40   35   30   25   Millions of years ago

Hippopotamus
Indohyus
Pakicetus
Ambulocetus
Kutchicetus
Rodhocetus
Dorudon
Toothed whales
Baleen whales

From land-dwelling mammals to modern dolphins and whales.

Adaptation Evidence

In his travels on the *HMS Beagle* to the Galapagos Islands, Darwin identified 13 different species of finches (Fig. 10.1). Each species had its own distinctive beak and ecological niche on the Galapagos Islands. Molecular studies (comparing mitochondrial DNA) suggest these finches probably evolved from an ancestral seed-eating, warbler-type, "dull-colored grassquit" that lived on the mainland several million years ago.

Adaptive radiation refers to the process by which a species or group of related species evolves rapidly into many different species that occupy new habitats or geographic zones. Darwin's finches diverged in response to the availability of food in the different habitats.

Within a taxon of organisms, some species are generalist. These organisms do not have specific adaptions for a particular trait. Other species are specialist and have specific adaptions of a particular trait. Although contrary to original thought in the game of survival, generalist species may have a survival advantage when conditions change and specialist species are unfit for the new environment.

Did you know . . .

Darwin's Finches

A particularly interesting example of contemporary evolution involves the 13 species of finches studied by Darwin on the Galapagos Islands, now known as Darwin's finches. A research group led by Peter and Rosemary Grant of Princeton University has shown that a single year of drought on the islands can drive evolutionary changes in the finches. Drought diminishes supplies of easily cracked nuts but permits the survival of plants that produce larger, tougher nuts. Drought thus favors birds with strong, wide beaks that can break these tougher seeds, producing populations of birds with these traits. The Grants (1991) estimated that if droughts occur about once every 10 years on the islands, a new species of finch might arise in only about 200 years.

Sources: Peter R. Grant, "Natural Selection and Darwin's Finches," *Scientific American* (1991): 82–87; Jonathan Weiner, *The Beak of the Finch* (New York: Alfred A. Knopf, 1994).

10

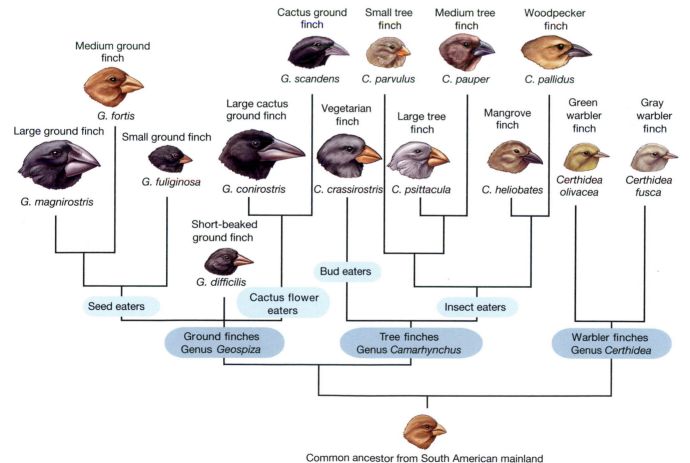

FIGURE **10.1** Cladogram of Darwin's finches showing the beak adaptations of the individual species.

Procedure 1

Natural Selection for Bird Beak Shapes

Materials
- Scissors
- Binder clip (2¼ in.)
- Spoon
- Small tweezers
- Paper clips
- Unpopped popcorn
- Rubber bands
- Lima beans
- Marbles
- Paper cup

As Darwin pointed out, the shape of a bird's beak is very important to survival. This activity is designed to illustrate Darwin's brilliant deduction about the shape of bird beaks and natural selection.* The activity begins on a hypothetical island known as Darwin Island. Inhabiting the island are four species of sparrows, each with a different beak modified to exploit a different food source. Common objects are assigned for the beaks of the four species of birds:

- Scissors for Huxley's sparrow.
- Binder clip for Henslow's sparrow.
- Spoon for Emma's sparrow.
- Tweezers for Hooker's sparrow.

On the island, food sources include paper clip fruit, unpopped popcorn fruit, rubber band fruit, lima bean fruit, and marble fruit. A paper cup will serve as the bird's stomach. Complete the activity, and answer the questions regarding the activity.

1 Before proceeding, write a simple hypothesis regarding "beak" shape (scissors, binder clip, spoon, tweezers) and the ability to gather food.

2 Identify the independent and dependent variable in this activity.

Independent variable: _____

Dependent variable: _____

3 Eight students will be chosen from the class: two will be Huxley's sparrows, two will be Henslow's sparrows, two will be Emma's sparrows, and two will be Hooker's sparrows. Each set of sparrows will procure the proper beak object. Each bird will also procure a paper cup to serve as the stomach. Students not serving as birds will distribute paper clips across several desktops.

4 After the birds are ready, and the food is distributed on the desktops, the instructor will give the command to feed. Each bird will have 20 seconds to pick up as much food as possible, and place the food in the upright paper cup. Make sure to hold the cup upright in one hand and the beak in the other. Only pick up the food with the beak! The instructor will give the command to stop feeding.

5 After the command to stop, the birds will return to their station and sort and count the food. Record your results in Table 10.1.

6 Repeat the feeding activity with the other food items, and fill in Table 10.2 accordingly.

7 Discuss your findings regarding feeding in the sparrows. Which sparrow's beak was best adapted to each particular type of food? Which sparrow's beak was least adapted to each particular type of food?

8 Clean the laboratory, and return the supplies to the designated area.

* This activity is adapted from Al Janulaw and Judy Scotchmoor's Clipbirds, http://www.ucmp.berkeley.edu/education/lessons/clipbirds and https://concord.org/sites/default/files/projects/er/materials/TeacherGuide_ClipBirds_TXMO-final.pdf.

TABLE **10.1** Individual Data

Type of Beak	Paper Clips	Unpopped Popcorn	Rubber Bands	Lima Beans	Marbles

TABLE **10.2** Group Data

Bird	Beak Type	Paper Clips	Unpopped Popcorn	Rubber Bands	Lima Beans	Marbles
Huxley's sparrow	Scissors					
Henslow's sparrow	Binder clip					
Emma's sparrow	Spoon					
Hooker's sparrow	Tweezers					

✔Check Your Understanding

1.1 In nature, why is being a generalist sometimes an advantage over being a specialist?

1.2 Which sparrow would have the advantage if all food sources were to disappear on the hypothetical island leaving only marbles? If the marble remained the only food source, which kind of beak would the sparrow population have in several years?

Paleontological Evidence

Stop! Don't throw that rock! You may be throwing away millions of years of geological and biological history. Your missile may contain a variety of fascinating fossils. A **fossil** is defined as the remains, or traces, of organisms that lived in the past, and the study of fossils is termed **paleontology**.

To be classified as a fossil, a specimen has to be at least 10,000 years old. Fossils may be complete organisms, parts of organisms, or even traces of organisms. Occasionally, a complete organism is found, such as fossils that have been preserved in ice. Most fossils, however, are parts of organisms, such as shells, bones, and teeth. Traces of life in the past include tracks, burrows, and fossilized excrement.

If a fossil could talk, it would tell an incredible tale. Paleontologists examine fossils, which are used to identify evolutionary trends in the morphology, anatomy, injuries, and diseases of organisms over time. Information about behavior can also be gathered from fossils that provides evidence of the characteristics of ancient environments and geography. In addition to being significant scientifically, many fossils are used for jewelry, for conversation pieces, in industry, and in the search for oil.

Formation of Fossils

In general, three major factors increase the chances of some-thing fossilizing: possession of hard parts, escape from immediate destruction, and rapid burial.

1. Organisms that have hard parts such as bones, teeth, shells, and woody tissue are more likely to fossilize than organisms with soft parts. Even delicate organisms, such as bacteria, sponges, worms, and jellyfish, however, may fossilize under the proper conditions.

2. The remains of most organisms are destroyed by mechani-cal forces, such as weathering, crushing, and wave action, or by biological actions, such as predators, scavengers, and mechanisms of decay. Paleontologists estimate that more than three-fourths of the life forms that ever lived did not form fossils.

3. Fossilization also can involve rapid burial in a medium capable of retarding decay. Organisms living in water or near water are likely to be buried in sediment that protects them from mechanical forces and decay. Tree sap, ice, volcanic ash, and even tar may preserve some organisms. Organisms living in arid regions may be pre-served through desiccation or mummification. A mummy of a dinosaur called a Trachodon (a species of duck-billed dinosaur) preserved in this manner indicated that these animals had leathery skin and webs between their toes on the front feet.

A tremendous fossil record has been unearthed through the years, and an immense number of fossils await future discovery. Fossilization can occur in many ways:

- Organisms preserved in ice provide paleontologists an ideal means of fossilization. Woolly mammoths recovered from Alaska and Siberia look as if they were just taken out of the freezer!

- In 1993, the movie *Jurassic Park* introduced movie-goers to the concept of fossils in amber. In this case, tree sap may have trapped insects, feathers, or other biological specimens and preserved them in a permanent encasement.

- The La Brea Tar Pits near Los Angeles have yielded an abundance of mammal fossils, including saber-toothed cats.

- The acidity (tannic acid) of peat bogs can mummify the soft tissues of plants and animals. This is sometimes called chemical preservation. Bog fossils are usually less than 15,000 years old.

- Carbonization may reduce plants and animals to shiny films actually finer than tissue paper. Many fossil fern leaves are carbonized films.

- Sedimentary rocks can contain fossils because they form at temperatures and pressures that do not destroy fossil remains. Dead organisms can become trapped in sediments that may, under the right condi-tions, become sedimentary rock. Limestone is a kind of sedimentary rock actually formed of shells largely reduced to calcium carbonate.

- Coal, petroleum, and natural gas are known as fossil fuels because they are the remains of plant and animal matter transformed by bacteria, heat, and pressure. Fossil plants and animals are often found embedded in seams of coal.

- Petrification is a common form of fossilization that results from mineral matter soaking into every cavity and pore of a specimen and replacing the living material. The hard parts, however, are not totally replaced.

- Unlike petrification, replacement involves total replacement of the original structures. Many times, replacement fossils lack details.

In addition to the actual remains of organisms being preserved, certain traces of organisms are considered fossils (**ichnofossils**). Tracks and trails of animals are one common type of trace fossils. Tracks of dinosaurs as well as trail marks of tiny worms have been collected. Fossilized excrement known as coprolites and gizzard stones (gastroliths) are extremely valuable in studying the life history of animals.

Molds and casts are two widespread types of trace fossils. Basically, molds are an impression of an organism or an organic structure in a hardened medium. Internal and external molds are common fossils. Casts result from a substance filling a mold. For example, clam and snail shells often dissolve, leaving a cast.

Determining Fossil Age

The age of fossils can be determined in two different ways: relative dating and absolute dating.

1. **Relative dating** involves age-dating fossils by association with known time periods grouped into four eras: the Precambrian, the Paleozoic, the Mesozoic, and the Cenozoic. The eras represent distinct ages in the history of Earth. Just as a movie can be age-dated by clothes, hairstyles, and cars, fossils can be age-dated by strata and sediments. Relative dating does not result in numerical ages; it simply orders the appearance of organisms. Fossils can be dated through **stratigraphy**, a method of placing fossils in relative sequence to each other based on their location in the sedimentary strata. In undisturbed beds of rock, the oldest rocks exist below the younger rock. The strata of one location can be correlated with another location by the presence of index fossils. These fossils are anatomically and morphologically distinct, common during a distinct period of time, widespread, and fossilized easily.

2. Although relative dating is important, **absolute dating** provides a more exact means of dating. Sometimes known as **radioactive dating**, absolute dating is based on the radioactive decay of certain isotopes. Every isotope has its own characteristic rate of decay. The time for one-half of an isotope to change to the more stable form is known as its **half-life**. The half-lives of isotopes vary from a few hours to millions of years. Because the half-life of an isotope does not vary, it is not influenced by variables such as temperature and pressure.

The age of a fossil is determined by comparing the percentages of the isotopes. For example, in carbon dating, the ratio of carbon 14 to carbon 12 is compared. Knowing that the half-life of carbon 14 is 5,730 years, we can determine the age of a fossil based upon calculations. Unfortunately, carbon dating is limited to specimens newer than 60,000 years. To date older specimens, other radioactive dating methods must be used. Argon to potassium to uranium dating has been used to date older specimens. New methods of dating have emerged in recent years. Fission tracking, paleomagnetism, dendrochronology, amino acid racemization, thermoluminescence, and electron spin resonance are becoming more popular.

Procedure 1

Classifying Fossils

Materials

❑ Model specimens (at least eight for each collection)
❑ Modeling clay

This activity asks students to identify, examine, and classify fossils. Working in groups of four, students should closely examine the fossils and answer the questions.

d. Is it a cast or a mold? _____

e. What is your fossil? _____

f. What are the closest living relatives to your fossil?

_____

1 Select a representative fossil or model provided by your instructor, and answer the following questions:

a. Is the fossil a complete organism, a part of an organism, or a trace of an organism?

_____

b. In what kind of environment might this organism have lived?

_____

c. How did fossilization occur?

2 Sketch your selected fossil.

---

3 Choose a model specimen from your collection and, using modeling clay, make an impression of your fossil. Examine the model and answer the following: Identify your specimen. Did your impression reveal any surface features? Did you create an internal or an external mold? How can similar exercises aid paleontologists?

_____

_____

_____

_____

_____

4 Although the type and number of fossils vary from region to region, the images in Figure 10.2 represent common fossils. Try to put these 15 images into the following categories:

a. Teeth (**Hint**: Look for pointy incisors and square-ish molars):

_____

b. Molluscs (**Hint**: Look for fossils with shells):

_____

c. Corals (**Hint**: Look for fossils with a honeycomb-like appearance):

_____

d. Marine creatures not listed in the above categories (**Hint**: Look for indentations of creatures):

_____

5 Now that you have categorized the fossils into different groups, try to match the fossil types in Figure 10.2 with the specific names below. Use all resources at your disposal: biology textbook, internet, etc.

a. Megalodon tooth _____

b. Trilobite _____

c. Bivalve (2 answers) _____

_____

d. Brachiopod _____

e. Horse tooth _____

f. Cephalopod (2 answers) _____

_____

g. Mastodon tooth _____

h. Crinoid _____

i. Crinoid crowns _____

j. Colonial coral (2 answers) _____

_____

k. Sea urchin _____

l. Fossil fish (_Pristacara liops_) _____

6 How many did you get right? Were you close on others? What does this exercise tell you about fossils?

_____

_____

10

FIGURE **10.2** Some common fossils.

✓ Check Your Understanding

2.1 _____ dating involves age-dating fossils with known time periods grouped into four eras that represent distinct ages in the history of Earth. (*Circle the correct answer.*)

a. Carbon

b. Absolute

c. Adaptive

d. Radioactive

e. Relative

2.2 Match the description or fossil example with the fossilization type. Some will have more than one answer.

_____ Replacement

_____ Carbonization

_____ Sedimentary rocks

_____ Chemical preservation

_____ Ice preservation

_____ Petrification

A. Mineral matter soaks into all the cavities and pores of the specimen, replacing living material; hard parts are not replaced.

B. Limestone is an example.

C. This process reduces plants and animals to shiny films.

D. This process results in the total replacement of the original structures and may lack detail.

E. Dead organisms become trapped in sediments.

F. Fossil ferns are examples.

G. The acidity of peat bogs mummifies the soft tissues of plants and animals.

H. Woolly mammoths have been found because of this process.

Did you know . . .

It's a Fish! It's an Amphibian! It's *Tiktaalik roseae*!

In 2004, a paleontological team led by Neil Shubin, Edward Daeschler, and Farish Jenkins discovered the fossilized remains of a 375 million year (Devonian era) transitional fossil in the Canadian Arctic. The fossil known as *Tiktaalik roseae* illustrates an important time in vertebrate evolution when lobe finned fishes (sarcopterygians) were transitioning into the first amphibians.

Called a "fishapod," *Tiktaalik* possessed several fishlike characteristics such as a tubular body, scales, gills, and fins. In addition, *Tiktaalik* possessed a flattened head, a neck, ribs, ear notches, a sturdy shoulder, a flexible wrist joint, and fin bones that would allow it to prop itself up out of the water like an amphibian. The presence of spiracles on the top of the head also suggests that *Tiktaalik* may have had lungs as well as gills. Many of the features of this amazing fossil were a mixture of fish and amphibian traits.

Tiktaalik represents a pivotal moment in geologic time when fishes were making the transition to land and is an ancient ancestor to other tetrapods (four legs) such as amphibians, reptiles, birds, and mammals. Although snakes, dolphins, birds, and humans do not have "four feet," they share a common ancestor with these ancient tetrapods.

Tiktaalik roseae.

Comparative Anatomical Evidence

In studying evolutionary relationships, scientists use a variety of methods to determine ancestry. Evolutionary biologists compare anatomical structures of organisms, embryological development, and the biochemical (macromolecular) makeup of organisms. Organisms in turn can be classified based on the extent of similarity or descent from a common ancestor.

Comparative anatomy of structures yields evidence that supports evolution of organisms from a common ancestor. When structures share similar anatomical features and embryological development, they are said to be **homologous**. For example, the wing of a bat, the flipper of a whale, and the arm of a human are homologous structures—having evolved from a common ancestor but not always performing the same function (Fig. 10.3).

In contrast, it is possible for organisms to share a similar structure but to have evolved via different pathways. For example, the wing of a bird and the wing of an insect have a similar function but differ in their evolutionary history. The wings are **analogous structures**.

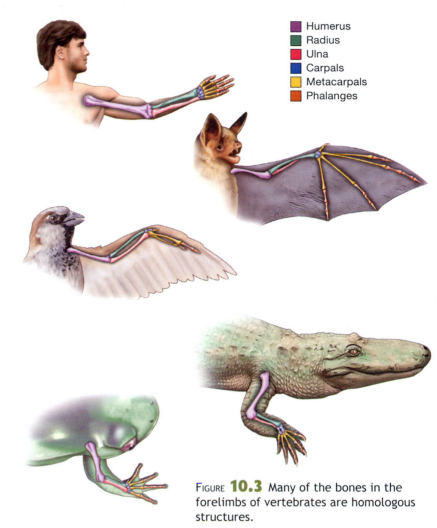

- ■ Humerus
- ■ Radius
- ■ Ulna
- ■ Carpals
- ■ Metacarpals
- ■ Phalanges

FIGURE **10.3** Many of the bones in the forelimbs of vertebrates are homologous structures.

10

Procedure 1

Homologous and Analogous Structures

Materials

None required

In this activity, students will refer to Figure 10.3 or given specimens, then conduct some research on the vertebrate forelimbs of two species.

1 Examine the bones of the forelimbs pictured. For each organism, name the bones that resemble a common ancestor and the bones that differ.

_____

_____

_____

_____

2 Explain how modifications in the structure of some of the bones were necessary for the existence of the organism.

_____

_____

_____

_____

3 Sharks and dolphins share many analogous structures that allow them to meet the demands of their habitat. Why are these similar structures analogous rather than homologous?

_____

_____

_____

_____

Vestigial structures in humans include the coccyx, or tailbone, the wisdom teeth, and the third eyelid (nictitating membrane). Some primates also have extrinsic ear muscles that serve no function in humans but do in monkeys (Fig. 10.4). Until recently, the appendix was thought to be a vestigial structure but now is thought to function as an "incubator" for good bacteria that colonize our colon. Vestigial structures can be found in other animals as well. Whales and dolphins have remnants of hind leg bones, and pythons have pelvic spurs, the remnants of legs. In contrast to vestigial structures, **atavisms** are ancestral traits that reappear in an organism after many generations of dormancy such as tails in human babies, extra toes on the modern horse, and a back pair of flippers on a dolphin.

When we are cold, goose bumps raise the body's hair to trap air to keep warm. This reflex is not considered vestigial, but when we get goose bumps that raise our hairs in response to stress, this is considered a vestigial reflex or behavior.

10

Humans, along with other organisms, might have structures that have lost their original function or have no apparent function. These structures, called **vestigial structures**, indicate common ancestry and thus can provide evidence for determining evolutionary pathways. Vestigial structures are often homologous to structures that function normally in other species. In his book, _The Descent of Man_, Darwin discussed a number of anatomical structures he considered useless or nearly useless.

FIGURE **10.4** Muscles connected to (**A**) human ears do not develop enough to afford the same mobility as in (**B**) monkeys' ears.

✔ Check Your Understanding

3.1 What is the relationship between goose bumps in humans and the hair standing up on the back of a startled cat or dog?

_____

_____

3.2 What vestigial structures can be found in humans?

_____

_____

3.3 Vestigial structures are often analogous structures that function normally in other species. (_Circle the correct answer._)

True / False

3.4 The wing of an insect and the wing of a bird have a similar function but differ in their evolutionary history. They are

considered _____ structures.

Embryological Evidence

Heinz Christian Pander (1794–1865) was a Russian embryologist who studied chick embryos. In 1817 he discovered the germ layers, or the three distinct regions of the embryo—ectoderm, endoderm, and mesoderm—that give rise to the specific organ systems.

Karl Ernst von Baer (1792–1876) was a German biologist who discovered the mammalian ovum. Through his observations of embryos of vertebrates (fishes, amphibians, reptiles, birds, and mammals), he concluded that all vertebrate embryos, especially in the early stages of development, are strikingly similar but that the adults are very different. Expanding on Pander's concept of germ layers in the chick embryo to include all vertebrates, von Baer laid the foundation for comparative embryology. Today, von Baer is known as the "father of embryology." The laws of von Baer state that the general characters of a group to which an embryo belongs appear in development earlier than the special characters and that the less general structural relations are formed after the more general ones, and so on, until the most specific appear.

Ernst Haeckel (1834–1919), a German zoologist, proposed the theory that ontogeny (changes in size and shape) recapitulates phylogeny (the evolutionary history of a species). He meant that advanced species repeated in their embryological development the stages more primitive species ventured through. Haeckel supported this theory with drawings of embryological stages of various vertebrate species, claiming that they closely resembled each other (Fig. 10.5). Modern-day embryologists agree that some of Haeckel's drawings were misleading, that many of the features were overly exaggerated, and that the suggestion that ontogeny recapitulates phylogeny could not be taken literally. Still, they agree that all vertebrates do share certain characteristics.

At some time during embryonic development, all have a postanal tail, somites (body segments), and paired pharyngeal pouches (gill slits). The shared structures usually appear early in embryonic development and differ in later stages of development. In a cow the postanal tail literally becomes a "flyswatter," in dolphins it becomes a fluke, and in humans it is not present. In a shark, the pharyngeal pouches and arches become functional gills, and in humans the first pair of pouches becomes the auditory tube and middle ear cavity, the second pair gives rise to the facial nerve and palatine tonsils, and the third and fourth pair give rise to parts of the thymus and parathyroid gland. The pharyngeal arches in humans contribute to the formation of the maxilla, mandible, bones of the ear, facial nerves, and facial and laryngeal muscles.

10

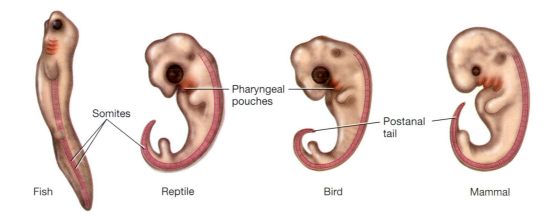

FIGURE **10.5** Similarity of vertebrate embryos.

Somites

Pharyngeal pouches

Postanal tail

Fish   Reptile   Bird   Mammal

Comparing Vertebrate Embryos

Materials
- ❏ Dissecting microscope
- ❏ Prepared slides of various embryos to be determined by the instructor, such as fish, reptile, bird, and mammal embryos

As the result of von Baer's work, scientists know vertebrate embryos share common embryological development. By comparing the comparable developmental stages of representative vertebrates, this relationship can be observed.

1 Procure a dissecting microscope and the slides of the vertebrate embryos.

2 Observe the given slides, and locate the somites, postanal tail, and pharyngeal pouches. Sketch and label these items in the space provided below.

10

Specimen _____   Specimen _____   Specimen _____

✔ Check Your Understanding

4.1 What is the fate of the pharyngeal pouches and postanal tail in an adult shark, turtle, cow, and human?

_____

_____

_____

_____

4.2 Which of the early biologists is credited with the theory that ontogeny recapitulates phylogeny? (*Circle the correct answer.*)

a. Heinz Christian Pander.

b. Ernest Haeckel.

c. Karl Ernst von Baer.

d. Stephen J. Gould.

e. T. H. Huxley.

Taxonomy

Look around—diversity abounds! Today, biologists face the awesome task of systematically identifying, studying, and developing the natural history of an estimated 5–30 million different species of organisms. Some microbiologists estimate possibly 1 billion species of bacteria exist. Zoologists already have classified more than 1.5 million species of animals, and approximately 13,000 new species are described yearly. Zoologists estimate the number of species of animals named so far represents 10% to 20% of all the living animals and less than 1% of all of the animals that have ever lived. Botanists have classified more than 300,000 species of plants, and many new species are described yearly.

Despite exciting times in taxonomy, many organisms will remain unknown forever. Some organisms are so minute and elusive that only good luck or random chance will place them in the taxonomists' hands. Others live in regions of the world not explored in depth. Still others are going extinct, especially in the tropical rainforests, faster than they can be classified.

Ancient people developed classification schemes based on the need to survive and on curiosity. Perhaps an ancient classification system was devised around the question, "Can I eat it, or will it eat me?" Early biologists realized if they were ever going to make sense of the diversity of life, they must devise a logical system of classifying organisms.

One of the early biologists, Carolus Linnaeus (1707–1778), known as the "father of taxonomy," developed a system of binomial nomenclature still used today. **Taxonomy** is defined as the study of the principles, procedures, and rules of scientific classification and the naming of organisms (Fig. 10.6). The Linnaean system of nomenclature will be discussed later in this chapter. It includes the application of distinctive names to each group of organisms.

The classification of living organisms is an arduous task. **Classification** is defined as the ordering of living organisms into groups, or **taxa,** on the basis of associations by contiguity, similarity, or both. Biological classification involves two major steps:

1. Defining and describing organisms; and
2. Arranging organisms into a logical classification scheme.

Classical taxonomy predates evolutionary biology, and many schemes of classical taxonomy are being challenged by newer theories. One of these is **systematics,** the scientific study of the kinds of organisms, their diversity, and their evolutionary relationships. Systematics examines organisms from many viewpoints with the aim of understanding the evolutionary relationships between organisms and the construction of a phylogenetic tree that relates organisms.

Don't be surprised if you see variations in classification schemes as you thumb through different texts. One relatively

FIGURE **10.6** Common and scientific names, respectively: (**A**) column stinkhorn, *Clathrus columnatus*; (**B**) baobab, *Adansonia digitata*; (**C**) nudibranch, *Berghia coerulescens*; (**D**) destroying angel mushroom, *Amanita virosa*; (**E**) pitcher plant, *Sarracenia leucophylla*; (**F**) bottlenose dolphin, *Tursiops truncatus*.

- Some organisms are named after their founder or a namesake. These names are called eponyms. As examples, the rhea *Rhea darwinii* is named after Charles Darwin. The southern magnolia *Magnolia grandiflora* is named after Pierre Magnol, and the trilobite *Struszia mccartney* is named after Paul McCartney.

- Some scientific names relate the organism to a geographical region, such as the American alligator *Alligator mississippiensis*, the white-tailed deer *Odocoileus virginianus*, and the Texas bluebonnet *Lupinus texensis*.

- Other scientific names are descriptive in nature, such as the mockingbird *Mimus polyglottis*, the flying squirrel *Glaucomys volans*, and the red maple *Acer rubrum*.

- Several scientific names are based in Greek mythology, such as the Louisiana flag iris *Iris versicolor*, named after Iris, goddess of the rainbow; *Zanthoxylem clavo-hercules*, Hercules club; and *Cassiopeia andromeda*, a jellyfish named after two Greek figures.

- Some scientific names are based on wordplay, puns, and humor, such as a dung beetle named *Ytu brutus*, a rhinoceros beetle named *Enema pan*, a yaupon tree named *Ilex vomitoria*, and a spider named *Darth-vaderum greensladeae*.

Many people ask why scientific names are necessary. Scientific names are universally accepted and serve to avoid confusion when discussing organisms. For example, bowfin, grindle, cypress trout, and choupique are colloquial names for the same fish, *Amia calva*.

Scientific names are relatively easy to find in biology books, on the internet, in field guides, and in natural history books. Learning these names requires study and, sometimes, making a game out of learning and remembering them. For instance, if your friend reminds you of a squirrel, you might call him *Sciurus carolinensis* after the gray squirrel.

In writing scientific names by hand, the genus and species names are underlined, and when typed, they are italicized. The genus name always begins with an uppercase letter, and the species epithet always begins with a lowercase letter. For example, the scientific name for a great white shark is *Carcharodon carcharias*, and a ginkgo tree is *Ginkgo biloba*. The genus name serves as a noun, and the species name usually is an adjective that must agree with the genus. In the scientific name for the domestic cat, *Felis domestica*, *Felis* refers to a genus of cats, including wild and domestic species, and *domestica* refers to the common domesticated cat.

Names of genera apply to specific groups of organisms only, whereas the species epithet may be used with different genera. As examples, the yellow-billed cuckoo is *Coccyzus americanus*, the black bear is *Euarctos americanus*, and the hookworm is *Necator americanus*. They all share the same

new theme you will discover in evolutionary taxonomy is termed **cladistics**, or **phylogenetic systematics**. This basically is a system of arranging taxa by analysis of primitive and derived characteristics so their arrangement will reflect phylogenetic relationships.

This lab experience has been designed to introduce students to the basic principles of taxonomy and cladistics. In addition, students will gain practical experience using and developing biological keys.

Binomial Nomenclature

In the mid-19th century, Swedish biologist Carolus Linnaeus devised a system of taxonomy based on two fundamental themes. First he assigned to each organism a two-part scientific name of Latin or Greek origin. This method was called binomial nomenclature. The first word of the name is the genus (plural, genera) to which a species belongs. The second word is specific to the organism and is called the species epithet. The genus and species epithet together make up the organism's scientific name.

In taxonomy, highly specific rules ensure reasonable uniformity and wide international acceptance of scientific names. Rigorous standards have been set for naming organisms. Whoever describes a genus or species has the honor of naming it.

10

species epithet. As another example, the house mouse, *Mus musculus*, shares its species epithet with the blue whale, *Balaenoptera musculus*.

Hierarchical Classification System

The second theme developed by Linnaeus is the hierarchical classification system. Basically, organisms are placed into taxa (major categories) and assigned a standard taxonomic rank. Taxonomists recognize more than 30 ranks in the animal kingdom. Today, animals are placed in eight mandatory ranks. These ranks, starting with general characteristics and ending with the species, are:

domain
    kingdom
        phylum
            class
                order
                    family
                        genus
                            species

The most general taxa is known as the domain and was introduced in 1977 by Carl Woese. Presently three domains are recognized: Archaea (bacteria living in extreme environments, such as methanogens, halophiles, and thermo-acidophiles); Bacteria (typical bacteria, such as cocci, spiro-chaetes, and bacilli); and Eukarya (protists, plants, fungi, and animals). In turn, the domains are comprised of kingdoms. For convenience, presently at least six kingdoms are recognized; they include Archaebacteria, Eubacteria, Protista,

Did you know . . .

Interesting Scientific Names

Sometimes whoever got to name the species had a little bit of scientific fun, as seen in the names below:

Preseucolia imallshookupus	gall wasp
Strigiphilus garylarsoni	biting louse on owls
Abracadabrella birdsville	jumping spider
Agra vation	Peruvian beetle
Bullisichthys caribbaeus	pugnose bass
Rhodophthalmokytodermogammarus cinnamomeus	amphipod
Aha ha	Australian wasp
Montypythonoides riversleighensis	extinct python
Desmodus draculae	vampire bat
La cucaracha	Bolivian moth
Zyzzyx chilensis	sand wasp
Ba humbugi	snail

Plantae, Fungi, and Animalia. Keep in mind kingdom Protista is a hodge-podge kingdom; it will be discussed in Chapter 15. In the near future, the current six kingdoms will be further divided into as many as 15 new kingdoms. Note the basic classification of humans in Table 10.3.

TABLE **10.3** Classification of Humans

Classification Level	Features	Example Organisms	Representative Illustrations
Domain Eukarya	Eukaryotic cells	Protists, plants, fungi, and animals	
Kingdom Animalia	Multicellular heterotrophs	Jellyfish, flatworms, roundworms, molluscs, insects, sea stars, humans, etc.	
Phylum Chordata	Presence of notochord, post-anal tail, dorsal hollow nerve cord, and gill slits	Sea squirts, amphioxus, fish, amphibians, reptiles, birds, mammals, etc.	
Subphylum Vertebrata	Spinal cord enclosed in vertebrae	Lampreys, sharks, perch, toads, sparrows, and humans	
Class Mammalia	Young nourished with milk, presence of hair, warm-blooded	Platypus, wombat, moles, bats, seals, dolphins, horses, lions, monkeys, and humans	
Order Primates	Tree dwellers or their descendants, fingers with flat nails, reduced olfaction	Lemurs, tarsiers, baboons, monkeys, orangutans, gorillas, chimps, and humans	
Family Hominidae	Flat face, eyes focusing forward, color vision, bipedal	Gorillas, chimps, and ancient and modern humans	
Genus Homo	Large brain, speech	*Homo erectus, Homo habilis, Homo ergaster, Homo neanderthalensis*, and *Homo sapiens*	
Species sapiens	Prominent chin, high forehead, sparse body hair	*Homo sapiens*	

Using a Dichotomous Key

With the great diversity of life, few biologists have the expertise to identify more than a small group of organisms. Usually biologists become familiar with the common species in a specific region or those species that have been a part of their research.

One simple means of identifying an organism is to compare the unknown organism with pictures and descriptions in a book. Various field guides and natural history books serve as an excellent means of identifying some unknown organisms. This method is often called "picture booking."

Another way of identifying organisms is to use a **taxonomical key**. The majority of keys, called **dichotomous keys**, are formatted as a series of paired choices. In each pair, only one choice describes the specimen. At the end of the correct choice, the user will find a reference number to the next set of choices. The choices will lead to identification of the unknown organism.

Procedure 1

Using a Key for Dinosaurs and Creating a Key for Insects

10

Materials
None required

Ever since the discovery of the first dinosaur fossil, these giant reptiles have occupied a major part of our thoughts, imaginations, and even movies. Dinosaurs were the dominant animals on our planet for more than 160 million years during the Mesozoic period. Although the giant dinosaurs get most of the attention, the average size of a dinosaur was about the size of a turkey. The fossil record has shown great diversity in these fascinating creatures.

Table 10.4 is designed to demonstrate the mechanics of a dichotomous key using a variety of dinosaurs. The examples in Figure 10.7 represent a few of the more popular dinosaurs.

1 Use the key in Table 10.4 to identify the dinosaurs in Figure 10.7. Write the correct name beside the dinosaur. You probably have known the name of these animals since you were in grade school.

TABLE **10.4** Simple Dichotomous Key of the Dinosaurs in Figure 10.7

1.	a.	Walks on two legs (bipedal)	go to 2
	b.	Walks on four legs (quadrupedal)	go to 3
2.	a.	Possesses short forelimbs	Allosaurus
	b.	Possesses large forelimbs and a large claw on the hindlimbs	Velociraptor
3.	a.	Possesses horns	go to 4
	b.	Does not possess horns	go to 5
4.	a.	Possesses one horn	Styracosaurus
	b.	Possesses three horns	Triceratops
5.	a.	Possesses plates or spines on dorsal side	go to 6
	b.	Does not possess plates or spines on dorsal surface	go to 7
6.	a.	Possesses two rows of plates on dorsal side	Stegosaurus
	b.	Possesses spines on dorsal side	Ankylosaurus
7.	a.	Forelimbs longer than hindlimbs	Brachiosaurus
	b.	Forelimbs and hindlimbs approximately of the same length	Apatosaurus

10

FIGURE **10.7** Dinosaurs identified using the dichotomous key.

2 In Table 10.5, construct a dichotomous key to the insects illustrated in Figure 10.8. When the table is complete, write the correct number and letter beside the insect.

TABLE **10.5** Key to Common Insects

FIGURE **10.8** Common insects: (**A**) grasshopper; (**B**) bee; (**C**) cockroach; (**D**) cicada; (**E**) fly; (**F**) ant; (**G**) moth; (**H**) butterfly; (**I**) beetle; (**J**) flea.

✔ Check Your Understanding

5.1 Name an animal and a plant with more than one colloquial or common name.

_____

_____

5.2 You are given the task of developing a classification scheme for 1,000 plants. What two things would you look at to start this process?

_____

_____

_____

_____

5.3 How might you classify life on Earth?

_____

_____

_____

_____

5.4 Identify the following taxonomic names for human. At each classification level, list one feature that would be found in an organism at that level.

Example Eukarya: eukaryotic cells

  a. Domain _____

  b. Kingdom _____

  c. Phylum _____

  d. Class _____

  e. Order _____

  f. Family _____

  g. Genus _____

  h. Species

5.5 What is a dichotomous key? How would you use it to identify a plant or animal?

_____

_____

_____

Constructing a Cladogram

Unlike Linnaean taxonomy, which classifies organisms in hierarchies (taxa) based on morphological similarities, cladistics (phylogenic systematics) groups organisms based on evolutionary ancestry. In 1950, Emil Hans Willi Henning (1913–1976), a German entomologist, pioneered the cladistics system of taxonomy. Cladists construct phylogenic trees called **cladograms** to discern evolutionary ties between species. Cladograms are treelike diagrams that depict hypothetical evolutionary processes gained from studies at the molecular level of the organism (DNA and RNA) and morphological similarities.

Cladistics is sometimes called **phylogenetics** because the cladograms are generated from similarities in the molecular structure of organisms. Cladistics is based on the principle that groups of organisms are descended from a common ancestor, and they do not mix ecology with phylogeny. The cladograms are branching diagrams showing evolutionary relationships among organisms. Each node (divergence of a population), or branching point, lists the derived characters of the descendants, **clades**, that follow the node. Clades are groups of organisms consisting of a common ancestor and all of its descendants. The derived characters (traits inherited from a common ancestor) are listed according to when cladists hypothesize they first appeared.

Cladists contend evolution is the result of modifications in characteristics over time. Cladistics recognizes only **monophyletic groups**, which include all of the organisms descended from a common ancestor. For example, traditional biologists have separate taxonomical groups for reptiles and birds. Although traditional taxonomists certainly would agree that birds have inherited scales, amniotic eggs, and lungs from the reptiles, they place the birds in a separate class, Aves, because of their differences from reptiles. In this cladogram, birds represent the ingroup (a set of taxa that are more closely related) whose evolutionary relationships are being investigated. Birds are warm-blooded and have specialized beaks and feet, adaptations they need to survive in their various habitats.

Mammals are also amniotes and represent an outgroup (a set of taxa that are less likely related). The outgroup is a group of species that diverged before the ingroup and lacks at least one of the shared derived characteristics of the ingroup. Figure 10.9 compares traditional taxonomy to that of the cladistics.

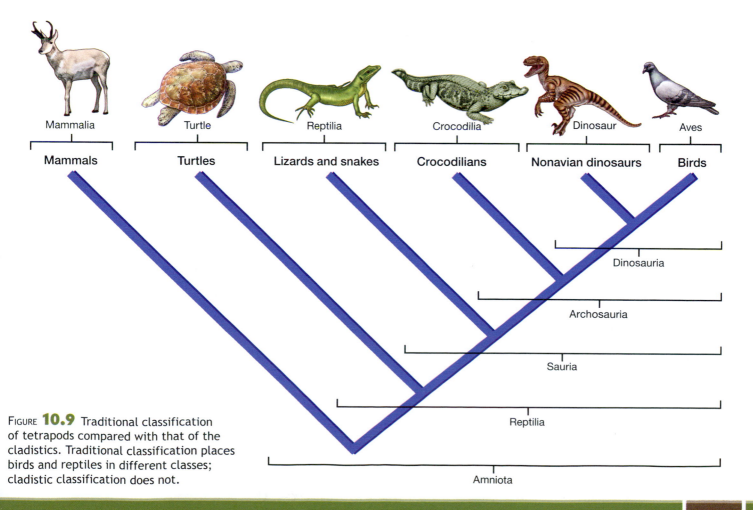

FIGURE 10.9 Traditional classification of tetrapods compared with that of the cladistics. Traditional classification places birds and reptiles in different classes; cladistic classification does not.

A Simple Cladogram

Materials

None required

Cladograms can become very complex. In order to fully understand how to construct a detailed cladogram, much experience is necessary. The following exercise will help you understand how to construct a simple cladogram.

1 Based upon Table 10.6, showing scrambled derived traits and sample organisms, place a plus in the boxes for the organism that has a particular trait and a minus if it does not have that trait. This will help you see relationships and aid in the construction of a simple cladogram.

2 Based upon your observations and Figures 10.10 and 10.11, construct a simple cladogram in the space provided in Figure 10.12, using the data from Table 10.6. Include in your cladogram the nodes

and clades, showing their relationships based on their shared derived characters. In the space provided, describe how and why you developed your cladogram.

TABLE **10.6** Constructing a Simple Cladogram

	Derived Trait					
	Placenta	Limbs	Hair	Segmented	Jaws	Multicellular
Catfish						
Earthworm						
Sponge						
Horse						
Gecko						
Paramecium						
Kangaroo						

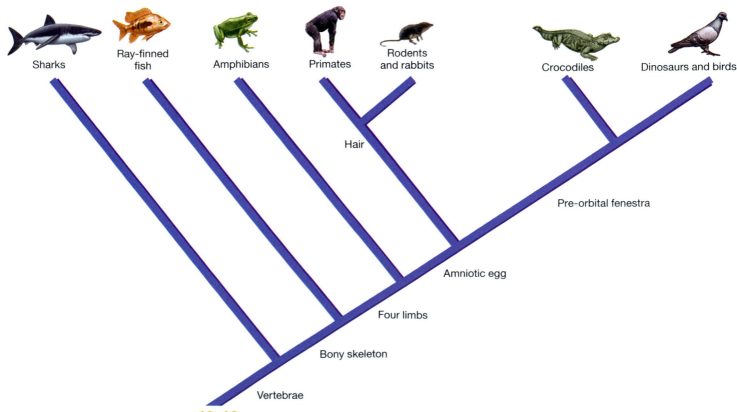

FIGURE **10.10** Cladogram depicting evolutionary relationships in vertebrates.

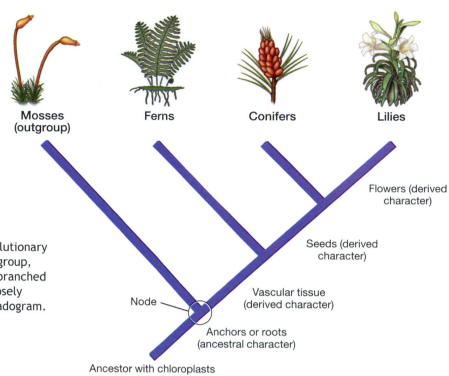

FIGURE **10.11** Cladogram of the evolutionary relationships of plants. Note the outgroup, which in cladistics is the group that branched from the ancestor first and is less closely related to the other groups in the cladogram.

Mosses (outgroup)

Ferns

Conifers

Lilies

Flowers (derived character)

Seeds (derived character)

Vascular tissue (derived character)

Node

Anchors or roots (ancestral character)

Ancestor with chloroplasts (ancestral character)

FIGURE **10.12** Cladogram.

✔ Check Your Understanding

6.1 Where would whales fit on the cladogram in Figure 10.10? Defend your answer. (*Hint:* Refer to Figure 10.3 to help you determine your answer.)

_____

_____

_____

_____

6.2 The hagfish and lamprey are considered ancestors of fish with jaws. In Figure 10.10, if a node were placed before the hagfish or lamprey, what would be a possible derived character?

_____

6.3 The system of cladistics _____. (*Circle the correct answer.*)

    a. classifies organisms into hierarchies based on their evolutionary ancestry

    b. organizes ordering of living organisms into groups or taxa

    c. uses branching diagrams showing evolutionary relationships

    d. classifies organisms based on morphological characteristics

    e. both a and c

10

Mystery of Mysteries
Understanding Evolution

MYTHBUSTING

Evolution of Thought

Debunk each of the following misconceptions by providing a scientific explanation. Write your answers on a separate sheet of paper.

1 Oh, evolution is "just a theory"!

2 Earth is only 10,000 years old.

3 There are no transitional fossils.

4 Evolution and Darwinism leads to immoral behavior.

5 Once established, scientific names do not change.

---

1 What are three matrices in which organisms may become fossilized?

2 Why are fossils important to scientists? What scientific evidence do fossils provide?

3 Compare and contrast artificial selection and natural selection.

4 Explain how microevolution and macroevolution are different.

5 Does Figure 10.13 show homologous or analogous structures?

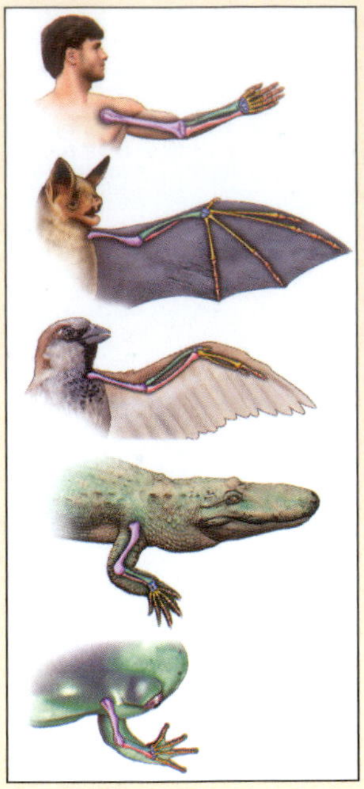

_____

6 Why do we use a two-part scientific name, or binomial nomenclature, rather than common names?

_____

_____

_____

_____

FIGURE **10.13** Bones in the forelimbs of various vertebrates.

7 What did Linnaeus contribute to taxonomy? (_Circle the correct answer._)

a. Classification system based on the ordering of living organisms into taxa.

b. Binomial nomenclature.

c. Classified organisms in hierarchies based on morphologic similarities.

d. Both b and c.

e. All of the above.

8 Match the following terms with the correct description. (_Terms may have more than one answer_.)

_____ Taxonomy

_____ Cladistics

_____ Classification

_____ Systematics

A. Scientific study of the kinds of organisms, their diversity, and their evolutionary relationships.

B. System of arranging taxa by analysis of primitive and derived characteristics.

C. Ordering of living things into groups, or taxa, on the basis of contiguity, similarity, or both.

D. Phylogenetic systematics.

E. Study of the principles, procedures, and rules of scientific classification and naming of organisms.

9 You want to determine the various species of birds in your backyard. Identify three sources that you could use to find their scientific names.

_____

_____

_____

10 The hierarchical system of classification classifies organisms based on _____ similarities, whereas cladistics groups organisms based on evolutionary ancestry. (_Circle the correct answer._)

a. phylogenetic

b. molecular level (DNA and RNA)

c. morphological

d. both a and c

e. all of the above

11 What is a monophyletic group?

_____

On the Edge of Life
Understanding Bacteria

The smallpox was always present, filling the churchyards with corpses, tormenting with constant fears all whom it had stricken, leaving on those whose lives it spared the hideous traces of its power, turning the babe into a changeling at which the mother shuddered, and making the eyes and cheeks of the bighearted maiden objects of horror to the lover.

— T. B. Macaulay (1800–1859)

Consider the following questions prior to coming to lab, and record your answers on a separate piece of paper.

1 I am planning a trip "out of the country." How can I find out what kind of pathogens and parasites are endemic to that region and what preparations should be taken?

2 Why have MRSA and VRSA become so powerful?

3 Why are some bacteria called the normal flora of the intestines?

4 What makes those hot springs in Yellowstone so colorful?

5 Are bacteria found in snowflakes?

Objectives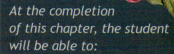

At the completion of this chapter, the student will be able to:

1. Discuss the roles of bacteria in the environment and in medicine.

2. Describe several significant bacterial species.

3. Describe the basic characteristics and morphology of bacteria.

4. Explain the difference between Archaea and Bacteria, and provide examples.

5. Demonstrate safe and sterile techniques in working with bacteria.

To a visitor from another world conducting a survey of life on Earth, our planet would be considered the planet of the bacteria. These prokaryotic organisms constitute an overwhelming percentage of the total biomass of life on Earth. Some soil samples contain more than 2 billion bacteria per gram, and a square centimeter of skin may contain nearly 100,000 bacteria. In fact, the total number of bacteria living on or in your body exceeds the total number of cells that make up your body.

Bacteria are the most cosmopolitan of life forms, living in diverse environments including in the clouds, in deep basalt deposits more than 4,500 feet underground, at the frigid poles, and in extremely hot, deep, hydrothermal vents in the oceans. In fact, one species best known as plant pathogen (*Pseudomonas syringae*) may be important in the formation of snow and rain. A protein found in this organism is used in making artificial snow.

Impact of Various Species

Unfortunately, as the result of disease-causing, or **pathogenic**, species, bacteria have acquired a bad reputation. The vast majority of bacteria are harmless, and many species are helpful. Below is a list of some of the many ways in which bacteria are helpful:

1. Bacteria are responsible for producing copious amounts of oxygen in the atmosphere and are vital members of the food chain.

2. *Rhizobium* spp. is an important bacterium that lives in a mutualistic, symbiotic relationship with the roots of plants known as legumes (clover, peanuts, etc.). *Rhizobium* bacteria fix atmospheric nitrogen, converting it to a water-soluble form the plants can use for various biosynthetic pathways (e.g., amino acid synthesis).

Chapter Photo
Thermophilic bacteria in Yellowstone National Park.

3. In industry, bacteria are necessary in the production of many products, including cheese, yogurt, wine, and several beneficial enzymes. Some bacteria are used to extend the shelf life of produce and others to control mosquito larvae.

4. Environmental applications of bacteria include their use in cleaning up oil spills (bioremediation), in breaking down toxic wastes and herbicides, and in the production of biodegradable plastics.

5. Recently, the species *Thiobacillus ferrooxidans* has been used in gold mining and initiated a new industry known as biomining.

6. In medicine, some bacteria are used in producing antibiotics.

7. The body harbors many harmless and beneficial bacteria, such as those that aid in the proper functioning of the digestive system. For example, one milliliter of saliva may contain more than 40 million bacterial cells!

However, some bacteria do cause problems. Bacteria cause a number of devastating diseases in plants and animals. Humans can fall victim to a number of bacterial infections. Some infections, such as the black plague and leprosy, have changed history. Unlike viruses, many bacterial diseases of animals can be transmitted to humans. These diseases, such as anthrax, are called **zoonotic infections**.

Bacterial Metabolism

Recall that bacteria are prokaryotic organisms (lacking a membrane-bound nucleus and organelles). Most prokaryotes are unicellular, although a few aggregate forms exist. In addition, compared with the cells of eukaryotes (protists, plants, fungi, and animals), the cells of prokaryotes are small, ranging from 0.5–5 nanometers in diameter. Prokaryotes display a number of nutritional modes, including **photo-autotrophism** (energy from sunlight, carbon from CO_2), **chemoautotrophism** (energy from inorganic chemicals, carbon from CO_2), **photoheterotrophism** (energy from sunlight, carbon from organic sources), and **chemoheterotrophism** (energy and carbon from organic molecules).

Oxygen requirements vary in prokaryotes. Some bacteria are **obligate aerobes**, requiring oxygen for survival; others, such as *Clostridium botulinum*, are **obligate anaerobes** that cannot live in an environment containing oxygen gas. Some bacteria, such as *Escherichia coli*, are **facultative anaerobes** that can survive in both worlds. The vast majority of prokaryotes reproduce asexually through binary fission; however, three forms of **genetic recombination**, or horizontal gene transfer, have been described: transformation, transduction, and conjugation. Recombination is the basis of **antibiotic resistance** in pathogenic bacteria as well as the evolution of prokaryotes.

Classification

The world of prokaryotes is diverse, and through the years, the classification of prokaryotes has been perplexing to scientists. Traditional classification is still important and is based upon observable characteristics such as shape, motility, and Gram staining. **Gram staining** is a procedure used to separate bacteria into two groups: Gram-positive cells have a thicker peptidoglycan cell wall and Gram-negative cells have a thinner peptidoglycan cell wall.

As the result of the contributions of molecular systematics, however, a new view of the prokaryotes is emerging. In 1977, based upon RNA sequencing technology, Carl Woese (1928–2012) proposed dividing the traditional bacterial kingdom Monera into two distinct domains—Archaea and Bacteria. Today, the domains Archaea and Bacteria stand along with domain Eukarya as the foundation of the new systematics.

The Archaea were first described in the 1970s. Initially, they were classified along with the typical bacteria, but further studies indicated they were worthy of their own domain. Archaea differ from Bacteria in the composition of their cell walls and plasma membranes as well as other features. Archaea also possess characteristics in common with domain Eukarya, such as the presence of histone proteins in their DNA. Although some Archaea rarely can be found in some subgingival dental plaques and in the human gut, no pathogenic Archaea have been described.

Domain Archaea

Archaea have been called **extremophiles** (Fig. 11.1) because the first studied specimens lived in harsh conditions, including:

1. Halophiles, living in extremely salty conditions, such as the Dead Sea;

2. Thermophiles, living in extremely hot temperatures, 60°C–90°C, such as in the hot springs in Yellowstone Park;

3. Hyperthermophiles, living in oceanic vents with temperatures exceeding 100°C;

4. Psychrophiles, living in extremely cold conditions, such as in the Antarctic;

5. Acidophiles, living in acid conditions, such as in pools of sulfuric acid; and

6. Alkaliphiles, living in extreme base conditions, such as slag dumps.

Some Archaea, however, can exist in soil, sediment, and other less extreme environments. These Archaea are called **mesophiles**.

One group of the Archaea, the **methanogens**, metabolize carbon dioxide to oxidize hydrogen and release methane gas as a waste product. Methanogens are decomposers in anaerobic, swampy soil (producing "marsh gas") and sewage

FIGURE **11.1** (**A**) Halophile; (**B**) thermophile; (**C**) acidophile.

treatment plants. Methanogens also can be found in the guts of several herbivorous insects and animals, including termites and cattle. In recent years, commercial methane digesters have been developed to use dung and compost to produce methane as an efficient source of fuel. But keep in mind that living next to a methane digester may be rather unpleasant!

Domain Bacteria

Most prokaryotes have been placed in domain Bacteria. This domain consists of many beneficial as well as pathogenic species. Bacteria are highly diverse and live in a variety of environments from the soil to your intestines. As systematics grows, so does the number of bacterial groups. Several well-known groups of bacteria are the cyanobacteria, proteobacteria, spirochaetes, chlamydias, and the Gram-positive bacteria. Table 11.1 lists some dangerous bacterial infections in humans.

1. The **cyanobacteria** (Fig. 11.2A) were once described as blue-green algae and can live as a single cell or in colonies. They live in aquatic environments and even on sidewalks, in trees, and on buildings. Cyanobacteria saturated the early atmosphere with oxygen and are still important oxygen producers today. Common cyanobacteria include *Oscillatoria*, *Anabaena*, and *Nostoc*.

2. The **proteobacteria** (Fig. 11.2B) are highly diverse but share many molecular traits. Mitochondria may have evolved from proteobacteria. The proteobacteria include the purple

TABLE **11.1** Review of Bacterial Pathogens Dangerous to Humans

Organism	Disease
Mycobacterium tuberculosis	Tuberculosis
Mycobacterium leprae	Hansen's disease, or leprosy
Neisseria gonorrhoeae	Gonorrhea
Neisseria meningitidis	Meningitis
Pseudomonas aeruginosa	Lung and bladder infections
Staphylococcus aureus	Pimples, boils, toxic shock syndrome, Methicillin-resistant *Staphylococcus aureus* (MRSA)
Streptococcus pyogenes	Strep throat
Corynebacterium diphtheriae	Diphtheria
Bacillus anthracis	Anthrax
Salmonella typhii	Typhoid fever
Shigella spp.	Bacterial dysentery
Escherichia coli	Gastrointestinal problems
Legionella pneumophila	Legionnaires' disease
Vibrio cholerae	Cholera
Vibrio vulnificus	Flesh-eating, intestinal problems
Yersinia pestis	Bubonic or black plague
Haemophilus influenzae	Meningitis, pinkeye, otitis media
Chlamydia trachomatis	Chlamydia
Clostridium perfringens	Gangrene
Clostridium tetani	Tetanus
Clostridium botulinum	Botulism
Helicobacter pylori	Ulcers
Leptospira interrogans	Leptospirosis
Treponema pallidum	Syphilis
Borrelia burgdorferi	Lyme disease
Mycoplasma pneumoniae	Atypical pneumonia
Rickettsia rickettsii	Rocky Mountain spotted fever
Rickettsia prowazekii	Epidemic typhus

11

sulfur bacteria, which use hydrogen sulfide, not water, as an electron donor in bacterial photosynthesis; the enteric bacteria, such as *E. coli* and *Salmonella*, which are inhabitants of the digestive tract of many animals; and the nitrogen-fixing bacteria, such as *Rhizobium*. Other proteobacteria include *Vibrio* spp., *Legionella* spp., *Yersinia pestis*, and *Neisseria* spp.

3. The **spirochaetes** (Fig. 11.2C) have spiral-shaped bodies and move in a corkscrew motion. Examples are *Treponema* spp., *Leptospira* spp., and *Borrelia* spp.

4. **Chlamydias** (Fig. 11.2D) are atypical bacteria that serve as intracellular parasites, such as *Chlamydia trachomatis*.

5. The **Gram-positive bacteria** (Fig. 11.2E) constitute another large, diverse group that includes *Bacillus anthracis*, *Clostridium* spp., *Staphylococcus* spp., *Streptococcus* spp., and *Mycobacterium* spp. Other Gram-positive bacteria are the **actinomycetes** that are found in the soil. Some actinomycetes are used in making antibiotics.

Identification by Shape

The shape of the cells of bacteria and their cellular arrangement are used in identification (Fig. 11.3).

1. Rod-shaped bacteria are known as the **bacilli** (singular, bacillus). Examples of bacilli are *E. coli* (Fig. 11.3A), *Bacillus anthracis*, and *Yersinia pestis*.

2. **Spirilli** (singular, spirillum) are spiral-shaped bacteria and include organisms such as *Leptospira interrogans* and *Treponema pallidum* (Fig. 11.3B). The sexually transmitted disease syphilis is caused by *Treponema pallidum*.

3. The spherical bacteria are called the **cocci** (singular, coccus). The cocci can be highly variable in arrangement:

   ♦ *Chlamydia trachomatis* exists as a singular coccus.

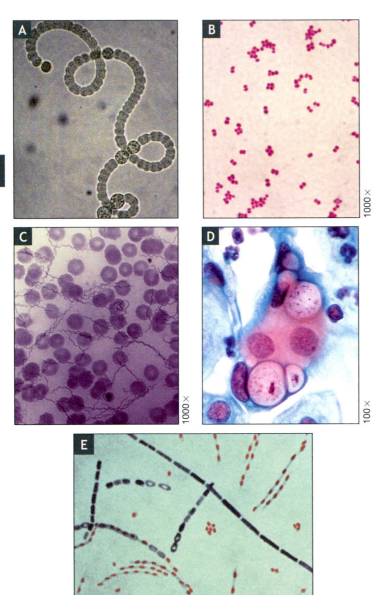

FIGURE **11.2** (**A**) Example of cyanobacteria: *Anabaena* sp.; (**B**) example of proteobacteria: *Neisseria gonorrhoeae*; (**C**) example of spirochaete: *Borella recurrentis*; (**D**) example of chlamydia: *Chlamydia trachomatis*; (**E**) example of Gram-positive bacteria: *Bacillus anthracis*.

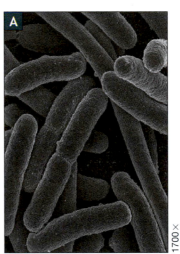

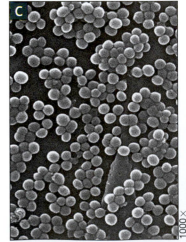

FIGURE **11.3** Micrographs of bacteria: (**A**) bacilli (rod-shaped), *Escherichia coli* (enteric bacteria); (**B**) spirilli (spiral-shaped), *Treponema pallidum* (syphilis); (**C**) cocci (spherical), *Staphylococcus aureus* (boils, MRSA).

- *Neisseria meningitidis* exists in groups of two cells known as diplococci.
- *Lactobacillus* spp. and *Streptococcus pyogenes* occur in chains and are called streptococci.
- *Staphylococcus aureus* (Fig. 11.3C) occurs in groups that look like bunches of grapes and is termed staphylococci.

Anatomical Structure

Even though the anatomy of a prokaryote is much simpler than that of an oak tree, mushroom, or dolphin, these unicellular organisms are capable of performing all of life's processes. The **cell wall** in prokaryotes is a means of protection, is involved in metabolic activity, and helps the cell maintain its shape. Archaea cell walls are different from those of Bacteria. In Archaea, the cell walls are composed of polysaccharides and proteins, whereas the cell walls of Bacteria are composed of peptidoglycans.

In 1884, Danish physician Hans Christian Gram (1853–1938) developed a simple laboratory technique to discriminate between various kinds of bacteria. Today, the Gram stain remains a standard procedure in the bacteriology laboratory. This staining technique is based on characteristics of the cell wall. Many species of bacteria have a capsule, glycocalyx, or slime layer composed of polysaccharides encasing the cell wall. This layer helps the bacterium attach to a substrate to keep from drying out, or in some species protects the bacterium from being destroyed by predators or the host's immune system.

Some bacteria possess **flagella** for motility. These flagella are simpler than those of eukaryotes. Several species of bacteria possess **pili,** hairlike structures extending from the cell wall. Those pili that attach the bacterium to a substrate are called **fimbriae,** and sex pili serve in a primitive type of sexual reproduction called **bacterial conjugation.** A bilayered **plasma membrane** underlies the cell wall in bacteria. This membrane serves as a barrier in regulating substances that move in and out of the cell and as a site of metabolic activities, such as photosynthesis. Folds in the plasma membrane called mesosomes increase the surface area and are integral in metabolism. In some bacteria, structures called magnetosomes serve as a compass and are found in the plasma membrane.

Bacteria exposed to harsh conditions may produce an **endospore.** This thick-walled structure aids them in surviving intolerable conditions (disinfectant, radiation, acid, and aridity) for long periods of time. Active endospores have been taken from the Egyptian pyramids and from a 7,500-year-old site in Minnesota. Examples of species that can form endospores are *Clostridium tetani* and *Bacillus anthracis*. *Clostridium tetani* is commonly found in the soil and can enter wounds potentially causing tetanus. Thus, it is imperative to thoroughly wash and clean all wounds.

Bacteria lack membrane-bound organelles, but still possess the means to perform various metabolic processes because of structures in their cytoplasm. Bacterial **ribosomes** serve as a site for building proteins. These ribosomes are different from eukaryotic ribosomes because the former are smaller and differ in their protein structure. Antibiotics such as tetracycline and streptomycin effectively inhibit bacterial growth because they inhibit the function of ribosomes.

In contrast to having as many chromosomes as a eukaryote does, bacterial genes usually are found in a single bacterial chromosome. This chromosome is housed in a nonmembranous region called the **nucleoid.** Bacteria also may possess one or more small, ringlike structures called plasmids that contain DNA and are integral in bacterial conjugation (Fig. 11.4).

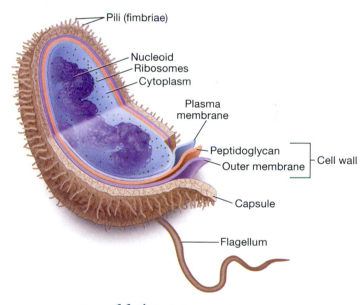

FIGURE **11.4** Basic bacterial anatomy.

Observing Bacteria

Procedure 1

Using Oil Immersion with Bacteria

Materials

❏ Compound microscope

❏ Immersion oil

❏ Prepared slides of *Oscillatoria* spp., *Bacillus* spp., *Pseudomonas* spp., *Streptococcus* spp., *Staphylococcus* spp., *Neisseria* spp., *Treponema* spp., and *Anabaena* spp.

❏ Colored pencils

Hints & Tips

Because bacteria are cosmopolitan and many species are potentially pathogenic, anyone working with bacteria must follow an aseptic laboratory technique and use common sense to reduce the risk of contamination of yourself and the environment. Although the bacteria used in the activities in this chapter are nonpathogenic, they should be treated with proper aseptic technique. The following aseptic tips should be strictly followed in the laboratory:

◗ Don't hesitate to ask questions regarding safety matters.

◗ Report all potentially hazardous conditions, including spills and potential contamination, to your instructor.

◗ Don't bring any food or drink into the laboratory. Store your book bags, coats, and other personal items in the designated area.

◗ Wear a lab coat or lab apron over your clothes. Use eye protection if required.

◗ Thoroughly wipe your laboratory table with the provided disinfectant before and after use.

◗ Thoroughly wash your hands prior to performing a procedure and at its completion. Any time you feel the urge to wash your hands—go for it!

◗ Always place disposable supplies in the proper biohazard container identified by your instructor. Never leave them lying around.

◗ Properly flame all qualified laboratory tools as indicated by your instructor.

◗ Report any injuries to your instructor.

◗ Don't remove material from the laboratory.

Bacteria exist in a variety of shapes and arrangements. This observational activity acquaints students with several species of bacteria and techniques for observing bacteria.

1 Before beginning, clean and disinfect your work area.

2 Procure a microscope, immersion oil, and the prepared slides.

3 Place the slide on the stage, and focus on the specimen using low power. Adjust the illumination if necessary. (Because bacteria are so small, only colored specks will be seen.) Rotate the high-power objective into place, and focus with the fine adjustment only. Rotate the oil-immersion objective so it is halfway in position.

4 Carefully place one drop of immersion oil on top of the specimen under the path of the light. Place the oil-immersion objective directly into the oil. Use only the fine adjustment to observe the specimen. Adjust the light if necessary.

5 In the space provided, sketch and label your specimen. In addition, briefly describe the basic morphology of the specimen, including the magnification of the specimen.

6 Repeat the process for each specimen.

7 Upon completion of the activity, carefully and thoroughly clean the oil from the slide and the oil-immersion objective. Return the materials.

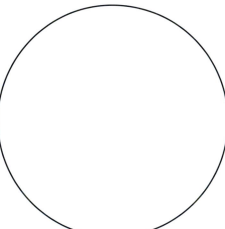

***Oscillatoria* spp.** Total magnification

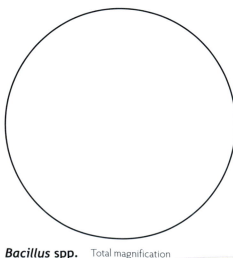

***Bacillus* spp.** Total magnification

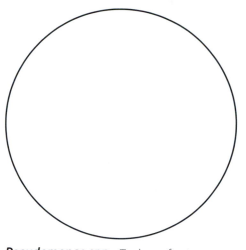

***Pseudomonas* spp.** Total magnification _____

***Streptococcus* spp.** Total magnification _____

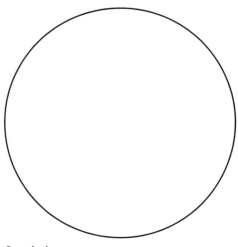

***Staphylococcus* spp.** Total magnification _____

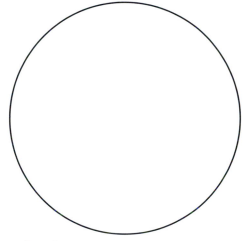

***Neisseria* spp.** Total magnification _____

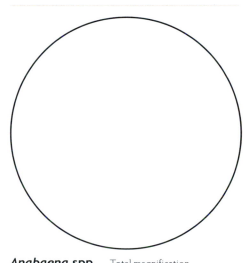

***Treponema* spp.** Total magnification _____

***Anabaena* spp.** Total magnification _____

✔ Check Your Understanding

1.1 What are the functions of pili in bacteria?

1.2 What is the function of an endospore in some bacteria?

1.3 Give an example of a disease-causing, spore-producing bacteria. How might this spore producer be introduced into the body?

1.4 Where is genetic material located in bacteria?

Finding Bacteria

Bacteria are inhabitants of nearly every environment. This activity asks students to detect whether bacteria, and perhaps fungi, can be found in the region around the laboratory and living in or on common objects. This activity is not designed to identify organisms but, instead, to examine the diversity of organisms that can be extracted from a specific environment.

Procedure 1

Bacteria

Materials

- ❏ Sterile petri dishes with growth medium (5)
- ❏ Sterile cotton swabs
- ❏ Loop
- ❏ Tape or Parafilm
- ❏ Wax pencil or Sharpie
- ❏ Colored pencils
- ❏ Bunsen burner

For this procedure, students should work in groups of two to four. Discuss various areas you would like to survey for bacterial growth. Some ideas: doorknob, desktop, candy bar, lipstick, eyeshadow, top of an open soda can or bottle, hand soap, antibacterial soap, soil, a puddle of water. Perhaps leave an open petri dish in a quiet corner of the room for 30 minutes. Consider swabbing your hands pre- and post-hand washing as well. Choose five areas for testing.

WARNING

Use aseptic techniques when collecting, culturing, and observing bacteria.

1 Procure materials to perform the activity, and disinfect your work area.

2 Using the wax pencil or Sharpie, place a name on the top and bottom of each petri dish, and number each plate. Record the number of the petri dish and the test area in Table 11.2.

3 Use a sterile cotton swab to rub over your study area (unless you choose to leave one dish exposed to air). Carefully open your sterile petri dish, and gently streak the cotton swab back and forth across the top of your dish. Place the used swab in the proper disposal container.

4 To spread the bacteria across the dish, flame your loop using the Bunsen burner, let the loop cool, and place your loop gently on the end of the swabbed region, spreading the bacteria as shown in the diagram in Figure 11.5. The loop should be flamed between streaks of different bacteria.

5 Tape the petri dish together as indicated by the instructor. Some instructors may prefer to seal the dish with Parafilm.

6 Repeat steps 3 through 5 for each sample.

TABLE **11.2** Bacteria Observations

Selected Environment	Colony Description and Numbers
1.	
2.	
3.	
4.	
5.	

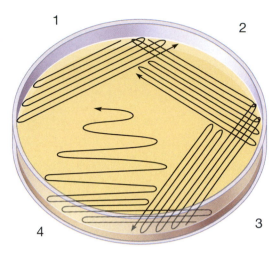

FIGURE **11.5** Quadrant streak pattern on nutrient agar.

7 Incubate the culture at 37°C in the designated area for one week.

8 Disinfect your work area, reflame the loop, and return the materials.

9 When you return to lab, observe the dishes. *Note:* Do not open the petri dishes.

10 Observe the number of colonies, their shape, and their color. Sketch the dishes and their colonies in the space provided using a colored pencil.

11 Record your results in Table 11.2.

12 Visit the other lab stations, and discuss your findings. Take notes on curious results.

13 Dispose of the petri dishes as instructed, and disinfect your area. Wash your hands thoroughly.

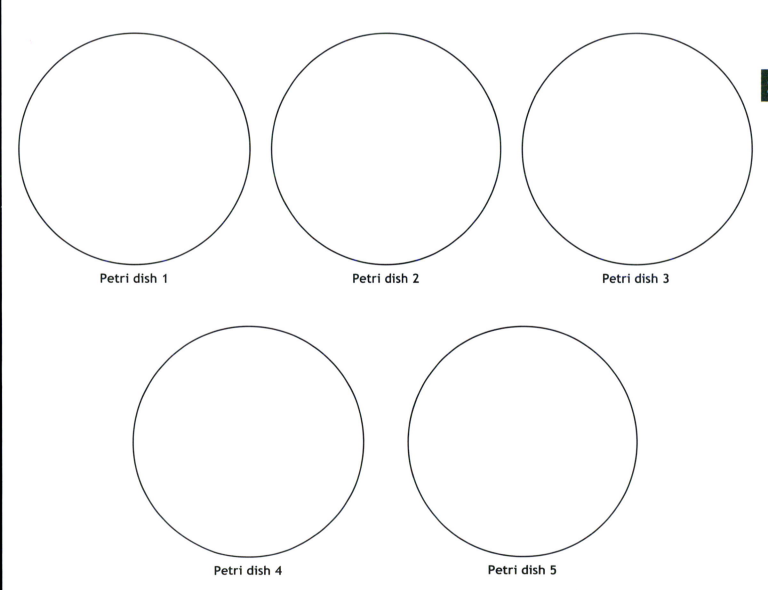

Petri dish 1

Petri dish 2

Petri dish 3

Petri dish 4

Petri dish 5

✔ Check Your Understanding

2.1 Which environment yielded the greatest number of colonies on the agar plates?

2.2 Which environment yielded the greatest diversity of colonies on the agar plates?

2.3 Describe two different colonies that you observed on your plates (color, etc.).

2.4 Were any fungal colonies noted on the plates? If any were seen, describe them.

11

2.5 Why is hand washing imperative to prevent disease? Does this activity make you more aware of the importance of hand washing?

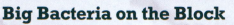

Did you know . . . ?

Big Bacteria on the Block

A giant prokaryote, *Thiomargarita namibiensis*, can be seen by the naked eye. It is a whopping 750 millimicrons in diameter. This giant is found in anaerobic, hydrogen sulfide-rich ocean floor sediments. The first forms were found off the coast of Namibia, but similar forms have been reported from the Gulf of Mexico. It is called the "sulfur pearl of Namibia."

MYTHBUSTING

Great Gruesome Germs!

Debunk each of the following misconceptions by providing a scientific explanation. Write your answers on a separate sheet of paper.

1 Antibiotics are effective against viruses.

2 Toilet seats are responsible for causing a host of maladies, including chlamydia.

3 The bactericide advertised on TV kills all "germs."

4 The bacterium that caused the plague was wiped out—don't worry.

5 Don't worry. The health profession is making a big deal about MRSA!

---

1 Archaea _____ . (*Circle the correct answer.*)

a. were first described in 1970s

b. are known as extremophiles because they are found living in extremely harsh environments

c. differ from bacteria in the composition of cell wall and plasma membrane

d. both a and b

e. all of the above

2 List three beneficial bacterial species and where are they found.

3 Which of the following is/are not true for bacteria? (*Circle the correct answer.*)

a. Bacteria are prokaryotic.

b. Bacteria have membrane-bound organelles.

c. Oxygen requirements vary from obligate aerobe to obligate anaerobe.

d. Bacteria exhibit a number of nutritional modes.

e. Both b and d.

4 Cyanobacteria _____ . (*Circle the correct answer.*)

a. formerly were described as blue-green algae and are single-celled or colonial

b. occur in aquatic environments, sidewalks, and trees; they also use an autotrophic mode of nutrition

c. are eukaryotic

d. both a and b

e. all of the above

5 What is the function of the following in bacteria?

    a. Cell wall _____

    _____

    b. Plasma membrane _____

    _____

    c. Plasmid _____

    _____

    d. Nucleoid _____

    _____

    e. Pili _____

    _____

6 Why are many scientists alarmed by the overuse of antibiotics? Provide an example of bacterial resistance along with its dangerous impact on health and medicine.

_____

_____

_____

_____

_____

7 Describe the three major shapes of bacteria, and give an example of each.

_____

_____

_____

_____

_____

_____

8 What are five safety factors to consider when working with bacteria?

_____

_____

_____

_____

_____

_____

This Fine Mess
Understanding the Protists

*Although traditional classifications of protists tended to have little respect for phylogeny—
neither in theory nor the data were available to make this possible—
they did have one great advantage. They were designed to be user-friendly.*

— Colin Tudge (1943–present)

Just wondering . . .

Consider the following questions prior to coming to lab, and record your answers on a separate piece of paper.

1 What are several important medical and commercial uses of algae?

2 How did a simple protist change the history of Ireland?

3 What is the importance of a protist known as *Globigerina*?

4 In the new classification scheme, what are the basal protists and their fundamental characteristics?

5 Did a protist change the life of Charles Darwin?

Objectives

At the completion of this chapter, the student will be able to:

1. Describe the general characteristics of the protists.

2. Discuss why protists present a unique problem for taxonomists.

3. Compare and contrast traditional and modern protist classification.

4. Describe the traditional organization of the plantlike protists.

5. Describe the basic biology of the plantlike protists.

6. Describe the traditional organization of the fungus-like protists.

7. Describe the basic biology of the fungus-like protists.

8. Describe the traditional organization of the animallike protists.

9. Describe the basic biology of the animallike protists.

10. Draw and label the protist examples used in this chapter.

Chapter Photo
Macrocystis sp., a type of brown algae.

In the classic 1930s comedy film, *Another Fine Mess*, Oliver Hardy's statement to his cohort Stanley Laurel, "What a fine mess you have gotten us into!" is also appropriate for trying to make sense of the hodgepodge group of organisms known as protists. The term *protist* describes microscopic, unicellular organisms, such as the amoeba, as well as gigantic multicellular organisms exceeding 600 feet in length, such as kelp.

Protists are ancient eukaryotes that first appeared in the Precambrian era nearly 2 billion years ago. A multitude of protists can live in a drop of water, in the terrestrial environment, or in the body of a host. In the environment, protists can function as autotrophs, heterotrophs, and even decomposers. Several protists are responsible for diabolical parasitic infections in humans.

The 100,000 named members of the protists are a polyphyletic group of organisms underlying the kingdoms Plantae, Fungi, and Animalia. This diversity is the main reason the kingdom "Protista" is no longer considered viable, though it is still a convenient, and widely understood, term to describe this diverse group.

Classically, the protists have been categorized as plantlike protists, fungus-like protists, and animallike protists in accordance with their roles in the environment. This scheme, although many consider it antiquated, is still a useful tool in developing a basic understanding of the protists.

Recently, studies using DNA sequencing and cytological analysis have painted a more complex picture of the protists. This new look at the protists suggests the once singular kingdom should be divided into several supergroups (Table 12.1 and Fig. 12.1) that can be divided further into kingdoms (as many as 30, according to some authors). As time passes, a clearer picture of protist classification will emerge.

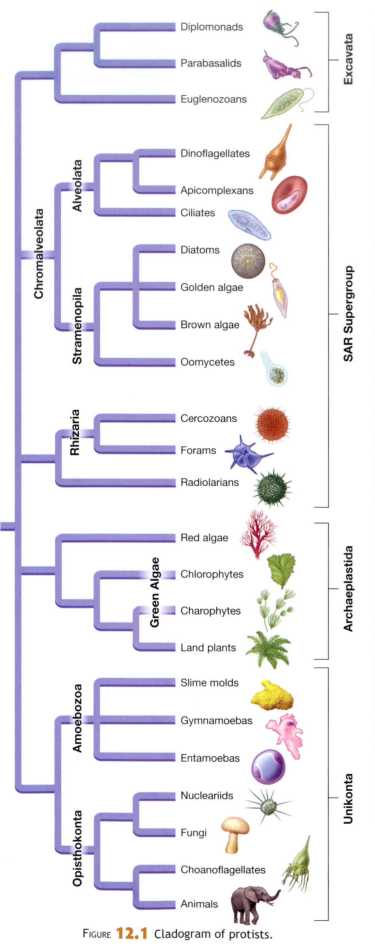

FIGURE **12.1** Cladogram of protists.

TABLE **12.1** New View of the Protists

Supergroup	Example	Representative Organism
Excavata		
	Diplomonadida Parabasala	*Giardia intestinalis* *Trichomonas vaginalis*
	Euglenozoa/ Discicristates (Euglenida and Kinetoplastea)	*Euglena deses* *Trypanosoma brucei*
SAR Supergroup		
Stramenopila (Heterokonta)	Bacillariophyta Oomycota Chrysophyta Phaeophyta	*Cytotella stelligera* *Phytophthora infestans* *Dinobryon sociale* *Macrocystis pyrifera*
Alveolata	Ciliophora Apicomplexa Dinoflagellata	*Paramecium caudatum* *Plasmodium vivax* *Gonyaulax catenella*
Rhizaria (Cercozoa)	Radiolaria Foraminifera Chlorarachniophyta	*Lychnaspis miranda* *Globigerina falconensis* *Chlorarachnion reptans*
Archaeplastida		
	Rhodophyta Chlorophyta Glaucophyta Vascular plants*	*Gelidium amansii* *Spirogyra crassa* *Cyanophora paradoxa* *Acer rubrum*
Unikonta		
Amoebozoa	Amoebas Cellular slime molds Plasmodial slime molds	*Amoeba proteus* *Dictyostelium discoideum* *Physarum polycephalum*
Opisthokonta	Choanomonada Fungi* Animalia*	*Monosiga brevicollis* *Amanita verna* *Homo sapiens*
* A separate kingdom		

Plantlike (Photosynthetic) Protists

The plantlike protists, commonly called **algae**, are extremely diverse, ranging from minute diatoms to giant multicellular kelp. These organisms exist as photoautotrophs possessing chlorophyll *a* and membrane-bound plastids. They produce copious oxygen and serve as the basis for the food chain. Traditionally, the names of the phyla were derived based upon accessory pigmentation. In this exercise, we will conduct procedures on specimens of phyla Chlorophyta, Rhodophyta, Bacillariophyta, and Euglenophyta. Phyla not included in this exercise but still considered protists are Phaeophyta (brown algae) and Pyrrophyta/Dinophyta (unicellular algae, or plankton).

Green Algae (Archaeplastida)

Phylum **Chlorophyta**, commonly known as green algae and placed in the major group Archaeplastida (Table 12.1), comprises about 10,000 unicellular and multicellular species (Fig. 12.2). Most green algae are aquatic, but some species can be found growing on tree trunks, sidewalks, buildings, unwashed cars, and even snow. One species even lives on the fur of tree sloths, imparting a green sheen to them. Green algae are thought to be the ancestors of modern plants. Not all green algae are green.

- *Volvox* sp. is a colonial freshwater green alga that resembles a basketball under the microscope. It may be composed of as many as 5,000 to 50,000 cells.

- Classically, *Spirogyra crassa* has been placed in phylum Chlorophyta, but current data suggests that it should be reclassified, with many leaning toward phylum Charophyta. This filamentous, freshwater green alga

can be found floating in mats on the surface of quiet ditches and ponds. Under the microscope, *Spirogyra crassa* appears ornamental with spiral chloroplasts occupying the inside of the filament.

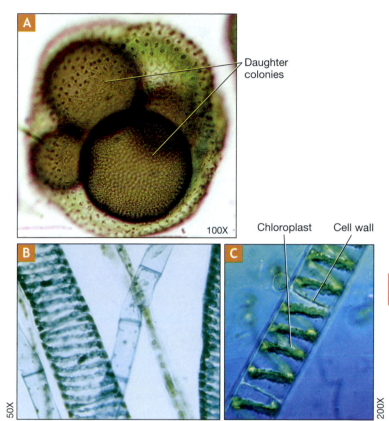

FIGURE **12.2** Examples from the phylum Chlorophyta: (**A**) *Volvox* sp.; (**B**) *Spirogyra* sp.; (**C**) magnified view of a single filament.

Procedure 1

Observing *Volvox*

Materials
- ❏ Compound microscope
- ❏ Prepared slides of *Volvox*
- ❏ Culture of *Volvox*
- ❏ Microscope slides and coverslips
- ❏ Eyedropper
- ❏ Forceps
- ❏ Paper towels
- ❏ Colored pencils

Volvox is a common colonial freshwater alga found in ponds, ditches, and puddles (Fig. 12.2A). The colony is composed of as many as 50,000 individual flagellated cells that form a hollow sphere. **Protoplasmates** (the substance of living cells, or strands of cytoplasm), connect adjacent cells. Each cell has an **eyespot** that helps locate the light necessary for photosynthesis.

Some colonies are asexual, composed of nonreproducing vegetative cells and gonidia that produce new **daughter colonies**. *Volvox* is capable of sexual reproduction, with male colonies producing and releasing microgametes and female colonies producing female sex cells, or macrogametes.

1 Procure the equipment and specimens.

2 Using the compound microscope and a prepared slide of *Volvox*, describe and sketch *Volvox* in the space provided.

3 Using the compound microscope, prepare a wet mount of *Volvox* from the provided culture. Note your observations and sketches in the space provided.

4 Make sure to sketch and label the following if you can see them: protoplasmate, eyespot, daughter colony.

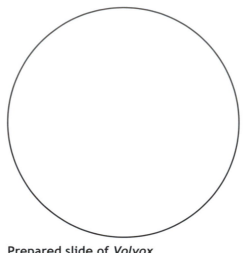

Prepared slide of *Volvox*

Total magnification _____

Wet mount of *Volvox*

Total magnification _____

Procedure 2

Observing *Spirogyra*

Materials

❏ Compound microscope
❏ Prepared slides of *Spirogyra* and conjugating *Spirogyra*
❏ Culture of *Spirogyra*
❏ Microscope slides and coverslips
❏ Forceps
❏ Eyedropper
❏ Paper towels
❏ Colored pencils

Spirogyra, sometimes called watersilk, is a filamentous freshwater species that possesses ornate spiral chloroplasts (Fig. 12.3). *Spirogyra* is capable of a type of sexual reproduction known as **conjugation**. During this process, a conjugation tube forms between adjacent filaments. **Protoplasts** (protoplasm of cell with cell wall removed) from one tube migrate via the tube to another filament to fuse with a waiting protoplast. The mobile protoplast is considered male, and the stationary protoplast is considered female. A thick-walled zygote results. Eventually the zygote undergoes meiosis, forming four haploid cells. Three disintegrate, and the remaining cell forms a new *Spirogyra* filament.

1 Procure the equipment and specimens.

2 Using the compound microscope and a prepared slide of *Spirogyra*, describe and sketch *Spirogyra* in the space provided.

3 Using the compound microscope and a prepared slide of *Spirogyra* undergoing conjugation, describe and sketch *Spirogyra* conjugation in the space provided.

4 Using the compound microscope, prepare a wet mount of *Spirogyra* from the provided culture. Note your observations and sketches in the space provided.

5 Make sure to sketch and label the following if you can see them: protoplast, chloroplast, pyrenoid.

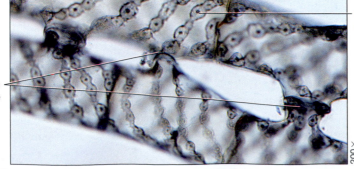

Pyrenoid in chloroplast

Conjugation tube

200×

FIGURE **12.3** Filaments of *Spirogyra* sp. showing initial contact of conjugation tubes.

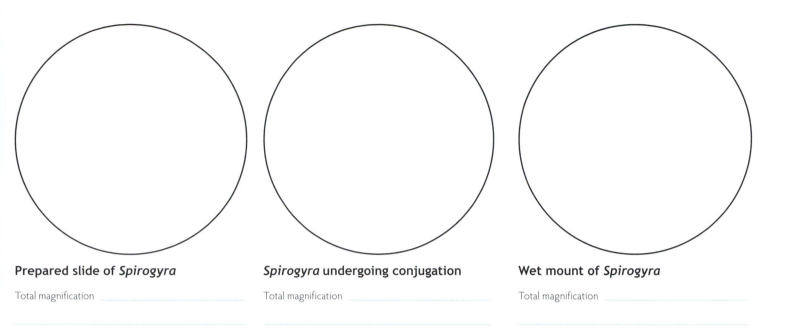

Prepared slide of *Spirogyra*

Total magnification _____

Spirogyra undergoing conjugation

Total magnification _____

Wet mount of *Spirogyra*

Total magnification _____

Diatoms (SAR Supergroup)

Phylum **Bacillariophyta** (in the major group Stramenopiles) consists of uniquely shaped algae called **diatoms** that possess a **test**, or shell, made up of two halves. These organisms are the most numerous unicellular algae found in marine and freshwater environments. There are more than 10,000 species, and some biologists estimate a million species. Diatoms are exquisite organisms, appearing in a variety of geometrical shapes.

A large component of the test is silica (SiO_2), a major component of glass incorporated into an organic mesh. Because their test is composed of silica, the fossil record of diatoms is well represented. They make up diatomaceous earth used in silverware polish, insulation, swimming pool filters, reflective paint, and even toothpaste. Some deposits of diatoms are more than 3,000 feet thick. A common diatom is *Cytotella stelligera*.

Procedure 3

Observing Diatoms

Materials

- ❏ Compound microscope
- ❏ Prepared slides of diatoms
- ❏ Culture of diatoms
- ❏ Microscope slides and coverslips
- ❏ Eyedropper
- ❏ Paper towels
- ❏ Forceps
- ❏ Colored pencils

Diatoms are unicellular algae that possess a unique silica cast composed of silicon dioxide and an organic matrix. The majority of diatoms live in aquatic and marine environments. Millions of diatoms may exist in a liter of water. Diatoms are known for their beauty (Fig. 12.4).

1 Procure the equipment and specimens.

2 Prepare a wet mount of diatoms from the culture provided.

3 Using the compound microscope, compare your slide with the prepared slide of diatoms. Note your observations and sketches in the space provided.

4 Make sure to sketch and label the following if you can see them: test, chloroplast, striae.

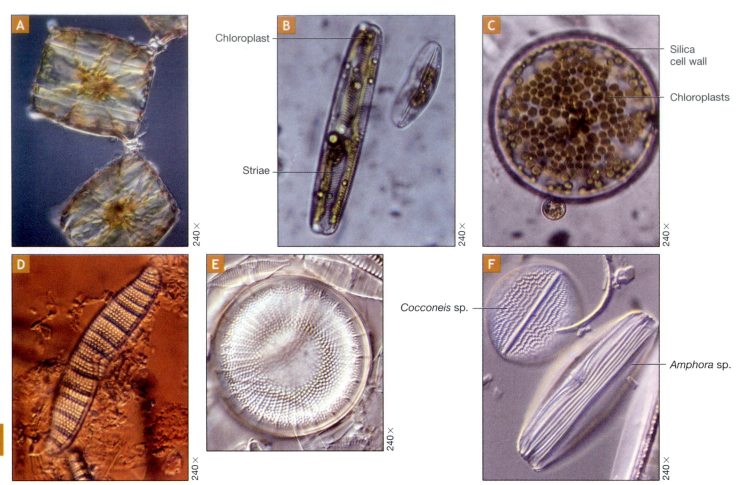

Chloroplast

Striae

Silica cell wall

Chloroplasts

Cocconeis sp.

Amphora sp.

240×

12

FIGURE **12.4** Several examples of diatoms: (**A**) *Biddulphia* sp., a colony forming colonies; (**B**) live specimens of pennate (bilaterally symmetrical) diatoms: *Navicula* sp. (left) and *Cymbella* sp. (right); (**C**) *Hyalodiscus* sp., a centric (radially symmetrical) diatom from a freshwater spring in Nevada; (**D**) *Epithemia* sp., a distinctive pennate freshwater diatom; (**E**) *Stephanodiscus* sp., a centric diatom; (**F**) two common freshwater diatoms.

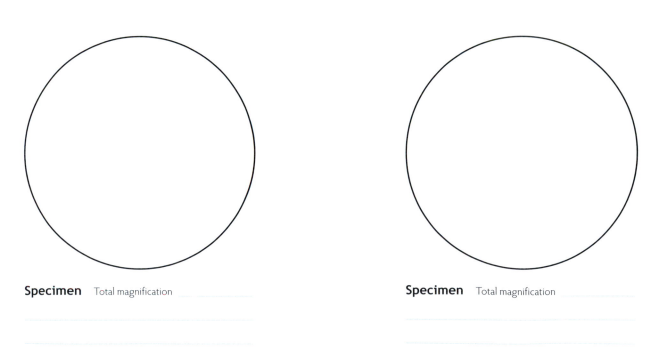

Specimen Total magnification

Specimen Total magnification

Euglenozoans (Excavata)

Phylum **Euglenophyta** (in the major group Excavata) includes approximately 1,000 species of unicellular flagellated freshwater species. Nearly 40 genera of euglenozoans have been described (Fig. 12.5). One-third of these genera possess chloroplasts, and the other two-thirds do not have chloroplasts. These nonphotosynthetic euglenozoans gather food by particle feeding and absorption, and we will discuss them in the exercise on animallike protists.

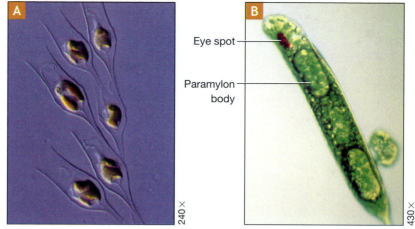

240×

430×

FIGURE **12.5** Examples from the phylum Euglenophyta include (**A**) *Dinobryon divergens* sp.; (**B**) *Euglena* sp.

Procedure 4

Observing *Euglena*

Materials

☐ Compound microscope
☐ Prepared slide of *Euglena*
☐ Culture of *Euglena*
☐ Microscope slides and coverslips
☐ Eyedropper
☐ Paper towels
☐ Forceps
☐ Colored pencils

Members of the genus *Euglena* are unicellular algae found primarily in freshwater environments (Fig. 12.6). They are capable of photosynthesis and can capture their own food—which presents a paradox for taxonomists. Chloroplasts serve as the site of photosynthesis. They are characterized by having a **pellicle** (helical protein bands that extend along the length of the cell beneath the plasma membrane) rather than a rigid cell wall. In addition, individuals possess two **flagella** (a short flagellum and a long flagellum used for locomotion; a paramylon body for starch storage; a gullet, through which food can be ingested; and a red stigma, or eyespot (aids in light detection). Some of these organisms are considered **mixotrophs** because they can undergo photosynthesis and can capture their own food.

1 Procure the equipment and specimens.

2 Using the compound microscope and a prepared slide of *Euglena*, describe and sketch *Euglena* in the space provided.

3 Using the compound microscope, prepare a wet mount of *Euglena* from the culture provided. Place your observations and sketches below.

4 Make sure to sketch and label the following in your drawings: pellicle, flagella, gullet, stigma (eyespot).

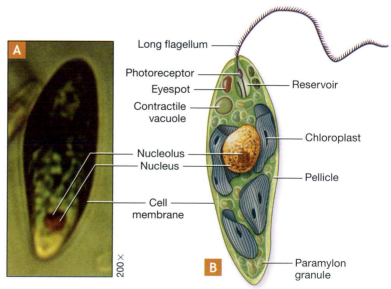

200×

FIGURE **12.6** Species of *Euglena* from a brackish lake in New Mexico: (**A**) micrograph; (**B**) illustration.

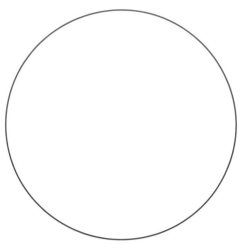

Prepared slide of *Euglena*

Total magnification _____

_____

_____

Wet mount of *Euglena*

Total magnification _____

_____

_____

✓ Check Your Understanding

1.1 Describe the movement of a *Volvox colony* and the unicellular *Euglena*. What structure(s) found on the *Euglena* is used for locomotion?

_____

_____

1.2 What is a mixotroph? Give an example of a plantlike protist that is considered a mixotroph.

_____

_____

1.3 Despite the fact that the various protists observed in this exercise belong to different eukaryotic lineages, what makes them all "plantlike" in nature?

_____

_____

_____

_____

Fungus-Like Protists

Once considered fungi, the **slime molds** are now considered protists. Fungus-like protists are decomposers in forests and woodlands. Another member of this group, the water molds, serves as decomposers in aquatic environments. Slime molds are motile organisms and produce spores on sporangia that in turn constitute fruiting bodies. The slime molds display complex life cycles in which they undergo remarkable morphological changes. Some slime molds measure 1 meter or more in diameter.

Phylum **Myxomycota** (in the supergroup Unikonta) consists of about 500 species known as the **plasmodial slime molds**. These organisms exist as a multinucleate mass covered in a sheath of slime. Many are colored bright yellow or orange and resemble a giant amoeba sliming ever so slowly across the forest floor. *Physarum* spp. is a common slime mold found in the woodlands of North America.

The **cellular slime molds** have been placed in phylum **Acrasiomycota** (also in the supergroup Unikonta). These organisms exist as solitary amoeboid cells in the soil. During adverse conditions, they aggregate, forming a mass called a pseudoplasmodium. This temporary stage gives rise to fruiting bodies that produce spores. *Dictyostelium discoideum* is a common cellular slime mold.

Water molds are placed in phylum **Oomycota** (in the major group Stramenopiles). They can be easily observed as a cotton-like filamentous mass growing on dying or dead fish. *Saprolegnia* spp. is a common water mold that helps decompose aquatic animals. The water molds have a cell wall composed of cellulose, unlike the fungi, which have a cell wall composed of chitin. One species, *Phytophthora infestans*, was responsible for the Irish potato famine in the 1840s. This organism had a direct impact on U.S. history because it forced a large number of Irish people to immigrate to the United States.

Procedure 1

Observing Plasmodial Slime Molds

Materials

- Sterile petri dish containing nutrient agar
- Fresh oatmeal flakes
- Culture of *Physarum polycephalum*
- Sterile forceps
- Sterile scalpel
- Appropriate marker for labeling petri dish
- Dissecting microscope
- Colored pencils
- Wax pencil

FIGURE **12.7** *Physarum* sp., a slime mold.

You will grow the plasmodium slime mold *Physarum polycephalum* on a petri dish with agar (growth media) and oatmeal flakes. Over a period of a week, you will be able to observe the active plasmodial stage of the slime mold, including protoplasmic streaming.

When nutrients (growth media/oatmeal flakes) are no longer available, the slime mold will enter into the sclerotia stage (Fig. 12.7). This is the inactive, or dormant, stage, and the mold can remain in this state for years. If nutrients and proper environmental conditions recur, fruiting bodies form within 12 hours, and sporulation will occur.

1. Procure the materials from your instructor.

2. Using a sterile scalpel, carefully cut a tiny cube of the *Physarum* culture from the stock dish and place it upside down on your sterile petri dish containing agar.

3. Using the sterile forceps, place two oatmeal flakes 1 in. away from the *Physarum* culture.

4. Place the lid back onto the dish, and with a wax pencil label your name/class section on a corner of the lid.

5. Each day, add two or three more oatmeal flakes, following the same procedure as in step 3.

6. Observe and record the growth and streaming of the plasmodium. Describe the growth of the slime mold over a period of a week. What color is the plasmodium? Describe the morphology of the slime mold.

_____

_____

_____

7. After viewing it with a dissecting microscope, describe the macroanatomy of your specimen. Record your observations in the space next to each specimen.

Physarum (First Day)

Physarum (Several Days Later)

Physarum (Week Later)

8 Clean up and dispose of your materials according to the instructor.

✔ Check Your Understanding

2.1 What is the significance of *Phytophthora infestans* to history?

2.2 Where can one find slime molds?

Did you know . . .

Did You Say "Dog Vomit?"

The plasmodium slime mold *Fuligo septica* is placed in the order Myxomycota. Even though it is bright yellow or orange in appearance, it gets the name "dog vomit slime mold" because it looks like a slimy mass of, you guessed it, dog vomit. *Fuligo* can appear almost overnight and can often be found living on bark mulch in urban areas after a heavy rain or excessive watering. It also readily grows on rotten wood and plant litter. *Fuligo* has a life cycle similar to other plasmodial slime molds.

The main vegetative phase of a plasmodial slime mold is the plasmodium, which serves as the active, streaming form of a plasmodial slime mold. The plasmodium form consists of intertwined networks of protoplasmic structures with many nuclei. During the plasmodial stage, the slime mold searches for food by creeping through the leaf litter. The plasmodium surrounds its food through phagocytosis and secretes digestive enzymes. Unfavorable dry environmental conditions cause the plasmodium to dry out and form a dry, multinucleate, dormant structure called a sclerotium. This stage can last a long time.

When favorable conditions return, the plasmodium reappears to continue feeding. As the food supplies dwindle, the plasmodium stops feeding and begins the reproductive phase of the life cycle. Stalks of sporangia form from the plasmodium. In the sporangia, meiosis occurs, and spores eventually form. Spores are released and spread by wind currents. Despite its very unpleasant name, *Fuligo* is completely harmless to humans, animals, and plants. In rare cases it may be linked to asthma problems.

EXERCISE 12.3

Animallike (Heterotrophic) Protists

Remember the first time you examined ditch water with a microscope? Wow! It was like viewing a miniature version of *Star Wars* with all of the tiny organisms gliding around in a thin film of water. Antony van Leeuwenhoek shared your fascination centuries before, terming these organisms "animacules and cavorting wee beasties." Today we know that many of Leeuwenhoek's wee beasties actually were animallike protists sometimes called protozoans (Fig. 12.8).

Classically, the animallike protists are divided into four categories:

1. Amoeboids (those that move by false feet);
2. Flagellates (those that move by flagella);
3. Ciliates (those that move by cilia); and
4. Apicomplexans (nonmotile forms).

Amoeboids (Unikonta and SAR Supergroups)

The animallike protist called the amoeba was first described in 1757 by German naturalist August Johann Rösel von Rosenhof (1705–1759). *Amoeba proteus* is a relatively common animallike protist that inhabits aquatic environments. Early naturalists called the amoeba *Proteus animalcule* after the Greek god Proteus, who could change his shape. The ability of some of the animallike protists to move via false feet, or **pseudopodia** (used for locomotion and for capturing food), makes them the ultimate shape-shifters of the protist world. In the amoebae, pseudopods are lobe-shaped and tipped by a **hyaline cap**, a clear space at the leading edge of the pseudopod.

Amoebas

Members of this phylum are found primarily in freshwater and marine environments, although several parasitic species exist. Classically, amoeboid protists, such as the common freshwater amoeba (*Amoeba proteus*), have been placed in phylum **Sarcodina** (**Rhizopoda**), which is in the supergroup Unikonta. The sarcodines do not have a wall or pellicle surrounding their plasma membrane. The majority of sarcodines reproduce asexually through fission. One species, *Entamoeba histolytica*, is an intestinal parasite responsible for amoebic dysentery spread in contaminated food.

In addition to locomotion, the pseudopodia are important in helping the amoeba surround its food through **phagocytosis**. The amoeba is surrounded by a plasma membrane that possesses two distinct parts. The **ectoplasm** is a thin, clear, nongranular region of cytoplasm directly beneath the plasma membrane, and the endoplasm is a granular region of cytoplasm that makes up the majority of the amoeba.

Food vacuoles (phagosomes) are also apparent in the amoeba (Fig. 12.9). They serve to hold food particles and fuse with a lysosome that

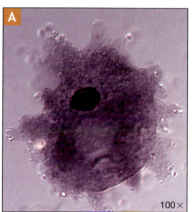 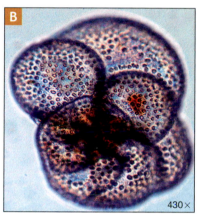 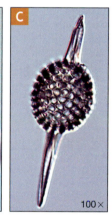

FIGURE **12.8** Animallike protists include: (**A**) *Amoeba proteus*; (**B**) *Globigerina* sp.; (**C**) *Stylatractus* sp.

Did you know . . .

Straight from Science Fiction!

Some science fiction movies feature a critter that eats the brain of its host. In the biological world, we don't have to look far to find a "brain-eating amoeba." *Naegleria fowleri* is an amoeba that inhabits the warm freshwaters of many southern states. It is responsible for the rare but fatal brain disorder primary amebic meningoencephalitis (PAM). It infects humans by entering the body through the nose while they are engaged in activities such as swimming or diving.

Naegleria travels up the nose to the brain and spinal cord, where it destroys nervous tissue. Early symptoms of PAM begin 1 to 14 days after *Naegleria* enters the brain and may include headache, fever, nausea, vomiting, and a stiff neck. As the disease progresses, symptoms include confusion, loss of balance, seizures, and hallucinations. After the start of symptoms, the disease progresses rapidly and usually causes death within 3 to 7 days.

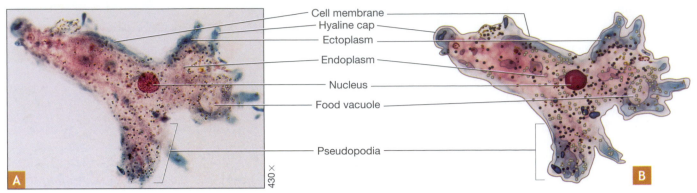

Figure **12.9** *Amoeba proteus:* (**A**) micrograph; (**B**) illustration.

Labels (top to bottom): Cell membrane, Hyaline cap, Ectoplasm, Endoplasm, Nucleus, Food vacuole, Pseudopodia

430×

aids in digesting the food. Waste is usually eliminated by exocytosis. The nucleus, or in some species several nuclei, is easily seen in amoebae. Many times the smaller nucleolus is also apparent within the nucleus. Occasionally, a **contractile vacuole** can be seen within an amoeba. It is important in osmoregulation.

Foraminiferans

Phylum Foraminifera (in the major group Rhizaria) is an intriguing group of aquatic and marine amoeboid protists with elaborate, colorful shells composed of calcium carbonate. These organisms have threadlike branched pseudopods that extrude from pores in the shell. Foraminiferans are found in tremendous numbers in the oceans. It is estimated about one-third of the sea floor consists of the shells, or casts, from dead foraminiferans, such as *Globigerina* sp. Fossil foraminiferans, or forams, are responsible for forming limestone. The white cliffs of Dover are composed primarily of the remains of foraminiferans.

Radiolarians

The radiolarians (**phylum Actinopoda** in the major group Rhizaria) possess an endoskeleton composed of silicon dioxide. Radiolarians, with their thin, relatively stiff pseudopods, are found in the marine environment. Many consider the radiolarians, with their geometric endoskeletons, the most beautiful organisms on Earth. Some radiolarians, such as *Lychnaspis miranda*, resemble snowflakes. In some parts of the ocean, radiolarian sediment on the floor can be more than 4,000 meters thick. Fossil foraminiferans and radiolarians such as *Stylatractus* sp. are helpful in correlating the age of various geologic strata.

Procedure 1
Observing Amoebas and Amoeba-Like Protists

Materials
- ❏ Compound microscope
- ❏ Prepared slide of *Amoeba* sp.
- ❏ Culture of *Amoeba* sp.
- ❏ Prepared slide of foraminiferans
- ❏ Prepared slide of radiolarians
- ❏ Prepared slide of *Entamoeba histolytica*
- ❏ Microscope slides and coverslips
- ❏ Eyedropper
- ❏ Paper towels
- ❏ Forceps
- ❏ Colored pencils

In this procedure, you will observe and record the basic anatomy of several animallike protists from prepared slides and live mounts. You should record your observations in the spaces provided and answer the questions in each section. Then, while referring to the descriptions of *Entamoeba histolytica*, foraminiferans, and radiolarians, you will observe each and describe the natural history of these organisms (Fig. 12.10).

1 Procure a slide of the provided species of amoeba, and record which species is used in the space provided.

2 Observe the specimen using low and high power. Locate the pseudopods, plasma membrane, hyaline cap, ectoplasm, endoplasm, nucleus, nucleolus, food vacuoles, and contractile vacuole.

3 Describe, sketch, and label your prepared specimen in the space provided.

4 Make a wet mount from the culture containing amoebae. Identify, describe, and sketch your specimen in the space provided.

5 Procure the following prepared slides: foraminiferans, radiolarians, and *Entamoeba histolytica*.

12

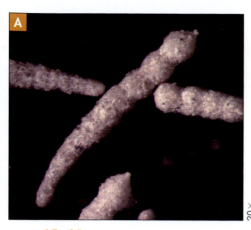

A

B

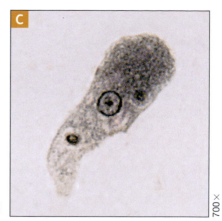

C

20×   100×   700×

FIGURE **12.10** Examples of some common amoeba-like organisms: (**A**) foraminiferan, *Arenaceous uniserial*; (**B**) radiolarian;
(**C**) *Entamoeba histolytica*.

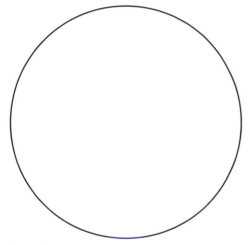

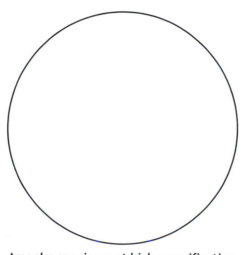

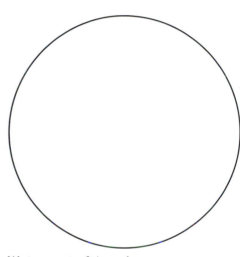

12

Amoeba specimen at low magnification

Total magnification _____

Amoeba specimen at high magnification

Total magnification _____

Wet mount of *Amoeba*

Total magnification _____

6 Observe the prepared slide of
foraminiferans on both low
and high power. Place your
observations and sketches in
the space provided.

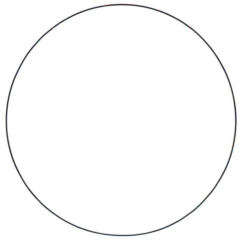

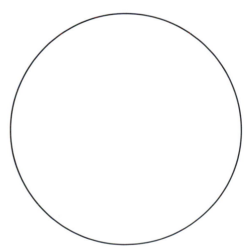

Foraminiferans at low magnification

Total magnification _____

Foraminiferans at high magnification

Total magnification _____

7 Observe the prepared slide of radiolarians on both low and high power. Place your observations and sketches in the space provided.

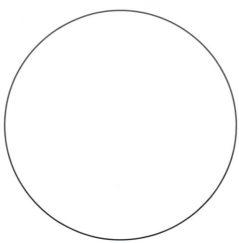

Radiolarians at low magnification

Total magnification _____

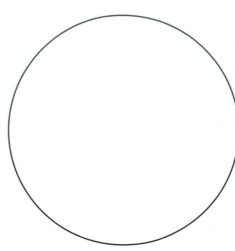

Radiolarians at high magnification

Total magnification _____

8 Observe the prepared slide of *Entamoeba histolytica* on both low and high power. Place your observations and sketches in the space provided.

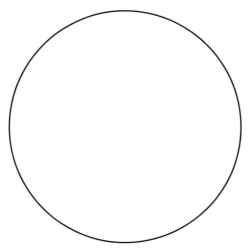

***Entamoeba histolytica* at low magnification**

Total magnification _____

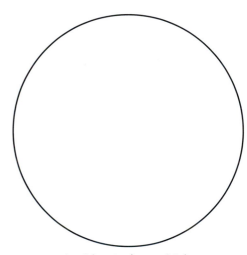

***Entamoeba histolytica* at high magnification**

Total magnification _____

9 Describe the natural history—the natural habitat of the organism, how it acquires nutrients and transforms energy, its unique structural features, type of movement, and its ecological role—of the organisms you observed.

a. *Amoeba* _____

b. Foraminiferans _____

c. Radiolarians _____

d. *Entamoeba histolytica* _____

Flagellates (Excavata and SAR Supergroup)

Perhaps the flagellated protists are the most confusing to taxonomists (Fig. 12.11). Classically, these are subdivided into the **phytomastigophorans** (photosynthetic flagellates) and the **zoomastigophorans** (animallike flagellates). For convenience, the well-known phylum **Zoomastigophora** (**Sarcomastigophora**) is used here to describe the animallike protists that move by flagella.

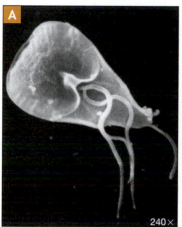 

FIGURE **12.11** Flagellated protists include: (**A**) *Giardia lamblia* (*intestinalis*); (**B**) *Gymnodinium* sp.

Dinoflagellates

Many biologists also consider the **dinoflagellates** to be zoomastigophorans, but several schemes place them in the plantlike protist phylum **Pyrrophyta**. The dinoflagellates are primarily marine, but a few freshwater species exist. These protists possess a faceted, hard, cellulose case and a single, long flagellum. Although some are capable of photosynthesis, most are heterotrophs.

Some, such as *Karenia brevis* and *Gonyaulax catenella*, are responsible for the devastating "red tide" in warm marine waters. *Gonyaulax* also can cause paralytic shellfish poisoning, which can be lethal to humans. *Pfiesteria piscicida*, living in the brackish water along the Atlantic coast, causes deadly infections in fish as well as humans. Discovered in 1988, *Pfiesteria* has a complex life cycle, with more than 20 body types. Symbiotic species (zooxanthellae) live in sponges, jellyfish, anemones, and coral and provide a supply of energy to their hosts. *Noctiluca scintillans* is a bioluminescent dinoflagellate that lights up marine waters at night.

Zooflagellates

The vast majority of zoomastigophorans are free-living, residing in freshwater, the marine environment, or in the soil. The species described here are well-known parasites that have been placed in the major group Excavata:

- The trypanosomes are protists responsible for a number of diseases in humans and other vertebrates. Trypanosomes are endoparasites transmitted by arthropod vectors, and they live in the blood plasma of their host, so they are called **hemoflagellates**. Trypanosomes are characterized anatomically as having an elongated shape with a flagellum supported by microtubules originating in the posterior region. *Trypanosoma brucei* is responsible for causing African sleeping sickness. The vector for this disease is the tsetse fly. *Trypanosoma cruzi* is the infective agent that causes Chagas disease in South and Central America. The vector is the triatomine, or kissing bug. Chagas disease kills approximately 50,000 people annually.

- Leishmaniasis is a serious infection transmitted by sand flies. *Leishmania donovani* causes "kala-azar," a lethal disease in India, China, Bangladesh, Ethiopia, and the Sudan. *Leishmania braziliensis*, a mucocutaneous form, causes grotesque facial deformities in people in South America. *Leishmania mexicana*, responsible for the chiclero ulcer or sores, is commonly contracted by unsuspecting tourists visiting Mexico, South America, and Central America.

- *Giardia lamblia (intestinalis)* causes giardiasis, an intestinal infection that causes intense diarrhea and dehydration. Giardiasis is a common waterborne infection worldwide. Many *Giardia* infections result from swimming in or drinking from contaminated waterways or even swimming pools. *Giardia* also is interesting in that it has modified mitochondria called mitosomes that do not generate ATP directly. In modern schemes, *Giardia* occupies its own kingdom known as **Diplomonadida**.

- Another unusual flagellate is *Trichomonas vaginalis*, a sexually transmitted parasite that causes urogenital infections. *Trichomonas* possesses modified mitochondria, termed hydrogenosomes, and is placed in its own kingdom called **Parabasala**.

- *Trichonympha* sp. is essential to termites. This flagellate, living in a symbiotic relationship with termites, helps them digest cellulose. If termites are available, trichonymphs can be easily observed in their gut.

12

Procedure 2

Observing Flagellates

Materials

- ❏ Compound microscope
- ❏ Prepared slide of *Trypanosoma* sp.
- ❏ Prepared slide dinoflagellates
- ❏ Prepared slide of *Giardia lamblia* (intestinalis)
- ❏ Termites
- ❏ Microscope slides and coverslips
- ❏ Colored pencils

In this procedure, you will observe representative flagellates from prepared slides and record your observations. Note that trypanosomes possess a kinetoplast (a mass of mitochondrial DNA close to the nucleus) near the kinetosome. A kinetosome is a self-duplicating structure at the base of the flagellum. It is derived from a large mitochondrion and forms the basis for the potentially new kingdom Kinetoplastea. In addition, an undulating membrane is prominent in trypanosomes (Fig. 12.12). *Giardia lamblia* (*intestinalis*) and *Trichonympha* sp. are other flagellates that you will observe in lab today (Figs. 12.13–12.15).

1   Procure a slide of the provided species of *Trypanosoma* sp., and in the space provided record which species is used. What disease is this specimen associated with?

2   Observe the specimen using low and high power. Locate the flagellum, undulating membrane, and nucleus. Draw and label your specimen in the space provided.

Trypanosoma sp.   Total magnification _____

_____

_____

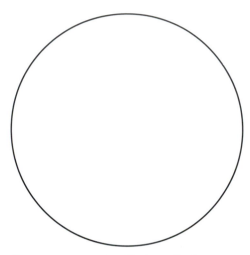

FIGURE **12.12** (**A**) Basic trypanosome anatomy; (**B**) trypanosomes in the blood plasma of a host.

Kinetoplast
Red blood cells
Nucleus   Kinetosome
Flagellum   Undulating membrane

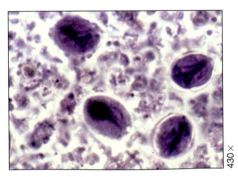

FIGURE **12.13** *Giardia lamblia* (*intestinalis*).

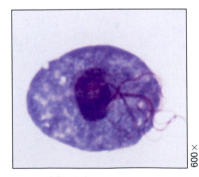

FIGURE **12.14** *Trichomonas vaginalis*.

FIGURE **12.15** *Trichonympha* sp.

3 Procure prepared slides of *dinoflagellates* and *Giardia lamblia (intestinalis)*. Observe the prepared slide of *dinoflagellates* on both low and high power. Make sure to locate and label the flagella and plates. Place your observations and sketches in the space provided.

4 Observe the prepared slide of *Giardia lamblia (intestinalis)* on both low and high power. Make sure to locate and label the nuclei, body, and flagella. Place your observations and sketches in the space provided.

5 Write a brief paragraph describing the natural history— the natural habitat of the organism, how it acquires nutrients and transforms energy, its unique structural features, type of movement, and its ecological role—of each organism you observed. This information will come in handy when answering Check Your Understanding question 4.1.

a. *Trypanosoma* sp. _____

_____

_____

b. Dinoflagellates _____

_____

_____

c. *Giardia lamblia (intestinalis)* _____

_____

_____

6 Procure the equipment and a termite from the instructor.

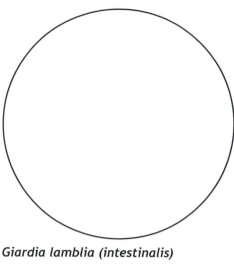

Giardia lamblia (intestinalis)

Total magnification _____

_____

_____

7 Place the termite on the center of a slide, and quickly place a coverslip over the termite. Carefully press down on the coverslip, squashing the termite.

8 Search through the remains of the termite. In the former gut region, you should see these unusual protists. Make sure to locate and label the body and flagella. Sketch *Trichonympha* sp., and place your observations in the space provided.

9 Clean your laboratory area, properly dispose of the slide and coverslip, and wash your hands thoroughly.

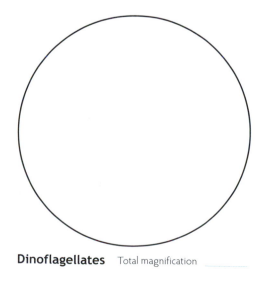

Dinoflagellates Total magnification _____

_____

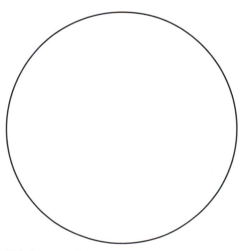

Trichonympha sp. Total magnification _____

_____

Ciliates (SAR Supergroup)

Phylum **Ciliophora** (in the major group Alveolata) consists of approximately 8,000 species of freshwater and marine organisms called **ciliates**. Their name is derived from the hairlike projections called **cilia** that cover their body surface or oral region. These structures are used in locomotion and feeding. The majority of ciliates are free-living, but there are several species of sessile (attached to a substrate), colonial, and parasitic forms.

The ciliates are considered the most structurally complex animallike protists. The outer covering, or pellicle, of ciliates is usually extremely tough. Most ciliates obtain their food using a **cytostome** (mouth) and an **oral groove**. The oral groove opens into a short canal, the cytopharynx, and ultimately into a region where food vacuoles are formed. The cytoproct, posterior to the oral groove, is responsible for elimination of fecal material. Contractile vacuoles are involved in osmoregulation. **Radiating canals** collect fluids and empty into the contractile vacuole.

Ciliates are multinucleate, possessing at least one **macronucleus** and one **micronucleus**. Macronuclei perform the basic nuclear duties, and the micronuclei are exchanged during conjugation, a form of sexual reproduction. Some ciliates possess trichocysts and toxicysts. Trichocysts lie beneath the pellicle and can be discharged like a harpoon for defense. Toxicysts are different from trichocysts in that they release a poison that can immobilize prey. Just think—if a paramecium were as big as a dolphin, it would be formidable!

The classic example of a free-living ciliate is *Paramecium caudatum*, a common inhabitant of freshwater environments. *Stentor* spp. is a sessile, vase-shaped ciliate found in the freshwater environment. Freshwater fishes may suffer from "ich" caused by the ciliate *Ichthyophthirius multifiliis* (Fig. 12.16).

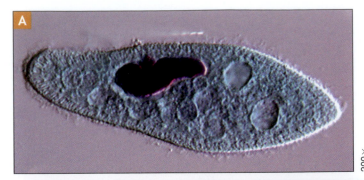

200×

FIGURE **12.16** Examples from phylum Ciliophora include: (**A**) *Paramecium caudatum*; (**B**) *Ichthyophthirius multifiliis*, known as "ich," seen here as small white specks on a freshwater fish.

Did you know ...

Distant Relatives?

The choanoflagellates are protists that might be the ancient ancestors of both kingdom Fungi and kingdom Animalia. Choanoflagellates can be single-celled or reside in colonies in freshwater and marine environments. Their basic anatomy is similar to the collar cells found in multicellular sponges. Choanoflagellates possess a single flagellum surrounded by a collar made of a number of cytoplasmic extensions called tentacles. They are placed in the supergroup Opisthokonta, which also includes fungi and animals.

Sphaeroeca sp., a choanoflagellate.

12

Procedure 3

Observing Ciliates

Materials

- Compound microscope
- Prepared slides of paramecia and conjugating paramecia
- Culture of *Paramecium* sp.
- Eyedropper
- Blank slides and coverslips
- Pipettes
- Vinegar
- India ink
- Toothpick
- Protoslo or methyl cellulose
- Probe or chemicals
- Colored pencils
- Paper towels
- Distilled water
- Forceps

In this procedure students will examine a common ciliate protozoan known as *Paramecium caudatum* (Fig. 12.17). These organisms are common in freshwater environments, such as ditches and ponds. Paramecia are more abundant in environments containing aquatic plants and decaying organic matter. The paramecium is slipper-shaped, transparent, and colorless. Paramecia appear highly active under the microscope, reaching speeds of 1 to 3 mm per second. These organisms have a pointed posterior end and a rather blunt anterior end. The oral groove is distinct, running backward to the ventral side. Paramecia can be observed using their cilia to channel food into the oral groove. The living pellicle is distinct and is covered with cilia.

In paramecia, the cilia near the oral groove create currents that bring water, and potentially food, into the oral groove. Trichocysts are used to gather food and for defense in paramecia. Contractile vacuoles in the paramecium act as an osmoregulatory apparatus, or bilge pump. Living

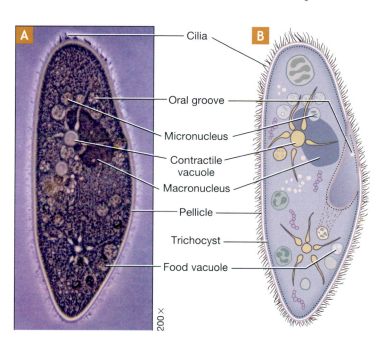

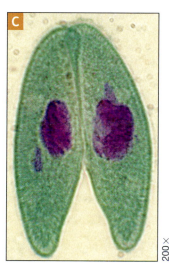

FIGURE **12.17** (A) *Paramecium caudatum*; (B) illustration of *Paramecium caudatum*; (C) *Paramecium* sp. in conjugation.

in a freshwater (hypotonic) environment, the organism has to remove excess water that diffuses through the cell membrane.

Stentor sp. and *Vorticella* sp. are two relatively common ciliates (Fig. 12.18). They can be found attached to a substrate in the freshwater environment. Many times *Vorticella* can be found living on the fins of fish.

1 Procure the equipment and material listed.

2 Examine the prepared slide of *Paramecium caudatum* under both low and high power. Draw and label your specimen in the space provided. Sketch and label the following structures if you can see them: pellicle, cilia, trichocyst, oral groove, macronucleus, micronucleus, and contractile vacuole.

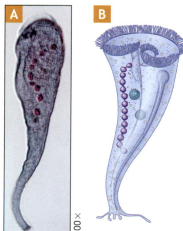

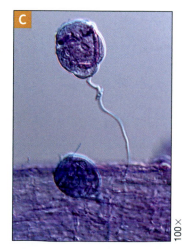

FIGURE **12.18** Common ciliates: (A) *Stentor* sp.; (B) illustration of *Stentor* sp.; and (C) *Vorticella* sp.

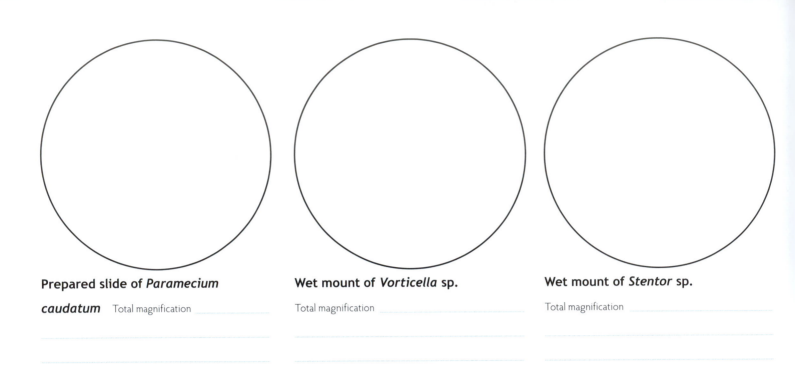

Prepared slide of *Paramecium*

caudatum   Total magnification _____

Wet mount of *Vorticella* sp.

Total magnification _____

Wet mount of *Stentor* sp.

Total magnification _____

3 Prepare a wet mount of paramecia, and describe their movement, the action of the contractile vacuole, and the organism in general. Paramecia are "speed demons," so Protoslo or methyl cellulose can be used to slow them down significantly.

4 Touch the end of toothpick into the India ink and then touch the toothpick near the oral groove of a paramecium slowed by Protoslo. Describe how the paramecium "tests" the water by drawing the ink into the oral groove.

5 Paramecia are sensitive to chemicals and objects in their environment. If a negative stimulus appears, the paramecium always turns to the left (the side away from the oral groove). Attempt to create a negative stimulus using a probe or chemicals, and describe the avoidance behavior in the paramecium.

6 Capture and place a single paramecium on a slide, and place a coverslip over the liquid. Then squeeze a small drop of vinegar along the edge of the coverslip. Describe in the space provided how the trichocysts react.

12

7 Count the number of times the contractile vacuole contracts in a slowed paramecium in its normal environment. After several trials, capture the paramecium with a pipette, and place it on a slide with distilled water. Count the number of times the contractile vacuole contracts in the distilled water. Capture another paramecium and repeat the process, placing the paramecium in a saltwater solution. Record your observations below, and explain why there is a difference in the number of contractions.

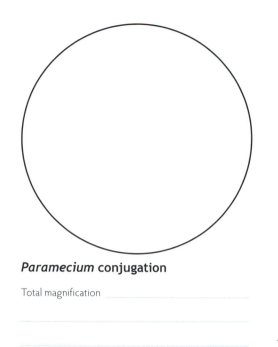

Paramecium **conjugation**

Total magnification _____

8 Paramecia are capable of conjugation, a crude means of exchange of genetic material between two individuals. Observe a prepared slide labeled *Paramecium* conjugation, and draw your specimen in the space provided.

9 Clean up your lab area, return the supplies, dispose of the material as instructed, and wash your hands.

12

✔ Check Your Understanding

3.1 Match the parasitic flagellate with the vector.

_____ Triatomine, or kissing bug

_____ Sand flies

_____ Tsetse fly

A. *Leishmania* sp.
B. *Trypanosoma cruzi*
C. *Trypanosoma brucei*

3.2 What is unique about *Giardia* and *Trichomonas*? How are they changing taxonomy?

3.3 Describe the action of the contractile vacuole of the *Paramecium* in a hypotonic solution.

3.4 Briefly describe the purpose of conjugation in paramecia.

3.5 Where on the cell are the conjugating *Paramecium caudatum* joined? (Refer to your drawings.)

3.6 Despite the fact that the various protists observed in this exercise belong to different eukaryotic lineages, what makes them all "animallike" in nature?

12

MYTHBUSTING

What a Mess!

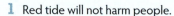

Debunk each of the following misconceptions by providing a scientific explanation. Write your answers on a separate sheet of paper.

1 Red tide will not harm people.

2 I would not worry about these so-called emerging viruses and other diseases!

3 Because cats can harbor *Toxoplasma*, if a woman is pregnant, she should give her cat away.

4 We should not worry about giardiasis in developed countries.

5 There are no pathogenic amoebas.

1 Traditionally, Protista has been called a kingdom, yet it is now recognized as a polyphyletic group. What is the likely future of the kingdom Protista?

2 How have the protists traditionally been classified?

3 Draw and label the anatomy of *Paramecium* spp. and *Euglena* sp.

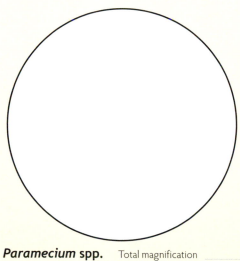

Paramecium spp.   Total magnification _____

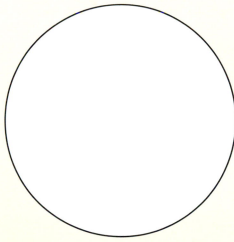

Euglena sp.   Total magnification _____

4 Which statement(s) or characteristic(s) is/are not true for the plantlike protists (algae)? (*Circle all correct answers.*)

a. They are a diverse group that includes diatoms to multicellular kelp.

b. They are photoautotrophs.

c. They have chlorophyll *a* and membrane-bound plastids.

d. Examples include *Spirogyra*, *Volvox*, *Polysiphonia*, and *Euglena*.

e. They do not undergo conjugation.

5 Identify the following as radiolarians, foraminiferans, or both, using the following key:

    R = Radiolarians   F = Foraminiferans   RF = Radiolarians and foraminiferans

_____ These have endoskeletons composed of silicon dioxide.

_____ *Globigerina* sp. is an example.

_____ These have colorful shells composed of calcium carbonate.

_____ Fossil species of these are helpful in correlating the age of geologic strata.

_____ These have threadlike, branched pseudopods.

_____ *Stylatractus* sp. is an example.

_____ These are fossil species that form limestone.

_____ Sediment of these organisms can be greater than 4,000 meters thick.

_____ These have thin, stiff pseudopods.

_____ These are found in marine environments.

6 What is a pseudopod?

_____

_____

7 Why should pregnant women be cautious of cats?

_____

_____

8 In the past, what were the main criteria by which protists were classified?

_____

_____

_____

9 Why is a mixotroph unique?

_____

_____

_____

10 What are the plans for classifying protists in the future?

_____

_____

_____

12

The Green Machine I
Understanding Basal Plants

13

*And Botany I rank with the most valuable sciences, whether we consider its subjects
as furnishing the principal subsistence of life to man and beast, delicious varieties for our tables,
refreshments from our orchards, the adornments of our flower-borders, shade and perfume of our groves,
materials for our buildings, or medicaments for our bodies.*

— Thomas Jefferson (1743–1826)

Just wondering...

*Consider the following questions prior to coming to lab,
and record your answers on a separate piece of paper.*

1 Are there any plants that have a parasitic lifestyle?

2 Are ferns commercially important?

3 What is the relationship between a seedless vascular plant
known as *Lycopodium* and the history of photography?

4 What was the world of ancient plants like during the
Carboniferous period?

5 What evolutionary adaptations helped ancient plants colonize
the land?

Objectives

*At the completion
of this chapter, the student
will be able to:*

1. Discuss the importance of
   and the origin of plants.

2. Compare and contrast
   nonvascular and vascular
   plants.

3. Explain the basic
   biology, life cycles, and
   reproduction of phyla
   Hepatophyta and
   Bryophyta.

4. Describe the process of
   alternation of generations.

5. State the characteristics
   and provide examples of
   nonvascular plants.

6. Describe how the vascular
   plants adapted to life on
   the land.

7. Discuss the functions of
   xylem, phloem, roots,
   stems, and leaves.

8. Explain reproduction in
   seedless vascular plants.

9. Name the characteristics
   and provide examples
   of phylum Lycophyta,
   Psilotophyta, Sphenophyta,
   and Pterophyta.

A wise man once declared, "We need the plants much more than they need us." This
statement is not only accurate but also meaningful. Have you ever stopped to thank
a plant? Think about it! Plants provide oxygen, food, shelter, shade, erosion control,
and commercial products for human uses, such as timber, medicine, and even the paper you
are looking at right now. In addition, many species of plants are aesthetically pleasing. As
Ralph Waldo Emerson said, "The Earth laughs with flowers." Remember—the next time you
are sitting under a majestic tree—say thank you!

Plants are a diverse group of eukaryotic, multicellular, photosynthetic autotrophs that
inhabit myriad environments from lush tropical rainforests to scorching deserts. Figure 13.1
is a cladogram of the plants. Biologists have identified approximately 300,000 species of
plants and estimate 400,000 species may exist. Plants vary in size from the smallest flowering
plant, *Wolffia angusta* (a duckweed), measuring less than 1 millimeter in diameter, to the
giant *Sequoia sempervirens*, which measures nearly 120 meters tall. The oldest plant in the
world is thought to be the King Holly (*Lomatia tasmanica*), a shrub that lives in Tasmania.
This remarkable plant is estimated to be more than 43,000 years old!

Paleobotanists believe plants evolved during the Paleozoic era from freshwater green
algae known as charophytes, approximately 450 million years ago. To make the transition
from the aquatic environment, ancestral plants had to evolve mechanisms that prevent
desiccation, anchor the plant body, transport water and nutrients, and ensure propagation
of the species.

The absence or presence of specialized conducting tissue known as **vascular tissue** is a
common way to distinguish plants. **Nonvascular plants** lack specialized conducting tissues to
transport water and nutrients throughout the plant's body. In addition, these plants lack true
roots, stems, and leaves. Examples of nonvascular plants are liverworts, hornworts, and true
mosses. Presently, about 25,000 species of nonvascular plants have been identified. Most

Chapter Photo
Water fern, or mosquito
fern, *Azolla* sp.

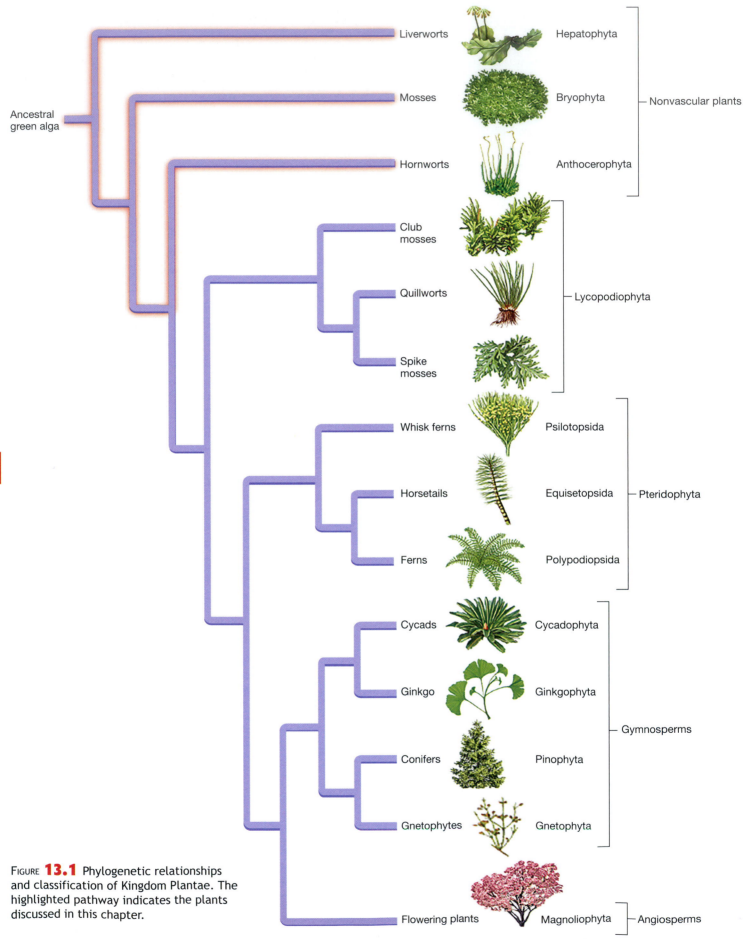

FIGURE **13.1** Phylogenetic relationships and classification of Kingdom Plantae. The highlighted pathway indicates the plants discussed in this chapter.

living plants are considered vascular plants because they possess an extensive conducting system composed of specialized tissues.

Vascular plants can be divided into the seedless vascular plants and the seed plants. The **seedless vascular plants** include the club mosses and the ferns. The **seed plants** are the largest group of vascular plants and include gymnosperms (such as ginkgos, cycads, and conifers) and angiosperms (flowering plants, such as zinnias and roses).

The life cycle of plants is characterized by an **alternation of generations** (Fig. 13.2). In this process, two distinct generations give rise to each other. The haploid (*n*) **gametophyte** generation is multicellular and is characterized by the production of male and female gametes through mitosis. The male and female gametes fuse during fertilization, forming a diploid **sporophyte**, which is also multicellular. The sporophyte generation is diploid (*2n*). It produces haploid spores that undergo mitosis to form a gametophyte. In nonvascular plants, the gametophyte generation is dominant, but in seedless vascular plants and seed plants, the sporophyte generation dominates. A fern, a pine tree, and a tulip are all examples of sporophytes.

■ Nonvascular Plants

In the history of plants on Earth, the nonvascular plants played an important role in establishing the transition from water to land and dominance of the gametophyte generation. The nonvascular plants are generally small and herbaceous (nonwoody). Although most of these plants are found in moist environments, some species can survive in arid environments. Three distinct phyla of nonvascular plants have been established:

1. Phylum Hepatophyta, the liverworts (not discussed in this chapter);
2. Phylum Anthocerophyta, the hornworts (not discussed in this chapter); and
3. Phylum Bryophyta, the true mosses.*

---

* The term *bryophyte* is often used to describe all of the nonvascular plants, even though it is the proper taxonomic name for the phylum in which the true mosses are classified.

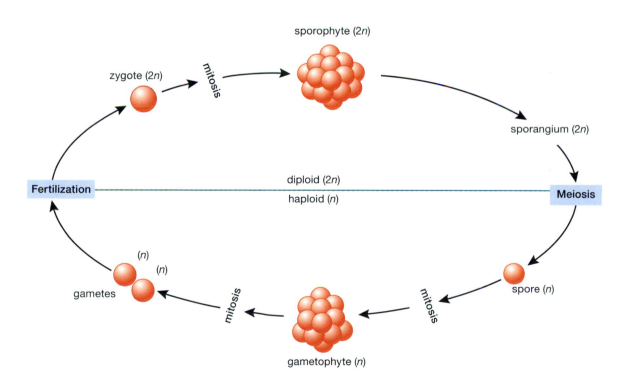

FIGURE **13.2** Model of alternation of generations.

Phylum Bryophyta

True mosses have been placed in phylum Bryophyta. Presently, nearly 15,000 species of mosses have been identified, but many small plants called mosses do not fall into this group. For instance, Spanish moss is actually a member of the pineapple family, reindeer moss is actually a lichen, Irish moss is a red algae, and club mosses are vascular plants.

The gametophyte stage of mosses consists of small, spirally arranged leaflike structures surrounding a central axis. The blades of the leaflike structure are one-cell-layer thick, lack vascular tissue and stomata, and surround a thickened midrib. Rhizoids anchor mosses to their substrate.

Mosses are capable of asexual reproduction through fragmentation, but they also undergo an alternation of generation with gametophyte and sporophyte stages. The "leafy" gametophytes are either male, bearing **antheridia**, or female, bearing **archegonia**. Flagellated sperm cells exit the antheridia and travel with the aid of water to the archegonia, where a single egg is fertilized. The zygote undergoes meiosis, forming spores housed in the sporophyte. The sporophyte appears as a tall stalk topped by a **sporangium**, or capsule. The calyptra protects the capsule. A foot connects the seta, or stalk, of the sporophyte to the leafy gametophyte. In some mosses, a single capsule may contain 50 million spores.

The tip of the capsule consists of a lid-like structure called the **operculum**. Teeth-like structures form the **peristome** (which means around the mouth) and lock the operculum to the capsule. During dry conditions, they unlock and allow spores to be carried by the wind. Through a hand lens, the peristome appears ornate and usually is orange or red. Immature spores land on a substrate and under suitable conditions develop into a filamentous **protonema**. The protonema eventually develops into the "leafy" gametophyte, and the cycle begins again (Fig. 13.3).

Most mosses live in moist environments of the temperate zone, although several species have been found living in the Arctic, Antarctic, and even deserts. Frequently, mosses can be seen growing on the trunks of trees and on the sides of buildings. After a fire or volcano, some mosses serve as pioneer species and help form soil.

Commercially, one of the most valuable species of moss is *Sphagnum*, or peat moss. In many regions, deposits of peat are mined for their use as packing material and fuel. Peat can absorb great amounts of water and is used in the gardening industry to enhance the water-holding capacity of soil and potted plants.

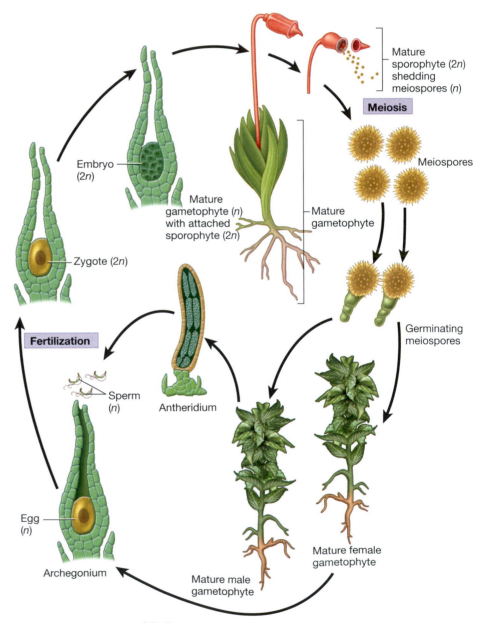

FIGURE **13.3** Life cycle of a moss (Bryophyta).

Labels in figure: Mature sporophyte (2n) shedding meiospores (n); Meiosis; Meiospores; Embryo (2n); Mature gametophyte (n) with attached sporophyte (2n); Mature gametophyte; Zygote (2n); Germinating meiospores; Fertilization; Sperm (n); Antheridium; Egg (n); Archegonium; Mature male gametophyte; Mature female gametophyte

Peat Moss in History

The use of peat moss, or *Sphagnum*, has been mentioned in folklore as far back as the 11th century. Native Americans used peat moss in diapers, and some ancient societies used it in menstrual pads. Medical texts in the 1800s mentioned that *Sphagnum* could be used as bandages or to pack abscesses. During the Russo-Japanese War from 1904 to 1905, *Sphagnum* was commonly used as bandage material. During World War I in a time of intense fighting in France, hospital staffs resorted to packing wounds and making bandages with *Sphagnum*. To their surprise, not only was the peat moss more absorbent than cotton but also the infection rate among the wounded was reduced significantly. The remarkable peat moss was used in World War II in a similar fashion. Peat moss has antibacterial properties as the result of the presence of phenol compounds, low pH, and the polysaccharide chitosan.

The moss *Sphagnum* was used as bandages during World War I.

Procedure 1

Macroanatomy of *Polytrichum*

Materials
- ❏ Dissecting microscope
- ❏ Living specimen of *Polytrichum*
- ❏ Colored pencils

Polytrichum sp., haircap moss, is a common moss found living in bogs. It is a relatively large moss, perhaps reaching 10 cm in length in certain environments. It is distinguished by having a tall sporangium with a golden calyptra sitting atop the seta and protecting the capsule (Fig. 13.4).

1 Procure the necessary supplies and equipment.

2 Using a dissecting microscope, draw and label the specimen of *Polytrichum*. Pay particular attention to the surface, rhizoids, stalk, calyptra, capsule, and reproductive structures. Place your labeled sketches and observations in the space provided.

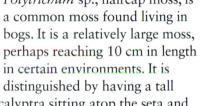

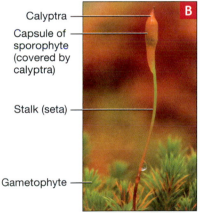

A

B
- Calyptra
- Capsule of sporophyte (covered by calyptra)
- Stalk (seta)
- Gametophyte

C
- Operculum
- Capsule of sporophyte (with calyptra absent)
- Stalk (seta)

FIGURE **13.4** (**A**) *Polytrichum* sp., a common moss often used in coursework; (**B**) gametophyte plants with sporophyte plant attached; (**C**) sporophyte plant and capsule.

Polytrichum

Sphagnum and Water Absorption

Materials

- ❏ 5 g of *Sphagnum*
- ❏ 5 g of a true sponge
- ❏ 5 g of an artificial sponge
- ❏ 5 g of a paper towel
- ❏ Scale
- ❏ 400 mL beakers (5)
- ❏ 100 mL graduated cylinder
- ❏ Wax pencil or Sharpie
- ❏ Timing device
- ❏ Tap water

Many gardeners use *Sphagnum* in their yards and plant containers because of its ability to absorb water (Fig. 13.5). In this procedure, students will compare the absorption rate of *Sphagnum* (peat moss) with equivalent masses of a true sponge, an artificial sponge, and a paper towel. When measuring the amount of water absorbed, 1 mL of water weighs 1 g. Write a hypothesis regarding this activity.

FIGURE **13.5** *Sphagnum* moss growing in the Pacific Northwest.

1. Weigh out 5 g portions each of *Sphagnum*, a true sponge, an artificial sponge, and a paper towel.

2. Label the beakers, one each: *Sphagnum*, true sponge, artificial sponge, and paper towel.

3. Add 250 mL of water to each of the labeled beakers.

4. Add to the appropriate beakers the 5 g portions of *Sphagnum*, a true sponge, an artificial sponge, and a paper towel.

5. Allow the materials to sit in the beakers for 2 min., and then remove the materials from each beaker.

6. Pour the water remaining in the beaker that contained Sphagnum into the 100 mL graduated cylinder, and record your measurement in Table 13.1. Empty the water from the graduated cylinder.

7. Repeat step 6 for the three remaining beakers of water.

8. Record your observations and discuss your findings in the space provided.

TABLE **13.1** Water Absorbency Results

Test Material	Water Remaining in Beaker (mL)	Water Absorbed by Test Material (mL)
Sphagnum		
True sponge		
Artificial sponge		
Paper towel		

✔Check Your Understanding

1.1 Describe the ways in which mosses reproduce asexually and sexually.

1.2 What unique characteristics made *Sphagnum* a successful bandage in World War I?

1.3 In the species *Polytrichum* and other mosses, the stalked capsule attached to the green "leafy" portion of the plant is the gametophyte stage. (*Circle the correct answer.*)

True / False

■ Seedless Vascular Plants

Imagine the world of the Silurian period 443 to 416 million years ago. The ancient oceans were teeming with life. Coral reefs were beginning to form, eurypterids (sea scorpions) cruised the seas, trilobites were abundant, and the ancestors of spiders and centipedes invaded the land. During this time, jawless fishes were common in the oceans, and fishes with jaws made their first appearance. Early in this period, the bryophytes lived near the water's edge, helping turn rock into soil. During the Silurian, plants took a giant leap toward conquering the land. Figure 13.6 is a cladogram of the plants.

Paleontologists have discovered that members of the extinct plant phylum Rhyniophyta, such as *Cooksonia*, had their humble origins during this time (Fig. 13.7). *Cooksonia*

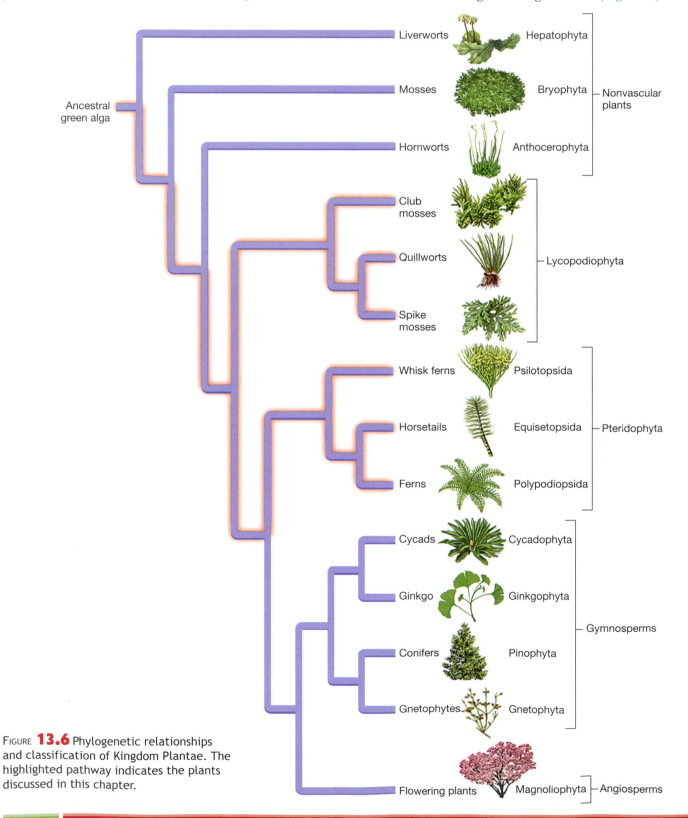

FIGURE **13.6** Phylogenetic relationships and classification of Kingdom Plantae. The highlighted pathway indicates the plants discussed in this chapter.

FIGURE **13.7** *Cooksonia* sp. is one of the oldest vascular land plants.

was a small plant only a few centimeters tall that possessed simple vascular tissues for conducting water and nutrients. It appeared as a branching stem with no roots or leaves. Spore-producing bodies known as sporangia appeared at the tips of the branches. The sporangia, like those of the bryophytes, produced only one type of spore, and thus were known as **homosporous**. *Cooksonia* and other primitive plants, such as *Rhynia*, established the foundations of the vascular plants.

Vascular plants possess specialized tissues for conducting water and nutrients throughout the plant. In modern vascular plants, **xylem** conducts water and dissolved minerals, and **phloem** conducts nutrients, such as sucrose, hormones, and other molecules. The cells responsible for conducting water are often strengthened by the polymer lignin, which allows the plant to grow tall. The central portion of the roots and stems of vascular plants has a **stele** composed of xylem and phloem.

Ancient vascular plants, such as lycophytes, possess a protostele in which a core of xylem is surrounded by phloem. More modern plants, such as sunflowers, possess a siphonostele in which a central spongy pith is surrounded by a ring of xylem and phloem. Other plants, such as conifers and the majority of flowering plants, have eusteles with xylem and phloem arranged into vascular bundles. In vascular plants, the sporophyte generation is dominant and possesses the vascular tissues. Vascular plants have a waxy **cuticle** for protection against desiccation as well as small openings called **stomata** on photosynthetic structures to allow for gas exchange. Vascular plants also possess true roots, stems, and leaves:

1. Roots are plant organs that absorb water and nutrients from the soil and anchor a plant.

2. Stems are vascular plant organs that support leaves and reproductive structures.

3. Leaves are the primary photosynthetic organs of plants.

The two major types of vascular plants are the seedless vascular plants and the seed plants. The seedless vascular plants, as the name indicates, do not produce seeds but instead reproduce through the production of spores like their ancestors. The seedless vascular plants dominated the landscape during the Devonian and Carboniferous periods. Unlike the relatively small seedless vascular plants of today, some species of seedless vascular plants, such as *Lepidodendron* (scale tree), reached heights exceeding 30 meters. The vast coal deposits of the Carboniferous period are the result of carbonization of seedless vascular plants that resided in giant swamp forests. Today, the seedless vascular plants are represented by the following phyla: Lycophyta (club mosses), Sphenophyta (horsetails), Psilotophyta (whisk ferns), and Pterophyta (ferns). Only phylum Pterophyta will be discussed in this chapter.

13

Phylum Pterophyta

Ferns placed in phylum Pterophyta (Polypodiophyta) are the most abundant seedless vascular plants. Most fern species live in moist, tropical regions of Earth, although some species reside in temperate regions as well as the Arctic Circle. Some ferns can even live in aquatic environments and dry areas. Ferns range in size from giant tropical tree ferns that can exceed 28 meters in height to *Azolla*, a diminutive aquatic fern that measures less than a centimeter in diameter (Fig. 13.8). Scientists believe ancestral ferns first appeared in the Devonian period approximately 375 million years ago. Today, plant taxonomists have identified approximately 11,000 species of ferns.

Ferns are cherished for their ornamental value. They are used for indoor as well as outdoor decorations and by florists to construct bouquets. The Environmental Protection Agency (EPA) suggests ferns are valuable in filtering formaldehyde and other toxins from the air. Fern rhizoids and fronds are foods in many cultures. Bracken fern fronds were used in the past to thatch roofs. Medicinally, ferns and their products have been used in the treatment of leprosy, parasitic worms, labor pains, sore throat, diabetes, dandruff, and many other maladies.

The leaves of ferns, known as **fronds**, arise from rhizomes. Immature fronds develop from the tip of a rhizome and appear as a tightly coiled and rolled-up structure called a **fiddlehead**. Compound frond ferns possess ornate leaflets, or **pinnae**. Simple frond ferns have leathery, broad, unbranched, strap-like fronds. The pinnae are attached to a midrib, sometimes called a rachis. A **petiole**, or stalk, attaches the pinnae to the rhizome. Branching roots also arise from the rhizomes.

The sporophyte stage is dominant in ferns. Most fern species are homosporous. The spores are produced in sporangia, appearing as distinct brown spots on the underside of the frond, called **sori**. To an untrained eye, the sori may appear like a fungus or insect eggs. In some species, the sori are protected by a colorless flap called an **indusium**. The **annulus**, a fuzzy region of the sorus, catapults the mature spores. Adder's tongue fern may produce more than 15,000 spores per sorus. The collective production of spores in some ferns exceeds 50 million.

The water fern *Marsilea* and several other species are heterosporous, producing spores in a **sporocarp**. Spores that

FIGURE 13.8 (**A**) Tree fern, *Cyathea* sp.; (**B**) aquatic fern, *Azolla* sp.

land in a favorable environment germinate, producing heart-shaped gametophytes, or **prothalli**. The gametophytes possess rhizoids that anchor them to their substrate. Flagellated sperm cells are produced in the antheridia of the gametophyte. The sperm are released and swim to the archegonium, where they fertilize the awaiting egg, forming a zygote. The zygote develops into the sporophyte generation, completing the life cycle of a fern (Figs. 13.9–13.13).

13

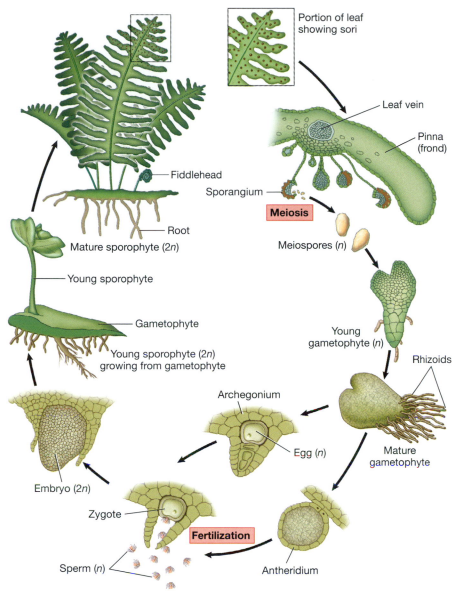

Fiddlehead

Root

Mature sporophyte (2n)

Young sporophyte

Gametophyte

Young sporophyte (2n) growing from gametophyte

Archegonium

Embryo (2n)

Zygote

Egg (n)

Sperm (n)

Fertilization

Antheridium

Portion of leaf showing sori

Leaf vein

Pinna (frond)

Sporangium

Meiosis

Meiospores (n)

Young gametophyte (n)

Rhizoids

Mature gametophyte

FIGURE **13.9** Life cycle of a fern.

FIGURE **13.10** (**A**) Compound and (**B**) simple fern leaf forming a fiddlehead.

FIGURE **13.11** Leaf of the fern *Polypodium virginianum*.

13

Pinna

Sori

FIGURE **13.12** Leaf of the fern *Polypodium virginianum* showing sori (groups of sporangia).

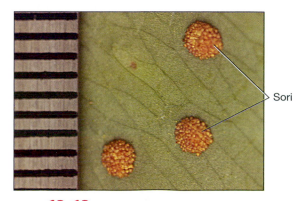

Sori

FIGURE **13.13** Closeup of the fern pinna of *Polypodium virginianum* (scale in mm).

Procedure 1

Macroanatomy of Ferns

Materials

- ❏ Dissecting microscope
- ❏ Living or preserved specimens of several species of ferns
- ❏ Scalpel
- ❏ Petri dishes or watch glasses (3)
- ❏ Tap water
- ❏ Dropper
- ❏ Microscope slides and coverslips
- ❏ Forceps
- ❏ Paper towels
- ❏ Acetone
- ❏ Ethyl alcohol
- ❏ Saltwater
- ❏ Distilled water
- ❏ Tap water
- ❏ Gloves
- ❏ Safety goggles
- ❏ Colored pencils

13

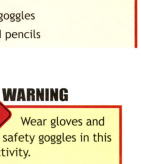

WARNING

Wear gloves and safety goggles in this activity.

1. Procure the necessary supplies and equipment.

2. Using a dissecting microscope, observe living or preserved fern specimens. Pay particular attention to the rhizome, roots, pinnae, midrib, petiole, sori, indusium, and annulus. Place your labeled sketches and observations in the space provided.

3. With the scalpel, remove a sorus from the underside of a frond. Place the sorus in a small amount of water in a petri dish or watch glass. Using a dissecting microscope, observe the anatomy of the sorus. Locate the sporangium, annulus, and indusium. Record your observations and labeled sketches in the space provided.

4. Using the scalpel, scrape a single sorus, dropping the spores into a drop of water on a blank slide. Cover your specimen with a coverslip. Record your observations and sketches in the space provided on the following page.

Specimen _____

Specimen _____

Specimen _____

Specimen _____

Specimen _____

_____

_____

Specimen _____

_____

_____

5 Remove another sorus from the pinnae with a scalpel, and place it in a petri dish or watch glass. Place 1 drop of acetone on the sorus, and observe it with a dissecting microscope. What happened?

_____

_____

6 Repeat step 5, using ethyl alcohol, saltwater, and distilled water. What happened?

_____

_____

_____

_____

7 Observe the fiddleheads of a fern, and place your observations and sketches below.

Specimen _____

_____

Specimen _____

_____

Bringing *Polypodium* Ferns Back to Life

Materials

- ❏ Dry *Polypodium* sp.
- ❏ Water
- ❏ Paper towels
- ❏ Plastic container

Consider photographing, using a webcam, or performing time-lapse photography in this activity.

1 Procure some living dry *Polypodium* sp. (Fig. 13.14).

2 Place the *Polypodium* sp. on a dry paper towel in a plastic container. Sprinkle the *Polypodium* sp. heavily with water.

3 Place the container in a safe place. Return the next day, and record your observations in the space provided.

4 If no results are noticed, repeat steps 2–3. Repeat up to three times if necessary.

FIGURE **13.14** Dry and green *Polypodium* sp.

Did you know . . .

Back to Life

Polypodium polypodioides is a common fern that grows on the trunks and branches of trees, particularly in the southeastern United States. An air fern, it receives its water and nutrients from the surface of the bark on the host plant. During periods of a long drought, the fronds turn brown and curl up, appearing dead. When they are exposed to water, they seem to return to life miraculously. Thus, this remarkable fern is called the *resurrection fern.*

✔ Check Your Understanding

2.1 What are fern sori and what is their function?

2.2 What is the common name of *Polypodium polypodioides*?

2.3 How does dried *Polypodium* respond to water in a plastic container and in nature?

2.4 Sketch and label the following structures on a fern frond: pinna and sori.

MYTHBUSTING

Ferns and Fronds

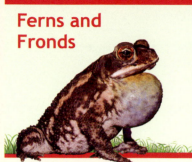

Debunk each of the following misconceptions by providing a scientific explanation. Write your answers on a separate sheet of paper.

1 Liverworts are good for liver ailments.

2 Spanish moss is a fine example of a true moss.

3 It is just a rumor that some ferns are poisonous.

4 Peat moss has no commercial value.

5 There were no giant seedless vascular plants in geologic history.

---

1 What are 10 reasons you should "thank a plant"? Think about the major impacts plants have on your life.

2 To conquer the land, what sort of adaptations did early land plants have to develop?

3 Which of the following statements correctly distinguish nonvascular and vascular plants? (*Circle the correct answer.*)

a. Nonvascular plants lack vascular tissue; vascular plants have specialized tissue for conducting water and nutrients within the plant.

b. Nonvascular plants lack true roots, stems, and leaves; vascular plants have true roots, stems, and leaves.

c. Vascular plants are divided into seeded and seedless plants with club mosses and ferns being examples of seeded plants; nonvascular plants are the liverworts, true mosses, and hornworts.

d. Both a and b.

e. All of the above.

4 What is true about nonvascular plants? (*Circle the correct answer.*)

a. They lack specialized conducting tissue for transport of water and nutrients.

b. They lack true roots, stems, and leaves.

c. The sporophyte generation is dominant.

d. Both a and b.

e. All of the above.

5 Distinguish between the sporophyte and the gametophyte generation in seedless vascular plants, making sure to discuss structures and ploidy.

_____

_____

_____

_____

6 Which of the following is not classified as a phylum of seedless vascular plants? (*Circle the correct answer.*)

a. Psilophyta.

b. Bryophyta.

c. Lycophyta.

d. Sphenophyta.

e. Both b and c.

7 Trace the life cycle of a typical fern in the space provided.

13

The Green Machine II
Understanding the Seed Plants

*For in the true nature of things, if we rightly consider, every green tree
is far more glorious than if it were made of gold and silver.*

— Martin Luther (1483–1546)

Just wondering . . .

*Consider the following questions prior to coming to lab,
and record your answers on a separate piece of paper.*

1 Are there any old wives' tales regarding plants that are true such as chewing willow bark will relieve a toothache?
2 Why does my hydrangea change colors?
3 What are five medically significant plants?
4 What is the main reason why sundews, pitcher plants, and other "carnivorous" plants consume insects?
5 What is catnip, and how does it work?

Objectives

*At the completion
of this chapter, the student
will be able to:*

1. Compare and contrast gymnosperms and angiosperms.
2. Explain the basic biology, natural history, and give examples of phylum Cycadophyta, Ginkgophyta, and Coniferophyta.
3. Distinguish springwood, summerwood, heartwood, and sapwood and determine a tree's age by annual rings.
4. Describe the evolution, organization, and fundamental characteristics of angiosperms.
5. Describe the life cycle of a typical angiosperm.
6. Compare and contrast basal dicots, eudicots, and monocots, and provide examples of each.
7. Define and describe the functions of a flower, a fruit, and a seed and describe flower and fruit types, providing examples of each.

Chapter Photo
Cross section of a pine limb, *Pinus* sp., filled with pine resin.

■ Gymnosperms

Just look around—most of the plants in the world are seed plants. When you are picnicking on soft grass, taking a stroll in the park admiring the ornamental azaleas and smelling the roses, walking through the majestic forest appreciating the splendor of the giant oaks and pines, and shopping for fruits and vegetables at the local market—seed plants are all around you. Today, the seed plants are the majority of plants on Earth, with an estimated 300,000 species.

The seed plants, **spermatophytes**, of today are divided into two major groups, the **gymnosperms** and the **angiosperms**. Figure 14.1 is a cladogram of the plants. The gymnosperm group consists of the following four living phyla:

1. Cycadophyta, the cycads and sago palms (not discussed in this chapter);
2. Ginkgophyta, only one living species, *Ginkgo biloba* (not discussed in this chapter);
3. Gnetophyta, three genera of plants including *Ephedra*, *Welwitschia*, and *Gnetum* (not discussed in this chapter); and
4. Coniferophyta, the largest phylum, including pine, spruce, sequoia, juniper, cedar, and cypress.

The angiosperms are flowering plants. Angiosperms constitute the majority of living plants on Earth, placed in phylum Magnoliophyta. Examples of angiosperms are duckweeds, cacti, oaks, grasses, tulips, sycamores, and magnolias.

Seed plants first appear in the fossil record in the late Devonian period, approximately 360 million years ago. The oldest known fossil is a "seed fern," *Elkinsia polymorpha* (Fig. 14.2). This ancient plant featured **ovules** (structures of seed plants containing the female sex cells with the potential to develop into seeds) at the tips of their slender branches. The tips formed cupules for the potential future development of seeds.

269

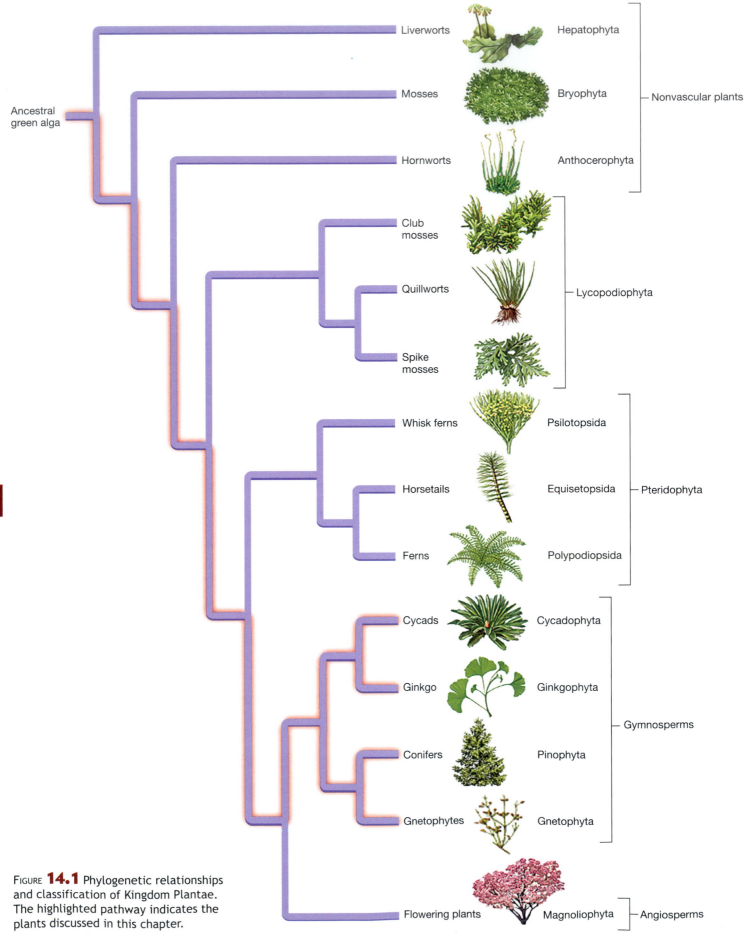

FIGURE **14.1** Phylogenetic relationships and classification of Kingdom Plantae. The highlighted pathway indicates the plants discussed in this chapter.

14

FIGURE **14.2** Fossil leaf of an Eocene gymnosperm.

These early seed plants, known as progymnosperms, did not have cones, flowers, or even seeds. They produced spores like ferns do. Another Devonian seed plant was *Archaeosperma arnoldii*, which possessed obvious cupules containing two ovules surrounded by prominent, clawlike appendages. Seed plants such as *Cordaites* continued to develop during the Carboniferous period but were overshadowed by the giant seedless vascular plants.

Paleobotanists agree that the gymnosperms, particularly the conifers, flourished in the Permian period. By the time of the Triassic period, all of the phyla of the seed plants were represented, with the exception of the flowering plants, the angiosperms. The flowering plants made their appearance about 140 million years ago, during the Cretaceous period, and became the dominant plants on Earth during the Paleocene epoch of the Cenozoic era, approximately 60 million years ago (see the inside back cover for a closer examination of the geologic time scale).

The seed plants feature a life cycle dominated by the sporophyte generation. Examples of this generation are the giant redwood and the tiny duckweed. The sporophyte produces two distinct types of gametophytes and is **heterosporous**. Multicellular male gametophytes (microspores) are called **pollen grains**. In nature, **pollination** occurs when pollen is carried from the male reproductive organs to the female reproductive organ in a number of ways, including wind, insects, and birds.

In seed plants a pollen tube forms, allowing the sperm in the pollen grain to unite with the egg in the ovule. The ovule is a sporangium enclosed by modified leaves called the **integument**. The fertilized egg becomes the embryo, and the ovule's integument forms a protective seed coat. The **seed** provides the embryonic plant essential nourishment and protection. Thus, the seed can withstand harsh conditions and stay dormant for many years.

Gymnosperm and angiosperm seeds are distinct (Fig. 14.3). The term *gymnosperm* literally means "naked seed." In these plants, the seeds are not enclosed in an ovule, and they mature on the surface of a cone scale, such as that of a pine cone. The nutritive material in gymnosperms accumulates prior to fertilization. In angiosperms, the nutritive material is stored only after fertilization. The gymnosperms generally lack flowers and fruits. The seeds of angiosperms are encased in a fruit. In both cases, the parental sporophyte generation provides nutrition to potential offspring, giving them a distinct advantage over those of seedless plants.

Evolution of the seed has changed the destiny of plants as well as humans. Seeds have allowed seed plants to become the dominant plants on Earth by allowing them to literally "get a head start" on life.

The gymnosperms first appear in the fossil record approximately 305 million years ago during the Carboniferous period. During the Mesozoic era, the gymnosperms dominated plant life. Although angiosperms dominate Earth today, the gymnosperms are still important plants in many ecosystems.

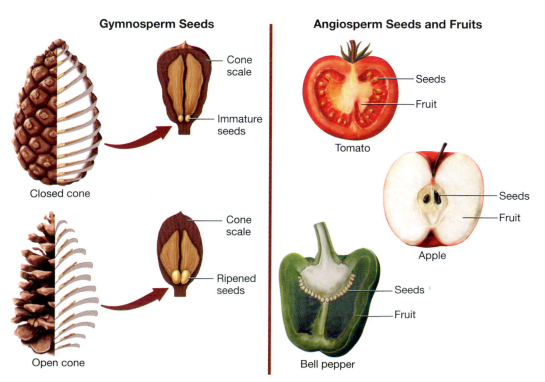

FIGURE **14.3** Comparison of gymnosperm and angiosperm seeds.

Phylum Coniferophyta (Pinophyta)

Pines, cypresses, spruces, redwoods, cedars, hemlocks, junipers, and yews are common gymnosperms placed in phylum Coniferophyta (Fig. 14.4). This phylum is composed of approximately 600 species of woody, mostly evergreen, cone-bearing plants most often found in cold and temperate climates.

Some conifers are record-setters. *Pinus longaeva*, the bristlecone pine, is the oldest known nonclone living organism on Earth. One specimen, "Methuselah," is nearly 5,000 years old. In California, a coastal redwood, *Sequoia sempervirens*, is the tallest tree in the world, measuring more than 110 meters in height. The "General Sherman," *Sequoiadendron giganteum*, a giant sequoia in California, measures more than 83 meters in height and has a circumference at the ground exceeding 31 meters. It is the largest tree on Earth by volume.

Some conifer species are sources of lumber, paper, wood alcohol, turpentine, and resin. Several species, such as juniper and yew, are ornamentals. In this regard, bonsai conifer plants are popular, along with spruce and fir Christmas trees. Oils from conifers are used in soaps and air fresheners. Humans eat some seeds, such as pine nuts. Several conifer products, such as taxol (a cancer treatment), are used in medicine.

In the seed cone, the ovules consist of **megasporangia** located on the upper surface of each of the individual scales. As the result of meiosis, four megaspores are produced in each megasporangium, but only one megaspore survives (Fig. 14.5). This megaspore undergoes mitosis and eventually forms a mature female gametophyte. The female gametophyte possesses between two and six **archegonia**. Each individual archegonium contains a large egg near the opening of the ovule.

Initial seed cones are small, scaly, and slightly opened to allow pollen to enter the ovule-bearing scales. Pollen enters the mature seed cone and forms sperm cells,

FIGURE **14.4** Various conifers: (**A**) bristlecone pine, *Pinus longaeva*; (**B**) bald cypress, *Taxodium distichum*; (**C**) Colorado blue spruce, *Picea pungens*.

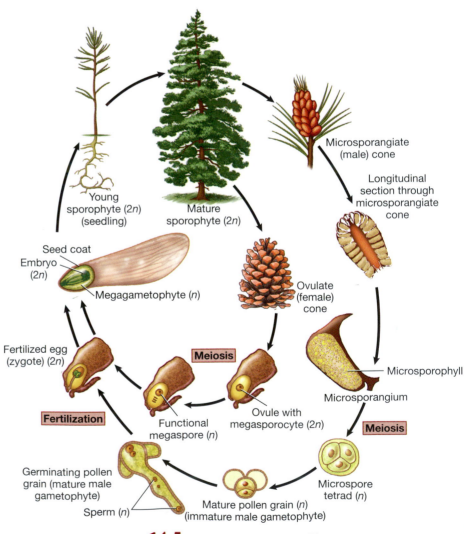

FIGURE **14.5** Life cycle of the pine, *Pinus* sp.

14

ultimately delivered to the waiting egg by means of a pollen tube. The egg is fertilized and forms a zygote. The seed cone closes and increases in size as the seeds develop.

Did you know . . .

Astonishing Numbers!

◆ Approximately 34.5 million Christmas trees are purchased in the United States, annually, costing more than 1.2 billion dollars!

◆ An average American uses about 500 pounds of paper annually.

Procedure 1

Macroanatomy of Conifers

Materials

❏ Dissecting microscope or hand lens
❏ Specimens of select conifers, such as pine, bald cypress, cedar, spruce, juniper, arborvitae, fir, and other specimens provided by the instructor
❏ Colored pencils

1 Procure needed equipment and supplies.

2 Observe the overall specimens, leaf arrangement, bark, cones, and distinguishing characteristics. Compare your observations with the images of conifers in this chapter. Place your observations and sketches in the space provided. Label as many structures as you can.

Specimen 1

Specimen 2

Specimen 3

Specimen 4

Specimen 5

Macroanatomy of Conifer Cones

Materials

- ❏ Dissecting microscope
- ❏ Colored pencils
- ❏ White paper
- ❏ Sterile microscope slides
- ❏ Pollen cones (male/staminate) from select conifers
- ❏ Seed cones (female/ovulate) from select conifers
- ❏ Seeds from select conifers

Many seed cones, such as those in pine, are woody (Figs. 14.6–14.10). Others, such as those in junipers, are fleshy. The seeds of conifers released from the seed cone are winged and require air dispersal. The dry, scaly, woody cones beneath a pine tree represent the spent seed cones. After a seed lands on a suitable substrate, it germinates and develops into a new sporophyte.

FIGURE **14.6** Images of the seed cones of conifers: (**A**) *Pinus* sp.; (**B**) *Abies* sp.; (**C**) *Picea* sp.; (**D**) *Taxodium* sp.; (**E**) *Taxus* sp.; (**F**) *Thuja* sp.

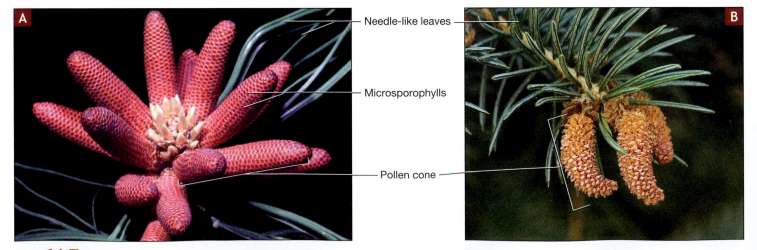

FIGURE **14.7** Microsporangiate cones of (**A**) *Pinus* sp. prior to the release of pollen; (**B**) *Picea pungens* after pollen has been released. The pollen cones are at the end of a branch.

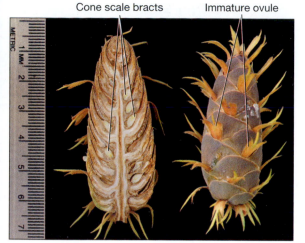

Cone scale bracts   Immature ovule

FIGURE **14.8** Transverse section through a first-year ovulate cone in *Pseudotsuga* sp.

1 Procure the needed equipment and supplies.

2 Examine pollen cones provided by the instructor, and record your observations in the space provided.

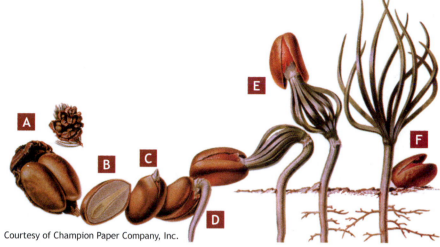

Courtesy of Champion Paper Company, Inc.

FIGURE **14.9** Pinyon pine seed germination producing a young sporophyte. (**A**) The seeds are protected inside the cone, two seeds formed on each scale. (**B**) A sectioned seed shows an embryo embedded in the female gametophyte tissue. (**C**) The growing embryo splits the shell of the seed, enabling the root to grow toward the soil. (**D**) As soon as the tiny root tip penetrates and anchors into the soil, water and nutrients are absorbed. (**E**) The cotyledons emerge from the seed coat and create a supply of chlorophyll. Now the sporophyte can manufacture its own food from water and nutrients in the soil and carbon dioxide in the air. (**F**) Growth occurs at the terminal buds at the base of the leaves.

3 Observe and compare seed cones from conifers provided by the instructor. Record your sketches and observations in the space provided.

4 If possible, collect and observe seeds from a seed cone (or the instructor will provide seeds for observation). Describe and sketch the seeds in the space provided on the following page, and note how they are distributed.

Seedling leaves (needles)

Young stem

Young roots

FIGURE **14.10** Young sporophyte (seedling) of a pine, *Pinus* sp. (scale in mm).

Specimen 1

Specimen 2

Specimen 3 _____
_____
_____

5 Drop a seed from *Pinus* sp. from
above your head. Describe the
action as it falls.

5 Drop a seed from *Pinus* sp. from
above your head. Describe the
action as it falls.

_____
_____
_____
_____
_____

Procedure 3

Conifer Seed Germination

Materials
- ❏ Pine seeds
- ❏ Sand
- ❏ Sectioned potting container with drains
- ❏ Potting soil
- ❏ Small pot
- ❏ Pencil with eraser

Consider photographing this activity.

1 Place sand in 20 sections of a well-drained potting container.

2 Using a pencil eraser, make a shallow hole in the center of each section.

3 Place a pine seed in each section.

4 Sprinkle water over Sections 1 through 10. Sprinkle nutrient solutions over Sections 11 through 20.

5 Place Sections 1 through 5 and 11 through 15 in a dark place. Place Sections 6 through 10 and 16 through 20 in a light place.

6 Every 2 days for the next 3 weeks, make detailed observations of all four sections, including height of shoot growth and depth of root growth. Record your observations in Table 14.1.

Did you know . . .

Of Knots and Knees

Look at a piece of flooring or furniture, and observe the knots. The pattern of knots seems to give wood its character. A knot is where the base of a branch has been overtaken by the lateral growth of the trunk. The bald cypress (*Taxodium distichum*) is characterized by the presence of knees, or pneumatophores. Although the exact function of cypress knees is unknown, they are thought to provide stability in wet soils and aerate the roots, providing oxygen.

14

TABLE **14.1** Pine Seed Growth Observations

	Sections 1 through 5: Seeds with Water and in Dark Place	Sections 6 through 10: Seeds with Water and in Light Place	Sections 11 through 15: Seeds with Nutrient Solution and in Dark Place	Sections 16 through 20: Seeds with Nutrient Solution and in Light Place
Day 0				
Day 2				
Day 4				
Day 6				
Day 8				
Day 10				
Day 12				
Day 14				
Day 16				
Day 18				
Day 20				

Procedure 4

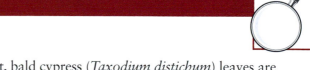

Macroanatomy of Conifer Needles and Leaves

Materials
- ❏ Dissecting microscope
- ❏ Selected specimens of conifer needles and leaves
- ❏ Colored pencils

The leaves of conifers are distinct. In pines, the leaf is called a needle. Pine needles reside in bundles called **fascicles**. In the majority of pine species, the fascicles contain between two and five needles. A fascicle is a short shoot with brown, nonphotosynthetic leaves at its base. Some conifers, such as firs, spruces, and redwoods, possess long, narrow leaves and do not have fascicles. In contrast, bald cypress (*Taxodium distichum*) leaves are feathery, and arborvitae (*Thuja* sp.) leaves are flattened. Most conifers are evergreen, slowly shedding their needles. Bald cypress and the dawn redwood are **deciduous** species, shedding their leaves yearly (Figs. 14.11 and 14.12).

Resin ducts are conspicuous structures in pine needles. Resin is a liquid containing terpenes, resin acids, and other compounds. It serves as a defense against insects and other animals that may want to eat the needle.

FIGURE **14.11** (**A**) Blue spruce, *Picea pungens*, like most conifer species, has needle-shaped leaves; (**B**) *Podocarpus* sp. has strap-shaped leaves; (**C**) *Araucaria heterophylla*, Norfolk Island pine, has awl-shaped leaves.

Needle

Fascicle

FIGURE **14.12** (**A**) Cluster of needles; (**B**) flat arborvitae leaf; (**C**) feathery bald cypress leaf.

1 Procure the needed equipment and supplies.

2 Examine specimens of conifer needles and leaves provided by the instructor. Note whether fascicles are present. If so, how many needles are associated with the fascicle? Record your sketches and observations in the space provided. Label as many structures as you can.

3 If your conifer needle is fresh, smell your specimen. Describe the smell. What is responsible for the distinct smell?

Specimen 1

Specimen 2

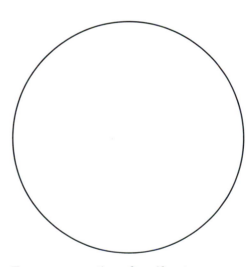

Procedure 5

Microanatomy of Conifer Stems

Materials

- ❏ Compound microscope
- ❏ Prepared microscope slide of transverse section through the stem of a young conifer, transverse section through the stem of *Pinus* sp., radial longitudinal section through the phloem of *Pinus* sp., and radial longitudinal section through the xylem of *Pinus* sp.
- ❏ Colored pencils

Xylem forms the wood in woody plants, such as conifers, and is responsible for conducting dissolved minerals and water throughout the plant. Xylem consists of two basic types of cells: tracheids and vessel elements. Tracheids are the only water-conducting cells of all gymnosperms with the exception of the gnetophytes. Tracheids appear as long, slender cells with tapered overlapping ends. Bordered pits in the cell wall allow for the passage of water.

Vessel elements are more advanced than tracheids and, with the exception of gnetophytes, are exclusive structures in angiosperms. Vessel elements are shorter and wider than tracheids and are stacked end to end to form vessels. Vessels are more efficient than tracheids in conducting water throughout the plant.

Phloem, responsible for transporting nutrients in plants, is composed of two types of cells, **sieve tube elements** and **companion cells**. Sieve tube elements are narrow tubes existing end to end that conduct nutrients. Porous sieve plates are found between adjacent sieve tubes. Narrow companion cells, adjacent to sieve tubes, help to control their function.

1 Procure the needed equipment and supplies.

2 Examine a prepared slide of a transverse section through the stem of a young conifer. Compare your slide with Figures 14.13 and 14.14. Sketch and label your specimen in the space provided below.

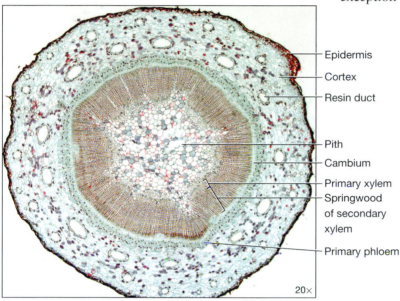

Epidermis
Cortex
Resin duct
Pith
Cambium
Primary xylem
Springwood of secondary xylem
Primary phloem

20×

FIGURE **14.13** Transverse section through the stem of a young conifer showing the arrangement of the tissue layers.

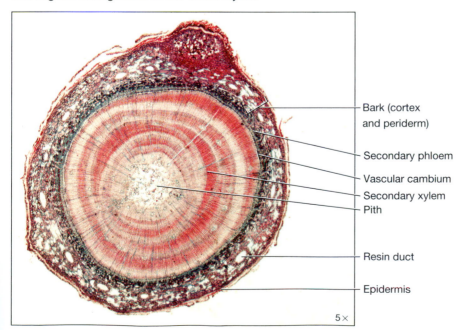

Bark (cortex and periderm)
Secondary phloem
Vascular cambium
Secondary xylem
Pith
Resin duct
Epidermis

5×

FIGURE **14.14** Transverse section through the stem of *Pinus* sp., showing secondary stem growth.

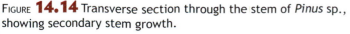

Transverse section of conifer stem

Total magnification

14

3 Examine a prepared microscope slide of the transverse section through the stem of *Pinus* sp. Compare your slide with Figure 14.15. Sketch and label your specimen in the space provided.

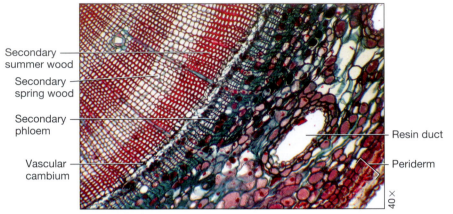

Secondary summer wood
Secondary spring wood
Secondary phloem
Vascular cambium
Resin duct
Periderm
40×

FIGURE **14.15** Enlarged view of the stem of *Pinus* sp. showing tissues following secondary growth.

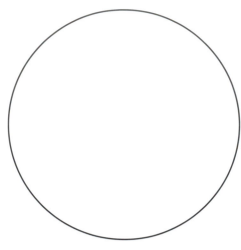

Transverse section of *Pinus* stem

Total magnification _____

_____

_____

4 Examine a prepared microscope slide of a radial longitudinal section through the phloem of *Pinus* sp. Compare your slide with Figure 14.16. Sketch and label your specimen in the space provided.

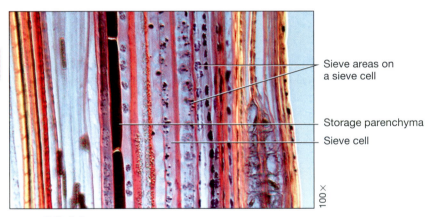

Sieve areas on a sieve cell
Storage parenchyma
Sieve cell
100×

FIGURE **14.16** Radial longitudinal section through the phloem of *Pinus* sp.

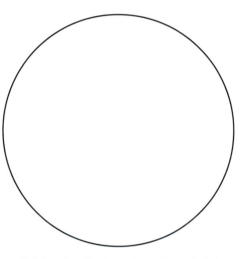

Radial longitudinal section through *Pinus*

phloem   Total magnification _____

_____

_____

5 Examine a prepared microscope of a radial longitudinal section through the xylem of *Pinus* sp. Compare your slide with Figure 14.17. Sketch and label your specimen in the space provided on the following page.

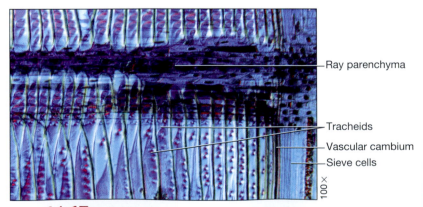

Ray parenchyma
Tracheids
Vascular cambium
Sieve cells
100×

FIGURE **14.17** Radial longitudinal section through a stem of *Pinus* sp., cut through the xylem tissue.

14

Radial longitudinal section through *Pinus*

xylem   Total magnification _____

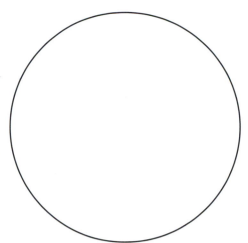

_____

_____

Procedure 6

Conifer Stem Aging

Materials

❏ Cross section of a tree trunk

The wood of conifers is considered **softwood**, and the wood of angiosperms is considered **hardwood**. Softwood is composed primarily of tracheids and rays (lateral conduction structures), whereas hardwoods have tracheids, rays, and vessel elements. As a result, softwoods are relatively light and usually less dense than hardwoods, with the exceptions of balsa and basswood, among several other types. Softwoods contain vertical resin canals that occur either naturally or as a result of injury. Generally, softwoods are easier to work with in the building and furniture industries. Softwoods also are used in the production of paper, medium density fiberboard (MDF), and the majority of plywood. Hardwood is used in fine furniture and flooring and typically is more expensive and durable than softwood.

The trunks (stems) of conifers and angiosperms share the same basic anatomy. The bark consists of the cork, cork cambium, and phloem. The visible outer bark, or periderm, protects the tree against water loss, extreme temperature, and infestations of insects and fungi. It is composed primarily of cork-producing cells of the cork cambium and nonliving cork cells. The cells of the phloem (inner bark) compress and become nonfunctional after a relatively short period. The inner bark consists of phloem that transports nutrients throughout the plant. The vascular cambium, located inner to the phloem, produces new phloem and xylem and produces the visible annual rings. The secondary xylem, beneath the vascular cambium, transports water and provides support (Fig. 14.18).

Annual rings are composed of a visible band of springwood and summerwood. **Springwood** is lighter in color, with larger cells. **Summerwood** is darker in color, with smaller cells. In many species, the age of a tree can be determined by counting the summerwood bands. In addition to displaying the age of a tree, the annual rings can tell the life story of a tree (Fig. 14.19). In many trunk cross sections, two distinct regions of xylem are present: sapwood is the lighter, outer xylem tissue, which actively transports water; **heartwood** in many species is the darker inner xylem tissue, which serves primarily as a reservoir for gum, resin, and tannin.

1 Procure a cross section of a tree trunk.

2 Identify the bark, rays, heartwood, sapwood, summerwood, and springwood.

3 Determine the age of your specimen.

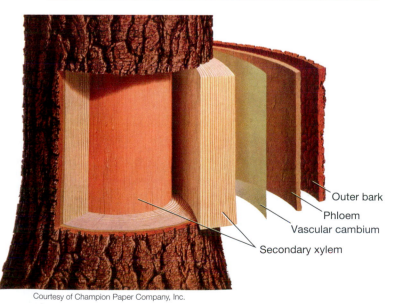

Courtesy of Champion Paper Company, Inc.

- Outer bark
- Phloem
- Vascular cambium
- Secondary xylem

FIGURE **14.18** Stem (trunk) tissues of a conifer.

14

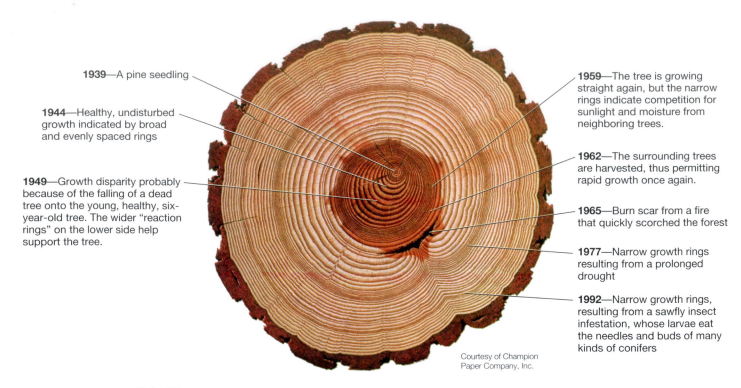

1939—A pine seedling

1944—Healthy, undisturbed growth indicated by broad and evenly spaced rings

1949—Growth disparity probably because of the falling of a dead tree onto the young, healthy, six-year-old tree. The wider "reaction rings" on the lower side help support the tree.

1959—The tree is growing straight again, but the narrow rings indicate competition for sunlight and moisture from neighboring trees.

1962—The surrounding trees are harvested, thus permitting rapid growth once again.

1965—Burn scar from a fire that quickly scorched the forest

1977—Narrow growth rings resulting from a prolonged drought

1992—Narrow growth rings, resulting from a sawfly insect infestation, whose larvae eat the needles and buds of many kinds of conifers

Courtesy of Champion Paper Company, Inc.

FIGURE **14.19** Stem (trunk) of a pine tree harvested in the year 2000 when the tree was 62 years old. The growth rings of a tree indicate environmental conditions that occurred during the tree's life.

4 Examine the annual rings. Was the tree exposed to any interesting conditions? How do you know, and when in the life of the tree did it occur?

14

✔ Check Your Understanding

1.1 Compare and contrast softwood and hardwood, and provide examples of each.

_____

_____

_____

1.2 In gymnosperms and angiosperms, the _____ is responsible for secondary growth that will develop into the epidermis in roots and stems.

1.3 Compare and contrast springwood and summerwood.

_____

_____

_____

1.4 Cross-section slides that you observed of young conifer stems show the location of both vascular tissues. The

_____ is found between the pith and cambium and the _____

is found between the cambium and the cortex.

1.5 List several uses of softwood and hardwood.

_____

_____

_____

14

Angiosperms

Charles Darwin (1809–1882) called the sudden appearance of modern flowers in the fossil record "an abominable mystery" because their abrupt emergence was difficult to explain. Since the time of Darwin, however, the fossil record has increased tremendously. The discovery in 1998 of an angiosperm fossil collected in China—*Archaefructus sinensis* from the Cretaceous period approximately 125 million years ago—and other discoveries have provided clues as to how angiosperms evolved.

These steps began in the Jurassic period, and today the flowering plants are the dominant plants on Earth, representing more than 90% of all living plant species and 18% of all species. Figure 14.20 is a cladogram of the plants highlighting the position of the angiosperms. Development of the angiosperms transformed the face of the planet. Today the landscape is painted with colorful flowers as plants advertise themselves to pollinators.

The **angiosperms** (Greek: *angio* = vessel, sperm = seed), known as the flowering plants, are the most diverse and numerous group of plants on Earth. Botanists have identified more than 280,000 species of angiosperms thus far. Angiosperms can be found in a number of environments. They range in size from duckweed *Wolffia angusta*, smaller than 1 millimeter in diameter, to Australia's 100 meter tall mountain ash tree, *Eucalyptus regnans*. Angiosperms vary in form, including delicate orchids, strange insectivorous sundews, succulent cacti, and the majestic baobab tree.

Most angiosperms are autotrophic. Several species of angiosperms are parasitic. Mistletoe (*Phoradendron* sp.) is a hemiparasite undergoing photosynthesis and parasitizing its host plant, and dodder (*Cuscuta* sp.) is a true parasitic plant. Indian pipe (*Monotropa uniflora*) and snowplant (*Sarcodes sanguinea*) are two of several saprophytic species of angiosperms. Spanish moss (*Tillandsia usneoides*) and some orchids, cacti, and ferns are epiphytes, or "air plants," that attach to a substrate, such as another plant or the side of a building. The angiosperms are important to humans as sources of food, medicine, aesthetic beauty, cotton, lumber, and other products.

The angiosperms are noted for their reproductive structures called **flowers**. Flowers are the exclusive reproductive organs of angiosperms. Although we cherish their beauty, they are elaborate reproductive structures solely programmed to propagate the species. Their color, texture, and nectar are designed to ensure pollination (Fig. 14.21). Evolutionarily speaking, the appearance of flowers during the Cretaceous period changed the face of the planet and opened the door for the coevolution of many pollinators. Most angiosperms are deciduous, losing their leaves during the winter or perhaps during a drought. Angiosperms such as peas are **herbaceous**, possessing little or no woody tissue, and others, such as an apple tree, are **woody**.

As in the gymnosperms, the sporophyte generation is the dominant portion of the life cycle of angiosperms. Flowering plants can cycle from germination to mature plant in less than a month or take as long as 150 years or more. In annuals, such as zinnias, the life cycle of a plant is completed in one season. Biennials, such as parsley, complete their life cycle in two growing seasons. Perennials, such as tulips, may take more than two seasons to complete their life cycle.

Fruits are the products of flowers and are exclusive to angiosperms. A fruit is a structure derived from the ovary of a plant and its accessory tissues. Fruits house, protect, nourish, and aid in the dissemination of seeds. Examples of fruits include strawberries, apples, oranges, peaches, tomatoes, cucumbers, pecans, rose hips, and beans. A **vegetable** is an edible part of a plant derived from petioles, leaves, specialized leaves, roots, stems, or flowers. Examples of vegetables are sweet potatoes, carrots, broccoli, turnips, and onions.

Darwin noted that organisms produce more young than naturally can be expected to survive. The vast numbers of seeds produced by a mustard plant, watermelon, or oak tree are examples of this principle. For example, of the thousands of acorns dropped by a typical mature oak tree, only one in 10,000 will become a tree. A **seed** is a ripened ovule of a plant that contains an embryo housed in a protective coat and nourished by stored food. Seeds have ensured the success

Did you know . . .

How about a Kiss under the Mistletoe?

In Anglo-Saxon, mistletoe literally means "dung on a twig." It was thought to appear through spontaneous generation from bird feces. The etymology does not exactly reflect the romantic reputation of the mistletoe plant! Mistletoe has been considered one of the most magical plants in European folklore. It was thought to bestow fertility and life to people, protect against poisons, and even serve as an aphrodisiac.

The first recorded tradition of kissing under the mistletoe is associated with the Greek festival of Saturnalia and early marriage rites. Mistletoe supposedly brought about fertility and long life. In Norse mythology, two enemies were said to find peace by kissing under the mistletoe. In Victorian England at Christmas, a young lady would stand under a kissing ball of mistletoe and await a kiss. If she was kissed, it could mean romance or close friendship. If she was not kissed, it meant that she would not marry during the next year. Today, the tradition of kissing under the mistletoe appears throughout the holiday season and signifies love or lasting friendship. Ironically, mistletoe berries are poisonous!

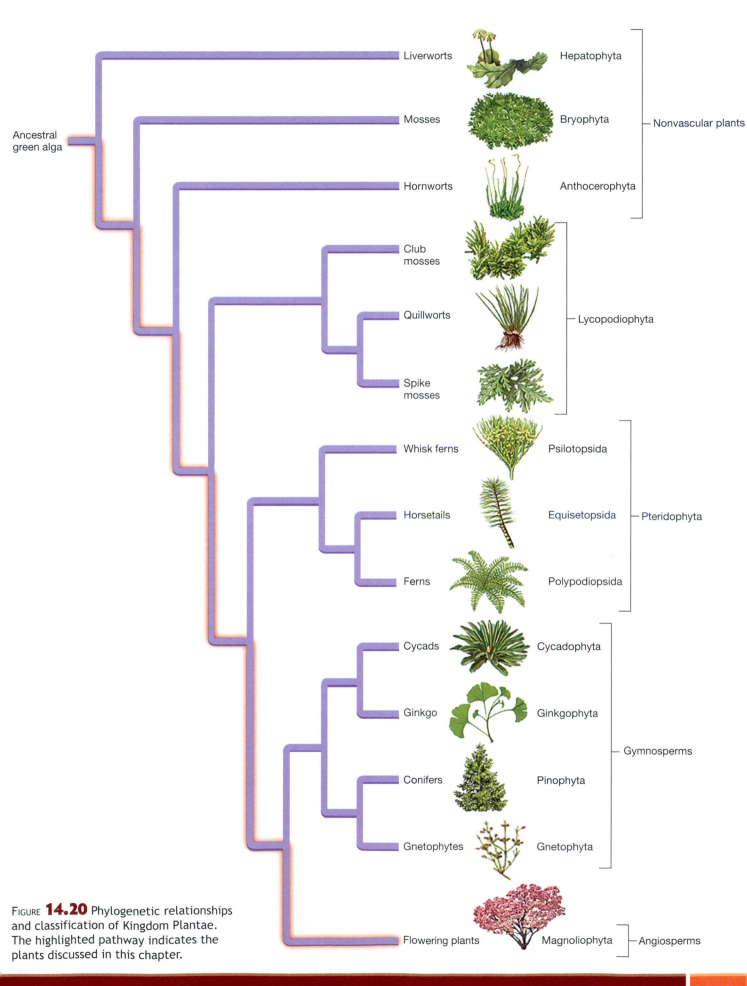

Liverworts — Hepatophyta ⎤
Mosses — Bryophyta ⎥ Nonvascular plants
Hornworts — Anthocerophyta ⎦

Club mosses ⎤
Quillworts ⎥ Lycopodiophyta
Spike mosses ⎦

Whisk ferns — Psilotopsida ⎤
Horsetails — Equisetopsida ⎥ Pteridophyta
Ferns — Polypodiopsida ⎦

Cycads — Cycadophyta ⎤
Ginkgo — Ginkgophyta ⎥ Gymnosperms
Conifers — Pinophyta ⎥
Gnetophytes — Gnetophyta ⎦

Flowering plants — Magnoliophyta — Angiosperms

Ancestral green alga

14

FIGURE **14.20** Phylogenetic relationships and classification of Kingdom Plantae. The highlighted pathway indicates the plants discussed in this chapter.

FIGURE **14.21** Flowers of many angiosperms are uniquely adapted for and rely on specific animals for pollination. Example animal pollinators include: (**A**) bee, *Anthophora urbana*; (**B**) broad-billed hummingbird (female), *Selasphorus platycercus*; (**C**) lesser long-nosed bat, *Leptonycteris yerbabuenae*.

of both gymnosperms and angiosperms in populating the planet. Angiosperm **ovules** are encased within diploid tissue (integuments) supplied by the parent plant. Angiosperms and their oldest living relatives, the gnetophytes, undergo double fertilization in which a fertilized egg and nutritive endosperm form. In addition to tracheids (see p. 279), the xylem of angiosperms possesses vessels that transport water efficiently throughout the plant.

14

Phylum Magnoliophyta

The angiosperms, or flowering plants, have been placed in phylum Magnoliophyta, also known as phylum Anthophyta. Traditionally, phylum Magnoliophyta has been divided into two major ranks, or classes: the dicotyledones (Magnoliopsida) and the monocotyledones (Liliopsida). Classically, dicotyledones, or **dicots**, are flowering plants whose seed contains two embryonic leaves, or **cotyledons**; and monocotyledones, or **monocots**, are flowering plants with a single cotyledon. A cotyledon is a seed leaf containing nutrients that nourish the developing embryonic plant. This is the first leaf evident upon germination. Although this scheme is user-friendly and is still used extensively, new molecular evidence and advanced observations are paving the way for a new classification of angiosperms.

Several newer schemes are attempting to make sense of the diversity of angiosperms. A commonly used scheme divides the angiosperms into the **basal dicots**, the eudicots, and the monocots. The basal dicots, or paleodicots, such as water lilies, avocado, and magnolias, possess **monosulcate pollen**, as do the monocots (Fig. 14.22). Monosulcate pollen has a linear, thin, furrow-like groove, or sulcus, on the surface of the grain and one pore. The true dicots, **eucotyledones**, or **eudicots**, sometimes called **tricolpates**, are the majority of angiosperms and possess **tricolpate pollen**. This pollen has three long, grooved apertures, or pores, on the surface. Example eudicots are roses, oak trees, sunflowers, and cabbage. The monocots are thought to have evolved from the dicots. Example monocots are cattails, corn, orchids, palms, and bananas.

Despite major problems and much confusion in organizing angiosperm taxa, the classical distinctions between dicots and monocots are helpful. Table 14.2 is based upon generalizations, with many exceptions.

In recent years the dicots have been split into the basal dicots and the eudicots. Basal dicots represent a small number of the total dicots and have a more ancient linage. The following are orders of basal dicots:

1. The Nymphaeales are aquatic plants, most of which have floating leaves, such as water lilies and lotus plants.

FIGURE 14.22 (**A**) Magnolia tree, *Magnolia* sp., is an example of a basal dicot; (**B**) American sycamore, *Platanus occidentalis,* is an example of a eudicot; and (**C**) coconut palm, *Cocos nucifera,* is an example of a monocot.

TABLE **14.2** Comparison of Dicots and Monocots

Eudicots	Monocots
Embryo with two cotyledons	Embryo with one cotyledon
Pollen with three apertures (except basal dicots)	Pollen with sulcus and one pore
Flower parts in multiples of four or five	Flower parts in multiples of three
Netted venation of leaf veins	Parallel venation of leaf veins
Vascular bundles arranged in a ring	Vascular bundles scattered
Secondary growth present	Secondary growth absent

14

2. The Piperales are a group of herbs, shrubs, and small trees, such as black pepper, lizard's tail, vine pepper, and wild ginger.

3. The Laurales, such as sassafras, cinnamon, and avocado, at one time were placed with magnolias.

4. The Magnoliales, the best-known group, consist of magnolia, sweet bay, nutmeg, and yellow poplars.

Several other basal dicot groups have been identified in recent years.

Most plants traditionally recognized as "dicots" are now considered eudicots. The eudicots consist of an extremely large, diverse group of angiosperms that has an incredible range of geographic distribution, variation within habitats, anatomy, morphology, and biochemistry. Table 14.3 lists common families of eudicots with examples of each.

Monocots constitute about one-fourth of all living angiosperms. The monocots diverged from the basal dicots approximately 90 million years ago. Monocots play a significant role in the floral and horticultural industries. Table 14.4 lists common families of monocots and gives examples of each.

TABLE **14.3** Representative Eudicots

Family	Examples	Family	Examples
Aceraceae	Sugar maple, red maple	Moraceae	Mulberry, fig
Anacardiaceae	Poison ivy, cashews, pistachios	Myrtaceae	Myrtle, bottlebrush, eucalyptus
Aquifoliaceae	American holly, English holly	Oleaceae	Olives, ashes, lilacs
Asteraceae	Zinnia, aster, sunflower, daisy, thistle, lettuce	Passifloraceae	Passion flower, love-in-a-mist
Brassicaceae	Radish, broccoli, cabbage, turnip	Platanaceae	Sycamore, plane tree
Cactaceae	Saguaro cactus, prickly pear cactus	Ranunculaceae	Buttercup, columbine, goldenseal
Convolvulaceae	Morning glory, sweet potato	Rosaceae	Roses, apples, strawberries, almonds
Cucurbitaceae	Pumpkin, squash, gourd	Rubiaceae	Coffee, chinchona, madder
Droseraceae	Sundew	Rutaceae	Orange, lime, lemon, grapefruit
Ericaceae	Azaleas, heath, huckleberry, mountain laurel	Salicaceae	Willow, poplar, cottonwood
Euphorbiaceae	Spurge, cassava, rubber plant, poinsettia	Sarraceniaceae	White-top pitcher plant, crimson pitcher plant
Fabaceae	Beans, peas, peanuts, clover, mimosa, licorice	Smilaceae	Green briar, kudzu, sarsaparilla
Fagaceae	Live oak, water oak, red oak, beech	Solanaceae	Potato, tomato, tobacco
Geraniaceae	Geranium, pelargonium	Theaceae	Camellia, stewartia
Juglandaceae	Walnut, hickory, pecan	Umbelliferae	Carrot, parsley
Meliaceae	Mahogany, chinaberry	Vitaceae	Grape, muscadine

Did you know . . .

Parlez-vous le francais?

The magnolia tree is named after French botanist Pierre Magnol (1638–1715).

14

TABLE **14.4** Representative Monocots

Family	Examples	Family	Examples
Agavaceae	Century plant, agave	Iridaceae	Iris, gladiolus, crocus
Aloaceae	Aloe	Lemnaceae	Duckweed, *Wolffia*
Amaryllidaceae	Spider lily, Amaryllis	Liliaceae	Lily, tulip, asparagus, onion, hyacinth
Araceae	Caladium, philodendron, skunk cabbage, peace lily, Jack-in-the-Pulpit	Musaceae	Banana, cannaplant, bird of paradise
Arecaceae	Coconut palm, palmetto	Poaceae	Corn, sugar cane, rice, bluegrass, wheat, oats, bamboo, crab grass, broom sedge, foxtail
Bromeliaceae	Pineapple, Spanish moss	Pontederiaceae	Pickerel weed, water hyacinth
Cyperaceae	Water chestnut, papyrus, spikerush, sawgrass	Orchidaceae	Orchids
Hydrocharitaceae	*Elodea*	Typhaceae	Cattail

Procedure 1

Macroanatomy of Basal Dicots, Eudicots, and Monocots

Materials

- ❏ Dissecting microscope
- ❏ Hand lens
- ❏ Specimens of basal dicots: stem, leaves, flowers, cones, and pollen from several basal dicots, such as a water lily, lizard's tail, sassafras, avocado, magnolia, tulip tree, or others
- ❏ Stems, leaves, flowers, fruits, and pollen from several eudicots, such as a rose, maple, oak, squash, sunflower, willow, or others
- ❏ Stems, leaves, flowers, and pollen from several monocots, such as corn, orchids, wheat, lilies, bananas, tulips, or others
- ❏ Colored pencils

1 Observe the images below and label each as either a monocot or a dicot in the space provided

2 Procure select specimens of the flowers and equipment provided by the instructor.

3 Using a dissecting microscope or hand lens, observe and draw the anatomical features (stems, leaves, flowers, roots, fruits) of the basal dicot specimens (Fig. 14.23). Place your labeled sketches and observations in the space provided. Remember to label your specimen.

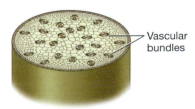
Vascular bundles

Vascular bundles

FIGURE **14.23** Examples of basal dicots: (**A**) water lily, *Nymphaea* sp.; (**B**) sassafras, *Sassafras* sp.; (**C**) lizard's tail, *Saururus* sp.; (**D**) tulip tree, *Liriodendron tulipifera*.

Basal dicot specimen 1

Basal dicot specimen 2

4 Using a dissecting microscope or hand lens, observe and draw the anatomical features (stems, leaves, flowers, roots, fruits) of the eudicot specimens (Fig. 14.24). Place your labeled sketches and observations in the space provided on the following page.

FIGURE **14.24** Examples of eudicots: (**A**) passion flower, *Passiflora* sp.; (**B**) Russian olive, *Elaeagnus angustifolia*; (**C**) rose, *Rosa* sp.

Eudicot specimen 1 _____

Eudicot specimen 2 _____

5 Using a dissecting microscope or hand lens, observe and draw the anatomical features (stems, leaves, flowers, roots, fruits) of the specimens (Fig. 14.25). Place your labeled sketches and observations in the space provided below.

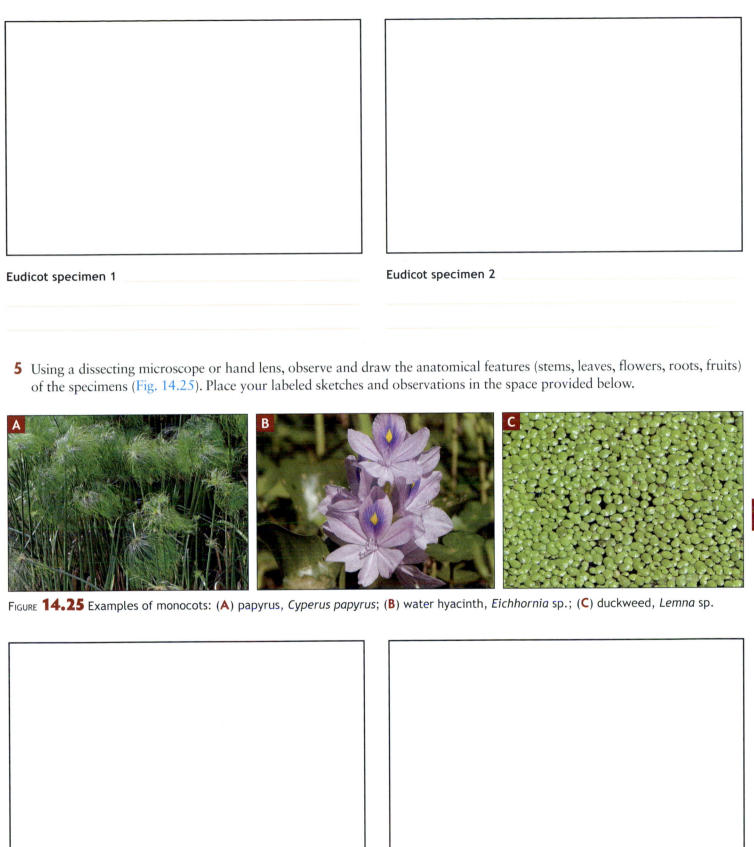

FIGURE **14.25** Examples of monocots: (**A**) papyrus, *Cyperus papyrus*; (**B**) water hyacinth, *Eichhornia* sp.; (**C**) duckweed, *Lemna* sp.

Monocot specimen 1 _____

Monocot specimen 2 _____

2.1 Compare and contrast basal dicots and eudicots.

2.2 Compare and contrast eudicots and monocots.

2.3 Identify each of the following plants as either basal dicots or eudicots.

_____ Water lily	_____ Willow
_____ Morning glory	_____ Poinsettia
_____ Sundew	_____ Tomato
_____ Sassafras	_____ Strawberry
_____ Mulberry	_____ Cotton

2.4 List three plants considered monocots that are used as food and three that are considered ornamental in nature.

MYTHBUSTING

Seeing the Forest for the Trees

Debunk each of the following misconceptions by providing a scientific explanation. Write your answers on a separate sheet of paper.

1 Lichens are parasites on trees.

2 All plants are terrestrial.

3 Cacti only thrive in deserts.

4 Poinsettias are highly poisonous to humans.

5 Pine trees all have softwood.

1 Which of the following are general characteristics of the gymnosperms? (*Circle the correct answer.*)

  a. They do not produce flowers.

  b. They do not produce seeds encased in a fruit.

  c. Seeds develop in association with a scale within a cone.

  d. Both a and b.

  e. All of the above.

2 When comparing the seeds of angiosperms and the seeds of gymnosperms, _____. (*Circle the correct answer.*)

  a. gymnosperm seeds mature on the surface of a cone scale while angiosperm seeds are enclosed in a fruit

  b. nutritive material accumulates prior to fertilization in angiosperms while nutritive material is stored after fertilization in gymnosperms

  c. nutritive material is stored after fertilization in angiosperms while nutritive material accumulates prior to fertilization in gymnosperms

  d. both a and b

  e. both a and c

3 Sketch and label the life cycle of a pine tree, and outline the steps in seed formation in the space provided.

4 Describe the life cycle differences in plants that are annuals, biennials, and perennials.

_____

_____

_____

5 Complete the following table:

	Basal Dicots	Eudicots	Monocots
Example organisms			
Support system			
Types of roots			
Transport system			
Type of transport tissue			
Type and arrangement of leaves			
Reproduction			
Type and arrangement of flowers			
Type of seeds			

14

6 Why was Darwin perplexed with the evolution of angiosperms?

_____

_____

7 Identify the following characteristics as either eudicot or monocot:

_____ Flower parts in multiples of threes

_____ Pollen with three apertures

_____ Scattered vascular bundles

_____ Embryo with two cotyledons

_____ Netted venation of leaf veins

_____ Secondary growth absent

_____ Embryo with one cotyledon

_____ Flower parts in multiples of fours or fives

_____ Parallel venation of leaf veins

_____ Secondary growth present

_____ Vascular bundles arranged in a ring

The Green Machine III
Understanding Plant Tissues, Roots, Stems, and Leaves

Nature is as delightful and abundant in its variations that there would not be one that resembles another, and not only plants as a whole, but among their branches, leaves, and fruit, will not be found one which is precisely like another.

— Leonardo da Vinci (1452–1519)

Just wondering . . .

Consider the following questions prior to coming to lab, and record your answers on a separate piece of paper.

1 Do some plants "communicate" when attacked by insects?

2 What is the significance of chinchona bark in medicine and history?

3 What kind of tissue allows many water plants to float?

4 Is the only function of spines on cacti defense?

5 Why do carnivorous plants "eat" insects?

Objectives

At the completion of this chapter, the student will be able to:

1. Describe basic plant tissues.

2. Compare and contrast herbaceous and woody plants.

3. Describe the function of roots, root hairs, stems, and leaves.

4. Compare and contrast the root systems, stems, and leaves of eudicots and monocots.

5. Identify and describe the anatomy and function of regular and specialized roots, woody stems, and leaves.

6. Locate and describe the function of stomata.

7. Classify leaves based upon phyllotaxy, shape, type of margin, type of apex, and type of base.

This chapter introduces students to the diversity of tissues found in plants. The study of tissues is termed *histology*. To colonize the terrestrial environment successfully, plants developed vascular tissues to carry water and nutrients throughout the body of the plant. The vascular plants of today possess a system composed of conducting tissues. **Xylem** is a specialized tissue that carries water, and **phloem** is a specialized conducting tissue that carries nutrients.

We will then discuss the fundamental vegetative organs: roots, stems, and leaves common to all flowering plants (Fig. 15.1). Working together, these three organs build the plant from the ground up. The basic tissues that make up the roots, stems, and leaves of the eudicots and monocots are arranged differently and provide a simple means of differentiating these two distinct groups of angiosperms. The roots and their ancillary structures constitute the root system of a plant, and the stems and leaves and their ancillary structures make up the shoot system of a plant.

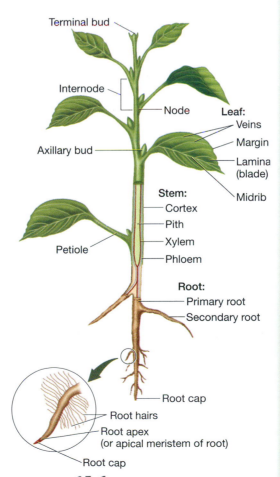

FIGURE **15.1** Basic organization of the vegetative organs of a generalized plant.

Chapter Photo
Begonia "escargot," or spiral begonia, *Begonia* sp.

Plant Histology

Plants have permanent regions of growth composed of **meristematic tissues**. In these tissues, cells are actively undergoing mitosis. The new cells resulting from cell division usually are small and six-sided with a prominent nucleus. As the newly formed cells mature, they begin to take on their characteristic size, shape, and function.

Meristematic Tissues

Meristematic tissues found at or near the tips of roots and stems make up the apical meristem. Growth of the apical meristem, known as primary growth, involves increasing the length of the root or stem. The apical meristem (Fig. 15.2) gives rise to three distinct regions: the protoderm, the ground meristem, and the procambium.

The lateral meristem provides the plant growth in girth, or secondary growth. Two derivatives of the lateral meristem are the vascular cambium (or simply cambium) and the cork cambium. The structures to which each region gives rise is listed in Table 15.1.

Grasses do not possess a vascular cambium or cork cambium, but they do have apical meristematic tissue called intercalary meristems near nodes (regions of leaf attachment) at intervals throughout the plant. Intercalary meristems allow grass to grow back quickly after being grazed by a cow or cut by a lawn mower.

Parenchyma Tissue

Parenchyma tissue (Fig. 15.3) is composed primarily of parenchyma cells, the most abundant and diverse type of cell in plants. Parenchyma cells vary in size and shape and tend to have large vacuoles. Parenchyma cells are involved in storage, photosynthesis, support, secretion, repair, and the movement of water and food in plants.

TABLE **15.1** Regions and Structures of Plant Growth

Region	Gives Rise to
Apical meristem	
Protoderm	Epidermis
Ground meristem	Building block tissue called parenchyma that usually exists between the epidermis and the vascular tissue
Procambium	Vascular tissue such as xylem and phloem
Lateral meristem	
Vascular cambium	Secondary vascular tissue important in support
Cork cambium (in woody plants)	Cork tissue that makes up the protective bark; the cork is impregnated with the waxy substance suberin, which makes the cells impenetrable to water

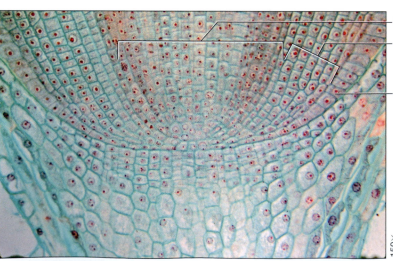

Procambium

Ground meristem

Protoderm

150×

FIGURE **15.2** Longitudinal section of a root of corn, *Zea mays*, showing primary meristems.

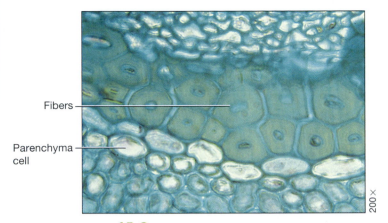

FIGURE **15.3** Transverse section through the stem of flax, *Linum* sp.; note the thick-walled fibers compared with the thin-walled parenchyma cells.

Fibers

Parenchyma cell

200×

The soft, edible parts of apples and other fruits consist mostly of parenchyma cells. In potatoes, parenchyma cells store starch. Parenchyma cells with numerous chloroplasts are sites of photosynthesis. These cells, chlorenchyma, are abundant in the leaves and stems of herbaceous plants.

Parenchyma cells with extensive air spaces found in water plants are known as aerenchyma. This tissue helps to support the plant and, when squeezed, is crunchy. Some mature parenchyma cells can divide when stimulated. When a plant is damaged, parenchyma cells are important in repair. When gardeners make cuttings, they take advantage of the growth of parenchyma cells.

Collenchyma Tissue

Collenchyma tissue (Fig. 15.4) is composed of elongated collenchyma cells. These cells develop thick, flexible walls that support young plants and specific plant structures such as leaves and flower parts. Collenchyma resides beneath the epidermis in stems. In a fresh specimen, collenchyma tissue glistens. This tissue borders the veins of leaves and makes up the "strings" in celery.

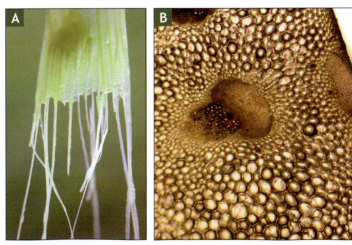

FIGURE **15.4** (A) Celery strings; (B) collenchyma tissue from celery.

Sclerenchyma Tissue

Sclerenchyma tissue is composed of thick sclerenchyma cells, often impregnated with the plant polymer **lignin**. Unlike parenchyma and collenchyma cells, sclerenchyma cells are dead at maturity and primarily provide support. The two types of sclerenchyma include the following:

1. **Fibers** are long, slender cells that occur in strands. They are commonly found in roots, stems, leaves, and fruits. Fibers are used in the manufacture of ropes, string, and canvas.

2. **Sclereids,** or stone cells (Fig. 15.5), are responsible for the gritty texture of plants, in which they may occur singly or in groups throughout; also, a major component of the shell of various nuts and the pit of a peach.

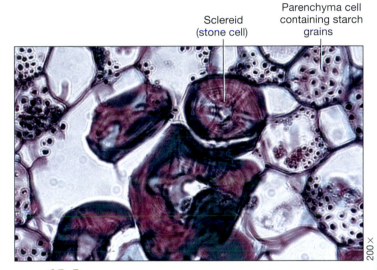

Sclereid (stone cell)

Parenchyma cell containing starch grains

200×

FIGURE **15.5** Stem of a wax plant, *Hoya carnosa*, showing thick-walled sclereids (stone cells).

Dermal Tissues

The **epidermis** (Fig. 15.6) constitutes the outermost layer of cells in plant structures, such as roots, stems, leaves, floral parts, fruits, and seeds. The epidermis generally is one cell layer thick and does not undergo photosynthesis. Because epidermal cells are in direct contact with the environment, they vary in form and function. The walls of many epidermal cells are covered with a waxy cuticle, minimizing water loss and protecting the plant against pathogens. The waxy cuticle can be easily observed on magnolia leaves.

Epidermal cells also can form **root and leaf hairs** that increase the surface area. Numerous small, pore-like structures are found primarily on the underside of leaves. The structures are called stomata (sing. = stoma), and they are bordered by a pair of guard cells. Stomata allow for gas exchange in plants.

In the roots and stems of woody plants, the epidermis is sloughed off and replaced by the **periderm**, which makes up the outer bark composed of box-shaped cork cells. Mature

15

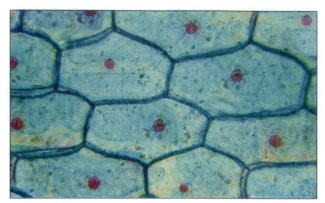

FIGURE **15.6** Epidermal cells from onion skin.

cork cells are dead. The fatty substance suberin is found in the walls of cork cells, providing protection from mechanical injury, desiccation, and extreme temperatures.

Vascular Tissues

Vascular tissue makes up the **vascular bundle**. In plants, xylem tissue is the primary water-conducting tissue and also serves in support, food storage, and the conduction of minerals. Xylem (Fig. 15.7) is composed of the following basic types of cells:

- Parenchyma cells serve in storage.
- Tracheids and vessel elements are the major conducting cells of the xylem.

- Ray cells serve in lateral conduction and storage.
- Fibers also can occur in xylem, adding support and storage.

Phloem tissue (Fig. 15.7) conducts dissolved food materials throughout the plant body. The food is composed primarily of sugars produced through photosynthesis. Phloem is composed of:

- Parenchyma cells, which provide storage;
- Sclerenchyma, which provide support;
- Sieve tube members, which provide conduction; and
- Companion cells, which also provide conduction.

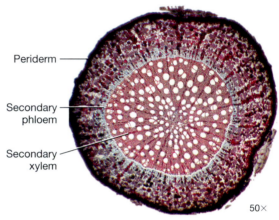

Periderm

Secondary phloem

Secondary xylem

50×

FIGURE **15.7** Transverse section of an older root of *Salix* sp.

15

Procedure 1

Microanatomy of Stems and Roots

Materials
- ❑ Compound microscope
- ❑ Prepared slides of longitudinal sections of stem tips and root tips, parenchyma tissue, collenchyma tissue, and sclerenchyma
- ❑ Living specimens: apple, celery, pear, leaf, and onion
- ❑ Blank slides and coverslips
- ❑ Forceps
- ❑ Paper towels
- ❑ Colored pencils

1 Acquire microscope slides of stem tip and root tip longitudinal sections and view them using a compound microscope. Record and label the following in the space provided: protoderm, ground meristem, procambium, cambium.

2 Acquire a slide of parenchyma tissue as well as an example from living apple tissue (flesh). Observe the slides using a compound microscope. Sketch the tissue in the space provided on the following page.

3 Acquire a slide of collenchyma tissue as well as a sample of celery. Observe the slide using a compound microscope. Remove and observe a string from a celery stalk. Sketch the tissues in the space provided on the following page.

4 Acquire a slide of sclerenchyma tissue as well as a section of living pear tissue (flesh). Sketch the tissues in the space provided on the following page.

5 Acquire a slide of epidermal tissue as well as a piece of onion skin. Observe the slide using a compound microscope. Create a wet mount using the onion skin and observe using a compound microscope. Sketch and label the tissues in the space provided on the following page.

6 Observe the underside of a leaf, and sketch and label the stomata and guard cells in the space provided on the following page.

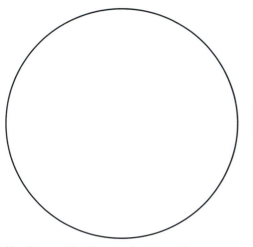

Meristematic tissues in stem tip longitudinal section

Total magnification _____

_____

_____

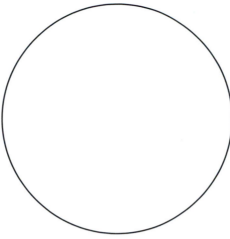

Meristematic tissues in root tip longitudinal section

Total magnification _____

_____

_____

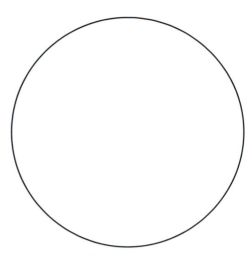

Parenchyma tissue

Total magnification _____

_____

_____

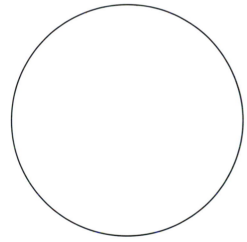

Collenchyma tissue

Total magnification _____

_____

_____

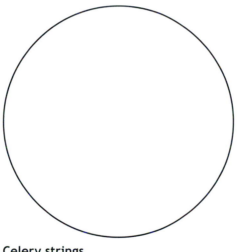

Celery strings

Total magnification _____

_____

_____

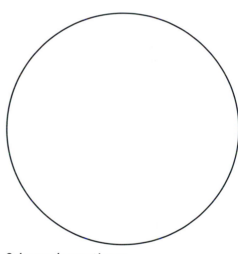

Sclerenchyma tissue

Total magnification _____

_____

_____

15

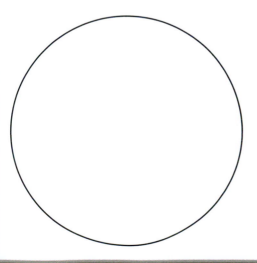

Epidermal tissue   Total magnification _____

_____

_____

Onion skin Total magnification _____

_____
_____

Stomata tissue Total magnification _____

_____
_____

Guard cells Total magnification _____

_____
_____

7 Acquire slides of the complex vascular tissue, xylem, and phloem. Sketch the tissues in the space provided.

8 Your instructor may have celery in colored water for you to observe. During the course of the lab, monitor the celery and note the changes you see taking place. Which tissue is involved in the transport of water through the stalk?

_____

15

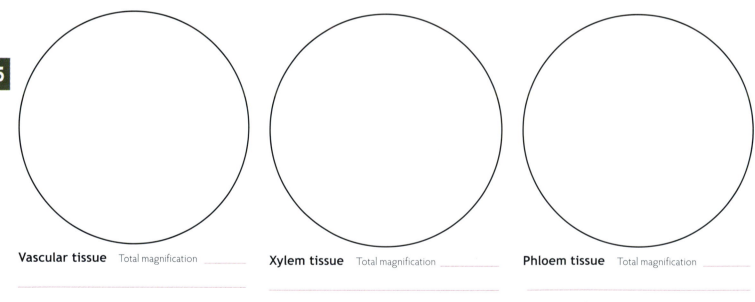

Vascular tissue Total magnification _____

_____
_____

Xylem tissue Total magnification _____

_____
_____

Phloem tissue Total magnification _____

_____
_____

✔Check Your Understanding

1.1 In vascular plants, _____ is the specialized tissue that carries

nutrients within the plants and _____ is the tissue that carries water.

1.2 In plants, what are the basic dermal tissues?

_____

_____

_____

1.3 What is the basic function of meristematic tissue?

_____

_____

_____

1.4 Label the following phloem tissues with the correct type of function.

_____ Parenchyma cells   A. Provides conduction

_____ Companion cells   B. Provides storage

_____ Sieve tube members   C. Provides support

_____ Sclerenchyma

1.5 In this exercise, you looked at longitudinal sections of stem and root tips, and the common celery stalk. What type of tissue makes up the "strings" in the celery? What is the function of this tissue?

_____

_____

_____

15

Roots

Roots are plant organs that anchor and support a plant, absorb water and necessary minerals, store food, and produce growth-stimulating hormones. The majority of the root system is found underground, although some plants, such as the tropical fig, possess extensive aerial roots. Generally, the extent of the root system is equal to or exceeds that of the shoot system.

In some plants, including many grasses, the root system is just a few millimeters beneath the soil, and in some dry-climate plants, such as *Juniperus monosperma*, the root system exceeds 20 meters in depth. The number of roots produced by a plant can be staggering. A single ryegrass plant, for example, may have up to 15 million roots and a surface area larger than a volleyball court.

Root Anatomy

A typical root has four distinct regions, or zones, which are shown in Figure 15.8 and discussed in Table 15.2.

Types of Roots

In plant development, when a seed germinates a small, root-like structure, the radicle, emerges from the embryo and forms the first root. The radicle usually gives rise to a single, tapered **taproot** with many small lateral branches. Taproot systems are common in conifers and many species of eudicots. Taproots anchor the plant and seek deep water supplies (Fig. 15.9). The fleshy portion of a carrot is an example of a taproot that stores food in the form of carbohydrates.

Monocots and some eudicots possess a fibrous root system. The stem or another plant part produces **adventitious roots**, such as the **prop roots** of corn. Prop roots help anchor and brace the plant against wind. Dodder (*Cuscuta* sp.) possesses parasitic roots called **haustoria** that parasitize a host plant. Some members of the pumpkin family (*Cucurbitaceae*) that live in dry climates produce large water storage roots. Certain species of figs and swamp trees, such as the tupelo and bald cypress, have expanded **buttress roots** for stability in wet environments (Fig. 15.10). Sweet potatoes are fleshy

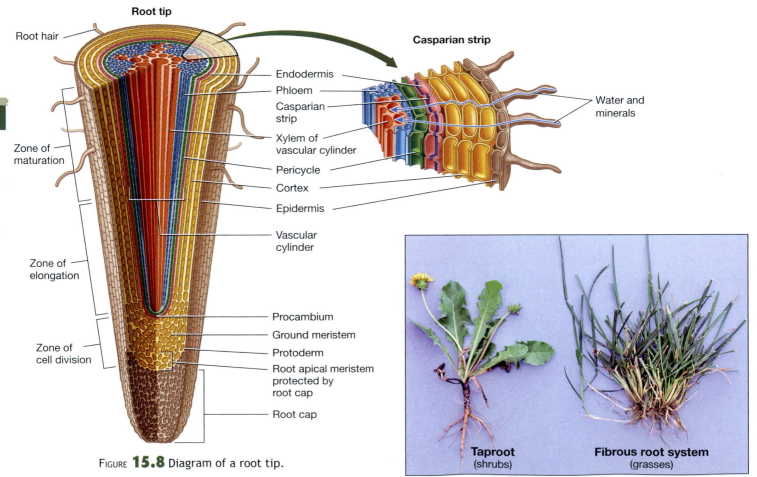

Root tip

Root hair

Casparian strip

Endodermis
Phloem
Casparian strip
Xylem of vascular cylinder
Pericycle
Cortex
Epidermis

Water and minerals

Zone of maturation

Vascular cylinder

Zone of elongation

Procambium
Ground meristem
Protoderm
Root apical meristem protected by root cap

Zone of cell division

Root cap

FIGURE **15.8** Diagram of a root tip.

Taproot
(shrubs)

Fibrous root system
(grasses)

FIGURE **15.9** Taproots and fibrous root systems anchor plants and absorb water and essential minerals.

TABLE **15.2** Zones of a Root

Zone	Where Found	Description
Zone of cell division (apical meristem or meristematic region)	Behind the root cap	Cells in this region are undergoing cell division at a high rate. Three distinct types of tissue are produced in the apical meristem: the protoderm, the ground meristem, the procambium.
Zone of elongation	Above the zone of cell division	In this region, the cells greatly increase in length and much less in width as small vacuoles merge, filling up more than 90% of the cell. The elongation in these cells helps to push the root cap through the soil. Some roots can push through the soil at a rate of 4 cm a day.
Zone maturation	Above the zone of elongation	The cells produced in the zone of elongation become differentiated and mature, forming the epidermis, cortex, endodermis, and vascular cylinder. The outermost cells differentiate into the single-layered epidermis. Interior to the epidermis is the cortex, composed of parenchyma cells. These cells possess starch granules and are used primarily in food storage.
Vascular cylinder	Center of the root, separated from the cortex by the endodermis (a single layer of cells that serves as the boundary)	The vascular cylinder has many parts. The primary cell walls of the endodermis contain lignin and suberin, which form an impermeable layer called the Casparian strip. The strip blocks the passage of water and minerals between adjacent cells and regulates the movement of water and minerals into the vascular cylinder. The vascular cylinder, which in eudicots sometimes is referred to as the stele, contains the xylem and phloem. The pericycle is the first layer of cells in the vascular cylinder. The cells of the pericycle can divide and initiate the development of lateral roots. In eudicots, the xylem is star-shaped with several radiating arms. The phloem is located between the radiating arms. In monocots, ground tissue forms the centrally located pith. Vascular tissue is located in bundles in a ring surrounding the pith, with the xylem oriented exteriorly and the phloem oriented interiorly. Roots can undergo primary growth and lengthen, and secondary growth can increase the diameter. Over time, eudicot roots may develop concentric rings of xylem, and in woody eudicot roots, everything outside the stele is replaced by bark.

FIGURE **15.10** Specialized roots: (**A**) prop root of corn; (**B**) haustoria of dodder; (**C**) bald cypress knees.

portions of a fibrous root system. Many species of mature plants have a combination of a taproot and fibrous roots. Roots also possess root hairs that increase the surface area of the root and, therefore, its ability to absorb water and nutrients. In removing plants, the root system should not be injured because the delicate roots and root hairs are vital features.

The **root cap** is a group of specialized cells at the tip of the root. Its major function is to protect the delicate inner root from abrasive soil. The cells also produce a muscilaginous lubricant that helps the root pass through the soil and aids the growth of nitrogen-fixing bacteria in some plants, such as clover, peas, and peanuts. Cells in the root cap also are thought to orient the root growth toward the center of gravity, called gravitropism.

Some food storage roots, such as horseradish, dandelion, beet, radish, turnip, and carrot, have expanded food storage capability. Lily bulbs have **contractile roots** that help to pull the bulb deeper into the soil. **Aerial roots** are diverse; examples are English ivy, Virginia creeper, and banyan trees. **Pneumatophores** are spongy roots that extend out of the water in swamp plants such as the black mangrove (*Avicennia nitida*). At one time, the knees of bald cypress (*Taxodium distichum*) were thought to be pneumatophores. Today, their complete function is unknown, but they are thought to be primarily for support.

Uses of Roots

Some roots display complex relationships with other organisms. Members of the legume family Fabaceae, such as peanuts, clover, beans, and peas, possess small root nodules containing nitrogen-fixing bacteria. The bacteria produce enzymes that convert atmospheric nitrogen into nitrates and nitrogenous substances that can be absorbed by the roots.

Root knots are swellings, found in tomatoes and several other plants, which may house parasitic roundworms, or nematodes. The roots of many plants have a mutualistic relationship with fungi known as mycorrhizae. The plant supplies the fungi sugars and amino acids, and the fungus helps the plant metabolize phosphorus.

In the environment, roots hold soil together. This is particularly important in coastal regions. Roots are used for food by animals and humans. Edible roots include carrot, sweet potato, cassava, beet, horseradish, turnip, radish, and rutabaga, among others. Several roots are used for spices, including licorice, ginger, and sassafras. Sugar beets are a primary source of sugar, and the roots of yams have been a source of estrogen compounds in making birth control pills. The roots of many plants, including ipecac, gentian, ginseng, and reserpine, are used in medicines. The insecticide rotenone is derived from the roots of *Lonchocarpus* sp. Other roots are used to weave baskets and in the souvenir industry.

Procedure 1

Macroanatomy of Roots

Materials

- ❏ Dissecting microscope or hand lens
- ❏ Scalpel
- ❏ Instructor's choice: root specimens, such as a taproot, a fibrous root system, specimen with numerous lateral roots and root hairs, various modified roots, prominent root nodules, root knots, and mycorrhizae
- ❏ Germinating seed with a prominent radicle
- ❏ Colored pencils

1 Procure the specimens and equipment. Follow your instructor's directions, and observe the specimens in detail using a dissecting microscope or a hand lens. Describe, draw, and label the features of your specimens in the space provided below.

2 Observe root nodules, root knots, and mycorrhizae. Record and sketch your observations in the space provided on the following page.

3 Procure a germinating seed with a prominent radicle. Record and sketch your observations in the space provided on the following page.

Specimen 1 _____

_____

_____

Specimen 2 _____

_____

_____

15

Specimen 3 _____

Specimen 4 _____

Specimen 5 _____

Radicle _____

Procedure 2

Microanatomy of Roots

Materials

- ❏ Compound microscope
- ❏ Slides and coverslips
- ❏ Forceps
- ❏ Probe
- ❏ Prepared slides of eudicot and monocot roots
- ❏ Prepared slides of lateral root growth
- ❏ Colored pencils
- ❏ Paper towels

1 Procure equipment and prepared slides of various eudicot and monocot roots.

2 Carefully crush the nodule, knot, and mycorrhizae from Procedure 1 with a probe. Make a wet mount, and observe the specimen with a compound microscope. Record and sketch your observations in the space provided.

3 Observe, sketch, and label the eudicot root slides in the space provided. Compare your observations with Figures 15.11–15.15.

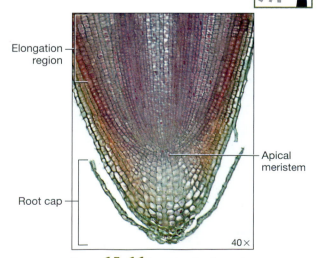

FIGURE **15.11** Longitudinal section of a pear root tip, *Pyrus* sp.

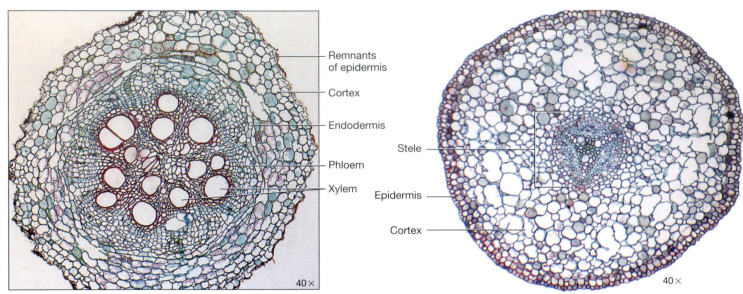

FIGURE **15.12** Transverse section of a sweet potato root, *Ipomaea* sp.

Labels: Remnants of epidermis, Cortex, Endodermis, Phloem, Xylem

40×

FIGURE **15.13** Transverse section of a young root of *Salix* sp.

Labels: Stele, Epidermis, Cortex

40×

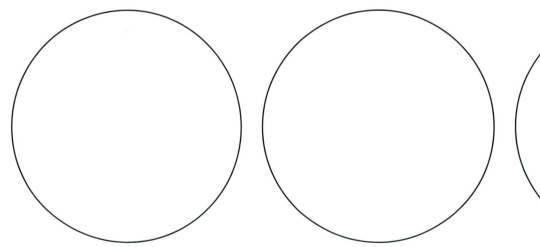

FIGURE **15.14** Transverse section of an older root of *Salix* sp. showing early secondary growth.

Labels: Epidermis, Cortex, Vascular tissue

120×

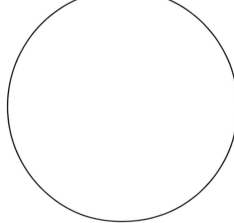

FIGURE **15.15** Transverse section of a young root of *Pyrus* sp.

Labels: Epidermis, Cortex, Primary phloem, Primary xylem, Endodermis

40×

15

Specimen 1 _____

Total magnification _____

Specimen 2 _____

Total magnification _____

Specimen 3 _____

Total magnification _____

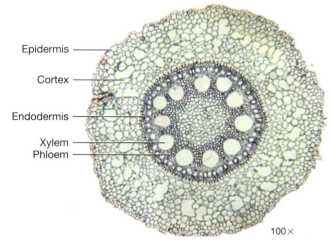

Specimen 4 _____

Total magnification _____

_____

_____

Specimen 5 _____

Total magnification _____

_____

_____

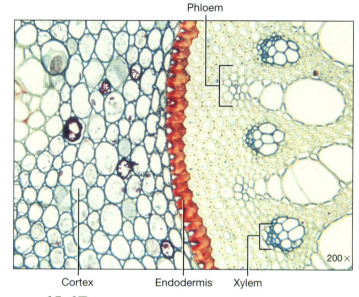

4 Observe, sketch, and label the monocot root slides in the space provided. Compare your observations with Figures 15.16 and 15.17.

Epidermis

Cortex

Endodermis

Xylem
Phloem

100×

FIGURE **15.16** Transverse section of the root of the monocot *Smilax* sp., low magnification.

Phloem

Cortex   Endodermis   Xylem   200×

FIGURE **15.17** High magnification of a root of the monocot *Smilax* sp.

15

5 Observe, sketch, and label your lateral root slides in the space provided. Compare your observations with Figures 15.18 and 15.19.

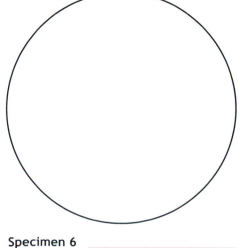

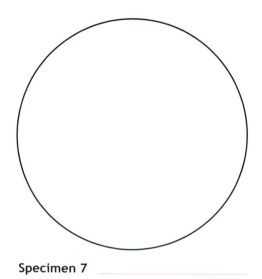

Specimen 6 _____

Total magnification _____

_____

_____

Specimen 7 _____

Total magnification _____

_____

_____

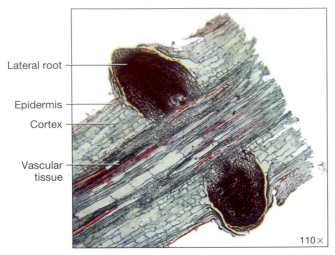

Lateral root

Epidermis

Cortex

Vascular tissue

110×

FIGURE **15.18** Longitudinal section of a willow species showing lateral root formation.

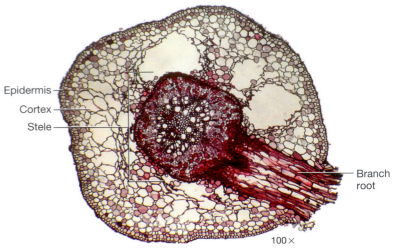

Epidermis

Cortex

Stele

Branch root

100×

FIGURE **15.19** Transverse section of the root of *Salix* sp. showing branch root development.

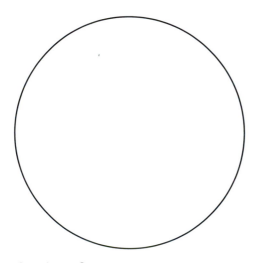

Specimen 8 _____

Total magnification _____

_____

_____

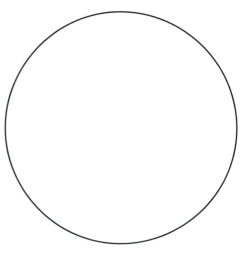

Specimen 9 _____

Total magnification _____

_____

_____

15

✔Check Your Understanding

2.1 Roots function to _____. (*Circle the correct answer.*)

   a. anchor and support a plant

   b. absorb water and nutrients

   c. store food

   d. produce growth-stimulating hormones

   e. all of the above

2.2 Fill in the table with the major zones of a typical root and their descriptions.

Zone	Description

2.3 Secondary growth involves the lengthening of a plant, and primary growth involves increasing the plant's diameter. (*Circle the correct answer.*)

True / False

2.4 Which of the following are considered specialized roots? (*Circle the correct answer.*)

   a. Buttress roots.

   b. Haustoria or parasitic roots.

   c. Pneumatophores.

   d. Both a and c.

   e. All of the above.

15

Stems

The trunk of a giant tree and all of its numerous branches and twigs are stems, as is the delicate body of a dandelion. Stems produce and support flowers and leaves, provide for the plant's growth, carry water and minerals up from the roots to the leaves to be used in photosynthesis, and carry food back down the plant to be stored and distributed as needed. **Herbaceous stems** are usually green, soft, and succulent compared with the harder, lignified, **woody stems**. The majority of monocots as well as several species of eudicots are herbaceous. Many species of eudicots possess woody stems.

Stem Anatomy

A simple twig can yield the basic external anatomy of a woody stem. The **terminal bud**, located at the tip of the twig, contains the tip of the shoot. The terminal bud is protected by modified leaves called **bud scales**. The apical meristem within the terminal bud is enveloped by immature leaves called **leaf primordia**. The bud scales associated with the terminal bud leave a distinct scar, the terminal bud scale scar. In many twigs, counting the number of terminal bud scale scars can denote the age of the twig.

Nodes mark the region of the stem where a leaf or leaves were attached by a stalk called a petiole (see Fig. 15.1, p. 295). Close examination of the nodes may yield bundle scars from past vascular tissue. The region between the nodes is the internode, which increases in length as a stem grows. **Axillary buds** that can give rise to new branches or flowers are located between the petiole and the stem. Close examination of a woody twig also may yield small, slightly raised structures on the twig known as **lenticels** (Figs. 15.20 and 15.21). These structures allow for gas exchange from the interior of the plant to the external environment.

When a twig breaks dormancy and begins to grow, cell division occurs within the apical meristem, giving rise to three types of primary meristem: protoderm, procambium, and ground meristem. The primary meristem adds to the length of a stem.

1. The protoderm, the outermost portion of the primary meristem, forms the epidermis. A waxy cuticle covers the epidermis in herbaceous plants.

2. The procambium, beneath the protoderm, forms primary xylem and phloem. A distinct band of cells between the primary xylem and phloem becomes the vascular cambium, which helps to form the girth of the twig.

3. The ground meristem forms the center of the stem, called the pith. In some plants the pith breaks down, forming a hollow cylinder, or in some woody plants is crushed as new tissues add girth to the stem. The cortex, also formed by the ground meristem, is used primarily for storage.

4. Cork cambium arises within the cortex or from the phloem or epidermis. It produces cork cells that, in turn, make up the bark.

Most conifers and flowering plants have eusteles, a structure in which the primary xylem and phloem exist in distinct vascular bundles separated by parenchyma tissue (Fig. 15.22). Ultimately, the arrangement of the vascular bundles depends upon the stem's ability to undergo secondary growth, or girth (Figs. 15.23 and 15.24). One of the characteristics that distinguish eudicots from monocots is the arrangement of the vascular bundles. In eudicots, the vascular bundles form a ring around the outside of the stem. In most monocots (except some grasses with a ground tissue cavity, such as wheat), the vascular bundles are spread throughout the ground tissue. In both the eudicots and the monocots, the phloem faces outward and the xylem faces inward. As discussed in Chapter 14, the annual patterns of vascular cambium form the growth or annual rings of a tree.

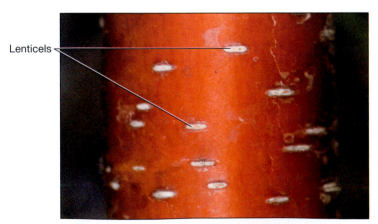

FIGURE **15.20** Bark of a birch tree, *Betula occidentalis*, showing lenticels.

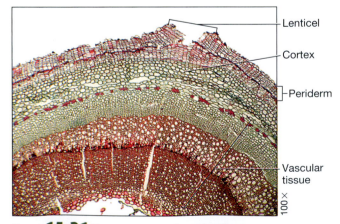

FIGURE **15.21** Transverse section of a dicot stem showing a lenticel and stem tissues.

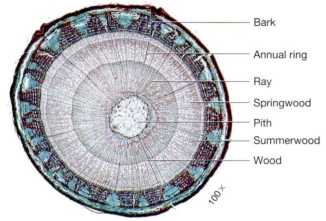

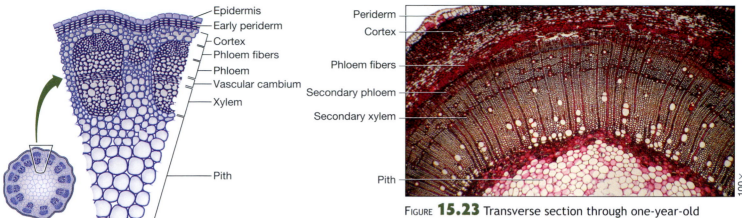

FIGURE **15.22** Diagram of collateral vascular bundles in the eustele of a dicot stem.

FIGURE **15.23** Transverse section through one-year-old *Fraxinus* sp. stem showing secondary growth.

FIGURE **15.24** Transverse section through a three-year-old *Tilia* sp. stem showing secondary growth.

Types of Stems

Many angiosperms possess modified stem systems that perform specialized tasks. Although the appearance of these stems may vary, they all have nodes, internodes, and axillary buds, which distinguishes them from roots.

1. **Rhizomes** are a type of modified stem in which the stem grows horizontally below the ground, resembling a root. The rhizome may be slender, as in some grasses and ferns, or a thick structure, as in some irises.

2. **Runners** are similar to rhizomes except they are above ground. Strawberry plants produce runners after they flower. In philodendrons, runners can be seen giving rise to new plants that trail down from a basket.

3. **Stolons** are runner-like, growing beneath the surface and in different directions. Some botanists do not distinguish between runners and stolons.

4. **Tubers** are swollen extensions of stolons modified to store carbohydrates. The Irish potato is an example of a tuber. The eyes of the potato are actually axillary buds.

5. **Bulbs,** as in onions, tulips, and lilies, possess small underground stems with large buds. Adventitious roots grow from the bottom of the stem, and fleshy leaves make up most of the bulb.

6. **Corms** look like bulbs, but they do not have fleshy leaves. The only leaves are thin, papery, brown structures on the outside of the corm. Examples of corms are gladiolus and crocus.

7. **Cladophylls** are flattened photosynthetic stems. In cacti, the cladophylls are the broadened green structure, and the spines are modified leaves.

8. **Tendrils** are common in climbing plants, such as grapes and greenbrier. Some tendrils, such as those of pumpkins and peas, are modified leaves.

9. The **thorns** of honey locust are modified stems.

Uses of Stems

Stems are integral in human civilization. Some stems are a food source for animals as well as humans. Stems are used in wood products from lumber to toothpicks. Wood is used in pulp for making paper, fibers, and even filler for ice cream and bread. Sugarcane is a major source of sugar in syrups, soft drinks, and other foods. Cinnamon spice is derived from the stem of *Cinnamomum* spp., and the antimalarial drug quinine is derived from the bark of *Cinchona* spp.

15

Macroanatomy of Stems

Materials

- ❑ Dissecting microscope or hand lens
- ❑ Instructors choice: stem specimens such as a woody twig, an herbaceous eudicot, a monocot, or cross section of a tree trunk
- ❑ Camera or camera phone (optional)
- ❑ Colored pencils

The stem of an angiosperm is often the ascending portion of the plant specialized to produce and support leaves and flowers, transport and store water and nutrients, and provide growth through cell division.

1 Procure the specimens and equipment.

2 Observe the specimens in detail using a dissecting microscope and a hand lens. Describe, draw, and label the anatomical features of your specimens in the space provided. Compare your observations with Figure 15.25.

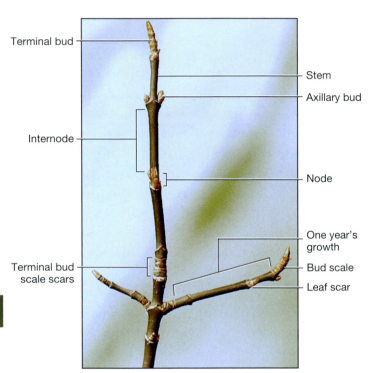

Terminal bud

Internode

Terminal bud scale scars

Stem

Axillary bud

Node

One year's growth

Bud scale

Leaf scar

FIGURE **15.25** External anatomy of a woody twig.

Woody twig

Herbaceous eudicot stem

Monocot stem

Cross section of tree trunk

15

Materials

- ❏ Dissecting microscope or hand lens
- ❏ Instructor's choice of modified stems, such as rhizome, runner, stolon, tuber, bulb, corm, cladophyll, tendril, thorn
- ❏ Colored pencils

Consider photographing this activity.

1 Procure the specimens and equipment.

2 Identify your specimen and the type of modified stem it has (Fig. 15.26).

3 Sketch each specimen, and discuss the function of each in the space provided.

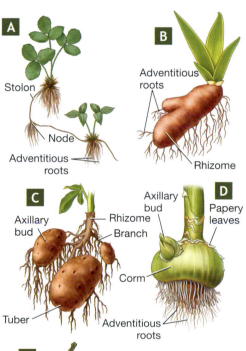

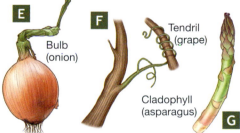

Specimen 1 _____

Specimen 2 _____

Specimen 3 _____

Specimen 4 _____

FIGURE **15.26** Examples of the variety and specialization of angiosperm stems: (**A**) runners; (**B**) rhizomes; (**C**) tubers; (**D**) corms; (**E**) bulbs; (**F**) tendrils; (**G**) cladophyll.

15

Microanatomy of Stems

Materials
- ❏ Compound microscope
- ❏ Prepared slides of eudicot and monocot stems
- ❏ Colored pencils

1 Procure equipment and prepared slides of various eudicot and monocot stems.

2 Observe, sketch, and label the eudicot stem slides in the space provided. Compare your observations with the eudicot stem images in Figures 15.27 and 15.28.

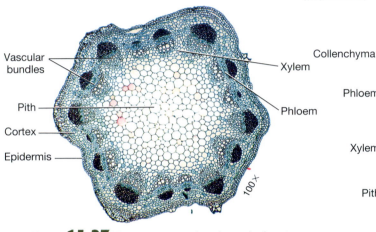

FIGURE **15.27** Transverse section through the stem of a young sunflower, *Helianthus* sp.

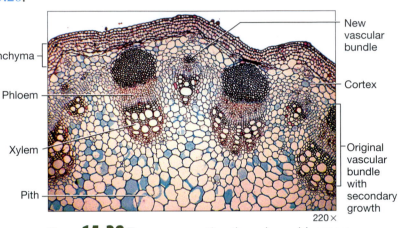

FIGURE **15.28** Transverse section through an older stem of a sunflower, *Helianthus* sp., at high magnification.

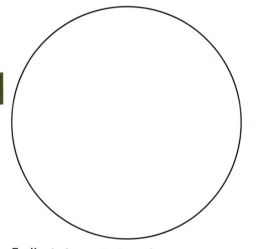

Eudicot stem   Total magnification _____

Eudicot stem   Total magnification _____

3 Observe, sketch, and label the monocot stem slides in the space provided. Compare your observations with the monocot stems in Figure 15.29.

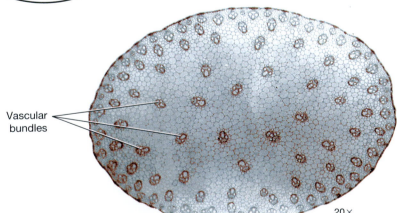

FIGURE **15.29** Transverse section from the stem of a monocot, *Zea mays* (corn) showing the vascular bundles.

15

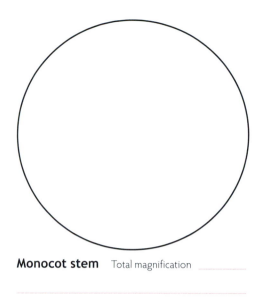

Monocot stem Total magnification _____

_____

_____

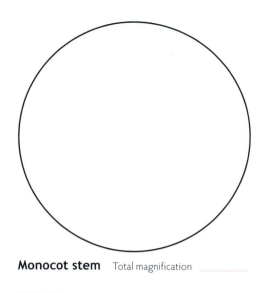

Monocot stem Total magnification _____

_____

_____

✔ Check Your Understanding

3.1 Which of the following are functions of plant stems? (*Circle the correct answer.*)

a. Carry food from the leaves down the plant to be stored.

b. Carry water and minerals up from roots to the leaves to be used in photosynthesis.

c. Support flowers and leaves.

d. Provide for the growth of the plant.

e. All of the above.

3.2 Identify the type of speciality stem described in each example.

_____ It is a swollen extension of a stolon modified to store carbohydrates.

_____ It is a flattened photosynthetic stem as seen in cacti.

_____ The stem grows horizontally below the ground, resembling a root.

_____ It is a small underground stem with large buds; adventitious roots grow from the bottom of the stem and it has fleshy leaves.

_____ It is runner-like; it grows beneath the surface in different directions.

_____ It is similar to a bulb, but has a thin, papery, brown structure on the outside, not fleshy leaves.

_____ Gladioli and crocuses are examples.

_____ The potato is an example.

_____ The iris is an example.

3.3 What is a lenticel, and what is its role in a plant?

_____

_____

15

Leaves

Albert Camus (1913–1960) once exclaimed, "Autumn is a second spring where every leaf is a flower." Leaves are the most conspicuous structures of a plant, and how and why the colors of autumn come about has always intrigued inquisitive minds. Most leaves are green as a result of chlorophyll; however, the brilliant colors of autumn leaves emerge when chlorophyll is broken down, and pigments such as carotene (orange), xanthophyll (yellow), and anthocyanin (red) show through. The leaves of oaks and some trees turn brown or tan as the result of a reaction between the tannin stored in vacuoles and the proteins in the leaf.

The process by which plants seasonally lose their leaves or lose their leaves from injury or drought is called abscission. This process begins in a specific region of the petiole called the abscission zone and is controlled by auxin and ethylene.

Leaves are the primary food factories in a typical plant. Powered by the sun's energy, leaves take in carbon dioxide and produce oxygen and glucose through photosynthesis. Because leaves require light to function, the leaves have to be efficient at gathering and using light energy. To capture sunlight, leaves have evolved many novel methods. Some leaves have attained tremendous size; examples are the aroid (giant elephant ear) of Borneo (its heart-shaped leaves are 10 meters across) and the raffia palm, *Raphia regalis*, of tropical Africa (with leaves up to 24 meters in length). Others, such as the sunflower, follow the sun during the course of the day. Some leaves have purple or red undersides to reflect the light back through the thickness of the leaf. In certain begonias, some surface cells are transparent, acting as tiny lenses to focus light into the plant.

Raking the lawn makes you realize that trees have the capability to produce many leaves. A mature oak tree may produce more than 500,000 leaves in a year and a mature elm several million. Each year, leaves produce an estimated 200 billion tons or more of sugar worldwide.

Leaf Anatomy

Leaves are attached to the stem at the node. The arrangement of leaves on a stem is the **phyllotaxy**. In plants with **opposite leaf attachment**, such as dogwood and maple, two leaves are attached at each node. In **alternate leaf arrangement**, such as in poplar and aspen, a single leaf appears at each node. The majority of plants have the alternate leaf arrangement. In **whorled leaf arrangement**, as in bedstraw and oleander, three or more leaves are attached at a single node (Fig. 15.30).

The veins of a leaf are composed of vascular tissue. In basal dicots and eudicots, the veins are arranged in a netted, or reticulate, pattern. **Pinnately veined** leaves, as in apple and cabbage, have one prominent primary vein, or midrib, and secondary veins branch off the midrib. In **palmately veined** leaves, as in sycamore and sweetgum, several primary veins branch out from a single point. Monocots, such as corn and sugarcane, have **parallel venation**, in which the veins are arranged nearly parallel to each other. Interestingly, *Ginkgo biloba* has no midrib, and the veins fork out from the base of the blade. *Ginkgo* exhibits **dichotomous venation** (Fig. 15.31).

FIGURE **15.30** Leaf arrangements on stems: (**A**) opposite; (**B**) alternate; (**C**) whorled.

FIGURE **15.31** Leaf venation: (**A**) pinnate; (**B**) parallel; (**C**) palmate; (**D**) dichotomous.

Types of Leaves

A leaf may be classified a **simple leaf** if it has a single blade, such as azalea and birch leaves, or a **compound leaf**, such as ash trees and pecans, if it is divided into smaller leaflets. In **pinnately compound** leaves, such as black walnut and locust, leaflets occur in pairs along the rachis (extension) of the petiole, and in **bipinnately compound** leaves, such as the silk tree and mimosa, the leaflets are subdivided into smaller leaflets. In **palmately compound** leaves, such as buckeye and Virginia creeper, all of the leaflets are attached at the same origin (Fig. 15.32).

Uses of Leaves

In addition to the fact that leaves are an essential source of oxygen and food for animals, humans use leaves and their products in a number of ways. Many medicinal products are derivatives of leaves. These include aloe for burns; digitalis, a heart stimulant from fox-glove; and atropine, which lowers parasympathetic activities, from belladonna, jimsonweed, and mandrake. Tobacco, marijuana, and cocaine also are leaf products. Prickly pear, bearberry, henna, sassafras, and indigo leaves are used for dyes. The leaves from some plants are used in beverages, including agave in tequila and camellia leaves in tea. Spices are derived from a number of leaves, including peppermint, oregano, spearmint, wintergreen, bay, and basil, among others.

FIGURE **15.32** Leaf complexity: (**A**) palmately compound; (**B**) simple; (**C**) pinnately compound; (**D**) bipinnately compound.

15

Did you know . . .

Kitty Craze!

Have you ever given your cat some catnip? When cats smell catnip, they display a variety of strange behaviors including head shaking, drooling, turning over, rubbing, and acting bizarre. An estimated 70% of cats have catnip receptors. Catnip is a weed-like perennial herb, *Nepeta cataria*, belonging to the mint family Labiatae. Nepetalactone, the active ingredient in catnip, is thought to mimic the effects of a pheromone that causes a variety of psychosexual behaviors. Catnip is commonly called "kitty cocaine." Catnip also has been used in human medicine for bronchitis, insomnia, diarrhea, and colic.

Macroanatomy of Leaves

Materials

❏ Dissecting microscope or hand lens
❏ Instructor's choice: specimens representing a variety of types of leaves discussed
❏ Colored pencils

Despite great diversity in size and shape, all leaves originate in the leaf primordia of a bud. Over time, the immature leaf takes the characteristic shape of its species. The majority of leaves are attached to a stem by a stalk-like petiole and possess a flattened blade, the **lamina**. The blade can vary in shape and size. Even within a tree leaf, size can vary; shade leaves are larger and thinner, and sun leaves are smaller and thicker. In some leaves, a pair of thornlike structures called stipules is present at the base of the petiole (Fig. 15.33).

FIGURE **15.33** Basic leaf anatomy of a cottonwood, *Populus* sp., leaf.

1 Procure specimens and equipment. Observe the specimens in detail using a dissecting microscope or a hand lens. Describe, draw, and label the anatomical features in the space provided. Remember to include both the scientific name and the common name of your specimen.

2 Using the dissecting microscope, make observations of the stomata and guard cells on the leaves, and sketch and label your observations in the space provided. Were the stomata open on your specimens?

Specimen 1 _____

Specimen 2 _____

Specimen 3 _____

Procedure 2

Leaf Classification

Materials

❏ Hand lens
❏ Instructor's choice: specimens representing a variety of leaf shapes and margins
❏ Colored pencils

Consider photographing this activity.

1 Procure the leaf specimens and equipment.

2 Using Figures 15.34 and 15.35 classify the leaves provided by the instructor. Record your observations in Table 15.3 on page 320.

Rhomboid
diamond-shaped

Acuminate
tapering to a long point

Ovate
egg-shaped, wide at base

Alternate
leaflets arranged alternately

Palmate
like a hand with fingers

Lanceolate
pointed at both ends

Spear-shaped
pointed, barbed base

Bipinnate
leaflets also pinnate

Peltate
stem attached centrally

Cordate
heart-shaped, stem in cleft

Lobed
deeply indented margins

Trifoliate/Ternate
leaflets in threes

Cuneate
wedge-shaped, acute base

Odd Pinnate
leaflets in rows, one at tip

Even Pinnate
leaflets in rows, two at tip

Deltoid
triangular

Truncate
squared-off apex

Digitate
with fingerlike lobes

Obtuse
bluntly tipped

Elliptic
oval-shaped, small or no point

Opposite
leaflets in adjacent pairs

Whorled
rings of three or more leaflets

FIGURE **15.34** Representative leaf shapes.

Ciliate
with fine hairs

Crenate
with rounded
teeth

Dentate
with symmetrical
teeth

Entire
even, smooth
throughout

Lobate
indented, but
not to midline

Serrate
teeth
forward-pointing

Spiny
with sharp,
stiff points

Sinuate
with wavelike
indentations

FIGURE **15.35** Representative leaf margins.

15

TABLE **15.3** Observations and Classification of Leaves

Specimen	Scientific Name	Common Name	Shape	Margin
1				
2				
3				
4				
5				
6				
7				
8				
9				
10				
11				
12				
13				
14				
15				
16				
17				
18				
19				
20				

15

Microanatomy of Leaves

Materials

- ❏ Compound microscope
- ❏ Prepared slides of various monocot and dicot leaves, such as transverse sections
- ❏ Colored pencils

The internal anatomy of a typical eudicot leaf is complex, consisting of three basic types of tissues: epidermal, ground, and vascular tissue. The epidermis of a typical leaf is present on both sides of the blade. It is transparent and does not undergo photosynthesis. In a horizontal leaf, a number of openings, or stomata, can be found mostly on the underside. An oak leaf may have nearly 60,000 stomata and a lettuce leaf nearly 11 million. The frequency and arrangement of stomata are determined genetically.

Stoma are essential in gas exchange. The opening of a stoma is regulated by a pair of adjacent guard cells. A waxy **cuticle** can be observed above the epidermis on many leaves. The wax cutin is important in water balance and in discouraging insect predation. **Leaf hairs**, a type of trichome, are extensions of the epidermis that increase the surface area of the leaf. Other trichomes may be glandular, producing oils.

The **mesophyll**, or ground tissue, in a typical leaf is composed of **palisade mesophyll** and **spongy mesophyll** (parenchyma). The palisade mesophyll is made of tightly packed parenchyma cells and the spongy mesophyll of loosely packed cells. The majority of photosynthesis occurs in the mesophyll. Veins held within bundle sheaths house the vascular tissue of a plant. The veins of a leaf are composed of xylem on the top side of a vein and phloem on the bottom side of a vein.

1 Procure the slides and equipment.

2 Observe the slide in detail using a compound microscope on both low and high power. Compare the microanatomy of your slides with the micrographs in Figures 15.36–15.42. Observe, draw, and label the anatomical features in the space provided.

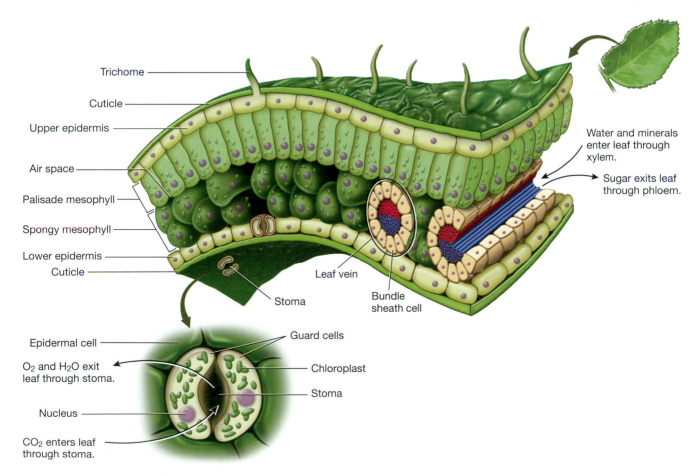

FIGURE **15.36** Cross section of a typical leaf. In this diagram, xylem is red and phloem is blue.

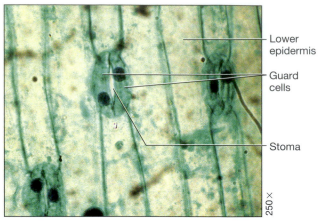

FIGURE **15.37** Face view of the epidermis of onion, *Allium* sp. Note the twin guard cells with the stoma opened.

Lower epidermis

Guard cells

Stoma

250×

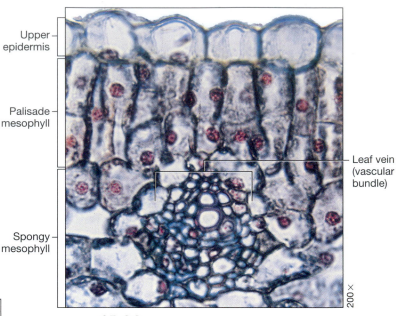

Upper epidermis

Palisade mesophyll

Spongy mesophyll

Leaf vein (vascular bundle)

200×

FIGURE **15.38** Transverse section of tomato leaf, *Lycopersicon* sp.

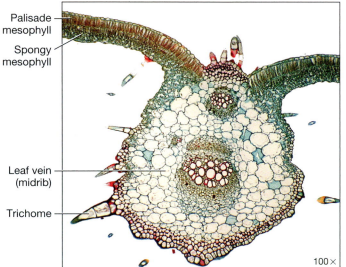

Palisade mesophyll

Spongy mesophyll

Leaf vein (midrib)

Trichome

100×

FIGURE **15.39** Transverse section through a cucumber leaf, *Cucurbita* sp.

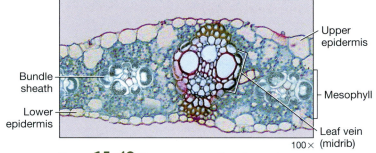

Upper epidermis

Bundle sheath

Lower epidermis

Mesophyll

Leaf vein (midrib)

100×

FIGURE **15.40** Transverse section through a corn leaf.

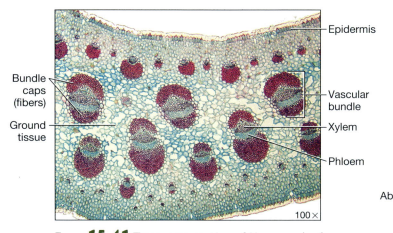

Epidermis

Bundle caps (fibers)

Vascular bundle

Ground tissue

Xylem

Phloem

100×

FIGURE **15.41** Transverse section of *Yucca* sp. leaf.

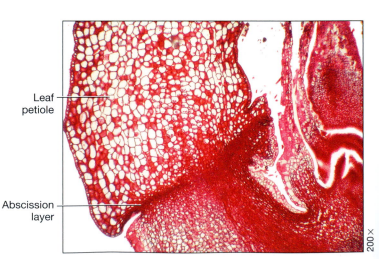

Leaf petiole

Abscission layer

200×

FIGURE **15.42** Longitudinal section of privet, *Ligustrum* sp., through the stem and petiole at the abscission layer.

15

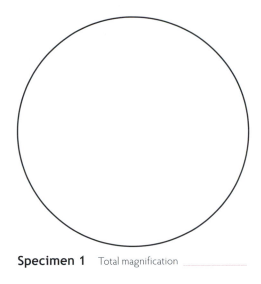

Specimen 1 Total magnification _____

_____

_____

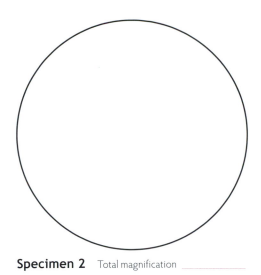

Specimen 2 Total magnification _____

_____

_____

Procedure 4

Macroanatomy of Specialized Leaves

Materials

- ❏ Hand lens
- ❏ Instructor's choice of select specialized leaves
- ❏ Colored pencils
- ❏ Camera or camera phone (optional)

Myriad specialized leaves are found in the plant world performing a number of duties. Thick, water-retaining succulent leaves, such as aloe, jade, and yuccas, can be found in plants from arid regions to salty seashores. The spines of cacti, effective defenses, are actually modified leaves that evolved to conserve water. Many plants living in aquatic regions, such as the water hyacinth, have large air spaces of aerenchyma tissue that allow them to float. The sensitive plant, *Mimosa pudica*, droops when touched. The drooping, or nastic movement, is the result of changes in turgor pressure. In some plants, such as the garden pea, leaves are modified to form tendrils that enable climbing.

Bracts are specialized leaves at the base of a flower. The brightly colored bracts of poinsettia are popular at Christmas. The obvious white or pink bracts of dogwood resemble flower petals. Approximately 200 species of plants have modified insect-trapping leaves; these include pitcher plants, sundews, and Venus flytraps.

1 Procure specimens and equipment.

2 Using a hand lens, observe the specimens in detail. Describe, draw, and label the specialized leaves in the space provided on the following page. Remember to include both the scientific name and the common name of your specimen and the function of the specialization. Compare your observations with Figure 15.43.

FIGURE **15.43** Examples of modified leaves: (**A**) succulent leaf of an aloe; (**B**) spines of a cactus; (**C**) sensitive plant; (**D**) floating leaves of a giant water lily; (**E**) tendrils of a garden pea; (**F**) insect-trapping leaves of a Venus flytrap.

15

Specimen 1 _____

Specimen 2 _____

Specimen 3 _____

✔ Check Your Understanding

4.1 Match the specialized leaf type of each plant with the descriptions. (Some descriptions may be used twice.)

_____ Dogwood

_____ Cacti

_____ *Mimosa pudica*

_____ Water hyacinth

_____ Garden pea

_____ Sundew

_____ Poinsettia

_____ Yucca

A. It has water-retaining succulent leaves and is found in arid regions.

B. Its leaves are sensitive to touch due to changes in turgor pressure.

C. Its tendrils are modified to enable climbing.

D. It has bracts (showy leaves at the base of a flower).

E. It has spiny structures that evolved to conserve water.

F. It has floating leaves with large air spaces of aerenchyma tissue.

G. It has insect-trapping leaves.

4.2 What is the function of leaf hairs?

_____

_____

_____

4.3 Sketch and label a typical transverse section of a dicot leaf.

15

MYTHBUSTING

Don't Bark Up the Wrong Tree!

Debunk each of the following misconceptions by providing a scientific explanation. Write your answers on a separate sheet of paper.

1 The best way to get rid of poison ivy is to burn it.

2 Trees do not have sun leaves and shade leaves.

3 When the weather gets cold, leaves do not receive as much sunlight and rain and soon die.

4 A potato is a big seed.

5 Plants are static and unresponsive to their environment.

1 Plant epidermal cells may form root and leaf hairs. What is the role of these structures?

2 To colonize terrestrial environments, plants developed many special adaptations. Discuss the importance of development of vascular tissue in this process.

3 Compare vascular bundles arrangement in eudicots and monocots. (*Circle the correct answer.*)

a. Vascular bundles form a ring around the outside of the stem in eudicots; vascular bundles are scattered throughout ground tissue in monocots.

b. Phloem faces inward and xylem outward in vascular bundles in both monocots and dicots.

c. Phloem faces outward and xylem inward in vascular bundles in both monocots and dicots.

d. Both a and b.

e. Both a and c.

4 Which of the following statements comparing roots, stems, and leaves in eudicots and monocots are true? (*Circle the correct answer.*)

a. Monocots have fibrous roots with a poorly developed primary root and well-developed secondary roots.
Eudicots have well-developed primary roots.

b. Monocots have scattered vascular bundles.
Eudicots have vascular bundles form a ring around the stem.

c. Monocots have netted venation.
Eudicots have parallel venation.

d. Both a and b.

e. All of the above.

5 Label the micrographs in Figure 15.44 of a (**A**) root; (**B**) stem; and (**C**) leaf.

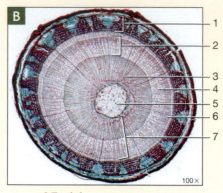

FIGURE **15.44** (**A**) Root micrograph.

FIGURE **15.44** (**B**) Stem micrograph.

FIGURE **15.44** (**C**) Leaf micrograph.

A	B	C
1. _____	1. _____	1. _____
2. _____	2. _____	2. _____
3. _____	3. _____	3. _____
4. _____	4. _____	4. _____
5. _____	5. _____	5. _____
	6. _____	6. _____
	7. _____	

6 Describe the orientation of xylem and phloem in monocot and eudicot roots, stems, and leaves.

Type	Part	Orientation of Xylem	Orientation of Phloem
Monocot	Root		
	Stem		
	Leaf		
Eudicot	Root		
	Stem		
	Leaf		

15

The Green Machine IV
Understanding Flowers, Fruits, and Seeds

16

From a moth's point of view, flowers that reliably provide nectar are like docile, productive milch cows. From the flowers' point of view, moths that reliably transport their pollen to other flowers of the same species are like a well-paid Federal Express service, or like well-trained homing pigeons."

— Richard Dawkins (1941–present)

Just wondering...

Consider the following questions prior to coming to lab, and record your answers on a separate piece of paper.

1 Why do many fruits taste so good?
2 What is kopi luwak, and why is it so unique?
3 What is the nutritional value of bananas?
4 What causes an apple to turn brown shortly after it is cut?
5 Why is wild thistle Scotland's national flower?

 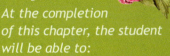
Flowers

In 1878, Italian botanist Odoardo Beccari (1843–1920) recorded finding a corpse flower (titan arum, or *Amorphophallus titanum*) in the rainforest of Sumatra with a circumference of more than 1.5 meters and a height exceeding 3 meters. The fragrance of the flower is like a rotting corpse—hence its common name. The corpse flower is the largest unbranched **inflorescence** (cluster of flowers) in the world. The largest *branched* inflorescence is produced by the talipot palm, *Corypha umbraculifera*, native to Southeast Asia. Its inflorescence can attain a length of 8 meters and have several million individual flowers.

The largest single flower is produced by *Rafflesia arnoldii*. It resembles a giant, reddish-brown mushroom and can attain a diameter of more than a meter and weigh 11 kilograms. Like the corpse flower, when in bloom its smell is similar to rotting meat, which attracts insects that pollinate the plant. It is a native of the rainforest of Indonesia. The smallest flower in the world is produced by *Wolffia globosa*, a type of duckweed. The mature plant weighs about the same as two grains of salt. A bouquet of a dozen of these tiny flowers would be about the size of the head of a pin.

Flower Anatomy

Despite the tremendous range in size, color, and shape, all flowers are made primarily of the same components (Fig. 16.1). Flowers begin to develop from a specialized stalk, the **peduncle**, or from several smaller stalks, **pedicels**. The peduncle or pedicels form the **receptacle**, a swollen region that contains the other floral parts arranged in **whorls**. Keep in mind that in eudicots, the flower parts are arranged in multiples of four or five, and in monocots in multiples of three.

FIGURE **16.1** Plumeria flower with peduncles and pedicels.

The outermost whorl is composed of leaflike, green **sepals**. The sepals, in turn, form the **calyx**. In many species the calyx serves to protect the flower while it develops within the bud. The **petals**, the most conspicuous parts of a flower, range in color, shape, size, and fragrance. The petals collectively form the **corolla**. The color of the petals and the shape of the corolla are significant in pollination. The calyx and corolla make up the **perianth**.

The "male" portion of the flower, the **stamen**, consists of a slender stalk, the **filament**, and the saclike **anther** where pollen is produced. Collectively, the stamens make up the **androecium**. Usually, anthers release pollen by splitting open (as in a daisy), but in some species (such as azaleas) the pollen is released from pores at the tip of the anther.

The most obvious female portion of the plant is the centrally located **pistil**. (Some texts refer to the pistil as a **carpel**.) It is composed of a sticky knob that receives pollen and is known as the **stigma**, which sits atop a slender tube called the **style**, which leads to the **ovary** (Fig. 16.2). The position of the ovary is significant in plant classification. If the calyx and corolla are attached to the receptacle at the

base of the ovary, the ovary is classified as superior, or **hypogynous**, as in grape, honeysuckle, and quince.

In an inferior, or **epigynous**, ovary, the calyx and corolla appear to be attached at the top of the ovary, as in blueberry, watermelon, and pear. In a semi-inferior, or **perigynous**, ovary, the calyx and corolla are found on a cup-shaped structure surrounding the receptacle, such as in crape myrtle, cherry, and *Pyracantha*. An ovary eventually forms the fruit. One or more carpels serve as the main portion of the ovary. The carpel or carpels are known collectively as the **gynoecium**. Ovules, produced on the carpels, contain the female gametophyte. Generally, the number of carpels is related to the number of divisions of the stigma. For example, each section of a tomato or a grapefruit represents a carpel.

Flower Classification

Flower terminology and arrangement is complex. **Complete flowers,** such as magnolias, tulips, apples, azaleas, and lilies, possess sepals, petals, stamens, and pistils. **Incomplete flowers** lack one or more sepals, petals, stamens, or pistils; examples are squash, begonia, oak, and walnut (Fig. 16.3). A **perfect flower,** such as a dandelion, lily, banana, or pea, possesses both stamens and pistils. An **imperfect flower** possesses only one sex because it lacks either the stamens or the pistil (Fig. 16.4). **Staminate flowers** have only stamens, and **pistillate flowers** have only pistils. Imperfect flowers may appear on separate plants, as in holly, mulberry, and persimmon, or on the same plant, as in cattail, oak, and corn.

Floral symmetry refers to the arrangement of flowers along a plane. **Actinomorphic** (radially symmetrical) flowers can be divided into symmetrical halves by more than one longitudinal plane passing through the axis. In these flowers the petals are similar in shape and size. Examples of actinomorphic flowers are azaleas, buttercups, and roses. **Zygomorphic** (bilaterally symmetrical) flowers can be divided by a single plane into two mirror-image halves (Fig. 16.5). Zygomorphic flowers generally have petals of two or more different shapes and sizes. Examples of zygomorphic flowers are the orchid, foxglove, and snapdragon.

Flowers may be **solitary,** as in a petunia or camellia, or appear in clusters known as an inflorescence, as in oak, sunflower, and willow (Fig. 16.6). Catkins (Dutch = kitten) are a drooping, slim inflorescence, lacking petals or having inconspicuous petals that resemble a kitten's tail. They contain many, usually unisex, flowers arranged closely along a central stem. In some plants, such as

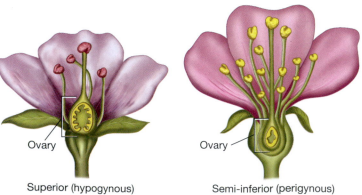

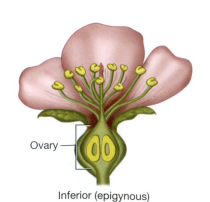

Superior (hypogynous) Semi-inferior (perigynous) Inferior (epigynous)

FIGURE **16.2** Position of the ovary in angiosperms.

16

FIGURE **16.3** Example of (**A**) complete flower, lily, *Lilium* sp.; and (**B**) incomplete flower, grass flowers.

FIGURE **16.4** Example of (**A**) perfect flower, hibiscus, *Hibiscus*, sp.; and (**B**) imperfect flower, pitcher plant, *Sarracenia* sp.

FIGURE **16.5** Example of (**A**) actinomorphic symmetry, daffodil, *Narcissus* sp.; (**B**) zygomorphic symmetry, pansy, *Viola tricolor*.

FIGURE **16.6** (**A**) Solitary flower, passionflower, *Passiflora* sp.; and inflorescent flowers, (**B**) sunflower, *Helianthus* sp.; (**C**) walnut catkin, *Juglans* sp.

willow, mulberry, and oak, only the male flowers form catkins and the female flowers are solitary. In other plants, such as poplar, both male and female flowers are borne in catkins.

Grasses

Many people do not realize grasses produce flowers. In the summer, just let your yard get out of control and notice the flowers and seed heads. The flowers of grasses are inflorescence. In bluegrass, wheat, rice, and other herbaceous grasses, each leaf consists of a **basal sheath** that encompasses the **culm** (grass stem) down to its point of origin, the **node**. The **internodes** of grasses typically are hollow, such as in bamboo. The leaf blade usually grows away from the culm. A membranous scale called the ligule can be found at the junction of the basal sheath and leaf blade. The tiny projections near the base of the leaf blade are known as auricles.

A **spikelet** is a conspicuous extension of the peduncle, consisting of many small florets (Fig. 16.7). A glume, designated as the first and second glume or protective husk, appears externally in the **floret**. A floret consists of two bracts—the

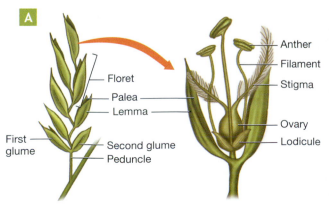

FIGURE **16.7** (**A**) Floral structure of grasses; (**B**) *Elymus flavescens*, showing spikelets with six florets.

lemma and the palea. Within the floret are one pistil, three stamens, the ovary, and the scale-like lodicule. After fertilization, the ovary develops into a one-seeded fruit, a grain, or caryopsis. When weeding the yard, if you cut the peduncle, the seed head will not repair.

Pollination

The sporophyte is the dominant generation in angiosperms (Fig. 16.8). In the male portion of the plant, anthers have four **pollen sacs** containing numerous microsporocytes. Each microsporocyte produces four haploid microspores. After cell division, the haploid nuclei of the microspores produce a pollen grain. In the female portion of the plant, one or several ovules can be found in the ovary. Within an ovule, four megaspores are formed. One of the megaspores becomes an **embryo sac,** or female gametophyte.

During pollination, a two-celled pollen grain lands on the stigma of the same species of plant; one cell forms a tube cell and the other a generative cell (Fig. 16.9). The tube cell will form the pollen tube, and the generative cell will produce two sperm cells. The pollen tube moves down the style to the ovary with an awaiting ovule. Of the two sperm cells, one fertilizes the egg and the other fuses with two polar nuclei to form a $3n$ endosperm. This is called **double fertilization**. The endosperm nourishes the developing embryo. The ovule develops into a seed containing the embryonic plant (sporophyte) and the endosperm.

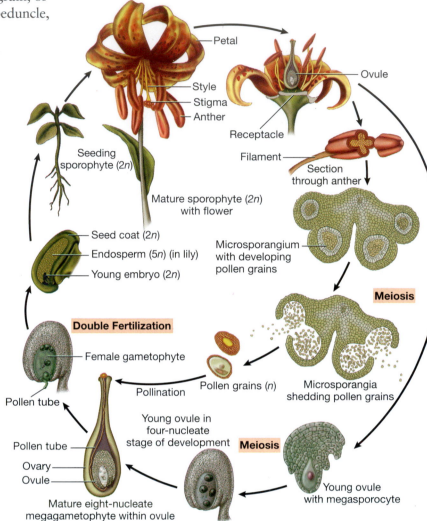

FIGURE **16.8** Life cycle of an angiosperm.

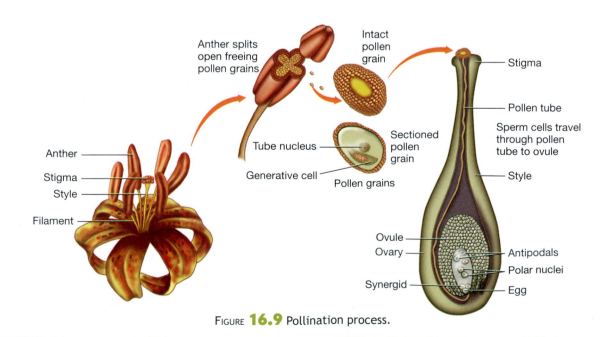

FIGURE **16.9** Pollination process.

Flowers

Procedure 1

Flower Dissection

Materials

- ❏ Dissecting microscope or hand lens
- ❏ Scalpel
- ❏ Dissecting tray
- ❏ Flowers provided by the instructor
- ❏ Colored pencils

1 Procure the specimens and equipment.

2 Describe the specimens in detail and, using a dissecting microscope or a hand lens, observe, draw, and label the anatomical features of your specimens in the space provided (refer to Figs. 16.10–16.13). Be sure to include the common name and the scientific name of the plant in your description. Also include whether the flower is a basal dicot, eudicot, or monocot. In addition, include whether the flower is complete or incomplete, perfect or imperfect, solitary or an inflorescence, actinomorphic or zygomorphic, and if it has a hypogynous, epigynous, or perigynous ovary.

FIGURE **16.11** Structure of a dissected cherry, *Prunus* sp., showing a perigynous flower.

FIGURE **16.10** Floral structure of a lily, *Lilium* sp.

FIGURE **16.12** Structure of a dissected pear, *Pyrus* sp., showing an epigynous flower.

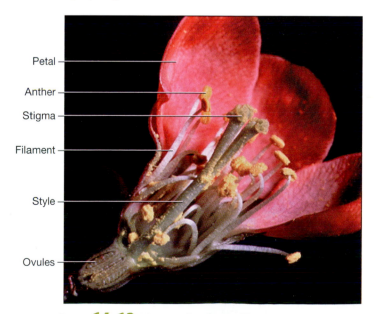

FIGURE **16.13** Dissected quince, *Chaenomeles japonica*, showing an epigynous flower.

Specimen 1

_____

_____

Specimen 2

_____

_____

3 Describe the smell of your flower.

_____

_____

_____

4 Carefully remove the sepals and petals with your fingers from a flower designated by the instructor. Closely observe the male and female reproductive parts, using a dissecting microscope.

5 Carefully cut the base of the pistil and ovary longitudinally, and record your observations in the space provided, using Figure 16.14 as a reference.

_____

_____

_____

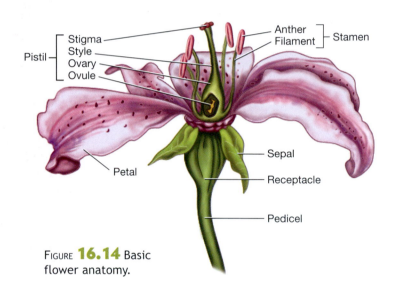

Stigma — Anther
Style — Filament — Stamen
Pistil — Ovary
Ovule
Sepal
Petal
Receptacle
Pedicel

FIGURE **16.14** Basic flower anatomy.

Procedure 2

Microanatomy of the Flower

Materials

❏ Compound microscope
❏ Scalpel
❏ Microscope slides and coverslips
❏ Forceps
❏ Paper towels
❏ Anther of a flower
❏ Colored pencils

1 Using the point of the scalpel, remove some pollen from the anther of a flower.

2 Make a wet mount of the pollen, and observe it under the compound microscope (Figs. 16.15). Record and illustrate your observations in the space provided.

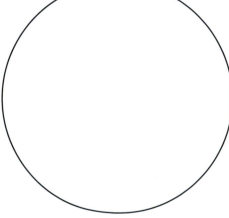

Pollen specimen   Total magnification _____

16

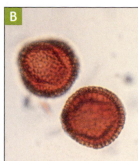
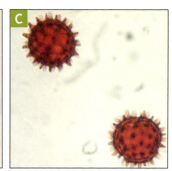
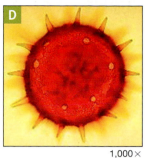

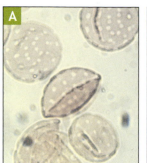

FIGURE **16.15** Examples of pollen grains: (**A**) pigweed, *Amaranthus* sp.; (**B**) lilac, *Syringa* sp.; (**C**) arrowroot, *Balsamorhiza* sp.; (**D**) hibiscus, *Hibiscus* sp.

1,000×

✔ Check Your Understanding

1.1 Monocot and eudicot flowers can be distinguished by looking at the arrangement of the floral parts.

Monocot flowers are arranged in _____ and eudicot flowers are arranged

in _____ .

1.2 Sketch a typical eudicot flower, and label the parts.

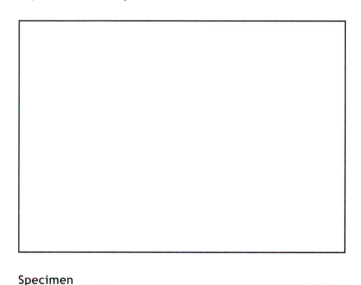

Specimen _____

_____

1.3 What is a catkin? Name several plants that produce catkins.

_____

_____

_____

16

Fruits are exclusive to angiosperms. All fruits are derivatives of the ovary or ovaries of a flower and associated structures such as the receptacle (Fig. 16.16). The diversity of fruits is astonishing, ranging from acorns to zucchini. Some fruits, such as the cultivated banana, wild parsnips, seedless watermelons, and seedless grapes, are **parthenocarpic**; they do not require fertilization to form a fruit. Fruits serve to nourish the seed and aid in seed dispersal.

Fruit Anatomy

Upon maturation, the ovary of a fleshy fruit usually has three regions (Fig. 16.17)—the **exocarp**, the **mesocarp**, and the **endocarp**. Because these regions may merge, it sometimes is difficult to distinguish between the regions. The three regions are known collectively as the **pericarp**. In dry fruits, the pericarp may be thin, as in the hull of a peanut.

1. The exocarp, which forms the skin, or peel, of a fruit, is variable in color and texture. With its associated glands, it is called a flavedo in a citrus fruit. As an orange ripens, the outside of the flavedo changes from green (chlorophyll) to orange (mostly xanthophyll) in color.

2. The mesocarp is the fleshy portion of the fruit between the exocarp and the endocarp. In citrus fruits, the whitish region just beneath the exocarp is actually the mesocarp, called the albedo.

3. The endocarp is the inside layer of the pericarp directly surrounding the seed. The endocarp may be papery as in apples, or stony as in a peach, or slimy as in a tomato, or a shell as in a pecan. In citrus fruits the endocarp is divided into distinct segments. **Juice vesicles** provide the treasured juice in a citrus fruit.

Fruit Classification

Fruit classification is based on several features, including whether the fruit is simple or compound, fleshy or dry, and whether other floral parts are present. This scheme is not exact, and arguments abound. In any case, Table 16.1 may be helpful in classifying fruits.

Simple fruits, such as grapes, beans, and hickory, are derivatives of a single ovary (Fig. 16.18). Many simple fruits, such as apples, oranges, and watermelons,

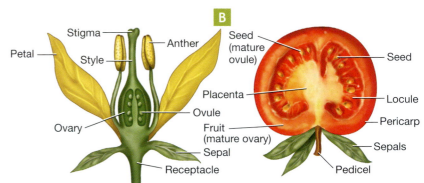

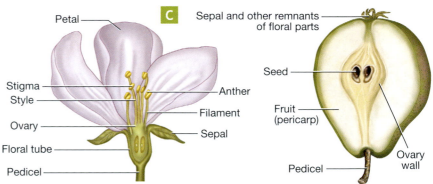

FIGURE **16.16** Flowers and fruits: (**A**) strawberry, *Fragaria* sp. (**B**) tomato, *Lycopersicon esculentum*; (**C**) pear, *Pyrus* sp.

FIGURE **16.17** Longitudinal section of a peach.

TABLE **16.1** Fruit Classification

Fruit Group	Type	Description	Examples
Simple Fruits			
Fleshy fruits	Berry	Ovary compound, skin from exocarp, fleshy pericarp	Tomato, grape, guava, kiwi, persimmon, papaya, pomegranate, avocado. *Note*: Bananas, cranberries, and blueberries are false berries.
	Pome	Accessory fruit, derived from several carpels, ovary (core) surrounded by fleshy receptacle tissue	Apple, pear, quince
	Hip	Accessory fruit, derived from several carpels, encloses achenes	Rose
	Pepo	Accessory fruit, berry with hard, thick rind, receptacle partially or completely encloses the ovary	Squash, watermelon, cantaloupe, gourd
	Drupe	Derived from a single carpel, possesses one seed, endocarp a stony pit, exocarp a thin skin	Peach, cherry, plum, olive, mango, pecan, walnut, coconut, almond, pistachio, cashew, macadamia
	Hesperidium	Berry with a leathery rind and juice sacs	Orange, lemon, lime, grapefruit, kumquat
Dry Fruits			
Dehiscent	Legume	Single carpel, pod splits along two sides	Peas, mimosa, bean, peanut, wisteria, redbud
	Follicle	Single carpel, splits along one side	Milkweed, oleander, columbine
	Silique	Two carpels that separate at maturity, leaving a permanent partition between them	Radish, mustard, cabbage, turnip
	Capsule	Composed of several carpels, separates in several ways	Okra, poppy, iris, yucca, cotton, orchid, sweet gum, agave, Mexican jumping bean, Brazil nut
Indehiscent	Achene	Simple ovary with pericarp is dry and free from the internal seed, except at the placental attachment	Sunflower, dandelion, buttercup, sycamore, buckwheat
	Samara	Simple ovary with a winged pericarp, produced in clusters	Maple, ash, elm
	Caryopsis (grain)	Simple ovary, one seed with the pericarp fused to the seed coat	Corn kernel, rice, oats, barley, wheat, Johnson grass, Bermuda grass
	Schizocarp	Two or more sections break apart at maturity, each with one seed	Carrot, fennel, celery, dill, puncture vine
	Nut	Single seed with hard pericarp surrounded by bracts and/or a receptacle	Oak (acorn), hickory, hazelnut, beech, chestnut
Compound Fruits			
Aggregate Fruits			
Fleshy fruits	Achenes	Consist of a number of matured ovaries from a single flower, arranged over the surface of a single receptacle; individual ovaries are called fruitlets	Strawberry, buttercup
	Drupes		Dewberry, blackberry, raspberry, boysenberry
Dry fruits	Follicles		Southern magnolia
	Samaras		Tulip (yellow) poplar
Multiple Fruits			
Fleshy fruits	Achenes	Collection of fruits produced by the grouping of many flowers crowded together in a single inflorescence, typically surrounding a fleshy stem axis	Fig
	Drupes		Breadfruit, mulberry, Osage orange
	Fused berries		Pineapple
Dry fruits	Achenes		Sycamore
	Capsules		Sweet gum
	Caryopsis		Corn cob and kernels

16

FIGURE **16.18** Examples of simple fruits: (**A**) peach; (**B**) grapes; (**C**) apple; (**D**) pea.

are classified as **fleshy fruits**. Others are classified as **dry fruits** (Fig. 16.19).

In dehiscent dry fruits, the pericarp is dry, and the fruit splits at maturity; these include peas, radishes, milkweed, and orchids. **Indehiscent dry fruits** do not split at maturity; examples are acorns, corn kernels, parsley, and rice.

Compound fruits, such as strawberries, blackberries, and figs, develop from several individual ovaries. **Aggregate compound fruits** are derived from a single flower with many pistils. In an aggregate fruit, the tiny fruitlets can be an achene or a drupe existing on a single receptacle. A strawberry is a fleshy fruit with achenes on the surface of the receptacle (the body). Look at a blackberry: Each tiny drupe came from an individual ovary. Strawberries and blackberries are **aggregate fleshy fruits** (Fig. 16.20). Aggregate fruits can also be dry, such as the fruits of magnolia and yellow poplar. Multiple fruits, such as figs, pineapples, and mulberries, are derived from several individual flowers in an inifloresence (Fig. 16.21).

Fruits that develop from tissues surrounded by the ovary are called **accessory fruits.** These generally develop from flowers with inferior ovaries, and the receptacle becomes a part of the fruit. Accessory fruits can be simple, aggregate, or multiple.

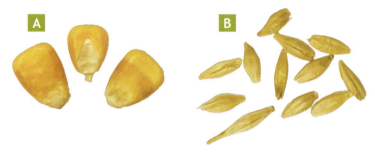

FIGURE **16.19** Examples of dry fruits: (**A**) corn; (**B**) oats.

Did you know . . .

Astonishing!

In the United States, 4,427,000 metric tons of apples are produced each year.

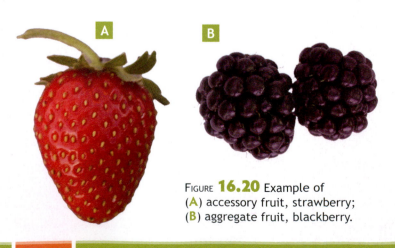

FIGURE **16.20** Example of (**A**) accessory fruit, strawberry; (**B**) aggregate fruit, blackberry.

FIGURE **16.21** Examples of multiple fruits: (**A**) pineapple; (**B**) fig.

Materials

- ❏ Dissecting microscope or hand lens
- ❏ Dissecting tray
- ❏ Dissecting needle
- ❏ Forceps
- ❏ Scalpel
- ❏ Fruits provided by the instructor, such as an apple, an orange, a peach, a bean, a strawberry
- ❏ Colored pencils

1 Procure the specimens and equipment.

2 Describe the specimens in detail and, using a dissecting microscope or a hand lens, observe, draw, and label the anatomical features of your specimens in the space provided. Be sure to include with your description both the common name and the scientific name of the plant as well as the type of fruit and what you infer about seed dispersal.

3 Dissect the fruits provided. Record your observations and labeled illustrations in the space provided. In your labeling, include as many structures as you can.

Specimen 1 _____

Specimen 2 _____

Specimen 3 _____

16

Specimen 4 _____

Specimen 5 _____

Specimen 6 _____

✔ Check Your Understanding

2.1 What is the function of a fruit?

_____

_____

_____

2.2 Provide the type of fruit for each description. (*Hint:* Refer to Table 16.1 for help.)

Example
Single carpel; splits along one side. Answer: follicle

_____ Berry with hard, thick rind

_____ Single carpel; pod splits along two sides

_____ Simple ovary; one seed with the pericarp fused to the seed coat

_____ Berry with leathery rind and juice sacs

_____ Accessory fruit derived from several carpels; ovary (core) surrounded by fleshy receptacle tissue

_____ Simple ovary with a winged pericarp; produced in clusters

_____ Fruit derived from a single carpel; possesses one seed; endocarp a stony pit; exocarp a thin skin

_____ Two carpels that separate at maturity, leaving a permanent partition between them

16

Did you know . . .

How to Grow a Pineapple

A green thumb isn't necessary to grow a pineapple—just patience. This can be done by following these steps:

1. Obtain a fresh pineapple with healthy green leaves.
2. Remove several of the lower leaves to expose the stem. Cut off the crown about 3 inches below the stem. Trim any tissue from around the rim.
3. Place the crown upside down in a cool, dry, insect-free place for one week.
4. Plant the crown in an 8 inch clay pot filled with light garden soil with a 30% blend of organic matter. Be sure to form the soil around the crown up to the base of the stem.
5. Place the plant in a humid, sunny environment. Lightly water the plant weekly. (Pineapples don't like to get wet!)
6. During the summer, lightly fertilize the pineapple monthly.
7. After 18–24 months, inspect the plant's development. The pineapple will produce a red cone surrounded by blue flowers. The flowers will drop, and the fruit will begin to develop.
8. If desired, force the fruit to develop by covering the entire plant with a clear polyethylene bag and placing two ripe apples in the pot. Ethylene gas produced by the apples encourages fruit development.
9. When the pineapple matures, photograph it, and then harvest it. *Bon appetit!*

Seeds

Seeds link the historical development of a species with the present and the infinite possibilities of the future. The seed is a structure formed by maturation of the ovule following fertilization. Seeds are the end products of sexual reproduction, and they house the embryo. Seeds protect, support, and nourish the embryonic plant until **germination** (resumption of growth and metabolic activity). Many plants have developed elaborate strategies to disperse the seeds into the environment (Fig. 16.22).

Seed Anatomy

All seeds (Fig. 16.23) are covered by a **testa** (seed coat) that protects the seed from drying out, extreme temperatures, bacteria, fungi, and predation. An opening called the **micropyle** is often visible on the seed coat as a small pore. The micropyle allows the pollen tube to enter the ovule to ensure fertilization. In addition, some seeds have a distinct scar, the **hilum**, left on the seed coat when the seed separates from

the supportive **funiculus**, or stalk. The nutritive endosperm, beneath the testa, contains copious starch that nourishes the seed after germination.

Angiosperm and gymnosperm seeds differ in the origin of their stored food. The female gametophyte in gymnosperms provides the food. In angiosperms, the food is supplied by cotyledons. In addition to the endosperm, an **embryo** can be found within the seed. The size of the embryo

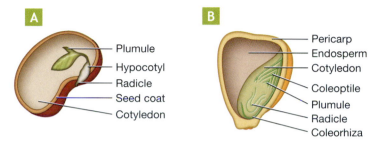

FIGURE **16.23** Seed structure: (**A**) red bean (eudicot); (**B**) corn kernel (monocot).

FIGURE **16.22** Methods of seed dispersal: (**A**) airborne, shown by maple, white pine, and willow seeds, which are light and often winged to enable dispersal by wind; (**B**) forceful discharge, shown by witch hazel seeds; (**C**) adherence to hair or fur, shown by beech seeds with spiny husks; (**D**) waterborne, shown by mangrove, coconut, and pecan seeds with protective husks; (**E**) burial by animal, shown by black walnut seeds or acorns; (**F**) ingestion by animal, shown by apple or cherry seeds, which are undigested and passed in feces to germinate in a new location.

varies with the plant species. The mature embryo consists of a stem-like axis bearing one (monocot) or two (eudicot) cotyledons. The cotyledons, or seed leaves, are the first leaves to appear in a new sporophyte. They serve as food storage organs for the seedling plant.

Upon examination, a bean has two distinct halves, each a cotyledon, and a corn kernel has a single cotyledon. In monocots the cotyledon may be called the **scutellum**. In monocots the scutellum is highly absorptive. At opposite ends of the plant embryo are the **apical meristem** of the shoot and the root. Many plants have a stem-like axis, the **epicotyl**, with one or more developing leaves above the cotyledon or cotyledons. The resulting embryonic shoot is called the **plumule**.

The stem-like portion beneath the cotyledon or cotyledons is called the **hypocotyl**. The embryonic root, or **radicle**, exists at the lower end of the hypocotyl. The radicle and the plumule are enclosed in sheath-like protective structures called the **coleorhiza** and the **coleoptile**, respectively. These structures protect the seed during germination.

Seed Germination and Growth

When environmental conditions are favorable, a seed breaks dormancy and germinates, forming a new generation of the plant (Fig. 16.24). The embryos of different species remain viable (capable of germination) for varying periods of time.

Seeds of one species of lotus germinated after they were discovered in a 3,000-year-old tomb!

Among the variables influencing germination are temperature, light, water, and scarification, the latter of which can be brought about by bacterial action, stomach acid, freezing, or fire. Usually, seeds germinate while under the surface of the soil. Although soil is the ideal environment for roots, shoots are poorly designed for growth under abrasive soil conditions. Fortunately, nature has provided several mechanisms for protecting the young shoot during its emergence from the soil.

In beans, after development of the root and anchorage in the soil, the hypocotyl grows toward the surface in the form of a hook, gently pulling the cotyledons upward. When the hypocotyl hook reaches the surface, light induces the tissue of the hook to straighten, bringing the cotyledons and the young shoot to the surface. During this process, the plumule is protected between the cotyledons.

In peas, the epicotyl elongates and forms a hook that otherwise is similar to the mechanisms of hypocotyl elongation in beans, but the cotyledons remain under the surface. In corn and in some grasses, the coleoptile protects the plumule. The coleoptile is a tough, protective sheath that completely surrounds the plumule. When the coleoptile reaches the surface, light induces it to split, and the plumule emerges.

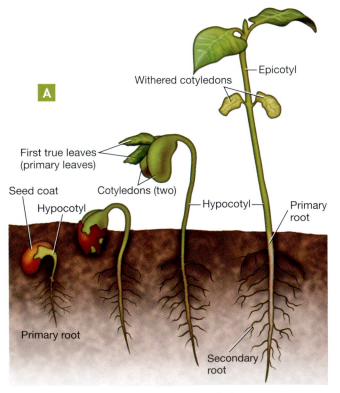

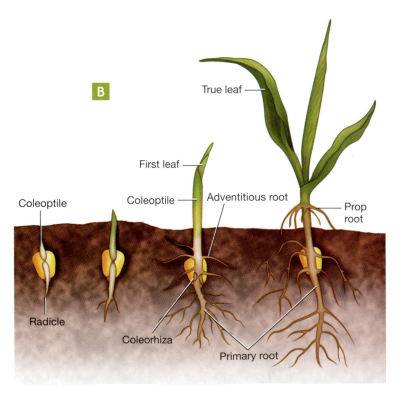

FIGURE **16.24** Germination: (**A**) red bean (eudicot); (**B**) corn kernel (monocot).

Procedure 1

Iodine and Starch

Materials

- ❏ Iodine solution or counterfeit pen
- ❏ Eyedropper
- ❏ Scalpel
- ❏ Dissecting tray
- ❏ Plastic petri dish
- ❏ Paper towels
- ❏ Forceps
- ❏ Paper currency (e.g., $1 bill)
- ❏ Cotton swab
- ❏ Corn kernel soaked in water for 24 hours
- ❏ Timing device

Why does the clerk mark your $20 bill with that "magic pen"? As you learned in Chapter 3, paper money contains many chemical safeguards to protect it from counterfeiters, including that of removing starch (*amylose*) from the paper. Counterfeiters have not discovered how to remove the amylose, and when counterfeit bills are marked with an iodine pen, the mark appears blue. Gotcha! Iodine reacts with starch, producing a bluish color. In this procedure, the cut surface of a corn kernel is treated with a drop of iodine solution. The storage tissues turn blue because of the presence of amylose.

1 Procure the material and equipment from the instructor.

2 Carefully cut the corn kernel in half longitudinally with the scalpel.

3 Place the kernel in the petri dish, and place one drop of iodine solution on the kernel.

4 Wait 2 minutes, and then describe what happened. What portion of the kernel is blue?

_____

_____

_____

_____

5 Lightly dip a cotton swab into the iodine solution, lightly rub the iodine solution on the paper money, and determine if your "bill" is counterfeit.

_____

Procedure 2

Macroanatomy of Seeds

Materials

- ❏ Dissecting microscope or hand lens
- ❏ Scalpel
- ❏ Dissecting tray
- ❏ Seeds provided by the instructor, such as a red bean, a peanut, a corn kernel, a true rice grain, and a watermelon seed
- ❏ Colored pencils

1 Procure the equipment and seeds.

2 Describe the specimens in detail and, using a dissecting microscope or a hand lens, observe, draw, and label the anatomical features of your specimens in the space provided. Use Figure 16.23 as a reference. In your description, be sure to include both the plant's common name and the scientific name.

3 Using a scalpel, carefully cut each seed longitudinally. Describe the specimens in detail and, using a dissecting microscope or a hand lens, observe, draw, and label the anatomical features of your specimens in the space provided.

Specimen 1 _____

_____

Specimen 2 _____

Specimen 3 _____

✔ Check Your Understanding _____

3.1 Adding a drop of iodine to a cut corn kernel turns part of the kernel blue. The reaction occurs because

of the presence of _____ .

3.2 How does a "real" $20 bill react to a drop of iodine? Why?

_____

_____

3.3 Embryonic seed leaves, or _____ , are the first leaves to appear in a new sporophyte.

These seed leaves function as _____ for the seedling.

3.4 What is the function of the radicle?

_____

_____

3.5 What is the main function of the coleoptile during germination?

_____

Did you know . . .

What Is a Mexican Jumping Bean?

A Mexican jumping bean is not a bean at all! It is a carpel of a seed capsule from the Mexican shrub *Sebastiana pavoniana* that houses the larva of a small gray moth called the jumping bean moth (*Laspeyresia saltitans*). While eating the nutritive material within the carpel, the larva wiggles, causing the jumping movements of the "bean."

MYTHBUSTING

Planting a Seed

Debunk each of the following misconceptions by providing a scientific explanation. Write your answers on a separate sheet of paper.

1 Rose hips (the berrylike fruit structure of a rose) are considered poisonous.

2 Seedless grapes are seedless because they are sprayed with hormones.

3 Flowers were placed on Earth for their beauty.

4 If you swallow a seed, it will germinate in your appendix.

5 The lotus flower provides eternal life.

1 What is the major function of a flower?

_____

_____

2 Describe the life cycle differences in plants that are annuals, biennials, and perennials.

_____

_____

_____

3 A flower that possesses both stamens and pistils is known as an imperfect flower. (*Circle the correct answer.*)

True / False

4 Label Figure 16.25.

1. _____
2. _____
3. _____
4. _____
5. _____
6. _____
7. _____
8. _____
9. _____
10. _____
11. _____
12. _____

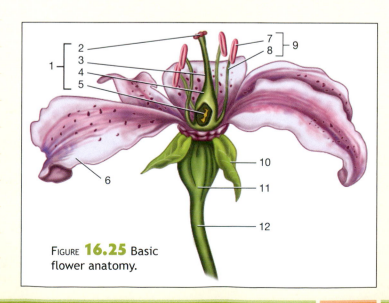

FIGURE **16.25** Basic flower anatomy.

5 Sketch, label, and describe the function of the anatomical features of a bean.

6 Draw a cross section of a peach, and label the structures. What parts of the flower give rise to these structures?

7 What is a fleshy fruit? List three categories of fleshy fruits along with an example of each.

16

8 What is a dry fruit? List three categories of dry fruits along with an example of each.

9 Compare aggregate and multiple fruits providing examples of each.

There's a Fungus among Us
Understanding Fungi

17

*Nature has been experimenting with fungi for a billion years, perfecting survival tools.
We can use these tools in fantastic ways- to retrieve damaged ecosystems,
to help offset global warming, and even to prevent disease.*

— Paul Stamets (1955–present)

*Consider the following questions prior to coming to lab,
and record your answers on a separate piece of paper.*

1 Why do some fungi smell like rotten flesh?

2 What is so special about a honey mushroom (*Armillaria ostoyae*) found in Malheur National Forest in eastern Oregon?

3 What is the nutritional value of mushrooms?

4 What is the relationship between bagpipes and *Aspergillus*?

5 Why are mycorrhizae important in the evolution of land plants?

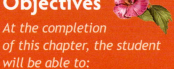

Objectives

At the completion of this chapter, the student will be able to:

1. Explain the ecological role and characteristics of kingdom Fungi.

2. Describe and identify the basic anatomical features of kingdom Fungi.

3. Discuss the taxonomical organization of kingdom Fungi.

4. Describe the characteristics of and biology of the phyla Zygomycota, Ascomycota, and Basidiomycota.

5. Describe and identify select examples of each phylum of fungi.

6. Trace the life cycle of *Rhizopus stolonifer*, a typical ascomycete, and *Agaricus bisporus*, a typical basidiomycete.

7. Label the parts of a typical mushroom.

8. Describe the biology of lichens and their three basic forms.

S ure, there are many species of fungi among us (Fig. 17.1)! Unfortunately, the term *fungus* evokes images of molds, mildew, rotting organic matter, spoiled food, and various maladies of plants, animals, and humans. Fungi do not limit their enzymatic attack to living things or dead things. Species of fungi attack plastic, leather, paint, petroleum products, film, and even the multicoating of optical equipment such as your digital camera.

Millions of dollars are spent yearly trying to control fungal diseases in plants, including Dutch elm disease, wheat rust, and corn smut, and in humans, diseases such as ringworm (red and itchy skin), coccidiomycosis (fever, rash, headache, joint pain, skin lesions, chronic pneumonia), and aspergillosis (allergic and lung infections). Some species of fungi produce powerful toxins, carcinogens, and hallucinogens.

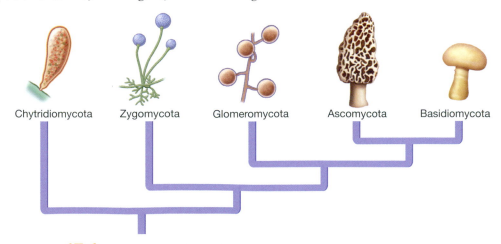

Chytridiomycota Zygomycota Glomeromycota Ascomycota Basidiomycota

Chapter Photo
Turkey tail fungi, *Trametes* sp.

FIGURE **17.1** Phylogenetic relationships and classification of major fungus lineages.

345

Fungi play a vital role in ecosystems and are economically essential (Fig. 17.2). Without certain species of fungi serving as decomposers, ecosystems would collapse. These decomposers break down dead organisms, leaves, feces, and organic matter and recycle their chemical components back into the environment. In addition, many species of plants depend upon mutualistic fungi to help their roots absorb minerals and water from the soil.

Animals and humans eat many species of fungi. Truffles and some species of morels and mushrooms are delicacies. Fungi also play a vital role in the bread, cheese, beer, and wine industries. Several species of fungi are used in the production of antibiotics, including penicillin and cyclosporine, and other beneficial medicines.

FIGURE **17.2** Examples of fungi: (**A**) yeast; (**B**) bread mold; (**C**) mushroom; (**D**) morel.

Anatomical Structure

Fungi are filamentous, spore-producing, heterotropic unicellular or multicellular eukaryotes. Fungi release digestive enzymes onto a food source, partially dissolving the source to make the essential nutrients available. Multicellular fungi are composed of numerous small filaments, known as **hyphae**, grouped together into a mass called a **mycelium**. For example, hyphae make up the body of a mushroom. Hyphae and mycelia can grow rapidly. Under ideal conditions, a single fungus might produce a kilometer of hyphal growth in one day. That is why a mushroom can appear overnight.

In the majority of fungi, the hyphae are divided into compartments by **septa** (cross walls) that allow for some structures, such as ribosomes and mitochondria, to pass through. **Coenocytic** fungi lack septa and appear as a cytoplasmic mass with perhaps thousands of nuclei. Various modifications of hyphae can be found. One modified hyphae, a haustoria, penetrates the tissues of a host. Other modified hyphae, called **rhizoids**, anchor fungi to a substrate. Some soil fungi have snare, or loop, hyphae to trap unsuspecting nematodes (roundworms) for future consumption. Fungi that have cross walls in their hyphae are connected to adjacent hyphae by tiny pores in the cross wall; in contrast, septa that separate reproductive cells have no pores.

Most fungi are **saprobes** that break down organic matter, but there are parasitic and mutualistic fungi as well. Many fungal species possess a cell wall composed of the polysaccharide chitin, and they store their energy in the form of glycogen. Fungi do not contain chlorophyll.

Most fungi are capable of undergoing asexual and sexual reproduction. Asexually, fungi can reproduce by budding, fragmentation, and spore formation. Spores can develop directly without uniting with another spore. Sexually, fungi produce gametes in specialized areas of the hyphae called **gametangia**. The gametes may be released to fuse into spores elsewhere, or the gametangia themselves may fuse. In the hyphae of some fungi, **dikaryons** (Greek, *di* = two, *karyon* = nucleus) form as the result of unspecialized hyphae fusing. In this case, their two nuclei remain distinct for a portion of the life cycle. When the two nuclei finally fuse, the zygote undergoes meiosis prior to spore formation. Upon germination, the spores form haploid hyphae.

A large mushroom can produce billions of spores. When a spore lands on a suitable substrate, it germinates and grows. That is why bread mold can appear mysteriously on a slice of bread you thought was pristine.

Classification

For many years, the fungi were classified as imperfect plants. Today, it is clear the fungi are not degenerate plants but, rather, unique eukaryotes deserving of their own kingdom. Newer classification schemes place fungi closer to the animal kingdom than to the plant kingdom. Mycologists (specialists in fungi) recognize nearly 100,000 species of fungi and predict this number could increase to 2,000,000 species. Representative fungi include mushrooms, puffballs, bread mold, morels, truffles, smuts, rusts, blight, mildew, and yeasts.

Members of kingdom Fungi are classified into several distinct phyla based on anatomy, types of hyphae, means of reproduction, and molecular biology.

1. Chytridiomycota is the most ancient group of fungi. Most chytrids are either aquatic decomposers feeding on dead plant or animal material in a pond or parasites living on water molds, insects, or snakes.

2. Zygomycota includes bread molds and *Pilobolus* spp., the "hat-throwing" fungus.

17

3. Glomeromycota comprises fungi that enter into a symbiotic relationship with the roots of plants, forming mycorrhizae.

4. Ascomycota, the largest phylum of fungi, includes organisms such as truffles, morels, and yeast.

5. Basidiomycota includes mushrooms, shelf fungi, and puffballs.

Microspora is a sixth phylum of kingdom Fungi. Once classified as protists, the microsporidia are a group of obligative intracellular parasites of invertebrate and vertebrate hosts. Clinical manifestations of microsporidians in humans include skin, eye, urinary, respiratory, gastrointestinal, muscle, and nervous system infections. The most common means of transmission are ingestion, inhalation, or coming into contact with spores. They commonly infect individuals suffering from compromised immune systems disorders such as HIV.

Phylum Zygomycota

Phylum Zygomycota includes 1,000 species of primarily terrestrial fungi known as the conjugating fungi. **Zygomycetes** commonly occur in soil, decaying organic matter (a white filamentous mass on decaying fruit), and feces. The hyphae of zygomycetes lack septa and are called coenocytic. A representative example of phylum Zygomycota is *Rhizopus stolonifer*, the common black bread mold. Three types of hyphae are found in *Rhizopus*:

1. Rhizoids (anchoring hyphae that penetrate the bread and have digestive enzymes);

2. Stolons (horizontal surface hyphae); and

3. Sporangiophores (reproductive hyphae).

Reproduction in *Rhizopus* can occur asexually or sexually. Asexually, when a **spore** (sporangiospore) lands on a suitable substrate, such as a slice of bread, it germinates and forms hyphae that soon form a mycelium. After the mycelium develops, it produces sporangiophores that rise above the surface and contain spore-containing **sporangia**. The sporangia release their spores and seek another supportive substrate.

Sexually, *Rhizopus* reproduces by conjugation (Fig. 17.3). *Rhizopus* produces two different hyphae (+ and − strains) that develop swollen **progametangia** on the ends facing each other. Eventually they touch, and a cross wall forms behind each tip. Next, a thick-walled **zygosporangium** forms, replacing the progametangia. The zygosporangium cracks open, forming sporangiophores and their associated sporangia. Meiosis occurs in the sporangia, producing **meiospores** then released to seek another substrate.

Several other notable species of zygomycetes are medically and ecologically significant. *Rhizopus nigricans*, which also can grow on bread, and *Mucor* spp., which can grow on stored foods and seeds and are commonly found in house dust, cause fungal sinusitis and allergies. Other species of *Rhizopus* and *Mucor* can cause serious and sometimes deadly infections (mucormycosis) that affect the skin, digestive tract, facial region, lungs, and brain. *Pilobolus*, discussed in the Nature's Cannon sidebar, is another zygomycete.

Did you know . . .

Nature's Cannon

Grazing animals such as cattle rarely graze near feces (dung pat). The zone of ungrazed grass around a dung pat is called the "ring of repugnance." Infective stages of several endoparasites are faced with the task of having to travel from the dung pat beyond the ring of repugnance to ungrazed grass.

The roundworm (nematode) *Dictyocaulus viviparus*, called the "cattle lungworm," has solved this problem in a unique way. The larvae of the cattle lungworm migrate up the sporangiophore of the saprobic fungus *Pilobolus* sp. and accumulate on the sporangium.

Pilobolus is known as the "shotgun fungus," or the "hat-throwing fungus," or the "cannon fungus," because it has explosive sporangia that can shoot spores beyond the ring of repugnance up to 2.5 meters toward light. The lungworm larvae hitch a ride on the spores and land beyond the ring of repugnance. When ingested by a cow, the larvae penetrate the wall of the cow's intestine and are carried by the lymphatic and circulatory systems to the lungs, where the adult worms develop.

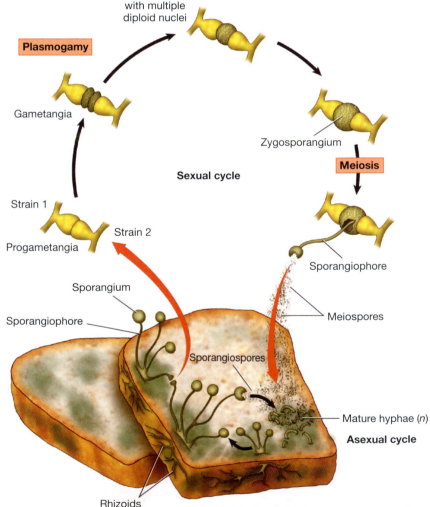

FIGURE **17.3** Life cycle of *Rhizopus stolonifer*.

17

Macroanatomy of *Rhizopus stolonifer* (bread mold)

Materials

- ❏ Dissecting microscope or hand lens
- ❏ Paper towels
- ❏ Bread contaminated with *Rhizopus stolonifer*
- ❏ Colored pencils

1 Procure the equipment and the bread mold from your instructor.

2 Place your specimen on a paper towel and using a dissecting microscope or hand lens, observe your specimen. Sketch, label as many structures as you can, and record your observations in the space provided.

Rhizopus stolonifer _____

_____

_____

Microanatomy of *Rhizopus stolonifer* (bread mold)

Materials

- ❏ Compound microscope
- ❏ Slides and coverslips
- ❏ Prepared slides of *Rhizopus stolonifer*
- ❏ Scalpel
- ❏ Forceps
- ❏ Paper towels
- ❏ Colored pencils

1 Procure a compound microscope and prepared slides of Rhizopus stolonifer.

2 Using the scalpel, remove a sporangium from the sporangiophore in Procedure 1. Place the sporangium on a microscope slide, and prepare a wet mount for observation.

3 Sketch, label as many structures as you can, and record your observations in the space provided.

4 Examine the prepared slides and compare them with Figures 17.4 and 17.5. Place your observations and labeled sketches in the space provided.

17

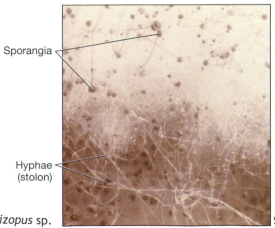

Sporangia

Hyphae (stolon)

15×

FIGURE **17.4** *Rhizopus* sp.

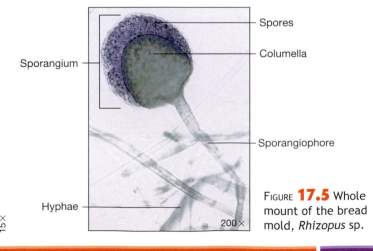

Spores

Columella

Sporangium

Sporangiophore

Hyphae

200×

FIGURE **17.5** Whole mount of the bread mold, *Rhizopus* sp.

Rhizopus stolonifer _____

Total magnification _____

_____

_____

Rhizopus stolonifer _____

Total magnification _____

_____

_____

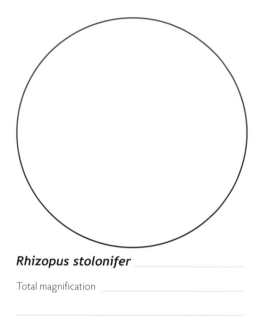

✔ Check Your Understanding

1.1 Describe three types of hyphae found in *Rhizopus stolonifer*.

_____

_____

_____

1.2 Match the type of hyphae found in *Rhizopus stolonifer* with the description:

_____ Stolons

_____ Rhizoids

_____ Sporangiophores

A. Reproductive hyphae

B. Anchoring hyphae that penetrate bread; produce digestive enzymes

C. Horizontal surface hyphae

1.3 The asexual structure produced in *Rhizopus stolonifer* and other zygomycetes that produce sporangiospores is called a/an _____ .

17

Phylum Ascomycota

Most fungal species (65,000) belong to phylum Ascomycota, the sac fungi (Fig. 17.6). Many species of **ascomycetes** are found in a symbiotic relationship with algae, forming **lichens**. Sac fungi live in a variety of marine, freshwater, and terrestrial habitats. These organisms range from unicellular to elaborate multicellular forms. Ascomycetes are responsible for various serious plant diseases such as powdery mildew, chestnut blight, and Dutch elm disease.

Coccidioidomycosis is a fungal disease in humans caused by an ascomycete, resulting in a rash and a potentially fatal pulmonary disease. *Cryptococcus neoformans* is an ascomycete associated with pigeon excretions and is potentially deadly to animals and humans. It is associated with meningitis and meningoencephalitis. *Histoplasma capsulatum*, another ascomycete, is responsible for the potentially lethal lung disease known as histoplasmosis. *Claviceps* spp. are associated with St. Anthony's fire in the Middle Ages, the madness of the witches of Salem in the witch trials, the hallucinogenic drug LSD, and other cases of ergotism throughout history.

The yeast *Saccharomyces cerevisiae*, an ascomycete, plays important roles in the brewing industry and in genetic research. Another yeast, *Candida albicans*, causes a variety of fungus infections, including oral thrush and vaginal infections. *Neurospora*, a type of bread mold, is also important in genetic studies. True morels are common woodland ascomycetes featuring a convoluted cap. Several species of morels are prized for their flavor and consistency. Although plain to the sight, truffles have an exquisite flavor and rival the cost of some precious metals per gram.

Aspergillus is a genus of green mold that can cause deadly respiratory infections. Some species of *Aspergillus* are used in producing soy sauce, inks, toothpaste, chewing gum, inks (especially black), and photograph-developing solutions. The *Aspergillus flavus* species that may grow on improperly stored grain produces potent carcinogenic substances that can cause liver cancer. *Stachybotrys chartarum* is a black mold responsible for "sick-building" syndrome, in which exposure to the spores can cause chronic sickness such as headaches, eye irritation, lung disease, rash, memory loss, and fever. This mold presented a major problem in New Orleans, Louisiana, and the Mississippi gulf coast (Gulfport and Biloxi) after Hurricane Katrina.

Penicillium chrysogenum, formerly called *Penicillium notatum*, is the source of penicillin, the antibiotic discovered fortuitously by Alexander Fleming in 1928. Several species of *Penicillium* are used in the production of gourmet cheeses, such as Brie, Camembert, Gorgonzola, and Roquefort. (What do you think the "blue stuff" in blue cheese is?) But

Did you know . . .

Ergot . . . Madness!

One of the most interesting ascomycetes, *Claviceps purpurea*, or ergot, grows on rye and similar plants. *Claviceps* is responsible for ergotism in humans and other animals that consume infected food. In the Middle Ages, the dreaded St. Anthony's fire was caused by ergot. Ever since the Middle Ages, ergot has been used to induce abortions and to stop maternal bleeding following childbirth. Ergot produces a chemical used to synthesize lysergic acid diethylamide, or LSD. Perhaps the "witches" of Salem, Massachusetts, prosecuted during the witch trials of 1692–1693, and other "possessed" and "mad" people in the past were merely "tripping out" as the result of ergotism.

17

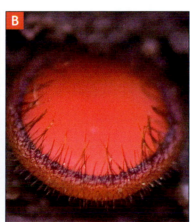

FIGURE 17.6 Fruiting bodies (ascocarps or ascoma) of common ascomycetes: (**A**) *Peziza repanda* is a common woodland cup fungus. (**B**) *Scutellinia scutellata* is commonly called the eyelash cup fungus. (**C**) *Morchella esculenta* is a common edible morel. (**D**) *Helvella* is sometimes known as a saddle fungus because the fruiting body is thought by some to resemble a saddle.

less-friendly species of ascomycetes, including *Trichophyton* sp., cause athlete's foot, ringworm, jock itch, and small non-pigmented splotches of skin called tinea versicolor.

Ascomycetes get their name from the **ascus**, a large, sac-like cell responsible for producing reproductive **ascospores**. The hyphae in ascomycetes are septate, but the cross walls are not complete. Fruiting bodies in ascomycetes are well developed and are called **ascocarps**. Sexual reproduction in the ascomycetes starts when hyphae with one nucleus of opposite mating strains come into contact (Fig. 17.7).

Each female gametangium, called an **ascogonium**, forms a **trichogyne** that grows toward the male gametangium, called the **antheridium**. After the trichogyne touches the antheridium, nuclei migrate from the antheridium to the female ascogonium. The ascogonium forms dikaryotic **ascogenous hyphae**. These hyphae form a crozier, or hook, and the nuclei fuse, forming a diploid nucleus. The nucleus undergoes meiosis, producing eight ascospores. Eventually, the ascospores forming in the ascocup are released. The asexual mitotic spores form singularly or in chains from **conidiophores** and are called **conidia**.

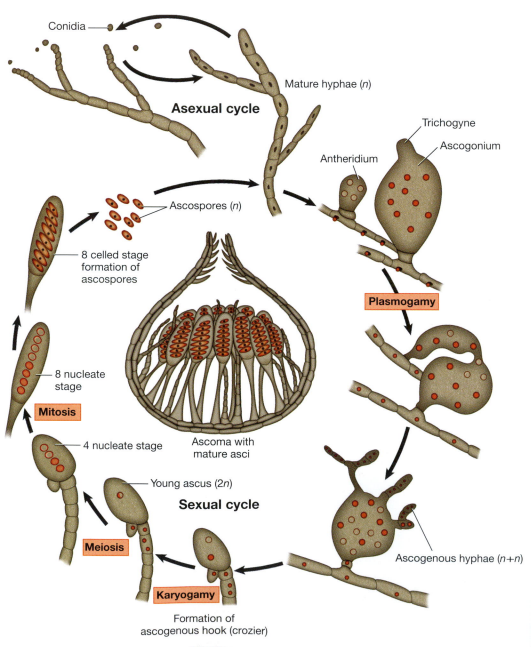

FIGURE **17.7** Life cycle of an ascomycete.

Procedure 1

Macroanatomy of Ascomycetes

Materials

- ❏ Dissecting microscope or hand lens
- ❏ Dissecting tray
- ❏ Several examples of true and false morels and plants infected with ascomycetes provided by the instructor
- ❏ Colored pencils

1 Procure the equipment and sample specimens.

2 Using a hand lens or dissecting microscope, observe your specimens. Record your observations and sketches in the space provided.

Specimen 1 _____

Specimen 2 _____

Specimen 3 _____

Specimen 4 _____

17

Procedure 2

Microanatomy of Ascomycetes

Materials

- ❏ Compound microscope
- ❏ Slides and coverslips
- ❏ Prepared slides of select ascomycetes, such as a morel (*Morchella* sp.), *Claviceps* sp., *Peziza* sp., *Penicillium* sp., or *Aspergillus* sp. provided by the instructor
- ❏ Scalpel
- ❏ Forceps
- ❏ Paper towels
- ❏ Ascomycete specimen
- ❏ Colored pencils

1 Procure a compound microscope, blank slides, and coverslips.

2 Carefully scrape away some of the tissue from an ascomycete specimen with a scalpel from Procedure 1, and prepare a slide with it. Observe, record, and sketch your observations (Figs. 17.8–17.10) in the space provided.

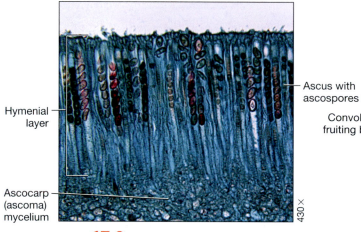

Perithecia

Stroma within multiple perithecia

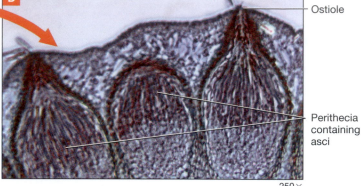

Ostiole

Perithecia containing asci

FIGURE **17.8** Ascomycete *Claviceps purpurea*: (**A**) longitudinal section through stoma showing ascocarps (ascoma); (**B**) enlargement of three perithecia.

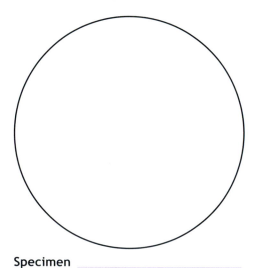

Hymenial layer

Ascus with ascospores

Ascocarp (ascoma) mycelium

FIGURE **17.9** Hymenial layer of the apothecium of *Peziza* sp. showing asci with ascospores.

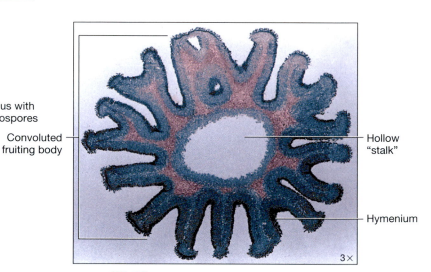

Convoluted fruiting body

Hollow "stalk"

Hymenium

FIGURE **17.10** Ascocarp (ascoma) of the morel *Morchella* sp. True morels are prized for their excellent flavor.

17

Specimen _____

Total magnification _____

3 Observe the prepared slides on both low and high power (Figs. 17.11–17.14). Record your observations and sketches in the space provided.

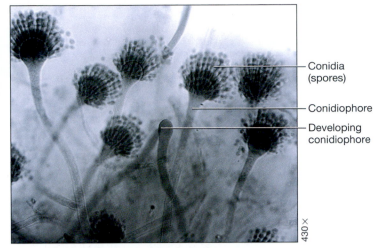

FIGURE **17.11** Close up of *Aspergillus* sp. sporangia.

Conidia (spores)

Conidiophore

Developing conidiophore

430×

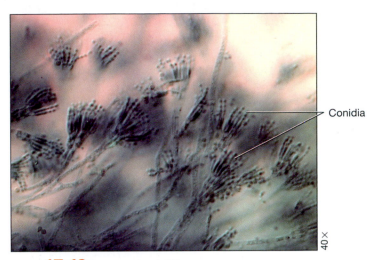

FIGURE **17.12** Fungus, *Penicillium* sp.

Conidia

40×

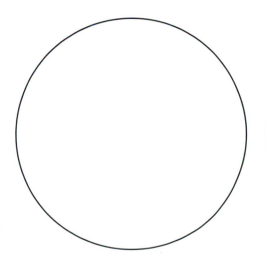

FIGURE **17.13** Blue mold, *Penicillium expansum*, growing on a rotten pear.

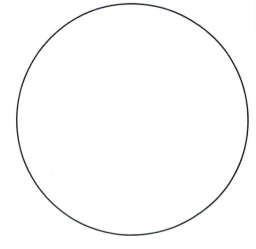

FIGURE **17.14** Ringworm, *Tinea corporis*.

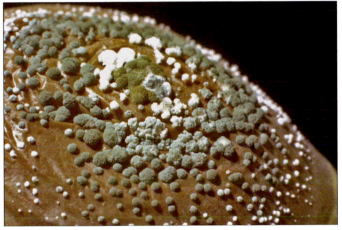

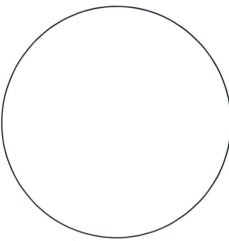

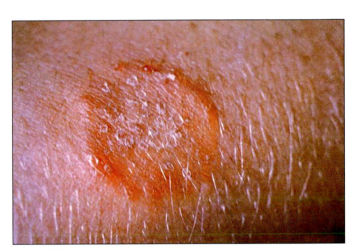

Specimen 1 _____

Total magnification _____

_____

_____

Specimen 2 _____

Total magnification _____

_____

_____

Specimen 3 _____

Total magnification _____

_____

_____

17

Procedure 3

Microanatomy of Yeast

Materials

- ❏ Compound microscope
- ❏ Slides and coverslips
- ❏ Dropper
- ❏ *Saccharomyces cerevisiae*
- ❏ Prepared solution containing baker's yeast
- ❏ Methylene blue
- ❏ Colored pencils
- ❏ Forceps
- ❏ Paper towels

1 Procure a compound microscope, blank slides, coverslips, a dropper, and a sample of *Saccharomyces cerevisiae* (Fig. 17.15).

2 Prepare a wet mount of the *Saccharomyces cerevisiae* (Fig. 17.16). Carefully place a drop of methylene blue on your slide to view the yeast more easily. Record your observations in the space provided, and label the stages of the yeast life cycle.

FIGURE **17.15** Yeast, *Saccharomyces cerevisiae*.

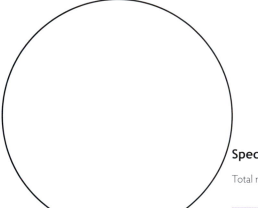

Specimen _____

Total magnification _____

_____

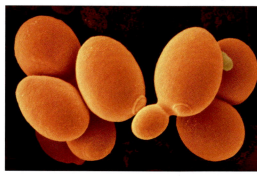

FIGURE **17.16** SEM of *Saccharomyces cerevisiae*.

✓ Check Your Understanding

17

2.1 Describe three species of ascomycetes that are beneficial to us economically. Describe three species that are potentially harmful.

_____

_____

_____

_____

2.2 How do the ascomycetes get their name?

_____

_____

2.3 Conidia are _____ spores produced by the ascomycetes, either singly or on chains produced

from _____ .

Phylum Basidiomycota

The best-known phylum of fungi is Basidiomycota, with more than 30,000 known species. Example **basidiomycetes** include mushrooms, puffballs, jelly fungi, earthstars, chanterelles, stinkhorns, rusts, and smuts (Fig. 17.17). Basidiomycetes are called "club fungi" because they produce spores, **basidiospores**, on a club-shaped structure, the **basidium**. Most basidiomycetes are saprobes living on dead or dying plants.

Mushrooms, the most obvious basidiomycetes, may be seen living singly, in groups, or in a circle called a **fairy ring**. Several species of mushrooms, including portabella (*Agaricus bisporus*), oyster mushroom (*Pleurotus ostreatus),* shiitake (*Lentinula edodes*), and chanterelles (*Cantharellus cibarius*), are edible and known for their delectable taste.

One has to be careful when collecting mushrooms for consumption because many poisonous mushrooms resemble edible species to the untrained eye. Some poisonous mushrooms, such as *Amanita* spp., are colorful and appealing, yet this species is termed the "death angel" or "death cap" because of its poison. Some mushrooms are hallucinogenic or psychedelic, such as the "magic mushroom," *Psilocybe* spp.

Puffballs are other common basidiomycetes. Puffballs disperse their reproductive spores when the peridium (covering) weathers or is compromised by rain, animals, or falling objects causing the spores to "puff" out of pores or a split. Shelf or bracket fungi resemble small shelves growing on the trunk of a tree.

Jelly fungi usually are colorful and feel cold, rubbery, or gelatinous to the touch. Stinkhorns are diverse, from orange, fingerlike structures erupting from the soil to something resembling a whiffle ball. If this were a scratch-and-sniff manual, everyone would agree the slimy covering of these fungi smells like rotting flesh.

Rusts such as cedar-apple rust and wheat rust (*Puccinia triticina*) are parasitic fungi devastating to wheat and rye crops. Smuts such as *Ustilago maydis* are parasitic fungi that attack corn, sugarcane, and other cereal crops, resulting in much devastation.

FIGURE **17.17** Representative basidiomycetes: (**A**) fly agaric, *Amanita muscaria*; (**B**) basidiomycete puffballs, *Calvatia* sp.; (**C**) jelly fungus, *Tremella* sp.; (**D**) earthstar, *Astreus* sp.; (**E**) golden chanterelle, *Cantharellus* sp.; (**F**) orange stinkhorn, *Clathrus* sp.; (**G**) wheat rust, *Puccinia podophylli*; (**H**) corn smut, *Ustilago maydis*.

The basidiomycetes reproduce through sexual reproduction. The life cycle of a mushroom is typical of most basidiomycetes (Fig. 17.18). When a spore lands on a suitable substrate, it germinates into a network of hyphae that form a mycelium beneath the surface. Haploid hyphae exist in several reproductive types. When two compatible types unite, they form a new dikaryotic mycelium. These mycelia can live for perhaps a hundred years and spread, forming the underground surface of a fairy ring. The mycelia eventually form a button that emerges from the soil. The button develops into a typical mushroom, sometimes called a basidiocarp, or basidioma.

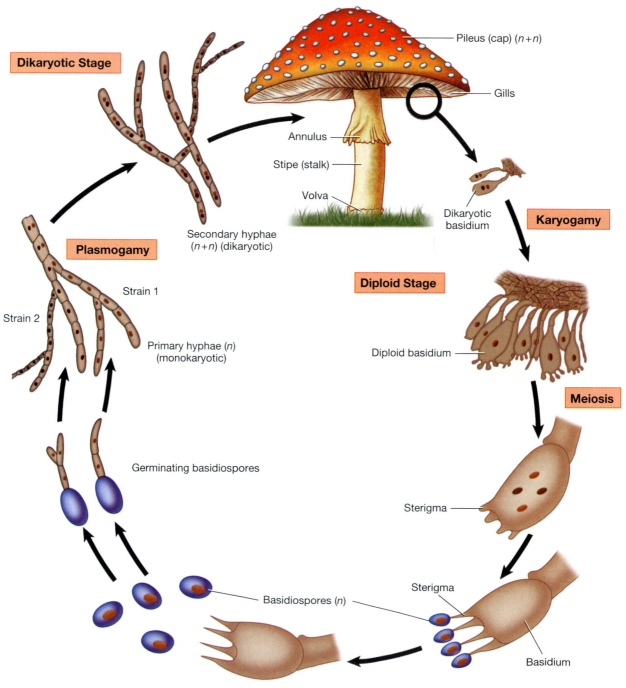

FIGURE **17.18** Life cycle of a "typical" basidiomycete (mushroom).

17

Macroanatomy of Basidiomycetes

Materials

- ❏ Dissecting microscope or hand lens
- ❏ Scalpel
- ❏ Paper towels
- ❏ Water
- ❏ Several examples of basidiomycetes, such as mushrooms and puffballs
- ❏ Colored pencils

A typical mushroom is composed of a cup-shaped **volva** at the base, a stalk-like structure called a **stipe**, a ring around the upper end of the stipe called an **annulus**, and a cap, or **pileus**. Beneath the cap are slit-like structures called **gills**, or they may be pore-like structures. The gills are composed of individual basidia. In immature mushrooms, a veil may cover the developing gills. The basidia mature, and the two nuclei fuse, forming a diploid nucleus that undergoes meiosis. The resulting four basidiospores can be found on peg-like **sterigma**. A large mushroom can produce several million basidiospores in a few days. The spores are released, and the cycle begins again. The student is encouraged to photograph basidiomycetes and bring the images to class for discussion.

1 Procure the equipment and sample specimens.

2 Using a hand lens or dissecting microscope, observe your specimens (Fig. 17.19). Record your observations and sketches in the space provided.

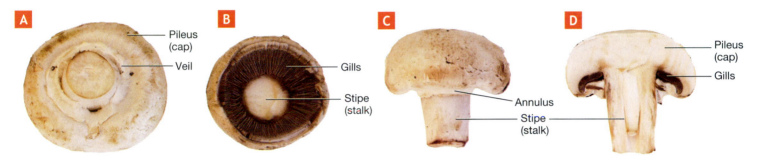

FIGURE **17.19** Structure of a mushroom: (**A**) inferior view with the annulus intact; (**B**) inferior view with the annulus removed to show the gills; (**C**) lateral view; (**D**) longitudinal section.

Specimen 1 _____

_____

_____

Specimen 2 _____

_____

_____

Procedure 2

Microanatomy of Basidiomycetes

Materials

- ❏ Compound microscope
- ❏ Slides and coverslips
- ❏ Prepared slides of select basidiomycetes, including a *Coprinus* sp. and *Puccinia* sp.
- ❏ Colored pencils

1. Procure a compound microscope and select slides.

2. Carefully scrape away some of the tissue from one of the basidiomycete specimens in Procedure 1, and prepare slides of the spores if possible. Record your observations and sketches in the space provided.

Specimen _____

Total magnification _____

_____

_____

3. Observe the prepared slides on both low and high power (Figs. 17.20–17.24). Record your observations and sketches in the space provided on the following page.

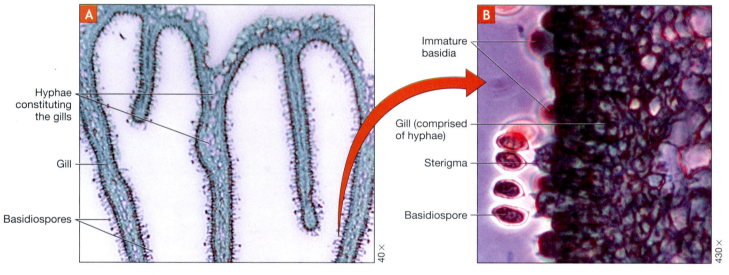

Hyphae constituting the gills

Gill

Basidiospores

40×

Immature basidia

Gill (comprised of hyphae)

Sterigma

Basidiospore

430×

FIGURE **17.20** Gills of the mushroom *Coprinus* sp.: (**A**) closeup of several gills; (**B**) closeup of a single gill.

FIGURE **17.21** Wheat rust, *Puccinia graminis*.

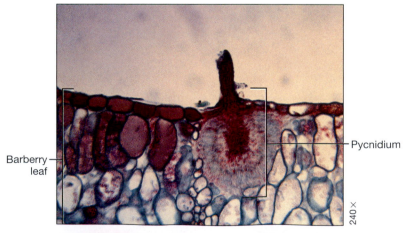

Barberry leaf

Pycnidium

240×

FIGURE **17.22** *Puccinia graminis*, pycnidium on barberry leaf.

17

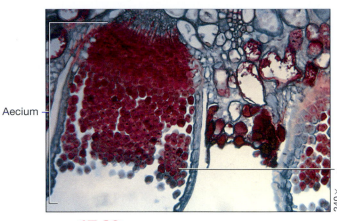

Aecium

Aeciospores

240×

FIGURE **17.23** *Puccinia graminis*, aecium on barberry leaf.

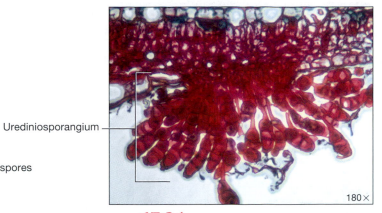

Urediniosporangium

180×

FIGURE **17.24** Urediniosporangium of *Puccinia* sp. on wheat leaf.

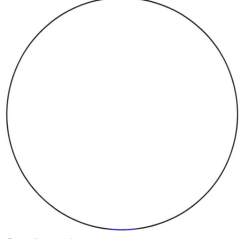

Specimen 1 _____

Total magnification _____

_____

_____

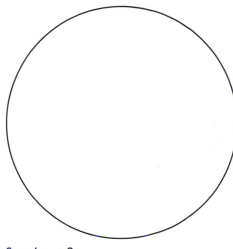

Specimen 2 _____

Total magnification _____

_____

_____

Specimen 3 _____

Total magnification _____

_____

_____

Specimen 4 _____

Total magnification _____

_____

_____

17

✔ Check Your Understanding

3.1 Describe five types of basidiomycetes.

_____

_____

_____

_____

_____

3.2 Basidiospores are asexual spores produced on peg-like sterigma on a club-shaped basidium. (*Circle the correct answer.*)

True / False

3.3 Cutting through a cap of a mushroom exposes gill tissue. What is seen when scrapings from gill tissue are placed on a slide under the microscope?

_____

_____

Lichens

Lichens are interesting symbionts consisting of a green algae or a cyanobacterium and, with the exception of a few species, ascomycetes. Algal cells or cyanobacteria are thought to provide food for both symbionts through photosynthesis, and the ascomycete retains water and minerals, anchors the organism, and protects the algae. Presently, nearly 20,000 species of lichen have been described. Scientific names are assigned to lichens like other species.

Lichens typically reside on trunks and branches of trees, bare rocks, and human-made structures, such as walls and gravestones. They also can survive in extreme conditions, such as the tundra (e.g., reindeer moss) and hot deserts. Lichens have been used to make dyes, litmus paper, bandages, antibiotics, packing material, decorations, and perfume. In the environment, pioneer lichens help build soil, and they provide food and habitat for small animals. Some species of lichen serve as environmental indicators of air pollution. Recently, several species of nitrogen-fixing lichens have been described.

The body, or **thallus,** of a lichen is usually derived from an ascomycete surrounding algal cells and enclosing them within complex fungal tissues. The thallus ranges in size from less than 1 millimeter to more than 2 meters in diameter. Lichens are noted for their longevity, perhaps living 4,500 years. Lichens vary in color from dull gray to bright red, green, and orange. Lichens primarily reproduce asexually.

Three basic types of lichens exist in nature:

1. **Crustose** lichens form brightly colored patches or crusts on rock or tree bark, without evident lower surfaces. This is often the start of primary succession (more on this in Chapter 25 on ecology, page 603).

2. **Foliose** lichens appear to have leaflike thalli that overlap, forming a scaly, lobed body. These lichens frequently are found on tree bark and on human-made structures.

3. **Fruticose** lichens may appear shrub-like or hanging mosslike on trees. Their thalli are either highly branched or cylindrical. Many people think lichens are parasites on trees but, with the exception of a few species, this is incorrect (Fig. 17.25).

Did you know . . .

Lichens in Space

In 2005, lichens were exposed to the harsh conditions of space aboard the BIOPAN-5 section of the European Space Agency facility for 16 days. Fungal and algal cells of lichens were found to survive in space after full exposure to massive UV and cosmic radiation—conditions proven to be lethal to bacteria and other microorganisms. In addition, after being dehydrated as the result of a vacuum, the lichens recovered within 24 hours.

FIGURE **17.25** Lichens are often categorized informally by their form: (**A**) gold cobblestone, *Pleopsidium* sp., a crustose lichen; (**B**) cabbage lungwort, *Lobaria* sp., a foliose lichen; (**C**) wolf, *Letharia* sp., a fruticose lichen.

Macroanatomy of Lichens

Materials

- ❏ Dissecting microscope or hand lens
- ❏ Dissecting tray
- ❏ Several examples of crustose, fruticose, and foliose lichens
- ❏ Colored pencils

Photograph lichens where you live or around your campus and bring the images to class for discussion.

1 Procure the equipment and specimens.

2 Place the specimen on a dissecting tray and using a dissecting microscope or hand lens, observe, describe, and sketch the lichens in the space provided.

Specimen 1 _____

Specimen 2 _____

Specimen 3 _____

✔ Check Your Understanding

4.1 Lichens are a symbiotic relationship between Ascomycetes and _____. (*Circle the correct answer.*)

- a. green algae
- b. green algae or cyanobacteria
- c. brown algae
- d. green algae or glomeromycetes
- e. brown algae or cyanobacteria

4.2 What role do lichens play in the environment?

_____

_____

4.3 Lichens _____. (*Circle the correct answer.*)

- a. are indicators of environmental pollution
- b. can survive in extreme conditions
- c. are known for their longevity
- d. both a and b
- e. all of the above

17

There's a Fungus among Us
Understanding Fungi

MYTHBUSTING

Break the Mold!

Debunk each of the following misconceptions by providing a scientific explanation. Write your answers on a separate sheet of paper.

1 If animals of the forest eat a particular mushroom, it is safe for me to eat.

2 I cannot get athlete's foot from a pair of used shoes that I buy online.

3 The presence of bread mold means that my house is filthy.

4 The false morel *Gyromitra*, or the beefsteak morel, contains jet fuel.

5 Ringworm is caused by parasitic worms.

---

1 What is the significance of the kingdom Fungi to the environment?

_____

_____

_____

_____

_____

2 Identify which phylum each of the examples belongs to, using the following key:

A = Ascomycota   B = Basidiomycota   Z = Zygomycota

_____ Rusts and smuts

_____ *Pilobolus* sp.

_____ *Penicillium* sp. and *Aspergillus* sp.

_____ Morels and *Peziza* sp.

_____ Earthstars, puffballs, and bracket fungi

_____ *Mucor* sp. and *Rhizopus* sp.

_____ *Candida albicans* and *Saccharomyces cerevisiae*

3 Sketch and label *Rhizopus* growing on bread in the space provided.

4 The ascomycetes produce sexual spores, or _____, within a saclike structure, or _____.

5 The ascomycete fruiting body is called a/an _____.

6 Trace the generalized life cycle of an ascomycete in the space provided.

7 What is the role of the fungus in the mycorrhizal relationship with the plant? (*Circle the correct answer.*)
   a. Provides carbohydrates to the plant.
   b. Provides water to the plant.
   c. Helps the plant take in phosphate and other minerals.
   d. Both a and c.
   e. Answers a, b, and c are all correct.

8 Which of the following characteristics of the kingdom Fungi is/are true? (*Circle the correct answer.*)
   a. They are eukaryotic cells with most fungal of the cell walls containing chitin.
   b. Fungi play an important role as decomposers; they release digestive enzymes into a food source, breaking it down.
   c. All produce haustoria.
   d. Both a and b.
   e. All of the above.

9 Trace the generalized life cycle of a basidiomycete in the space provided.

17

10 Draw and label a typical mushroom with gills in the space provided.

11 Match the following descriptions with the correct type(s) of lichen. (Some answers may be used twice.)

_____ Crustose   A. Forms a body or thallus

_____ Foliose   B. Leaflike thalli that overlap, forming a scaly lobed body

_____ Fruticose   C. Forms crusty patches on rocks and on tree bark

                           D. Is shrub-like or hangs mosslike from trees

Animal Planet I
Understanding Animal Histology and Basal Animals

18

I traveled among cells, watched their functioning ... and realized that within myself was a grand assemblage of living organisms, all of which added up to be me.

— John C. Lilly (1915–2001)

Just wondering...

Consider the following questions prior to coming to lab, and record your answers on a separate piece of paper.

1. Why are the cilia in pseudostratified columnar epithelia important to lung health?
2. Why does skeletal muscle have striations?
3. What is the general first aid for a jellyfish sting?
4. What is meant by reconstitution in sponges?
5. Are members of Class Cubozoa that venomous?

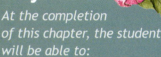

Objectives

At the completion of this chapter, the student will be able to:

1. Define the term *histology*.
2. Discuss the classification and basic characteristics of common human tissues.
3. State the location and function of the tissues used in the exercises in this chapter.
4. Identify the tissues examined under a compound microscope.
5. Discuss the major characteristics of animals.
6. Describe the fundamental characteristics and natural history of phylum Porifera, Cnidaria, and Ctenophora and distinguish between the organisms in its classes.
7. Record and sketch macroscopic and microscopic anatomy of phylum Porifera, Cnidaria, and Ctenophora.

Chapter Photo
Green striped mushroom coral, *Actinodiscus* sp.

Could you imagine what life would be like on Earth if multicellularity had not developed? Life on this planet would be extremely different. Multicellular organisms are made up of many cells and cell types capable of carrying on specialized functions. Evolutionarily, multicellularity arose independently in many separate lineages of eukaryotic organisms, such as plants and animals. Interestingly, forms of multicellularity also arose in fungi, green algae, brown algae, red algae, slime molds, and some ciliates.

Cells similar in structure and function compose tissues. Both plants and animals are made up of several basic tissue types. Animal tissues are generally classified into one of four primary types: epithelial, connective, muscular, and nervous. These tissues resemble each other only to the extent that they are composed of cells and intercellular substances. There are many different types of primary tissues, each with morphological and functional modifications. Organs comprise several tissue types. In fact, all four basic tissue types exist in the heart.

To understand how animals work, one must be able to recognize and comprehend the functions of basic tissues. Histologists (scientists who study tissues) are important in our understanding and diagnosis of disease, for example. In this chapter on histology, you will take a closer look at the four types of animal tissue.

Hints & Tips

1. Read the description of the tissue, and study the pictures thoroughly.
2. Do not become dependent on the color of the tissue.
3. View the tissue using the microscope powers suggested by your instructor.
4. In viewing, scan different parts of the slide and use different depths of field.
5. Accurately draw and color what you view.
6. Spend quality time viewing the specimens. *Don't rush!*

Animal Histology

Epithelial Tissue

Epithelial tissue, or epithelium, refers to tissue that covers body surfaces, lines body cavities, and forms glands. Usually this tissue is attached to connective tissue by a basement membrane and has one surface exposed to the environment. Epithelial tissue prevents most objects and substances from the outside environment from entering the body, and they keep most of the internal material within the body. Epithelial tissue is highly modified for absorption, excretion, and secretion.

Fundamental characteristics of epithelial tissue are the following:

1. The cells that make up epithelial tissue are relatively regular in shape (Fig. 18.1A–C).

2. Because this tissue is tightly packed in single or multi-layered sheets (Fig. 18.1D–F), there is little to no inter-cellular material between adjacent cells.

3. Epithelial tissue is primarily avascular (blood vessels are located in underlying vascularized tissue).

4. Epithelial tissues usually have several types of surface specializations. Although some epithelial tissues have smooth surfaces, most tissues have many complex folds, or **microvilli**. The microvilli on the free surface of epithelial cells is termed **brush border**. The primary function of microvilli is to increase surface area. **Cilia** occur in some epithelial cells, such as those lining the trachea.

Simple squamous epithelium occurs as a single layer of flattened cells tightly held together (Fig. 18.2). The nuclei appear broad and thin and are parallel to the surface. This tissue is thin and highly adapted for osmosis, diffusion, and filtration. Simple squamous epithelium is found in regions of little wear and tear.

Simple cuboidal epithelium consists of a single layer of cube-shaped cells. When viewed from above, the cells appear as polygons. The large nuclei are centrally located and are rounded (Fig. 18.3). This tissue is found in the lining of many glands and their ducts, the surface of the ovaries, the inner surface of the lens of the eye, the pigmented epithelia

Single layer of flattened cells

300×

FIGURE **18.2** Simple squamous epithelium.

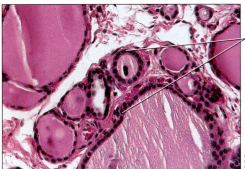

Single layer of cells with round nuclei

300×

FIGURE **18.3** Simple cuboidal epithelium.

FIGURE **18.1** Some common shapes and arrangements of tissues: (**A**) squamous; (**B**) cuboidal; (**C**) columnar; (**D**) simple; (**E**) stratified; (**F**) pseudostratified.

Hints & Tips

1. **Simple:** a single layer.

2. **Stratified:** two or more layers.

3. **Pseudostratified:** appears to have many layers because the cells vary in height and shape; the nuclei appearing at different heights give an impression of false layers.

4. **Squamous:** flat or scalelike cells forming a mosaic pattern.

5. **Cuboidal:** cells appear to be cube-like in cross section.

6. **Columnar:** cells are long and cylindrical like a column.

18

of the retina of the eye, and some kidney tubules. Simple cuboidal epithelium is active in absorption and secretion.

Simple columnar epithelium consists of a single layer of long, column-shaped cells. The nuclei are large and oval-shaped, usually located near the base of the cell (Fig. 18.4). Simple columnar epithelium serves in protection, secretion, absorption, and initiating movement. This tissue can be ciliated or nonciliated, depending on its location and function.

Ciliated tissue can be found in the oviduct, where it helps to sweep the egg cell toward the uterus after it leaves the ovary. Nonciliated cells can be found in the stomach, digestive glands, and gallbladder, where they protect the delicate linings and function in absorption and secretion. In the intestines, modified cells called **goblet cells** are interspersed in the columnar cells and secrete mucus that protects and lubricates the walls of the digestive tract.

Pseudostratified columnar epithelium is a simple epithelium as well. It consists of a single layer of cells of varying height and shape; all of the cells are attached to a basement membrane, but not all reach the free surface (Fig. 18.5). Because of the differences in the height of the cells, the nuclei appear at several different levels, giving the erroneous impression of stratification. When cilia are found on the free surface of the cells, the tissue is called **pseudostratified ciliated columnar epithelium**.

Stratified squamous epithelium usually consists of several layers of cells, but only the superficial layer consists of squamous cells (Fig. 18.6). The underlying basal cells are modified columnar and cuboidal cells. The basal cells have the ability to replace the superficial squamous cells as they become damaged or worn. Because of its regenerative powers, stratified squamous epithelium appears in areas of drying, wear, injury, and friction. Some stratified squamous epithelium is associated with a protective protein called **keratin**. Keratinized tissue can be found in the epidermis. Non-keratinized tissue can be found in the mouth, esophagus, and vagina.

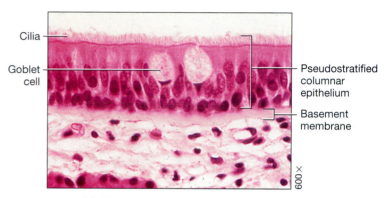

FIGURE **18.5** Pseudostratified columnar epithelium.

Cilia

Goblet cell

Pseudostratified columnar epithelium

Basement membrane

600×

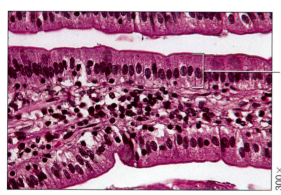

FIGURE **18.4** Simple columnar epithelium.

Single layer of cells with oval nuclei

300×

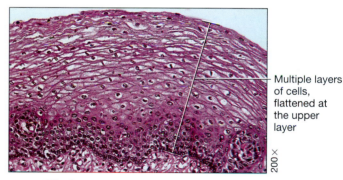

FIGURE **18.6** Stratified squamous epithelium.

Multiple layers of cells, flattened at the upper layer

200×

18

Procedure 1

Identifying Epithelial Tissues

Materials

❑ Compound microscope
❑ Prepared slides labeled as simple squamous, cuboidal, columnar, pseudostratified columnar, and stratified squamous epithelium
❑ Colored pencils

1 View the slide labeled "Simple Squamous Epithelium" under both low and high power. Sketch the tissue in the space provided

2 View the slide labeled "Cuboidal Epithelium, or "Kidney Section," or "Thyroid Gland" under both low and high power. Sketch the tissue in the space provided.

3 View the slide labeled "Columnar Epithelium" or "Frog Intestine" under both low and high power. In the slide, note that the columnar epithelium and goblet cells can be found in the **villi** (fingerlike projections) of the small intestine. In addition, note the presence of smooth muscle in this slide. Sketch the tissue in the space provided.

4 View the slide labeled "Pseudostratified Ciliated Columnar Epithelium" or "Trachea" under both low and high power. In addition, note the presence of adipose tissue and hyaline cartilage in this slide. Sketch the tissue in the space provided.

5 View the slide labeled "Stratified Squamous Epithelium" under both low and high power. Sketch the tissue in the space provided.

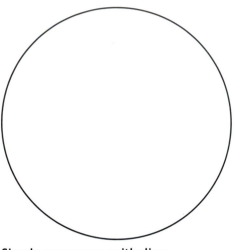

Simple squamous epithelium

Total magnification _____

_____

_____

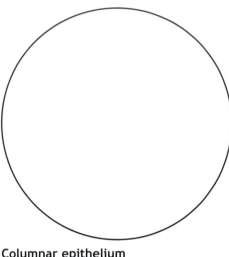

Cuboidal epithelium

Total magnification _____

_____

_____

Columnar epithelium

Total magnification _____

_____

_____

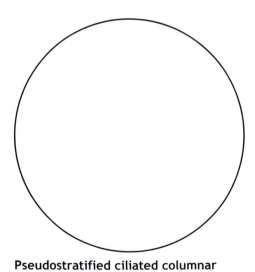

Pseudostratified ciliated columnar

epithelium   Total magnification _____

_____

_____

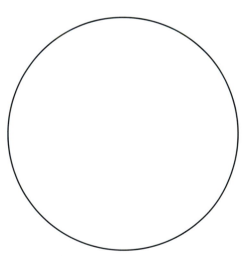

Stratified squamous epithelium

Total magnification _____

_____

_____

18

Connective Tissue

Connective tissue is the most widely dispersed and abundant type of tissue in the body. As a general rule, connective tissues have abundant extracellular fibrous material that supports the cells of other tissues. Connective tissues perform a variety of functions, but they primarily protect, support, and bind together other tissues. Other functions include insulation and cushioning, storage of fat, repair of body tissues, and production of **red blood cells (RBCs)**. These tissues vary in their morphology and anatomy.

In this laboratory, several representative types of connective tissue will be studied. The tissues explored next are adipose tissue, **hyaline cartilage**, and bone. Blood cells and blood-forming tissues are included because they have the same embryonic origin (mesenchyme) as the connective tissues.

Characteristics of connective tissue are as follows:

1. Connective tissue is well vascularized, with the exception of cartilage, tendons, and ligaments.
2. The extracellular fibers and ground substances make up the nonliving **matrix**.
3. The ground substance is a homogeneous, extracellular material that ranges from a semifluid to a thick gel in consistency. The ground substance provides a suitable medium for the passage of nutrients and waste products between the cells and the bloodstream.

Adipose Tissue

Adipose tissue consists of numerous adipocytes, or fat cells, and a small amount of reticular matrix. Adipose cells may appear singly or in clusters. A single adipocyte appears as a large, clear lipid droplet with a large nucleus flattened against the side. This tissue resembles signet rings or fishnet. Adipose tissue functions in insulation, cushioning, and the storage of energy. This tissue is found beneath the epidermis and surrounding organs and throughout the body (Fig. 18.7).

Cartilage

Cartilage is a type of connective tissue that provides support and aids in movement. Cartilage is avascular. Oxygen, nutrients, and waste products diffuse through the cartilage matrix. The three types of cartilage are:

1. Hyaline cartilage (studied here);
2. Fibrocartilage, somewhat flexible and capable of withstanding pressure (found in the pubic symphysis, intervertebral disks, knee joints, and temporomandibular joints); and
3. Elastic cartilage, which provides rigidity and great flexibility (found in the epiglottis, external ear, and eustachian tubes).

Cartilage cells are called **chondrocytes** and are embedded in small cavities within the matrix. The small cavities are known as **lacunae**.

Hyaline cartilage is the most common and rigid type of cartilage. The collagenous fibers are scattered in a network completely filled with ground substance. It usually is enclosed in a fibrous covering called the perichondrium. Hyaline cartilage forms a major part of the skeleton of embryos and reinforces respiratory passageways in the trachea, larynx, and bronchi. At the ends of long bones it is called articular cartilage, and at the distal ends of ribs it is called costal cartilage (Fig. 18.8).

Bone

Bone tissue, or **osseous tissue**, is a hard, connective tissue that consists of living cells dispersed in an organic and mineralized matrix. The organic portion of the matrix contains collagen fibers and other organic molecules. The mineral part of the matrix contains tricalcium phosphate crystals called hydroxyapatite and calcium carbonate. The human body is about 62% water, but bone tissue contains about 20% water. As a result of the minerals and lack of water, bone tissue is stronger and more durable than other tissues.

18

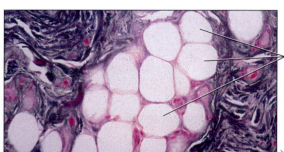

200×

Figure **18.7** Adipose connective tissue.

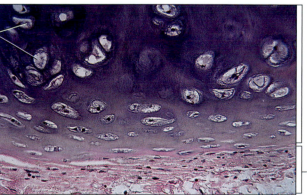

Chondrocytes within lacunae

Hyaline cartilage

Perichondrium

200×

Figure **18.8** Hyaline cartilage.

Bone serves in protection, support, movement, production of red blood cells, mineral homeostasis, storage of energy, and storage of calcium.

The two types of bone are as follows:

1. **Spongy bone** has spaces between the plates (trabeculae) of bone. The spaces between the trabeculae of some bones are filled with red bone marrow. Spongy bone tissue in the sternum, vertebrae, ribs, hip bones, and near the ends of long bones is involved in production of red blood cells. Spongy bone makes up most of the bone tissue in short, flat, and irregular bones.

2. **Compact bone** contains few spaces and also is known as dense bone. It constitutes the external layer of all bones and makes up the bulk of the shaft of long bones. Compact bone tissue provides support and protection and helps the long bones resist stress.

Compact bone contains cylinders of calcified bone known as **osteons**, or **Haversian systems** (Fig. 18.9). These cylinders consist of 4 to 20 concentric rings of bone called **lamellae**. The lamellae contain numerous lacunae, each housing a bone cell or **osteocyte**. Radiating out from each lacuna are **canaliculi** that channel nutrients and wastes by diffusion into and out of the blood vessels in the **Haversian canal** (central canal), the most prominent portion of the osteon. These longitudinal channels contain nerves and blood vessels.

Connected to and running at a right angle to the Haversian canals are Volkmann's canals (perforating canals). The Volkmann's canals extend the nerves and vessels outward to the periosteum (outer covering) and endosteum (inner lining) of the bone marrow cavity. The osteon complex provides the strength necessary to resist everyday stress.

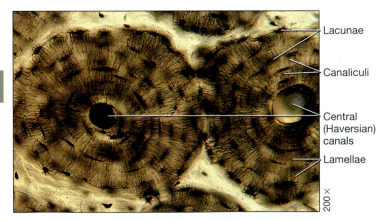

FIGURE **18.9** Cross section of two osteons in bone tissue.

Blood

Blood is classified as a specialized kind of fluid connective tissue. Blood contains a fluid matrix called plasma and formed elements called **erythrocytes**, leukocytes, and **platelets**. The functions of blood include transportation of respiratory gases, nutrients, enzymes, hormones, and waste products;

regulation of acid-base balance; regulation of body temperature; regulation of electrolytes; and defense against toxins and harmful microorganisms (Fig. 18.10).

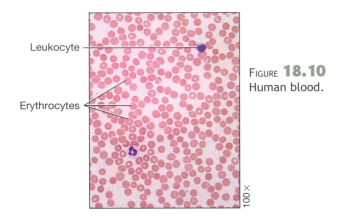

FIGURE **18.10** Human blood.

Blood has the following characteristics:

1. The straw-colored liquid portion of the blood, called plasma, makes up about 55% of the total blood volume. Plasma is 92% water and 8% dissolved or suspended molecules, such as plasma proteins, gases, cellular waste products, hormones, and ions.

2. Red blood cells (RBCs) are termed *erythrocytes*. These cells make up 99% of the formed elements of blood; the human body has about 25 trillion erythrocytes. Mature red blood cells in humans are anucleated and appear as biconcave disks that have more surface area for diffusion and are flexible so they can more readily pass through blood vessels. The surface antigens of RBCs are responsible for the various blood groups, such as the ABO group and Rh group.

3. White blood cells (WBCs) are termed *leukocytes*. Adults have about 1,000 erythrocytes for every leukocyte. The five basic kinds of leukocytes are neutrophils, eosinophils, basophils, lymphocytes, and monocytes. WBCs are classified into two major groups based on the staining properties of their cytoplasmic granules:

    a. Granular leukocytes (granulocytes) have conspicuous granules in the cytoplasm and a lobed nucleus. The granulocytes include neutrophils, eosinophils, and basophils. The neutrophils, the most common leukocyte in the blood, constitute 60%–70% of the total white blood count. Neutrophils have nuclei with two to six lobes (polymorphonucleated). The granules appear pale and lilac-colored. They destroy microorganisms and foreign particles and are a major component of pus.

       The eosinophils make up approximately 2%–4% of the total WBC count. The nucleus of an eosinophil is usually bilobed. The granules in eosinophils range from red to red-orange in acid stain. Common in allergic reactions and parasitic infections, eosinophils

18

can phagocytize antigen-antibody complexes formed during the allergic reaction.

The basophils make up less than 1% of the total white blood count. The nucleus is bilobed or irregular in shape, often in the form of an S. The granules are large, stain blue-black to red-purple, and often obscure the nucleus. These cells contain the anticoagulant heparin. In addition, they liberate histamine and serotonin to intensify the inflammatory response.

b. Agranular leukocytes (agranulocytes) show no cytoplasmic granules under the light microscope. The two kinds of agranulocytes are the lymphocytes and monocytes. Lymphocytes, the smallest of the leukocytes, constitute approximately 20%–30% of the total white blood count and encompass a number of different types. Lymphocytes have a round, dark-stained nucleus, and the cytoplasm appears as a thin ring around the nucleus. These cells are abundant in lymphoid tissue and play a major role in immunity and antibody production.

The largest of the leukocytes, monocytes make up 2%–8% of the total white count. They have a dark-stained, large, kidney-shaped nucleus, and the cytoplasm appears bluish and foamy. Monocytes generally remain in circulation for three days, leave the circulation, and become migrating macrophages.

They phagocytize bacteria, dead cells, cell fragments, and other debris.

4. Platelets are disk-shaped cell fragments of large multi-nucleate cells (megakaryocytes) formed in the bone marrow. When a blood vessel is injured, platelets move to the site and begin to clump together, attaching themselves to the damaged area. If the injury is slight, the platelets form a platelet plug to stop bleeding. If the damage is more extensive, the platelet plug is reinforced by fibrin threads that form as a result of activation of one of the blood coagulation pathways.

Procedure 2

Identifying Connective Tissues

Materials
- ❏ Compound microscope
- ❏ Prepared slides labeled as pseudostratified ciliated columnar epithelium, hyaline cartilage, ground bone, human blood, and amphibian blood
- ❏ Colored pencils

1 To find adipose tissue, view the slide labeled "Pseudostratified Ciliated Columnar Epithelium" or "Trachea" under both low and high power. In addition, note the presence of pseudostratified ciliated columnar epithelium tissue and hyaline cartilage. Sketch the tissue in the space provided.

2 View the slide labeled "Hyaline Cartilage" under both low and high power. Sketch the tissue in the space provided. Label the matrix, lacunae, and chondrocytes.

3 View the slide labeled "Ground Bone" under both low and high power. Sketch the tissue in the space provided. Label the matrix, Haversian canal, lamellae, lacunae, and canaliculi.

4 View the slide labeled "Human Blood" under high power. Note that the mature human RBCs are anucleated. Identify erythrocytes, leukocytes, and platelets. Your instructor may want you to identify the different kinds of leukocytes. Sketch the slide in the space provided.

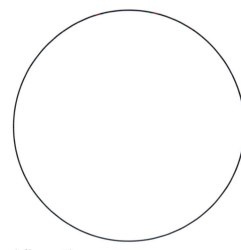

Adipose tissue

Total magnification _____

18

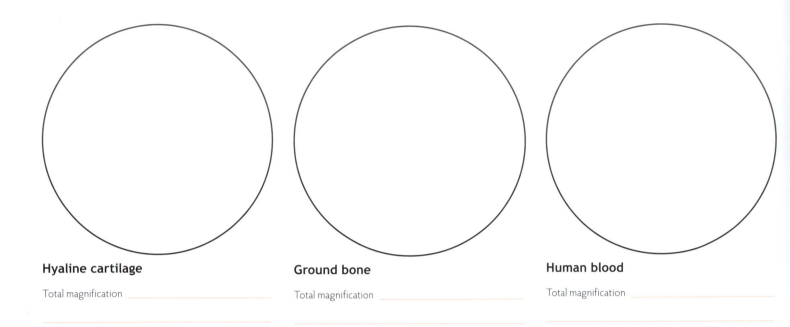

Hyaline cartilage

Total magnification _____

_____

_____

Ground bone

Total magnification _____

_____

_____

Human blood

Total magnification _____

_____

_____

5 View the slide labeled "Amphibian Blood," "Frog Blood," or "Amphiuma Blood" under both low and high power (Fig. 18.11). Note that amphibian RBCs are nucleated. In addition, the amphiuma has the largest erythrocytes in the animal kingdom. Identify erythrocytes, leukocytes, and platelets. Your instructor may want you to identify the different types of leukocytes. Sketch the slide in the space provided.

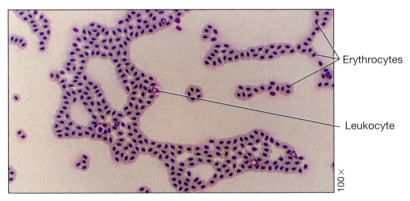

FIGURE **18.11** Amphiuma blood.

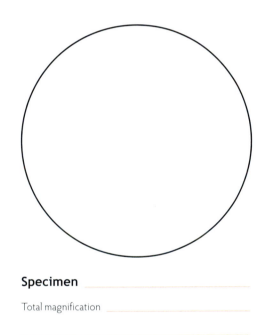

Specimen _____

Total magnification _____

_____

_____

Muscular Tissue

Muscle tissue is specialized to generate force, perform work, generate heat, maintain posture, and provide movement. The three major types of muscle tissue are smooth muscle, skeletal muscle, and cardiac muscle. These three types of muscle differ from each other in their microscopic anatomy, location, and control by the nervous and endocrine systems.

Muscle tissue has the following characteristics:

1. Muscle tissue exhibits contractility; as muscle tissue contracts, it generates force to do work.

2. Muscle tissue exhibits excitability, the ability to receive and respond to stimuli. Muscle tissue responds to neurotransmitters released by **neurons** or hormones distributed by the blood.

3. Muscle tissue exhibits extensibility, the ability to be stretched.

4. Muscle tissue exhibits elasticity, the ability to return to its original shape after constriction or extension.

Smooth muscle is considered involuntary muscle because it is controlled by the autonomic (involuntary) division of

18

the nervous system. Under the microscope, the muscle appears to lack the striations characteristic of skeletal and cardiac muscle. Instead, this muscle comprises fibers that bulge in the center and are tapered at both ends (Fig. 18.12). A single oval nucleus is located within each fiber of this muscle. Smooth muscle is not connected to bone. Smooth muscle is distributed throughout the body and is more variable in function than other types of muscle.

Skeletal, or striated, muscle is associated with voluntary movement and thus makes locomotion possible. In addition, it generates heat and guards the entrances and exits of the respiratory, digestive, and urinary tracts. Skeletal muscle is described as striated because of the alternating light and dark bands of the proteins **actin** and **myosin** (Fig. 18.13).

As the name indicates, cardiac muscle is found exclusively in the heart. Cardiac muscle pumps blood throughout the body. It contains the same type of myofibril and protein composition as skeletal muscle. Although this muscle is closely packed, each cell is separate and has its own nucleus. Like skeletal muscle cells, cardiac muscle cells are striated.

Unlike skeletal muscle cells, which are long and tubular, cardiac muscle cells are relatively short and branched.

Cardiac muscle cells are linked end to end at **intercalated disks**, which appear as dark bands running perpendicular to the cardiac muscle (Fig. 18.14). The functions of intercalated disks are to help impulses pass quickly from one cell to the next, strengthen the junction between cells, and separate the cells within a muscle fiber.

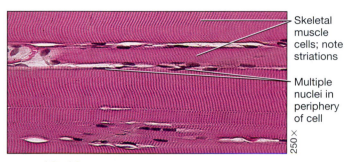

Skeletal muscle cells; note striations

Multiple nuclei in periphery of cell

250×

FIGURE **18.13** Longitudinal section of skeletal muscle tissue.

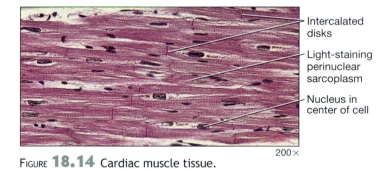

Intercalated disks

Light-staining perinuclear sarcoplasm

Nucleus in center of cell

200×

FIGURE **18.14** Cardiac muscle tissue.

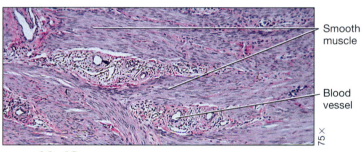

Smooth muscle

Blood vessel

75×

FIGURE **18.12** Smooth muscle tissue.

Procedure 3

Identifying Muscle Tissues

Materials

❏ Compound microscope
❏ Prepared slides labeled as smooth, striated, and cardiac muscle
❏ Colored pencils

1 View the slide labeled "Smooth Muscle" or "Small Intestine" (cross section) under both low and high power. Sketch the tissue in the space provided.

2 View the slide labeled "Striated Muscle," "Skeletal Muscle," or "Voluntary Muscle" under both low and high power. In the slide, note the striations and prominent nuclei. Sketch the tissue in the space provided.

3 View the slide labeled "Cardiac Muscle" under both low and high power. On the slide, note the striations, prominent nuclei, and intercalated disks. Sketch the tissue in the space provided.

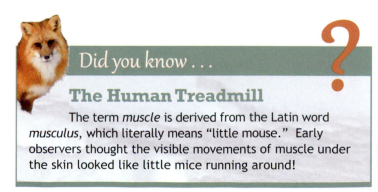

Did you know . . .

The Human Treadmill

The term *muscle* is derived from the Latin word *musculus*, which literally means "little mouse." Early observers thought the visible movements of muscle under the skin looked like little mice running around!

Smooth muscle

Total magnification _____

_____

_____

Striated muscle

Total magnification _____

_____

_____

Cardiac muscle

Total magnification _____

_____

_____

Nervous Tissue

The nervous system is responsible for integrating and coordinating the other systems of the body, sensing stimuli, and continuously monitoring the external and internal environment. The nervous system is made up of more than 100 billion nerve cells called neurons and other cells called neuroglia ("nerve glue") that serve as supportive cells.

Characteristics of **nervous tissue** include the following:

1. Neurons are one of the most specialized types of cells. These cells display great diversity in size and shape. The cell bodies of neurons range in diameter from 5 microns (smaller than a RBC) to 135 microns (large enough to be seen by the unaided eye). Some neurons extend more than 3 feet. Neurons vary in shape from star-shaped to oval.

2. A typical neuron (Fig. 18.15) has three parts: a cell body, dendrites, and an axon.

    a. The **cell body**, or **soma**, consists of varying amounts of cytoplasm with a prominent nucleus and nucleolus. A variety of organelles resides within the cytoplasm.

    b. **Dendrites** are neuron processes that conduct impulses toward the neuron cell body. The dendrites usually are unmyelinated, tapered, and branched. These structures are generally extensions of the cell body and share organelles with the cell body. Some neurons have more than 200 dendrites.

    c. **Axons** are long, thin, cylindrical neuron processes that carry impulses away from the cell body. Generally, neurons possess one axon that may branch into collaterals.

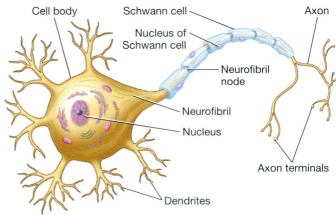

FIGURE **18.15** Typical neuron.

3. Neurons also can be classified according to their functions. Multipolar neurons possess several dendrites and only one axon. This type of neuron can be found in the brain and spinal cord.

4. Neurons may be classified by the number of processes attached to the cell body. There are three basic types of neurons:

    a. Unipolar neurons;

    b. Bipolar neurons; and

    c. Multipolar neurons.

18

Identifying Nervous Tissue

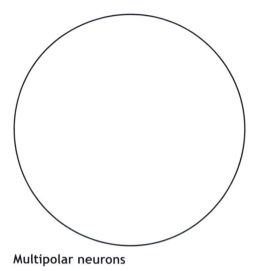

Materials

- ❏ Compound microscope
- ❏ Prepared slides labeled as multipolar neurons
- ❏ Colored pencils

1 View the slide labeled "Multipolar Neurons" or "Ox Spinal Smear" under both low and high power (Fig. 18.16).

2 Sketch the tissue in the space provided.

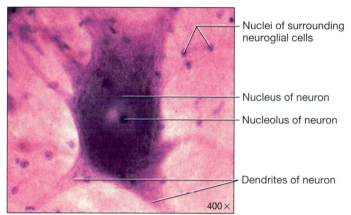

- Nuclei of surrounding neuroglial cells
- Nucleus of neuron
- Nucleolus of neuron
- Dendrites of neuron

400×

FIGURE **18.16** Neuron smear.

Multipolar neurons

Total magnification _____

_____

_____

✔ Check Your Understanding

1.1 What is the function of the basement membranes of epithelial tissues?

_____

_____

1.2 Simple columnar and pseudostratified columnar epithelium both are considered simple epithelium that consist of column-shaped cells. Simple columnar cells line up next to each other whereas there is an irregular arrangement with cells varying in height and shape in pseudostratified columnar epithelium, giving the impression of stratification. (*Circle the correct answer.*)

True / False

1.3 _____ cells are modified cells found in the intestines interspersed within columnar cells. They secrete mucus, which protects and lubricates the walls of the digestive tract.

1.4 Match the type of connective tissue with the correct function. (Some will have more than one answer.)

_____ Bone

_____ Blood

_____ Adipose tissue

_____ Cartilage

A. Transportation of respiratory gases, nutrients, and waste products
B. Support and hematopoiesis
C. Energy storage, insulation, and cushioning
D. Regulation of acid-base balance and body temperature
E. Support and movement
F. Regulation of electrolytes and defense against toxins and harmful microorganisms

(continues)

1.5 What is the function of lacunae?

_____

_____

1.6 Compact bone contains cylinders of calcified bone called _____. These cylinders consist of 4 to 20 concentric rings of bone known as _____. (*Circle the correct answer.*)

    a. osteons, canaliculi

    b. lamellae, canaliculi

    c. osteons, lamellae

    d. Haversian canals, lamellae

    e. none of the above

1.7 Observing a blood smear that has red blood cells with no nuclei would indicate that the specimen came from what type of organism?

_____

1.8 Describe the four general characteristics of muscle tissue.

_____

_____

_____

_____

1.9 Match each major type of muscle with the correct description:

_____ Smooth   A. Striated; cells are short and branched

_____ Cardiac   B. Has fibers that bulge in the center and are tapered at both ends; lacks striations

_____ Skeletal   C. Striated; cells are long and tubular

1.10 Compare and contrast the three types of muscle in regard to shape, striations, special features, and control.

_____

_____

_____

_____

_____

1.11 _____ carry impulses away from the neuron cell body and _____ conduct impulses toward the cell body.

1.12 What is the difference between a sensory and a motor neuron?

_____

_____

_____

_____

18

■ Basal Animals

From trilobites etched in stone to the mighty *Tyrannosaurus rex* and from simple sponges to the great blue whale, the great diversity of animals captures our curiosity and imagination (Fig. 18.17). Kingdom Animalia conservatively consists of approximately 1.5 million extant organisms and many more extinct forms. Animals are eukaryotic, multicellular heterotrophs that exist in marine, aquatic, and terrestrial environments. Many animals—frogs and cats, for example— are free-living, whereas several species, including barnacles

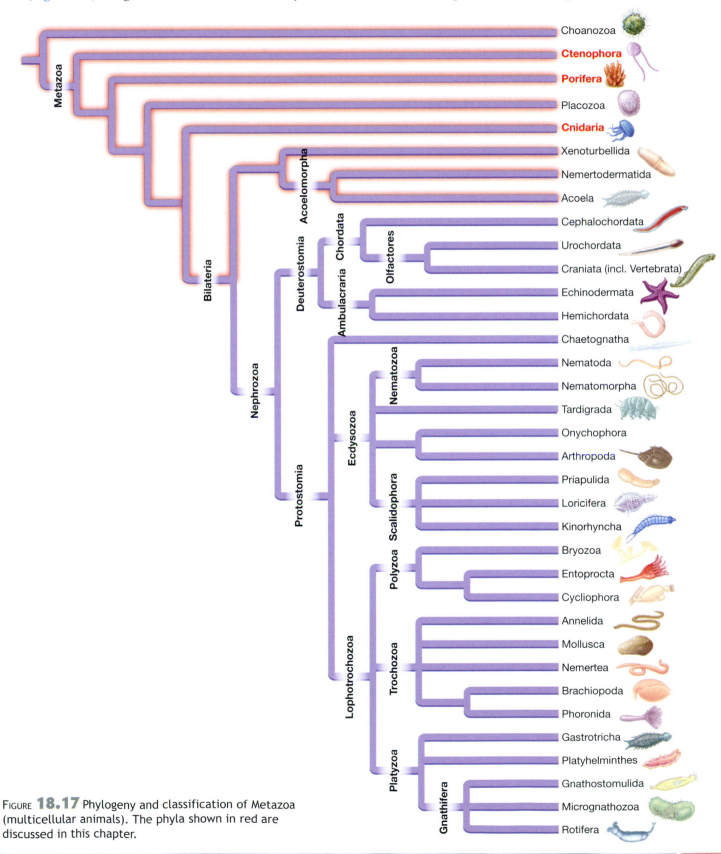

FIGURE **18.17** Phylogeny and classification of Metazoa (multicellular animals). The phyla shown in red are discussed in this chapter.

and others, are sessile (attached to a substrate). Many animal species exist in complex symbiotic relationships, such as mutualism (sea anemone and clown fish) and parasitism (flea and dog).

Animal cells do not have cell walls and are organized into complex tissue types, including muscle, nerve, and others. The muscular system, working with the nervous system, results in most animals being motile. Sexual reproduction is the primary means of reproduction, and the diploid stage dominates the life cycle. Some animals, such as jellyfishes and adult sea stars, exhibit **radial symmetry**; others, ranging from tapeworms to wasps to humans, exhibit **bilateral symmetry**. In radial symmetry the body parts are arranged around a central axis to radiate out like the spokes of a wheel. In bilateral symmetry, the body can be divided into two mirror images. In evolution, bilateral symmetry is tied closely to development of the head region, called cephalization, and forward movement such as swimming and running.

Although the shape and size of animals vary tremendously, only a few basic body plans exist in nature. Some animals, for examples jellyfishes, tapeworms, and flukes, have a saclike body plan. These animals have an **incomplete digestive tract**—what goes in the mouth goes back out the mouth. Other animals, for example squids and lions, have a tube-within-a-tube body plan, possessing a **complete digestive tract**. Jellyfishes and their relatives are diploblastic, composed of two germ layers, the ectoderm and the endoderm. Animals such as earthworms, beetles, snakes, and eagles are triploblastic—having three germ layers, the ectoderm, mesoderm, and endoderm (Fig. 18.18).

In gastrulation, the fate of the blastopore has resulted in two major groups of animals. In **protostomes**, such as snails, leeches, and ants, the blastopore gives rise to the mouth. In **deuterostomes**, such as sea stars, stingrays, pelicans, and monkeys, the blastopore gives rise to the anal opening.

Earthworms, insects, and vertebrates are segmented animals in which body parts are repeated along the length of the animal's body. *Hox* genes influence the embryological patterning of the body plan, including the body axis and arrangement of animal body parts. The study of these genes is providing scientists with a better understanding of the organization and evolution of the animal kingdom.

The most likely candidate for the ancestor to animals is a colonial flagellated protist that probably was similar to today's choanoflagellates. Between 600 and 550 million years ago, complex, soft-bodied, multicellular animals first appeared in the fossil record.

Keep in mind that the classification of animals, as well as the classification of eukaryotes as a whole, is in a state of transition. Traditional classification was based primarily on body plans, whereas modern schemes are more molecular in origin. Because traditional views have been followed for more than a century with wide acceptance, coupled with the understanding that modern views are not complete, taxonomy is confusing to experts and novices alike. Many biologists and medical professionals are taking a more user-friendly approach to classification based upon a synthesis of classical and modern concepts until newer schemes are solidified.

For convenience, the animal kingdom is divided here into four chapters. This chapter covers the poriferans, cnidarians (jellyfish and coral), and ctenophores (comb jellies). Chapter 19 describes the lophotrochozoans, including platyhelminths (flatworms), rotifers, molluscs (snails, octopi), and annelids

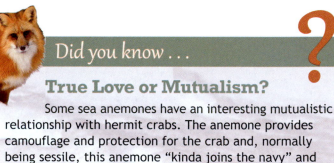

Did you know . . .

True Love or Mutualism?

Some sea anemones have an interesting mutualistic relationship with hermit crabs. The anemone provides camouflage and protection for the crab and, normally being sessile, this anemone "kinda joins the navy" and gets to travel, meeting other anemones and sampling new foods!

FIGURE **18.18** Body plans of triploblastic animals.

Labels: Gut, Internal organ, Ectoderm, Mesoderm, Endoderm, Internal organ, Gut, Pseudocoelom, Mesentery, Internal organ, Endoderm, Coelom (segmented worms). In Chapter 20, the ecdysozoans, including the nematodes (roundworms) and arthropods (insects and arachnids), are discussed. Chapter 21 covers the deuterostomes, including the echinoderms (starfishes) and chordates (fishes, amphibians, reptiles, birds, and mammals). Table 18.1 and Figure 18.19 outline some important directional terms you will need to use as you study the anatomy of animals.

TABLE **18.1** Helpful Anatomical Orientation Terms

Term	Description	Term	Description
Dorsal	Pertaining to back	Cephalic	Head
Ventral	Pertaining to underside	Celiac	Abdomen
Lateral	To the side	Caudal	Tail
Anterior	Front end	Crural	Leg
Posterior	Rear end	Oral	Mouth
Superior	Above another part or closer to head	Pedal	Foot
Inferior	Below another part or toward feet	**Body Planes**	
Medial	Toward imaginary midline	Sagittal plane (medial)	Lengthwise cut that divides the body into left and right halves.
Central	Middle	Transverse plane	Divides the body into superior and inferior sections
Peripheral	Near the surface or outside of	Frontal plane (coronal)	Divides the body into anterior and posterior portions

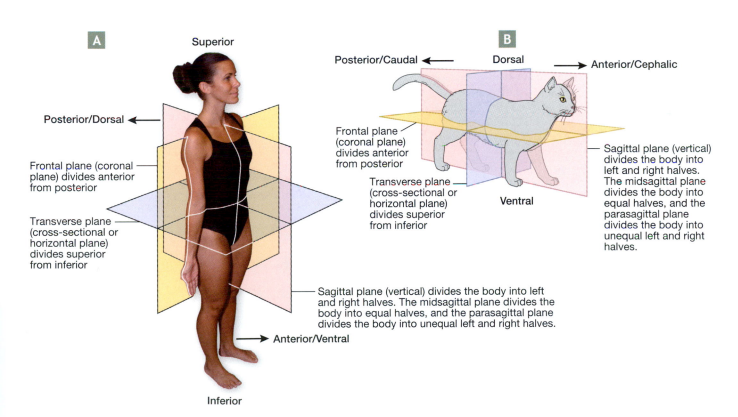

FIGURE **18.19** Body planes and basic anatomical orientation terms in a (**A**) bipedal vertebrate; (**B**) quadrupedal vertebrate.

18

Phylum Porifera

Typically, when you are thinking of a sponge, visions of bathing or washing the car enter your mind. The term *sponge*, however, has different connotations in biology. Upon examining a living or dried sponge, some people are amazed that sponges are actually the simplest of the multicellular animals. Many scientists think sponges evolved from a group of flagellated aquatic eukaryotes known as the choanoflagellates (Fig. 18.20).

Phylum Porifera (from Latin = pore-bearer) includes approximately 10,000 species of parazoan animals known as sponges (Fig. 18.21). Although sponges are multicellular, they are phylogenetically distinct from other metazoans (multicellular animals) because they do not have tissues or organs. Fewer than 200 species of sponges live in freshwater environments. The vast majority of sponges are sessile marine organisms. Despite the adult sponge being sessile, larval sponges are free-swimming.

Sponges vary in size from a few millimeters to 2 meters across. The body of a sponge is organized around a system of water canals and chambers. Many species are brightly colored (red, yellow, orange, purple, or green). Sponges vary from radially symmetrical to irregularly shaped. Some sponges bore holes in shells and rocks, and others stand erect or form low masses on a substrate (Fig. 18.22). The skeletal structure of sponges consists of fibrous collagen and calcareous or siliceous crystalline **spicules** (Fig. 18.23). These structures are associated with **spongin**, a collagenous protein in many species.

Excretion and respiration in sponges occur through diffusion. Digestion in sponges is intracellular. Sponges

Nucleus

Body

Collar cell

Flagellum

FIGURE 18.20 Choanoflagellates might be the ancestors of sponges and other animals.

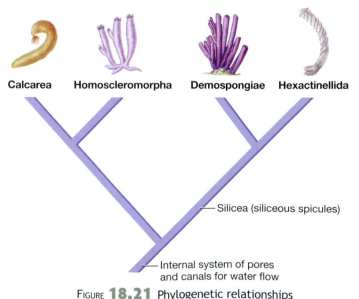

Calcarea   Homoscleromorpha   Demospongiae   Hexactinellida

Silicea (siliceous spicules)

Internal system of pores
and canals for water flow

FIGURE 18.21 Phylogenetic relationships and classification of Porifera.

FIGURE 18.22 Sponges come in many colors and shapes.

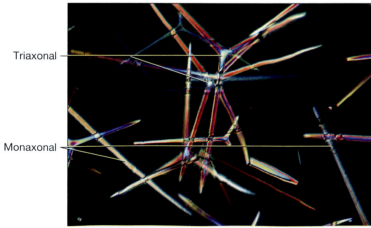

Triaxonal

Monaxonal

FIGURE 18.23 Branched silica spicules of a freshwater sponge.

reproduce asexually by budding or by forming gemmules, structures that help them survive harsh conditions. Under favorable conditions, gemmules form new sponges. Sexually, sponges produce sperm and eggs that unite to form free-swimming larvae. Most sponges are monoecious, having both sexes in the same organism.

Although the body of a sponge may vary in shape and size, the general anatomical features are similar. Many sponges are shaped like a porous vase. The pores, or **ostia**, allow water into the interior of the sponge. The central cavity, or **spongocoel**, is lined with flagellated, collar-shaped cells known as **choanocytes**. Water is eliminated from the sponge by way of the **osculum**. The flow of water through the sponge allows food to be taken in and circulated within the sponge and also enables the intake of sperm. The cells of sponges are arranged in a gelatinous matrix called **mesohyl**.

In sponges, flat, thin cells called pinacocytes cover the exterior, forming a layer called the pinacoderm. Another type of cell, called **amoebocyte,** or archaeocyte, moves about in the mesohyl and absorbs, digests, and transports food. Amoebocytes also are involved in the formation of spicules and spongin. Spicules are either siliceous or calcareous supportive (skeletal) structures. They can vary in shape and are important in sponge classification. Some sponges do not possess spicules. Spongin serves to provide sponges support.

Most species of sponge have one of three types of canal systems: asconoid, syconoid, or leuconoid (Fig. 18.24).

1. Sponges that have an asconoid canal system are generally small and tube-shaped. Water enters these sponges via tiny ostia in the dermis and makes its way to a large cavity called a spongocoel lined with choanocytes. The water is filtered and exits via a large opening called an osculum. Asconoid sponges are placed in class Calcarea.

2. Syconoid sponges resemble large versions of asconoid sponges. These sponges possess a tubular body with a single prominent osculum. Syconoid sponges, however, have a more complex canal system than asconoid sponges. The choanocytes are found in numerous radial canals that empty into the spongocoel lined with epithelia-like cells in syconoid sponges. The water with its nutrients enters the sponge through a large number of ostia into an incurrent canal. The water then passes through prosopyles into the radial canals, where the food is ingested by the choanocytes. The flagella of the choanocytes force the water through apopyles into the spongocoel. Finally, filtered water exits the osculum. Syconoid bodies are found in classes Calcarea and Hexactinellida.

3. Leuconoid sponges, the most common and complex type of sponge, generally form large masses, each member having its own osculum. Clusters of flagellated chambers receive water from incurrent canals, and discharged water exits via the excurrent canals and eventually to the osculum. One species of leuconoid sponge has been estimated to have several million flagellated chambers.

Traditionally, zoologists have recognized three classes of sponge:

1. Class Calcarea consists of small marine sponges with spicules composed of calcium carbonate. The spicules vary from

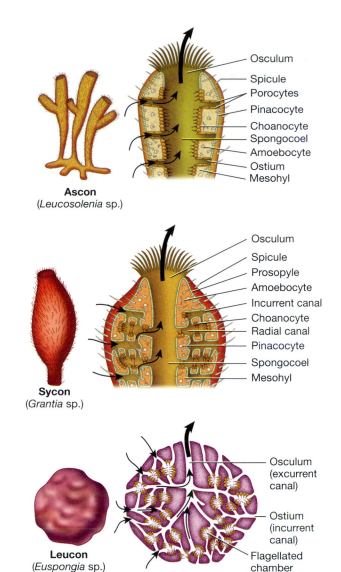

FIGURE **18.24** Examples of sponge body types. A diagrammatic representative of each of the three types depicts with arrows the flow of water through the body of the sponge.

monaxonal (needle-shaped) to triaxonal (three rays), tetraxonal (four rays), and six-rayed. The majority of sponges in this phylum are vase-shaped and drab in color, although a few bright yellow, lavender, red, and green species exist. Asconoid, syconoid, and leuconoid body forms are found in class Calcarea. *Scypha* (*Grantia*) is a vase-shaped syconoid sponge that can live in colonies. It is only 1 to 3 centimeters long and possesses a group of monaxonal spicules at the entrance of the osculum. *Leucosolenia* is a small, branched asconoid sponge (Fig. 18.25).

2. Class Demospongiae is the largest class of sponges. They are usually brilliantly colored with monaxonal or tetraxonal siliceous spicules sometimes bound together by spongin. Members

of this class have leuconoid canal systems. One family of this class lives in freshwater habitats. Examples are the bath sponge (*Spongilla* spp.) and the barrel sponge (*Xestospongia testudinaria*).

3. Class Hexactinellida is referred to as the "glass sponge" because of the six-rayed siliceous spicules that are fused into an intricate glass-like lattice. Members of this class of sponges are primarily deep-water marine forms. The body of these sponges is usually cylindrical or funnel-shaped. The flagellated chambers can be simple syconoid or leuconoid. Some attain lengths of 1.3 meters. The Venus flower basket (*Euplectella* sp.) is a beautiful member of this class (Fig. 18.26).

Osculum

FIGURE **18.25** *Leucosolenia* sp., a member of class Calcarea, a sponge with an ascon body type.

FIGURE **18.26** Venus flower basket, *Euplectella* sp.

Procedure 1

Macroanatomy of Classes Calcarea, Demospongiae, and Hexactinellida

Materials
❏ Dissecting microscope or hand lens
❏ Selected specimens of sponges, such as *Scypha* (*Grantia*) sp., *Spongilla* sp., or *Euplectella* sp.
❏ Colored pencils

1 Procure the needed equipment and specimens.

2 Observe the specimens with a dissecting microscope and magnifying glass. Record your observations and sketches in the space provided.

18

Specimen 1 _____

Specimen 2 _____

Specimen 3 _____

Microanatomy of Classes Calcarea, Demospongiae, and Hexactinellida

Materials

- ❏ Compound microscope
- ❏ Specimen of *Scypha (Grantia)*
- ❏ Microscope slides and coverslips
- ❏ Teasing needle
- ❏ Water
- ❏ Bleach
- ❏ Dropper
- ❏ Lab coat or apron
- ❏ Safety glasses
- ❏ Gloves
- ❏ Selected prepared slides of spicules, *Leucosolenia* sp., *Scypha (Grantia)* sp., *Spongilla* sp., or *Euplectella* sp.
- ❏ Colored pencils
- ❏ Paper towels
- ❏ Forceps

1 Procure the equipment, prepared slides, and specimens.

2 Observe the prepared slides with a compound microscope on both low and high power (Figs. 18.27–18.31). Record your observations and sketches in the space provided.

3 Tease (gently tear into small pieces) a small section of *Scypha (Grantia)* sp. onto a microscope slide, and make a wet mount, being sure to crush the sponge. Record your observations and sketches in the space provided.

4 Follow the directions in step 3, but place two drops of bleach on the sponge specimen and gently stir. The bleach should dissolve the spongin and make the spicules easier to see. Record your observations and sketches in the space provided.

⚠ WARNING

Remember to wear safety glasses, gloves, and a lab coat when handling bleach.

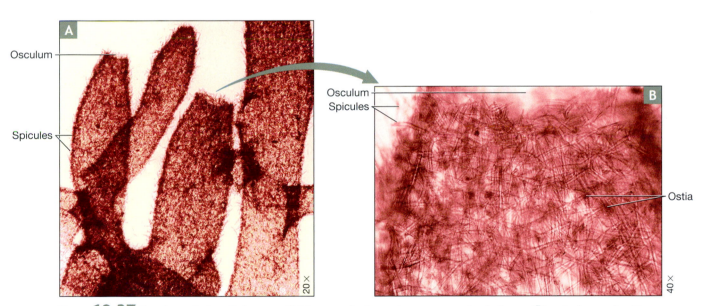

FIGURE **18.27** Member of class Calcarea: (**A**) *Leucosolenia* sp. has an ascon body type; (**B**) spicules and ostia.

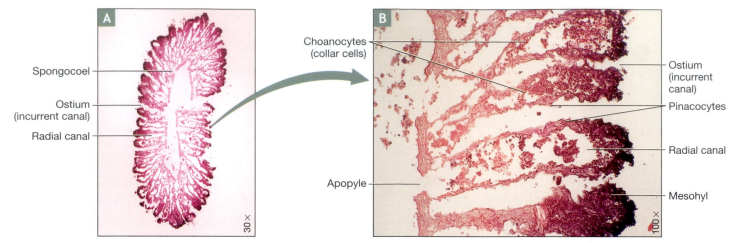

FIGURE **18.28** Transverse sections of the sponge *Scypha* (*Grantia*) sp., a member of the class Calcarea: (**A**) low magnification; (**B**) high magnification.

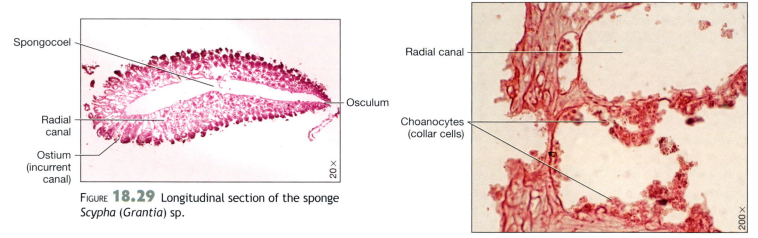

FIGURE **18.29** Longitudinal section of the sponge *Scypha* (*Grantia*) sp.

FIGURE **18.30** Transverse section of the sponge *Scypha* (*Grantia*) sp. showing collar cells.

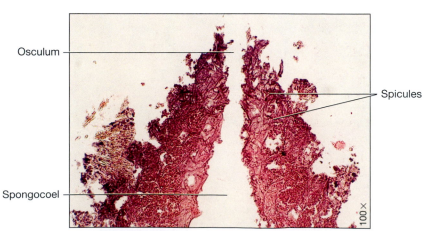

FIGURE **18.31** Longitudinal section of the sponge *Scypha* (*Grantia*) sp. showing magnified view of osculum.

18

Specimen 1 _____

Total magnification _____

_____

_____

Specimen 2 _____

Total magnification _____

_____

_____

Specimen 3 _____

Total magnification _____

_____

_____

Specimen 4 _____

Total magnification _____

_____

_____

Scypha (Grantia)

Total magnification _____

_____

_____

Scypha (Grantia)

Total magnification _____

_____

_____

Scypha (Grantia)

Total magnification _____

_____

_____

18

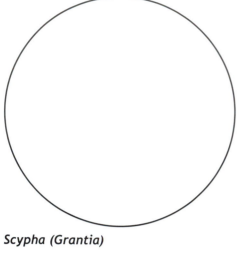

Scypha (Grantia)

Total magnification _____

_____

_____

_____

Scypha (Grantia)

Total magnification _____

_____

_____

_____

Scypha (Grantia)

Total magnification _____

_____

_____

_____

✔ Check Your Understanding

2.1 What is a choanoflagellate, and what is its evolutionary significance?

_____

_____

_____

_____

2.2 Match each type of sponge body form with the correct description.

_____ Asconoid

_____ Syconoid

_____ Leuconoid

A. Complex sponge that forms large masses; each member has its own osculum

B. Small, tube-shaped sponge; water enters via ostia and exits via an osculum

C. Large version of asconoid sponge; has tubular body with a single prominent osculum and a complex canal system

2.3 Viewing a longitudinal section of *Scypha* or *Grantia*, the central portion of the sponge is the _____, which is lined with _____. (*Circle the correct answer.*)

a. spongocoel, choanocytes

b. osculum, choanocytes

c. gastrovascular cavity, choanocytes

d. canal system, choanocytes

e. canal system, spicules

18

Phylum Cnidaria

"Ouch! I was just taking a dip in the ocean when I suddenly felt something stinging my leg! When I looked down, I saw this blob of jelly floating near my aching leg." Sound familiar? You may have had similar encounters with one of its classical members, the jellyfish, at the beach. Welcome to the incredible world of phylum Cnidaria!

The Radiata are eumetazoans represented by two distinct phyla, Cnidaria and Ctenophora (Fig. 18.32). Phylum Cnidaria contains approximately 10,000 species of primarily marine invertebrates, including many bizarre and beautiful forms, such as sea anemones, jellyfishes, Portuguese man-of-war, coral, and the freshwater *Hydra*.

The majority of species of cnidarians are sessile, although many floating or free-swimming forms exist. The colonial cnidarian Portuguese man-of-war is an excellent example of a cnidarian that has its own sail used for wind locomotion. The cnidarians get their name from cells called **cnidoblasts** (cnidocytes) that contain stinging cells, or **nematocysts**, found in members of this phylum. Nematocysts aid in food gathering and defense (Fig. 18.33).

The cnidarians exhibit two body forms, termed *dimorphism*. The **medusa** form resembles an upside-down cup with tentacles, and the **polyp** form consists of a tubular sessile body. Jellyfishes represent the medusa form, and coral represent the polyp form. The life cycle of several species of cnidarians, including *Obelia* spp., consists of both a medusa and a polyp generation. Cnidarians can vary in size from less than 1 millimeter to longer than 70 meters including the tentacles.

Cnidarians exhibit radial symmetry and are diploblastic. Because these organisms do not have mesoderm, the muscular system is made from contractile ectodermal and endodermal, or epitheliomuscular, cells. A gelatinous nonliving substance called **mesoglea** exists between the epidermis and the gastrodermis, or endodermis.

Amoebocytes within the mesoglea aid in digestion, transport, storage, repair, and defense against bacteria. Cnidarians have a single opening leading into their gastrovascular cavity (digestive system) and tentacles surround the mouth. Their digestion is extracellular, and they have no coelom. They do not have a respiratory system; gas exchange occurs through diffusion. In these organisms, individual cells perform excretion. Cnidarians possess a nerve net composed of neurites and sensory organs (for example, photosensitive organs for light detection and statocysts for balance).

Several species of cnidarians are bioluminescent; they can produce their own light. Interestingly, scientists have isolated this gene and inserted it into the embryos of animals, such as mice and pigs, to have a visible marker for gene transfer. Asexual reproduction in some members of this phylum can take place through **budding**, and in some colonial forms the life cycle includes an asexual part. Many cnidarians are dioecious: male and female gametes are produced in separate individuals. The gametes combine to form an embryo that develops into a ciliated **planula** larva (Fig. 18.34). The larva eventually attaches to a substrate and develops into the polyp form.

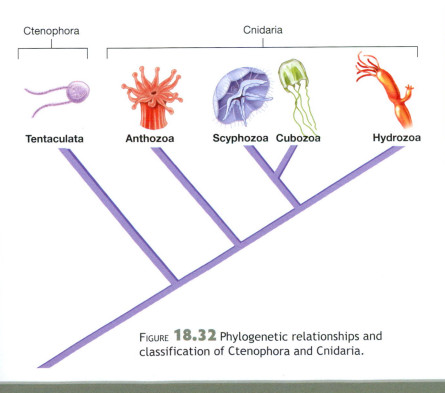

FIGURE **18.32** Phylogenetic relationships and classification of Ctenophora and Cnidaria.

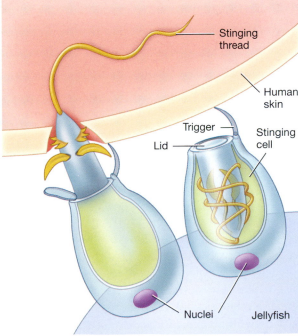

FIGURE **18.33** Nematocyst.

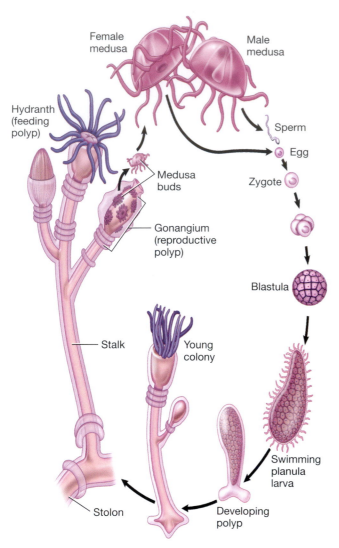

FIGURE **18.34** Life cycle of *Obelia*.

The life cycle of most hydrozoans consists of an asexual polyp and a sexual medusa stage. One of the most common hydrozoans is a small (25 millimeter) freshwater species, *Hydra*. This interesting organism is common in cool, clean, freshwater pools and streams throughout the world. The hydrozoan *Obelia* is a typical member of class Hydrozoa. *Obelia* can exist in both the asexual polyp form and the sexual form of the medusa. This organism is one of many colonial hydroids attached to rocks, pilings, and shells in the brackish and marine environment.

The body of *Hydra* (Fig. 18.35) appears to be a cylindrical tube, with its **aboral** (away from the mouth) end forming a slender stalk ending in a **basal disk** for attachment. The basal disk contains specialized gland cells that allow it to attach to a substrate (perhaps a lily pad). In addition, it allows the organism to form a gas bubble for floating. The mouth of *Hydra* is located on an elevated portion of the oral end, the **hypostome**. The hypostome is encircled by hollow **tentacles** (6–10 in number). The tentacles help to capture food, such as small insect larvae, crustaceans, and worms. The mouth itself opens into a **gastrovascular cavity** continuous with the tentacles. Upon close examination, testes or ovaries, when present, appear as rounded structures on the body. *Hydra* can reproduce asexually by budding. Many times, a bud projects from the side of the animal. The epidermis of the *Hydra* contains specialized cells, including the nematocysts.

Obelia attaches to the substrate via a rootlike structure, the **stolon**, which gives rise to various **stalks** (see Fig. 18.34). The stalk is protected by a chitinous sheath, the **perisarc**. The

Phylum Cnidaria comprises four classes:

1. Class Hydrozoa is made up of the hydras and many colonial species collectively called hydroids.

2. Members of class Scyphozoa are called the true jellyfishes.

3. Class Cubozoa once was considered an order of class Scyphozoa. These are the box jellyfishes.

4. Members of class Anthozoa are the "flower animals."

Class Hydrozoa

In class Hydrozoa, the polyp form is dominant. Most hydrozoans are marine, but a few are freshwater species. Representative hydrozoans include *Hydra* spp., *Obelia* spp., and Portuguese man-of-war, *Physalia physalis*. Each summer on the U.S. East Coast, the Portuguese man-of-war inflicts up to half a million stings. A large *Physalia* may have tentacles exceeding 50 meters. *Physalia* can easily be identified by its purple, sail-shaped, gas-filled pneumatophore.

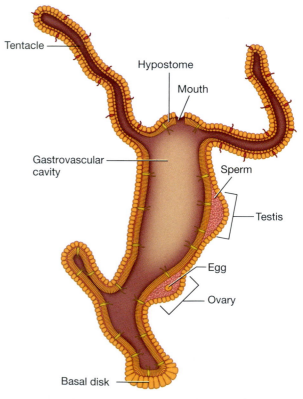

FIGURE **18.35** Basic anatomy of *Hydra* sp.

18

individual polyps also are attached to the stalk. Most polyps, also called **zooids**, are used for feeding and are called **hydranths**. In addition to the feeding polyps, reproductive polyps, individually called the **gonangium**, are attached to the stalk. Dioecious medusae bud from the gonangium. The free-swimming medusae mature and form gametes. After fertilization, a free-swimming planula larva develops and finds a new substrate. Upon settling, a new *Obelia* colony is established.

Class Scyphozoa

Class Scyphozoa (true jellyfishes) consists of solitary organisms in which a polyp stage is reduced or absent, and the dominant stage is a bell-shaped medusa (Fig. 18.36). The edge of the bell, the umbrella, has eight notches with sense organs. The umbrella of some jellyfishes exceeds 2 meters across. Scyphozoa is derived from the Greek, meaning a kind of drinking cup and referring to the cup shape of the organism. Examples of true jellyfishes are *Aurelia* spp., *Chrysaora* spp., and *Cassiopeia* spp.

In the life cycle of *Aurelia*, male and female medusae produce their respective gametes, which undergo fertilization to form a **zygote** that may be retained on the oral arms of the medusa and eventually becomes a ciliated planula larva (Fig. 18.37). The larva lands on a suitable substrate and forms a **scyphistoma** that grows perhaps asexually, buds and forms an asexual **strobila**. The strobila gives rise to swimming **ephyra** that eventually develop into a medusa.

Class Anthozoa

Members of Class Anthozoa (flower animals) exist as polyps only. Anthozoans can be colonial or solitary marine organisms. The pharynx leads into a gastrovascular cavity divided by eight or more septa. Examples of anthozoans are sea anemones, sea fans, sea pens, sea pansies, and corals (Figs. 18.38–18.41). More than 6,000 species have been described, and great numbers of fossil forms have been found. Anthozoans occur from the intertidal zone of the ocean to the depths of the great marine trenches (6,000 meters).

Sea anemones occur in warm coastal waters worldwide. They are sessile, attaching by their pedal disk to a suitable substrate. Several species can burrow into the sand or mud. Sea anemones are cylindrically shaped with a crown of tentacles surrounding the mouth. Some anemones have separate sexes, and others are monoecious. Anemones feed primarily upon fishes. Asexual reproduction can occur as fragments of the pedal disk break off (pedal laceration), transverse fission, or budding.

FIGURE **18.36** Red-striped jellyfish, *Chrysaora melanaster*.

Some anemones have complex symbiotic relationships with other organisms, such as algae and fishes.

The corals are sessile marine invertebrates that first appeared in the Cambrian period over 540 million years ago. They typically live in colonies made of genetically identical polyps living in an exoskeleton made of calcium carbonate. Most of the energy used by corals results from their mutualistic relationship with photosynthetic algae. Coral can also feed upon invertebrates and small fishes that can be trapped by their tentacles with nematocysts. Most corals undergo sexual reproduction but are capable of asexual budding. Groups of corals comprise a coral reef such as the Great Barrier Reef near Australia. Today coral reefs are under stress worldwide from changes in temperature and oceanic acidification. They are also threatened by human activities such as pollution and collecting.

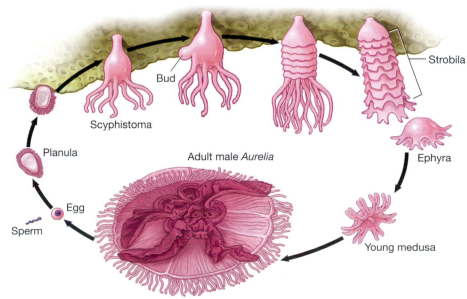

FIGURE **18.37** Life cycle of *Aurelia* sp.

FIGURE **18.38** Sunburst anemone, *Anthopleura sola.*

FIGURE **18.39** Firecracker coral, *Dendrophyllia* sp.

FIGURE **18.40** Tube anemone, *Pachycerianthus fimbriatus.*

FIGURE **18.41** Sea pen, *Ptilosarcus gurneyi.*

Did you know . . .

As Seen From Space

Off the coast of Queensland, Australia, you can see the Great Barrier Reef. It is composed of nearly 3,000 individual reefs and over 900 islands. Approximately 400 species of coral comprise the reef. The reef covers nearly 345,000 square kilometers providing habitat for myriad marine organisms including more than 1,500 species of fishes. It is so massive that it can be seen from space and is the largest structure formed by organisms.

Procedure 1

Macroanatomy of Classes Hydrozoa, Scyphozoa, and Anthozoa

Materials

- ❏ Dissecting microscope or hand lens
- ❏ Petri dish or depression slide
- ❏ Dissecting tray
- ❏ Probe
- ❏ 5% vinegar or Congo red solution
- ❏ Scalpel
- ❏ Living *Hydra* sp., preserved *Obelia* sp., *Physalia* sp., and other select hydrozoans, such as *Gonionemus* sp., *Daphnia* sp., or *Artemia* sp.
- ❏ Preserved specimens of *Aurelia* sp. and select scyphozoans
- ❏ Preserved specimens of *Metridium* sp., corals, sea fans, and other select anthozoans
- ❏ Colored pencils

1 Procure the needed equipment and supplies.

2 Obtain a living *Hydra*, and place it in a small petri dish or on a depression slide. Allow the *Hydra* a few minutes to acclimate to the new conditions. Place the petri dish or slide under the dissecting microscope. Sketch the specimen, label your sketch, and record your observations in the space provided.

Hydra

18

3 Tap the dish, or gently touch the *Hydra* with a probe. Record the response of the *Hydra* in the space provided.

_____

_____

_____

4 Place some *Daphnia* or *Artemia* in the petri dish, and observe the feeding behavior of *Hydra*. You may have to gently nudge them toward the tentacles of the *Hydra* with a probe. If a solution of 5% vinegar or Congo red is available, place a few drops in the petri dish and record the reaction of the *Hydra* in the space provided.

_____

_____

_____

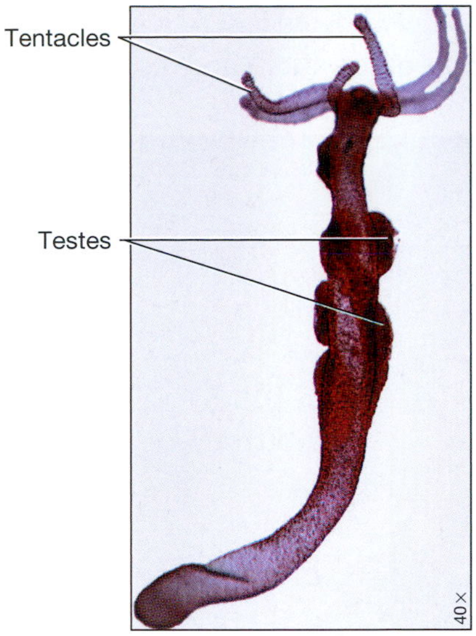

FIGURE **18.42** Male *Hydra* sp.

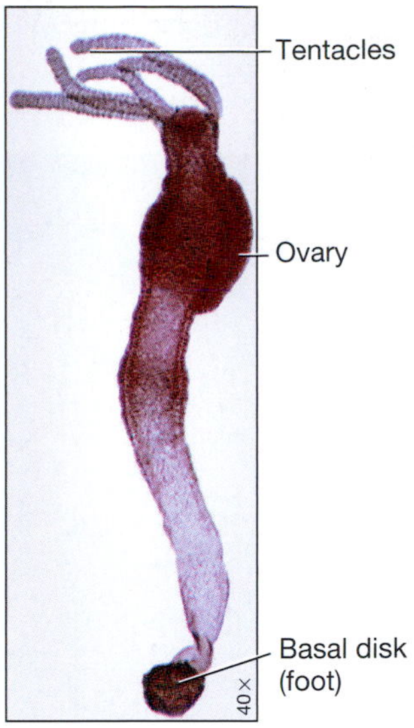

FIGURE **18.43** Female *Hydra* sp.

5 Using a dissecting micro-scope or hand lens, observe the anatomical features of *Obelia*, *Physalia*, and select hydrozoans (Figs. 18.42–18.45). Sketch, label, and describe your specimens in the space provided.

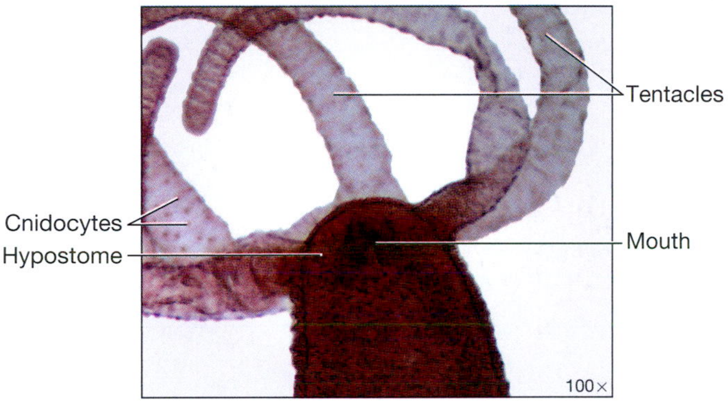

FIGURE **18.44** Anterior end of *Hydra* sp.

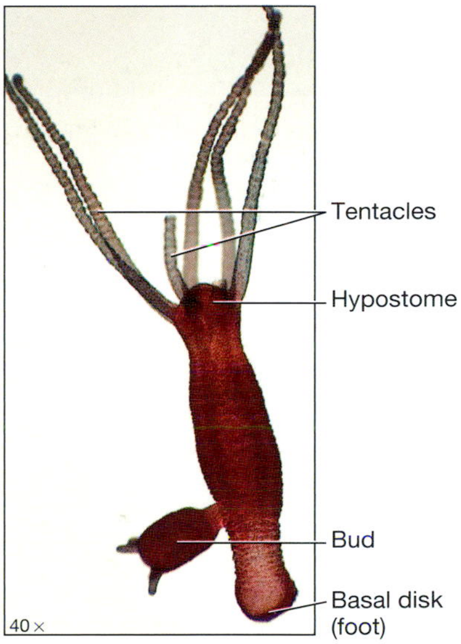

FIGURE **18.45** *Hydra* sp. budding.

Specimen 1 _____

_____

_____

Specimen 2 _____

_____

_____

Specimen 3 _____

_____

_____

6 Using a dissecting microscope or hand lens, observe the anatomical features of *Aurelia* and selected scyphozoans (Figs. 18.46–18.47). Record your observations and detailed sketches in the space provided.

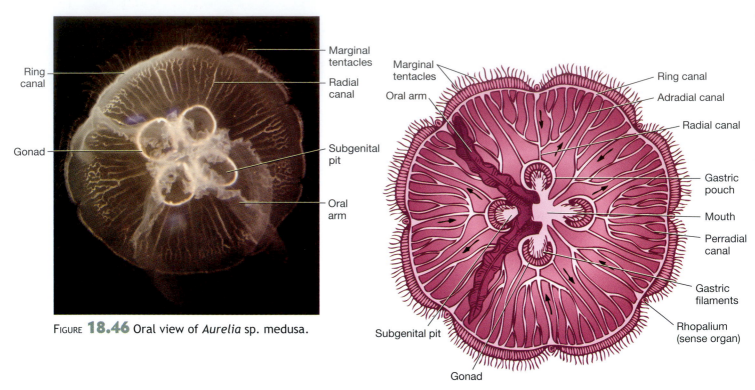

FIGURE **18.46** Oral view of *Aurelia* sp. medusa.

FIGURE **18.47** Oral view of *Aurelia* sp. medusa. In this diagram, the right oral arms have been removed. The arrows depict circulation through the canal system.

Specimen 1 _____

Specimen 2 _____

Specimen 3 _____

7 Place a specimen of the sea anemone *Metridium* on a dissecting tray, and examine it with a dissecting microscope or hand lens. Using the scalpel, make a longitudinal cut through *Metridium*. Locate the structures found in Figure 18.48. Record your observations, label, and sketch the organism in the space provided.

18

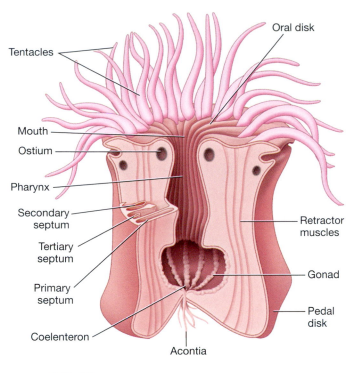

Tentacles

Oral disk

Mouth

Ostium

Pharynx

Secondary
septum

Tertiary
septum

Primary
septum

Coelenteron

Acontia

Retractor
muscles

Gonad

Pedal
disk

FIGURE 18.48 Partially dissected sea anemone, *Metridium*.

Metridium

8 Observe and sketch the various forms of coral, sea pens, and sea fans in the space provided.

9 Return or dispose of your dissected specimens as directed by your instructor, and clean your equipment.

Coral

Sea pen

Sea fan

18

Microanatomy of Classes Hydrozoa and Scyphozoa

Materials

- ❏ Compound microscope
- ❏ Select microscope slides of:
  - *Hydra* whole mount
  - *Hydra* longitudinal section
  - *Hydra* cross section
  - *Hydra* budding
  - *Obelia* colony
  - *Obelia* medusa
- ❏ Select microscope slides of *Aurelia* planula larva, scyphistoma, strobila, and ephyra
- ❏ Colored pencils

1 Procure the compound microscope and select microscope slides.

2 Observe the *Hydra* whole mount slide on low power and high power (Fig. 18.49). Sketch and label the anatomical features of *Hydra* in the space provided.

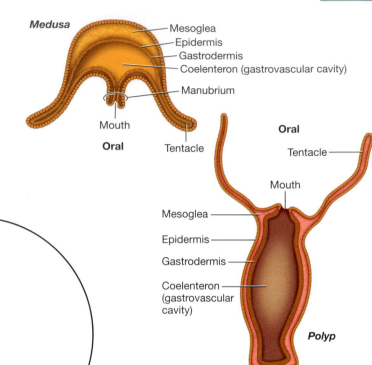

FIGURE **18.49** Cnidarian.

Whole mount of *Hydra*

Total magnification _____

_____

_____

_____

3 Observe the *Hydra* longitudinal section, cross section, and budding slides on low and high power. Sketch and label them in the space provided.

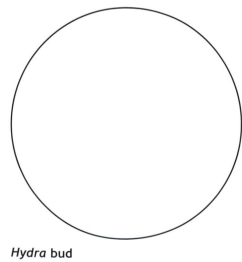

Longitudinal section of *Hydra*

Total magnification _____

_____

_____

Cross section of *Hydra*

Total magnification _____

_____

_____

***Hydra* bud**

Total magnification _____

_____

_____

18

4 Observe the *Obelia* colony and medusa slides on low and high power (Figs. 18.50–18.52). Sketch and label your observations in the space provided.

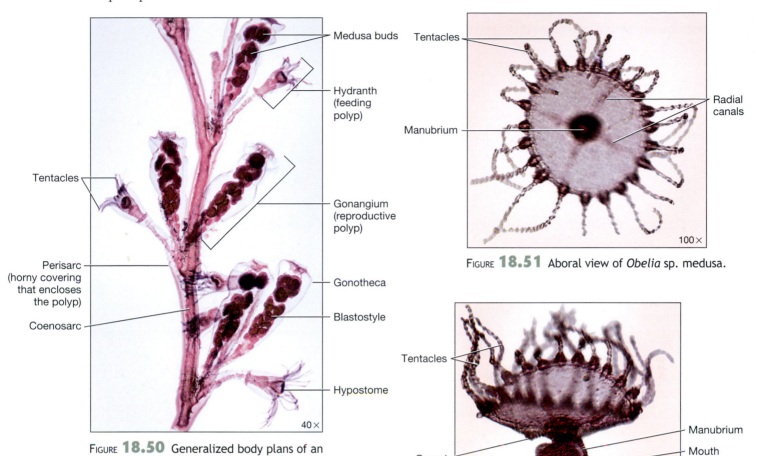

FIGURE **18.50** Generalized body plans of an *Obelia* sp. colony.

Labels (Figure 18.50): Medusa buds, Hydranth (feeding polyp), Tentacles, Gonangium (reproductive polyp), Gonotheca, Blastostyle, Hypostome, Perisarc (horny covering that encloses the polyp), Coenosarc

40×

FIGURE **18.51** Aboral view of *Obelia* sp. medusa.

Labels (Figure 18.51): Tentacles, Radial canals, Manubrium

100×

FIGURE **18.52** *Obelia* sp. medusa in feeding position.

Labels (Figure 18.52): Tentacles, Gonad, Manubrium, Mouth

100×

Obelia

Total magnification _____

Obelia

Total magnification _____

18

✔ Check Your Understanding

3.1 Describe the polyp body form using the living *Hydra* you observed in this exercise.

_____

_____

3.2 Describe the movement of the *Hydra* when it was initially placed in the petri dish, before adding the *Daphnia* or vinegar.

_____

_____

3.3 The presence of two different body forms, or dimorphism, is seen in the phlyum Cnidaria. The _____

form resembles an "upside-down cup" with tentacles, and the _____ body form consists of a tubular

sessile body. *Hydra* is an example of the _____ body form, and a jellyfish is an example of the

_____ body form.

Phylum Ctenophora

Any discussion of phylum Cnidaria should briefly mention phylum Ctenophora. Often, a small, clear, walnut-shaped blob washes up on the beach. Chances are it is a ctenophore. Members of phylum Ctenophora are known as comb jellies. People sometimes call these solitary, harmless, marine, jellyfish-like animals sea walnuts or sea gooseberries. Presently, there are approximately 150 described species of ctenophores, ranging in size from 1 centimeter to 1.5 meters. Ctenophores are exclusively marine, living in warm water worldwide. They exist as a medusa only.

Ctenophores have adhesive cells called **colloblasts** to capture food and do not possess nematocysts. The tentacles of ctenophores are solid, consisting of epidermis only. Ctenophores swim by means of rows of fused cilia, **comb plates**. The majority of ctenophores are monoecious, reproducing only by sexual means. Many ctenophores are bioluminescent. Common examples of ctenophores are *Pleurobrachia* spp. and *Mnemiopsis* spp. (Fig. 18.53). Another interesting ctenophore is the ribbon shaped Venus girdle (*Cestum veneris*).

FIGURE **18.53** Ctenophore, *Mnemiopsis* sp.

Procedure 1

Macroanatomy of a Ctenophore

Materials
- ❏ Dissecting microscope or hand lens
- ❏ Preserved specimens of *Pleurobrachia* spp. and *Mnemiopsis* spp.
- ❏ Colored pencils

1 Procure the needed equipment, supplies, and selected specimens.

2 Using a dissecting microscope or hand lens, examine the specimens and record your observations and sketches in the space provided below and on the following page.

Specimen 1 _____

Specimen 2 _____

18

✔Check Your Understanding

4.1 _____ use adhesive cells, or colloblasts, on their tentacles to procure food, whereas

cnidarians use _____ for obtaining food.

4.2 Members of the phylum Ctenophora are known as sea walnuts or sea gooseberries, but most commonly are called comb jellies. What is the background for this name?

_____

_____

4.3 Which of the following is not a characteristic of a member of the phylum Ctenophora? (*Circle the correct answer.*)
   a. Is bioluminescent.
   b. Swims by means of rows of fused cilia or comb plates.
   c. Exists in two body forms: polyp and medusa.
   d. Uses colloblasts for food capture.
   e. Is monoecious, reproducing only by sexual means.

MYTHBUSTING

At Sea

Debunk each of the following misconceptions by providing a scientific explanation. Write your answers on a separate sheet of paper.

1 Your white blood cells are only important for defense.

2 The Portuguese man-of-war is a true jellyfish.

3 Rubbing sand on a jellyfish sting makes it feel better.

4 While at the beach, I was attacked by a ferocious jellyfish!

5 Sponges are only found in marine waters.

1 Describe the arrangement and general function of simple squamous epithelium.

2 Both _____ muscle types are described as striated because of the alternating light and dark bands of the proteins _____ . (*Circle the correct answer.*)

  a. skeletal and cardiac, actin and myosin

  b. skeletal and smooth, actin and myosin

  c. cardiac and smooth, actin and myosin

  d. skeletal and cardiac, actin and myofibril

  e. none of the above

3 What is the function of goblet cells?

4 What is a major component of pus?

5 What are five characteristics of cnidarians?

6 Match the class of sponge with the following descriptions and examples. (Some will have more than one answer.)

_____ Calcarea

_____ Demospongiae

_____ Hexactinellida

A. "Glass sponge" with 6-rayed siliceous spicules fused into an intricate, glass-like lattice
B. *Spongilla*
C. Small marine sponges with spicules composed of calcium carbonate
D. Leuconoid body form
E. *Euplectella*
F. Brilliantly colored sponges with monaxonal or tetraxonal silicoeous spicules that may be bound together with spongin
G. Asconoid, syconoid, and leuconoid body forms
H. Syconoid or leuconoid body forms
I. *Scypha*

7 In the life cycle of a jellyfish, a ciliated _____ lands on the substrate and forms a _____, which eventually may reproduce asexually by forming a _____. This stage gives rise to swimming _____, eventually developing into a free-swimming medusa.

8 Label the *Obelia* sp. colony (Fig. 18.54).

1. _____

2. _____

3. _____

4. _____

5. _____

6. _____

7. _____

8. _____

9. _____

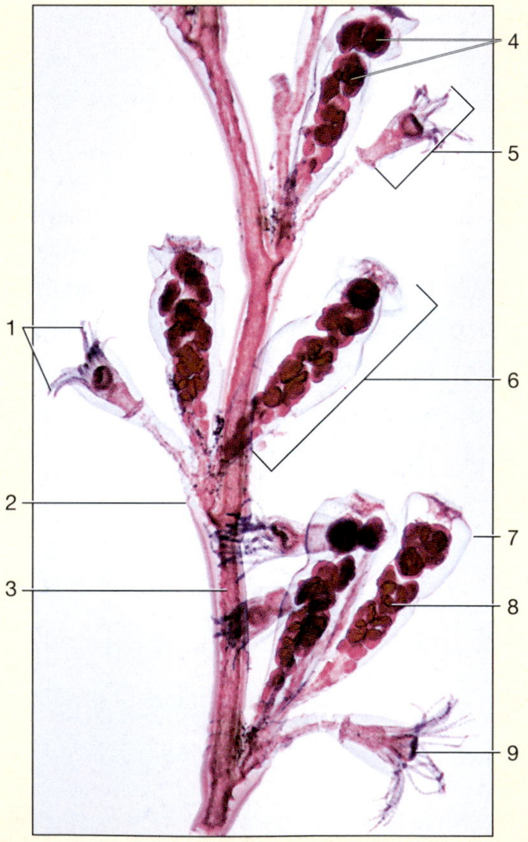

9 Which of the following characteristics are unique to ctenophores? (*Circle the correct answer.*)
a. Do not possess nematocysts; swim by using comb plates.
b. Possess nematocysts; swim by using comb plates.
c. Have adhesive cells, or colloblasts, for capturing food.
d. Both a and c.
e. Both b and c.

FIGURE **18.54** *Obelia* sp. colony.

18

Animal Planet II
Understanding the Lophotrochozoans

19

The winds, the sea, and the moving tides are what they are. If there is wonder and beauty and majesty in them, science will discover these qualities. . . . If there is poetry in my book about the sea, it is not because I deliberately put it there, but because no one could write truthfully about the sea and leave out the poetry.

— Rachel Carson (1907–1964)

Just wondering . . .

Consider the following questions prior to coming to lab, and record your answers on a separate piece of paper.

1 Why are scientists so interested in the geographic cone snail *Conus geographus*?

2 What are some signs and symptoms of a tapeworm infection in cats and dogs?

3 How are leeches used in medicine?

4 What were Charles Darwin's observations on earthworms?

5 Why are cephalopods considered the "Albert Einsteins" of the invertebrate world?

Objectives

At the completion of this chapter, the student will be able to:

1. Describe the characteristics of Lophotrochozoa.

2. Describe the characteristics, natural history, and organization of phyla Platyhelminthes, Rotifera, Mollusca, and Annelida.

3. Describe and compare and contrast the characteristics and anatomical structures of organisms in various classes of phyla Platyhelminthes, Rotifera, Mollusca, and Annelida.

4. Dissect and identify the anatomical features of a squid.

5. Dissect and identify the anatomical features of an earthworm.

In that great tree of life, where should flukes, snails, crabs, and worms be placed? The debate on how to classify bilaterally symmetrical animals is ongoing with no clear immediate resolution. Classically, the bilaterally symmetrical animals (protostomes and deuterostomes) were divided into acoelomates, pseudocoelomates, and coelomates. In recent years, incorporating data from molecular studies, scientists have reorganized the protostomes into two distinct clades and kept the deuterostomes in a separate clade. The two clades of protostomes are Lophotrochozoa and Ecdysozoa.

Lophotrochozoa is a clade consisting of several phyla (Fig. 19.1). The best-known phyla within this clade are Platyhelminthes (flatworms), Rotifera (rotifers), Mollusca (snails, oysters, and squid), and Annelida (segmented worms), the phyla we will study in depth in this chapter. Several lesser-known phyla are Acanthocephala (thorny-headed worms), Gastrotricha (spiny aquatic organisms), Bryozoa (moss animals), Entoprocta (entoprocts), Brachiopoda (lampshells), and Nemertea (ribbon worms; Fig. 19.2). The two defining characteristics of Lophotrochozoa are:

1. The presence of a horseshoe-shaped crown of **ciliated tentacles** (lophophores); and

2. The minute, translucent top-shaped **ciliated larvae** (trochophores).

Phylum Platyhelminthes consists of approximately 20,000 species collectively called the flatworms. Common representatives of this phylum are planarians, flukes, and tapeworms. Flatworms vary in size from shorter than 1 millimeter to longer than 10 meters (a species of tapeworm). Many species of platyhelminths are free-living and are found in terrestrial, freshwater, and marine environments. Others, such as tapeworms and flukes, are parasitic.

Phylum Rotifera consists of approximately 1,900 species of primarily freshwater organisms. Rotifers range in size from 0.1 to 1.0 millimeter and have an elongated, saclike

Chapter Photo
Sandworm, *Nereis virens*.

403

FIGURE **19.1** Phylogeny and classification of Metazoa (multicellular animals). The phyla shown in red are discussed in this chapter.

FIGURE **19.2** Examples of lophotrochozoans: (**A**) tapeworm, *Taenia* sp.; (**B**) snail, *Cornu* sp.; (**C**) leech, *Macrobdella* sp.; (**D**) lampshell, *Lingula* sp.

body. Commercially and medically, rotifers are not significant. However, biologists study rotifers to gain a better understanding of evolution, diversity, regeneration, ecology, and behavior.

Malacologists study members of phylum Mollusca. This phylum consists of 90,000 living species plus more than 70,000 fossil species. The molluscs inhabit a variety of environments, including marine, freshwater, and terrestrial habitats. This diverse phylum encompasses chitons, limpets, slugs, snails, whelks, abalones, nudibranchs, oysters, scallops, and octopi. Molluscs vary in size from almost microscopic to gigantic. The giant squid can attain lengths of more than 20 meters and weigh more than 454 kilograms (1,000 pounds). The giant clam can reach 1.5 meters in length and weigh more than 227 kilograms (500 pounds). Phylum Mollusca includes herbivores, carnivores, filter feeders, detritus feeders, and even parasites.

Perhaps you already have been introduced to a member of **phylum Annelida** on a fishing trip. The earthworm, or night crawler, you used for bait is a common representative of phylum Annelida, which includes approximately 15,000 species of segmented worms. In contrast to your likely image of worms, the annelids are highly diverse, and many species are quite attractive. In addition to earthworms, this phylum includes a variety of other worms, such as leeches, tubiflex worms, sandworms, parchment worms, and bloodworms. They live in terrestrial, freshwater, and marine environments. Annelids can vary in size from less than a millimeter to one species of tropical earthworm 4 meters long.

19

Phylum Platyhelminthes

The bodies of platyhelminths characteristically are flattened dorsoventrally and are ribbonlike, ensuring a large surface area. Flatworms are bilaterally symmetrical, triploblastic acoelomates. Some species are dull in coloration, and others are brightly colored. Many platyhelminths are monoecious but practice cross-fertilization. In addition, many parasitic flatworms have complex life cycles.

The platyhelminths lack specialized respiratory and circulatory systems. As a result, they exchange gases through diffusion. The digestive system of flatworms is incomplete, with only one opening to the exterior. Many flatworms possess a **mouth** connected to the **gastrovascular cavity** by a muscular **pharynx**. In larger flatworms, the gastrovascular cavity is branched within the body. The main structures in the excretory system are the **protonephridia**, capped by **flame cells.**

Parasitic flatworms are covered with a protective syncytial tegument (a body covering composed of multinucleate tissue with no clear cell boundaries). Platyhelminths exhibit cephalization. A pair of cerebral ganglia receives sensory information from the environment. **Eyespots** are present in some species. Each ganglion is connected to a nerve cord that runs the length of the body.

Phylum Platyhelminthes contains four classes of flatworms (Fig. 19.3), but only the first three are discussed here:

1. Class Turbellaria (Fig. 19.4A) consists of more than 3,000 mostly free-living flatworms, such as the planaria (*Dugesia* and *Bipalium*).

2. Class Trematoda (Fig. 19.4B) consists of flukes. Approximately 11,000 species of trematodes have been described, all of which are parasitic. *Clonorchis sinensis* and *Schistosoma spp.* are typical trematodes.

3. Class Cestoda (Fig. 19.4C) includes approximately 3,500 species of parasitic tapeworms, such as *Dipylidium caninum* and *Taenia spp.*

4. Class Monogenea contains ectoparasitic flatworms usually found on the skin and gills of fishes. Approximately 1,400 species have been identified, examples of which are *Gyrodactylus cylindriformis* and *Polystoma intergerrimum.* These organisms are not discussed in this chapter.

Class Turbellaria

Class Turbellaria mostly lives in marine environments. Members of class Turbellaria range in size from less than 5 millimeters to more than 60 centimeters. These organisms are covered by a ciliated epidermis that can range in color from shades of gray or brown to bright, rich colors. Turbellarians usually swim or crawl along the bottom of an aquatic or terrestrial environment by ciliary propulsion. Some species move by means of undulating waves of muscle contractions.

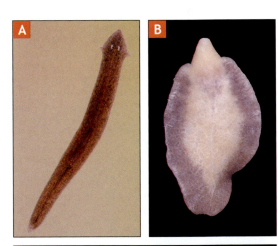

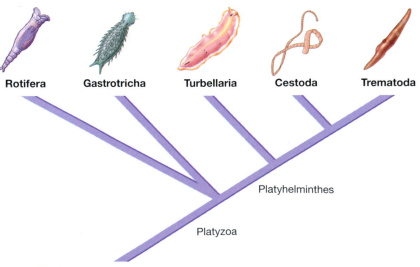

FIGURE **19.3** Phylogenetic relationships and classification of representative flatworms.

FIGURE **19.4** Examples of platyhelminths: (**A**) planarian *Dugesia* sp.; (**B**) liver fluke, *Fasciola* sp.; (**C**) tapeworm, *Taenia* sp.

Glands in the epidermis and underlying tissue secrete a mucous film to help the organism glide over the environment.

Most turbellarians are carnivores, feeding on other small invertebrates. The mouth is located along the midventral line of the organism. The gut consists of the pharynx and the intestinal sac. Movement of food materials in flatworms occurs primarily through diffusion. Paired protonephridia with flame cells function in removing nitrogenous wastes and in osmoregulation. These organisms do not have an anus; they eject digestive wastes through their mouths. In turbellarians, gas exchange occurs at the surface of the body. The flattened body increases the surface area, thus increasing the rate and efficiency of gas exchange.

In many turbellarians, the neurons exist in longitudinal bundles located beneath the epidermis. The connecting lateral nerve cords form a characteristic ladderlike pattern. The brain appears as a bilobed mass of ganglion cells at the anterior end of the organism. Some members of this class have light-sensitive eyespots, or **ocelli**. In many species, chemoreception and tactile reception are well developed. Small tentacles are present in some species. The **auricles** (lobes) on the side of a planarian's head are associated with tactile reception and chemoreception.

Many turbellarians reproduce asexually through fission. Being monoecious, they also are capable of sexual reproduction. Although these animals are **hermaphroditic**, they do not exhibit self-fertilization. Turbellarians generally reproduce by mutual fertilization, eventually resulting in cocoons laid in jellylike masses. Turbellarians exhibit amazing regenerative powers.

A common turbellarian found in many gardens and greenhouses is *Bipalium kewense*. A native of Indo-China, it was accidentally introduced into the United States more than a century ago. *Bipalium* is photonegative and can be found in dark, cool, moist areas. *Bipalium* is slender, about 25 centimeters in length, and usually brown in color with dark longitudinal stripes. The head is shovel-shaped, and eyespots are obvious.

The best-known turbellarian is the planarian *Dugesia* sp., a common inhabitant of freshwater environments that lives on plants, under rocks, and in debris. *Dugesia* is 3–15 millimeters long and brown to gray in color. *Dugesia* feeds primarily upon other invertebrates. The large mouth and pharynx are in the middle of its body. The auricles and eyespots are prominent on its anterior end.

Class Cestoda

Class Cestoda consists of approximately 3,500 species known as tapeworms. The tapeworms are **endoparasites** of humans and other vertebrates (Fig. 19.5). Most tapeworms require two hosts. Adult tapeworms usually are present in the digestive tract of the final vertebrate host. Some tapeworms reach lengths of more than 10 meters and perhaps up to 25 meters in their final host and can live up to 20 years. Perhaps at least 135 million people worldwide have a tapeworm infection at any given moment!

Cestodes are highly adapted for the life of a parasite. They are totally dependent on the host for nutrition because they lack a digestive tract. The tegument of the tapeworm permits nutrients to enter the body while protecting the body against alkaline substances and digestive enzymes in the host.

Cestodes possess an anterior region called a **scolex** that contains **hooks** and **suckers** for attachment. The hooks usually encircle a crown, or **rostellum**. Behind the scolex is a series

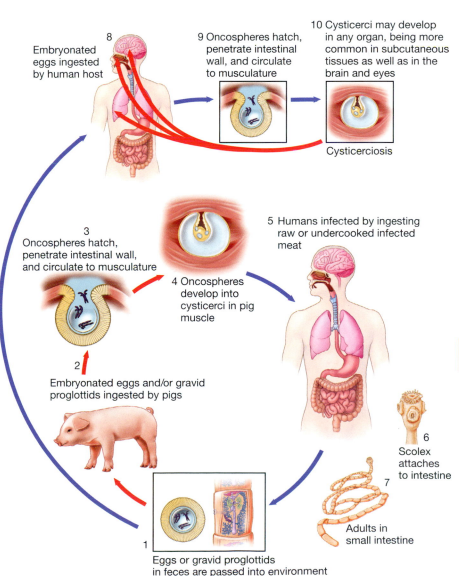

8 Embryonated eggs ingested by human host

9 Oncospheres hatch, penetrate intestinal wall, and circulate to musculature

10 Cysticerci may develop in any organ, being more common in subcutaneous tissues as well as in the brain and eyes

Cysticerciosis

3 Oncospheres hatch, penetrate intestinal wall, and circulate to musculature

4 Oncospheres develop into cysticerci in pig muscle

5 Humans infected by ingesting raw or undercooked infected meat

2 Embryonated eggs and/or gravid proglottids ingested by pigs

6 Scolex attaches to intestine

7 Adults in small intestine

1 Eggs or gravid proglottids in feces are passed into environment

FIGURE **19.5** Life cycle of the pork tapeworm, *Taenia solium*.

of subunits called **proglottids**. The **strobila**, the main mass of the tapeworm, is composed of proglottids. Immediately behind the scolex are germinative proglottids, in the process of asexually forming new proglottids. The proglottids mature as they are pushed posteriorly.

Following the germinative proglottids, a series of mature proglottids can be found. Mature proglottids can contain numerous testes and ovaries. Each proglottid is a hermaphroditic individual, and any two proglottids on either the same or different tapeworms are capable of exchanging sperm. Usually, proglottids do not self-fertilize. Tapeworm eggs containing embryos are surrounded by protective shields. Some eggs exit the proglottid via the **gonopore** and enter the host's intestine, and others are stored in the uterus. **Gravid** (egg-bearing) proglottids often break away from the strobila and rupture in the host's intestine, or they may exit the host with the feces.

A gravid proglottid can contain more than 100,000 eggs. After the eggs are released, they must be ingested by an **intermediate host** to hatch. The hatched eggs form larvae that bore through the intestinal wall, where they are picked up by the circulatory system and make their way to striated muscle. In the muscles, the larvae develop into a **cysticercus stage** (a fluid-filled cyst). If the cysticercus is eaten in raw or undercooked cooked meat, it develops into an adult tapeworm in the intestine of the final host.

Class Trematoda

Members of class Trematoda are parasitic flukes. As adults, most trematodes are endoparasites of various vertebrates; however, some ectoparasitic species exist. Almost 11,000 species of flukes have been described, many of which have great economic and medical importance.

Most trematodes reside in the lung, liver, bile ducts, pancreatic ducts, intestines, and blood. The flukes are flattened and leaflike in appearance. The body of a fluke is covered by a nonciliated syncytial tegument that lacks cell membranes between the nuclei.

Being parasitic, flukes have a variety of general and species-specific adaptations. General adaptations include:

1. The presence of various specialty glands for penetration or cyst formation; suckers and hooks for attachment.

2. A mouth at the anterior end of the organism.

3. The ability to produce a tremendous number of offspring.

The trematodes share with the turbellarians a well-developed alimentary canal, similar musculature, and similar body systems. In trematodes, the sense organs are poorly developed.

A representative fluke is *Clonorchis sinensis*, better known as the Chinese liver fluke (Fig. 19.6). This parasite is common in many regions of Asia, where it parasitizes dogs, cats, pigs, and humans. The adult varies in length from 10 to 70 millimeters. *Clonorchis* possesses an oral and a ventral sucker, and the digestive system consists of an **esophagus** and two long, unbranched intestinal **caeca**. Two protonephridial tubules unite to form a median bladder that points to the outside of the organism. The sense organs are degenerate, and the nervous system is similar to that of turbellarians.

Fertilized eggs exit the body of an infected host and, ideally, land in water and are ingested by a snail. In the snail, the egg hatches and forms a **miracidium larva**. At this point, the snail is the first intermediate host. The miracidium develops within the snail's body and forms a **sporocyst**. The sporocyst asexually produces thousands of larvae, the **rediae**. Each redia in turn reproduces asexually to produce up to 50 **cercariae** that emerge through the epidermis of the snail and enter the watery environment.

The cercariae are free-swimming and make their way to a fish, where they burrow into the muscle beneath the scales.

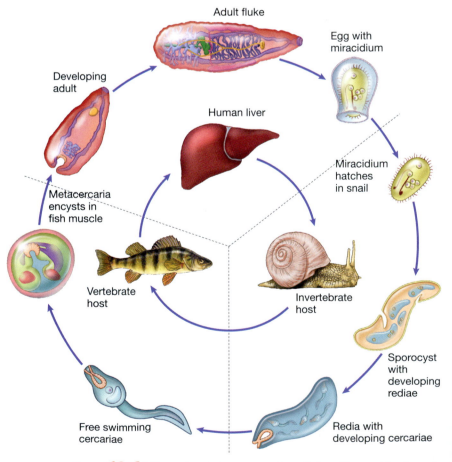

Adult fluke · **Egg with miracidium** · **Human liver** · **Miracidium hatches in snail** · **Developing adult** · **Metacercaria encysts in fish muscle** · **Vertebrate host** · **Invertebrate host** · **Sporocyst with developing rediae** · **Free swimming cercariae** · **Redia with developing cercariae**

FIGURE **19.6** Life cycle of the human liver fluke, *Clonorchis sinensis*.

19

The infected fish is the second intermediate host. The cercariae encyst in the fish and are called metacercariae. Next, the fish is eaten by the final host (perhaps a human) and the parasite makes its way to the bile ducts, where it parasitizes the host. It reaches sexual maturity and produces eggs to start the cycle over again.

Schistosoma spp., commonly called blood flukes, are found mostly in Africa (Fig. 19.7). Approximately 200 million people are infected with schistosomiasis. Because it is difficult to find a mate for copulation in the miles of vessels, the female lives in a groove in the male's body (Fig. 19.8).

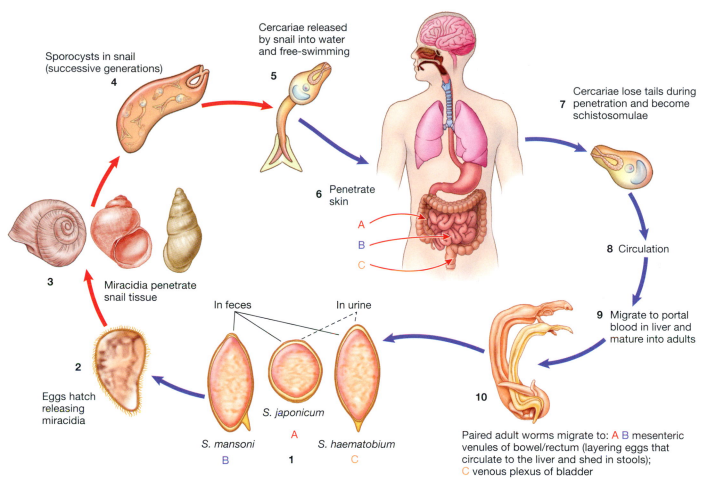

Sporocysts in snail (successive generations) **4**

Cercariae released by snail into water and free-swimming **5**

Cercariae lose tails during penetration and become schistosomulae **7**

6 Penetrate skin

A
B
C

8 Circulation

Miracidia penetrate snail tissue **3**

In feces

In urine

9 Migrate to portal blood in liver and mature into adults

2 Eggs hatch releasing miracidia

S. mansoni
B

S. japonicum
A

1

S. haematobium
C

10

Paired adult worms migrate to: A B mesenteric venules of bowel/rectum (layering eggs that circulate to the liver and shed in stools); C venous plexus of bladder

FIGURE **19.7** Generalized life cycle of *Schistosoma* spp.

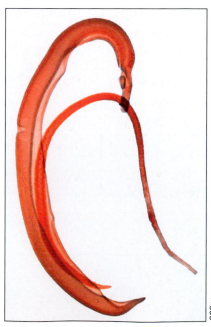

FIGURE **19.8** Schistosome with the female within the groove of the male.

200×

19

Procedure 1

Macroanatomy of Classes Turbellaria and Cestoda

Materials

- ❏ Dissecting microscope or hand lens
- ❏ Petri dish, jar, or watch glass
- ❏ Water
- ❏ Probe
- ❏ Pipette
- ❏ Living specimens of *Dugesia* sp.
- ❏ Hard-boiled egg yolk
- ❏ Instructor's choice of preserved specimens of various tapeworms
- ❏ Colored pencils

1 Procure the needed equipment and specimens.

2 With a pipette, transfer *Dugesia* sp. to a petri dish or watch glass. Do not let the planarian sit in the pipette very long because it will attach to the sides and be extremely hard to expel. After placing the *Dugesia* sp. into a petri dish or watch glass, allow the organism a few minutes to acclimate.

3 Using a hand lens or dissecting microscope, observe the specimen. Pay attention to locomotion and behavior. Nudge the specimen with a probe, and record your results.

_____

_____

4 Using transmitted light, observe the internal structures. Record your observations, sketches, and labels in the space provided.

5 Using a probe or pipette, place a small piece of egg yolk next to the planarian. If possible, turn down the lights on your scope and in the room. (Planaria do not like

to eat in bright light.) Record your observations in the space provided.

_____

_____

_____

6 Observe tapeworm specimens in a jar or in a petri dish. Record your observations and labeled sketches in the space provided.

7 In a jar or a petri dish, observe a fluke with the dissecting microscope or hand lens. *Be sure to observe the internal structures with transmitted light.* Record your observations and labeled sketches in the space provided.

8 Follow the instructor's directions regarding clean up and storage of materials and specimens.

Did you know . . . ?

Strange but True!

Tapeworms have been the subjects of lore and old wives' tales in the past. Tapeworm eggs actually have been given to patients to help them lose weight. Not good! In lore, if a long tapeworm were passed by a person, it was thought to be a "bosom serpent." Not really. Several other intestinal parasites also share this distinction.

19

Dugesia _____

Dugesia _____

Dugesia _____

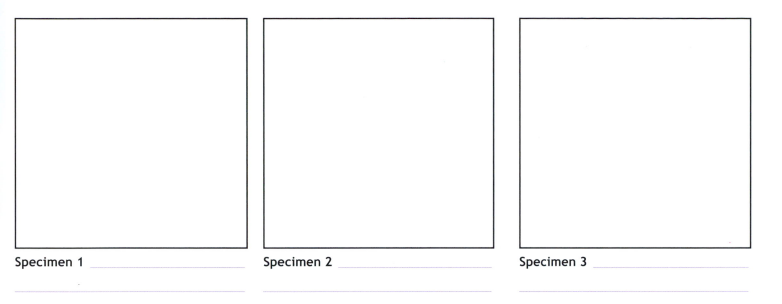

Specimen 1 _____

Specimen 2 _____

Specimen 3 _____

Procedure 2

Microanatomy of Classes Turbellaria and Cestoda

Materials

- ❑ Compound microscope
- ❑ Prepared slide of *Dugesia*
- ❑ Prepared slide of various tapeworms, such as *Taenia* spp., *Diphyllobothrium latum*, or *Dipylidium caninum*
- ❑ Colored pencils

Be sure to observe these slides using low power. Using a high power objective may cause the coverslip to crack and cause damage to the lens.

1 Procure the microscope and selected slides.

2 Using the compound microscope on scanning power or low power. Observe the whole mount of *Dugesia* sp. Compare your observations with Figures 19.9 and 19.10. Record your observations and labeled sketch in the space provided.

3 Using the compound microscope on scanning power or low power, observe a composite tapeworm (Figs. 19.11–19.15). Record your observations and labeled sketches in the space provided.

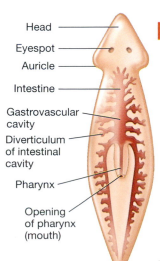

Head

Eyespot

Auricle

Intestine

Gastrovascular cavity

Diverticulum of intestinal cavity

Pharynx

Opening of pharynx (mouth)

A

B

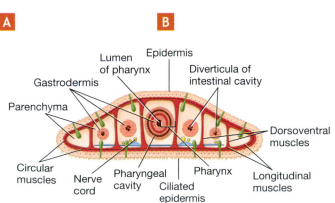

Lumen of pharynx

Epidermis

Gastrodermis

Diverticula of intestinal cavity

Parenchyma

Circular muscles

Nerve cord

Pharyngeal cavity

Ciliated epidermis

Pharynx

Dorsoventral muscles

Longitudinal muscles

FIGURE **19.9** Internal anatomy of *Dugesia*: (**A**) longitudinal section; (**B**) transverse section through the pharyngeal region (parenchyma are undifferentiated cells).

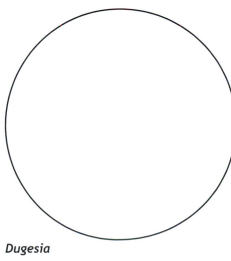

Dugesia

Total magnification _____

_____

_____

19

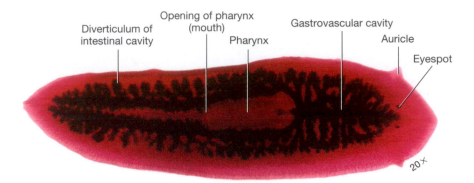

FIGURE **19.10** Stained and prepared specimen of planarian, *Dugesia*.

Diverticulum of intestinal cavity

Opening of pharynx (mouth)

Pharynx

Gastrovascular cavity

Auricle

Eyespot

20×

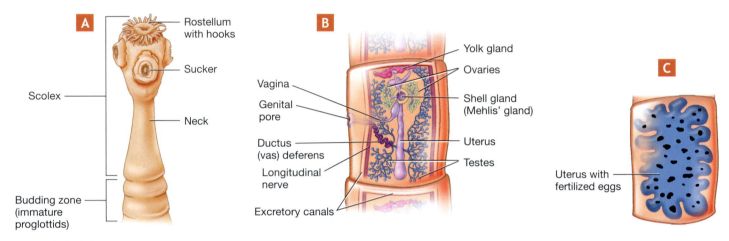

FIGURE **19.11** Diagrams of a parasitic tapeworm, *Taenia pisiformis*: (**A**) anterior end; (**B**) mature proglottids; (**C**) ripe proglottid.

A

Rostellum with hooks

Sucker

Scolex

Neck

Budding zone (immature proglottids)

B

Yolk gland

Ovaries

Shell gland (Mehlis' gland)

Vagina

Genital pore

Ductus (vas) deferens

Longitudinal nerve

Excretory canals

Uterus

Testes

C

Uterus with fertilized eggs

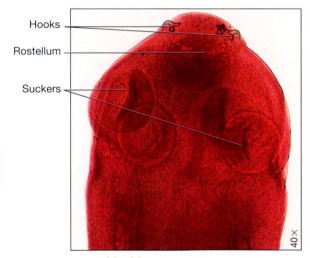

Hooks

Rostellum

Suckers

40×

FIGURE **19.12** Scolex of *Taenia pisiformis*.

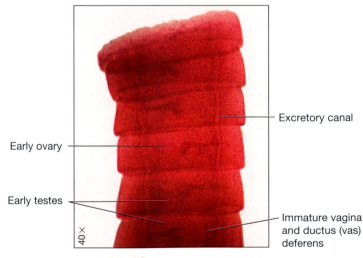

Excretory canal

Early ovary

Early testes

Immature vagina and ductus (vas) deferens

40×

FIGURE **19.13** Immature proglottids of *Taenia pisiformis*.

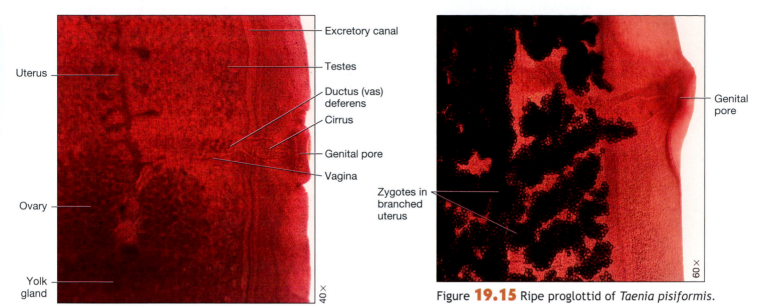

Uterus

Ovary

Yolk
gland

Excretory canal

Testes

Ductus (vas)
deferens

Cirrus

Genital pore

Vagina

40×

FIGURE **19.14** Mature proglottid of *Taenia pisiformis*.

Genital
pore

Zygotes in
branched
uterus

60×

Figure **19.15** Ripe proglottid of *Taenia pisiformis*.

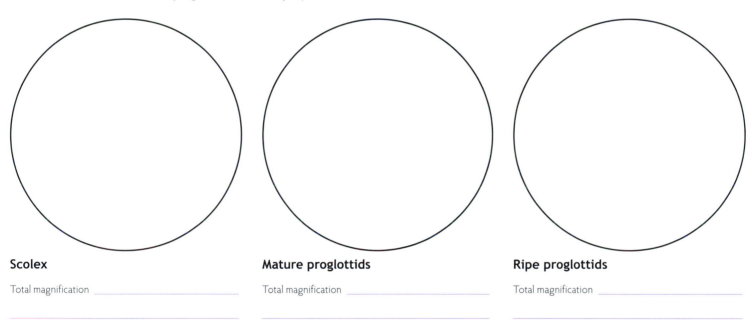

Scolex

Total magnification _____

_____

_____

Mature proglottids

Total magnification _____

_____

_____

Ripe proglottids

Total magnification _____

_____

_____

Procedure 3

Regeneration of *Dugesia*

19

Materials

❏ Dissecting microscope or hand lens
❏ Living specimens of the planaria *Dugesia* sp.
❏ Petri dishes and lids
❏ Scalpel
❏ Sharpie pen
❏ Pipette or probe
❏ Colored pencils

1 Procure the equipment and a specimen of *Dugesia* sp., and place it in a petri dish with water.

2 Using a sharp scalpel, cut the planarian into several cross sections.

3 Place the lid on the petri dish, and label it with a Sharpie. Put your petri dish in a designated area.

4 Each week for 5 weeks, observe your planarian (Fig. 19.16). Ensure the petri dish does not dry out.

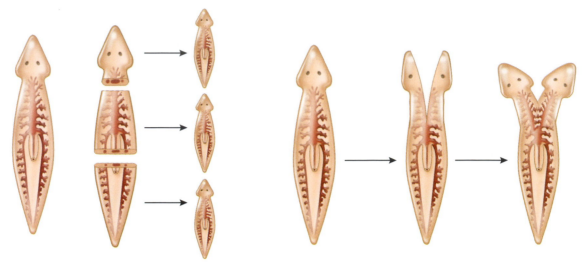

FIGURE **19.16** Planarians are capable of regeneration.

5 Using a pipette or probe, place a tiny piece of egg yolk in the petri dish several times over the 5-week period. Record your observations and sketches in the space provided.

Dugesia   Week _____   *Dugesia*   Week _____   *Dugesia*   Week _____

Did you know...

Ever Seen One of These?

Occasionally when walking on a moist sidewalk one comes across a long, striped, odd-looking flatworm with a shovel-shaped head. Chances are it is a yard planarian or hammerhead flatworm scientifically known as *Bipalium kewense*. Today, these animals are cosmopolitan but originated in Southeast Asia and are thought to have been introduced to many regions by plant trade. *Bipalium* is an extraordinary predator of earthworms. It produces a neurotoxin that it uses to paralyze its prey. If you try to pick it up, it is slimy and easily fragments into many pieces.

Bipalium adventitium.

✔ Check Your Understanding

1.1 Label the descriptions with the correct class of flatworm, using the following key:

Tu = Turbellaria   Ce = Cestoda   Tr = Trematoda

_____ *Dugesia* sp. is an example.

_____ *Fasciola hepatica* is an example.

_____ *Taenia* sp. is an example.

_____ These are parasitic flatworms composed of hooks, scolex, segments (proglottids); most require two hosts.

_____ These glide along in their environment due to secretion of a mucus film; their mouths exist on their ventral surfaces.

_____ Most are endoparasites of vertebrates with suckers and hooks for attachment; some species have oral and ventral suckers.

_____ These are free-living flatworms found in an aquatic or terrestrial environment.

_____ Adults reside in the digestive tract of the final vertebrate host where it may reach length in excess of 10 m.

_____ These have a flattened, leaflike appearance, reside in lung, liver, bile and pancreatic ducts, intestines, and blood.

1.2 How are intestinal platyhelminths protected from harsh digestive juices?

_____

_____

1.3 What is unique about where a mature female schistosome lives? Why?

_____

_____

1.4 What is a tapeworm proglottid?

_____

_____

1.5 What structures make up the scolex of a tapeworm?

_____

_____

_____

Phylum Rotifera

Antonie van Leeuwenhoek termed a curious group of aquatic creatures the "wheel animalcules." These tiny animals were so named because of a crown of cilia on their head. The cilia in motion resembled a spinning wheel. Van Leeuwenhoek's enthusiasm can be shared by examining a bottom sample of a nearby ditch or pond with a compound microscope. Today, wheel animalcules are placed in phylum Rotifera. The rotifers are pseudocoelomates.

Ecologically, rotifers consume small invertebrates and are eaten by other animals. The anterior portion of a rotifer bears a **corona**, or crown, with numerous cilia arranged on two disks that beat in a circular motion in opposite directions. These cilia are used in feeding and locomotion. Sensory bristles near the head region are called papillae. In some rotifers the **cuticle**, or covering on the trunk of the body, forms an armor-like girdle called a **lorica** that may bear **spines**. The posterior portion of a rotifer, the **foot**, contains adhesive glands that open to the exterior via spurs, or toes.

Being transparent, rotifers provide the observer an open window to their internal anatomy. A mouth can be seen beneath the ciliated corona. Behind the mouth in several species is a unique pharyngeal apparatus called the mastax that possesses grinding jaws, or tropi. In living rotifers,

the tropi easily can be seen grinding up algae and smaller invertebrates.

Rotifers have a complete digestive system starting with a mouth and ending in an anus. They have a large stomach and a short intestine. The terminal portion of the intestines is called the **cloaca** because it receives solid wastes, liquid wastes from the bladder, and sex cells such as eggs from the oviducts. Rotifers lack a circulatory system; they respire through their body surface. The nervous system of rotifers consists of a bilobed brain that sends paired nerves to various organs. Sensory organs in rotifers include papillae, ciliated pits, dorsal antennae, and paired eyespots in several species.

Rotifers are dioecious. Generally, female rotifers are larger than the males. Males either insert their penis into the female's cloaca or stab the female with the penis to insert sperm (hypodermic impregnation). Rotifers can produce thin-shelled, fast-hatching eggs or thick-shelled, dormant eggs. Rotifers are tiny organisms that live in environments susceptible to drying up. To cope with these conditions, rotifers can enter an arrested state of biological activity termed *cryptobiosis*. Rotifers have been known to stay in this state for up to 4 years.

Procedure 1

Macroanatomy of a Rotifer

Materials

❑ Compound microscope
❑ Microscope slides and coverslips
❑ Water
❑ Dropper
❑ Live rotifer specimens
❑ Paper towels
❑ Selected prepared slides of rotifers
❑ Colored pencils

1 Procure the needed equipment and culture (Figs. 19.17 and 19.18).

2 Prepare a wet mount of the rotifers and observe using a compound microscope.

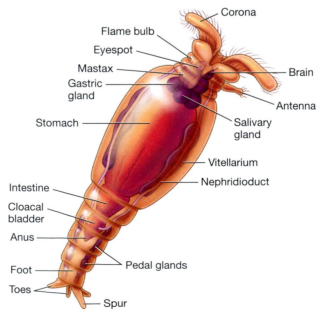

FIGURE **19.17** Rotifer, illustration.

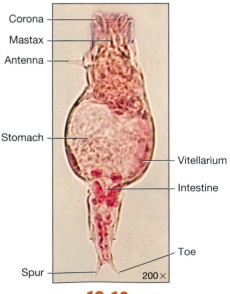

FIGURE **19.18** Rotifer, *Philodina* sp.

3 Observe prepared slides of rotifers using the compound microscope.

4 Record your observations and labeled sketch in the space provided.

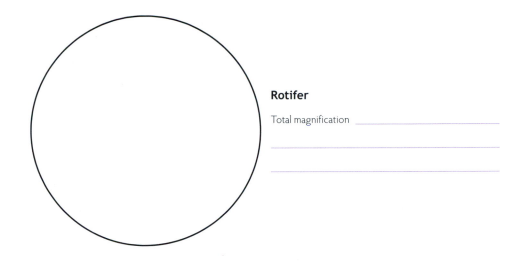

Rotifer

Total magnification _____

_____

_____

Procedure 2

Microanatomy of a Rotifer

Materials

❏ Compound microscope
❏ Microscope slides and coverslips
❏ Pond water containing rotifers or culture of rotifers
❏ Dropper
❏ Selected prepared slides of rotifers
❏ Paper towels
❏ Colored pencils

1 Procure the needed equipment and slides.

2 Using the compound microscope, examine the prepared slides of rotifers. Record your observations and labeled sketch in the space provided.

3 Using the dropper, place a living rotifer on a microscope slide, and prepare a wet mount.

4 Observe the behavior, feeding activities, and anatomy of the rotifer. Record your observations, prepare a sketch, and record the locomotion and activities of the rotifer in the space provided.

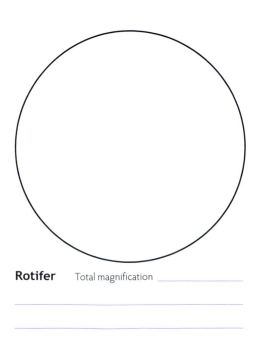

Rotifer   Total magnification _____

_____

_____

✔ Check Your Understanding

2.1 In what environment would you look for rotifers?

_____

_____

2.2 Which of the following characteristics are found in rotifers? (*Circle the correct answer.*)
   a. They are pseudocoelomates with complete digestive systems.
   b. They are coelomates with complete digestive systems.
   c. They are dioecious.
   d. Both a and c.
   e. Both b and c.

2.3 Describe the features found in rotifers that led van Leeuwenhoek to refer to them as "wheel animalcules."

_____

_____

_____

_____

2.4 What is the function of the mastax found in rotifers?

_____

_____

_____

2.5 What is cryptobiosis? What advantages might this give to rotifers?

_____

_____

_____

Did you know . . .

How Is a Pearl Made?

Buying pearl jewelry can damage a budget! Pearls can occur naturally when an irritant such as a parasite enters an oyster or mussel, and the mantle tissue secretes calcium carbonate from the nacre over the irritant. In cultured marine species, a mother of pearl "seed" is placed in the oyster or mussel. In freshwater species, a tiny "seed" of mantle material is placed in the mussel to initiate pearl formation.

Formation of pearls can be natural or artificially induced.

Phylum Mollusca

Although the molluscs exhibit great diversity, they share a bilaterally symmetrical body plan (Fig. 19.19). The molluscs are considered coelomates even though the coelom is limited to the space around the heart. The basic body plan of molluscs consists of two major portions: the **head-foot region** and the **visceral-mass region**. The head-foot region contains the cephalic portions of the organism as well as the feeding and locomotor structures. The visceral-mass region contains the digestive, respiratory, circulatory, and reproductive systems.

Molluscs can possess a heart, vessels, and sinuses, but the majority have an **open circulatory system** in which blood is not contained entirely in vessels. Most cephalopods have a **closed circulatory system** in which the blood is contained in vessels. **Gills** or lungs are responsible for gas exchange. The digestive tract is complete and highly specialized. Most molluscs possess a pair of **metanephridia**, or kidneys, that open into the coelom through a **nephrostome**. In many molluscs, the kidney ducts also discharge sperm and eggs.

Most molluscs have a well-developed head region that bears the mouth and sensory organs. Within the mouth of most molluscs known as gastropods (snails) is a unique structure called the **radula**, a protrusible, tonguelike organ used for rasping (Fig. 19.20). The radula can contain up to 250,000 "teeth" that serve to scrape, pierce, and cut. Observe a snail crawling up an aquarium glass, and notice the radula rasping the algae. Also, while at the beach, notice perfectly round holes in a shell. These holes were done by a mollusc's radula, such as that of an oyster drill.

The nervous system is composed of several pair of ganglia and their associated nerve cords. The nervous system in molluscs, especially the cephalopods (squid and octopi), is well developed. Cephalopods exhibit problem-solving ability. Sensory organs of vision, touch, smell, taste, and equilibrium vary in molluscs. The eyes of the cephalopods are particularly well developed.

The **mantle** in molluscs is a sheath of skin extending from the visceral mass and hanging down each side of the body, protecting the soft parts of the organism. Between the soft parts of the organism is a **mantle cavity**. The outermost surface of the mantle is responsible for secreting and lining the **shell** of some molluscs. Typically, the shell has three distinct layers. The outer layer of the shell, the **periostracum**, serves to protect the inner layers. The middle portion, called the **prismatic layer**, is composed of calcium carbonate and a protein matrix. The innermost layer, the **nacreous layer** of the shell, is produced by the adjacent portions of the mantle surface. This is the iridescent mother-of-pearl layer visible in many shells. Many molluscs secrete nacre around foreign or induced particles, producing pearls. The mantle cavity houses the respiratory organs of molluscs.

Many molluscs are dioecious. Many molluscs undergo **external fertilization**, but a few species undergo internal fertilization. Some molluscs and the segmented worms (annelids) have free-swimming, ciliated larvae called **trochophore larvae**. In others, such as many gastropods and **bivalves**, the trochophore larvae develop into larvae with the beginnings of feet, shells, and mantles, called **veliger larvae**. The cephalopods and some other molluscs produce juveniles that hatch directly from the egg.

The molluscs are generally placed into six major classes. We will discuss four of these classes:

1. Class Polyplacophora contains the chitons. (Not covered in this manual.)

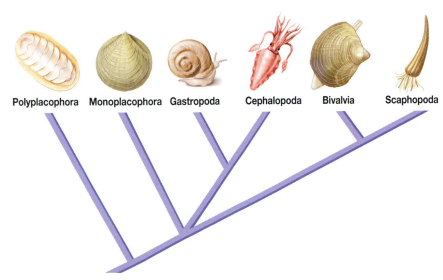

Polyplacophora Monoplacophora Gastropoda Cephalopoda Bivalvia Scaphopoda

FIGURE **19.19** Phylogenetic relationships and classification of Mollusca.

FIGURE **19.20** Snail radula.

2. Class Monoplacophora contains organisms with a cap-like shell. (Not covered in this manual.)

3. Class Gastropoda consists of snails, abalone, whelks, limpets, slugs, and nudibranchs.

4. Class Cephalopoda contains octopi, squid, cuttlefish, nautiluses, and fossil ammonites.

5. Class Bivalvia features nearly 25,000 species of clams, oysters, mussels, scallops, and many others.

6. Class Scaphopoda consists of the tusk shells. (Not covered in this manual.)

Class Gastropoda

Class Gastropoda is the most diverse class of molluscs (Fig. 19.21), with more than 70,000 identified species. Gastropods live in marine, freshwater, and terrestrial environments. Some gastropods, such as slugs, do not have a shell. In gastropods that have a shell, it is a one-piece **univalve**. Common examples of molluscs with univalves include snails, whelks, periwinkles, and conches. The end of the shell is called the **apex**. As the animal grows, the successive whorls increase in size. The central axis is called the **columnella**. The opening of the shell is called the **aperture**, and in many species a protective **operculum** covers the aperture.

Gastropods have well-developed cephalic regions. The head usually is characterized by the presence of two tentacles with an eye at the end of each. Gastropods use a muscular foot for locomotion. In air-breathing snails and slugs, a breathing pore or pneumostome is present on the right side of the mantle. Both monoecious and dioecious gastropods exist. Eggs of marine species are enclosed in egg cases, commonly found by beachcombers.

Class Cephalopoda

Class Cephalopoda (Fig. 19.22) is composed of approximately 800 living species. Cephalopods possess a modified foot in the head region that appears as a funnel-shaped structure, called a siphon, used to expel water from the mantle cavity. This structure is surrounded by tentacles with suckers. The siphon is used in a form of jet propulsion,

FIGURE **19.21** Examples of gastropods: (**A**) Venus comb murex, *Murex pecten*; (**B**) slug, *Deroceras* sp.; (**C**) banana slug, *Ariolimax californicus*; (**D**) sea hare, *Aplysia californica*; (**E**) keyhole limpet, *Megathura crenulata*; (**F**) snail, *Achatina fulica*.

19

expelling water during locomotion. Cephalopods range in size from 1 centimeter in length to *Architeuthis*, the giant squid, which may exceed 12 meters in length.

Nautiloids such as the chambered nautilus and the paper nautilus have an external chambered shell and, as the organism grows, new chambers are added (Fig. 19.23). Cuttlefish have a small, curved shell enclosed by the mantle, called the **cuttlebone**. This hard yet brittle structure is a chambered, gas-filled shell that keeps the animal buoyant. Interestingly, the cuttlebone can be used by caged birds to sharpen their beaks. It can be obtained at pet stores as a source of calcium for birds, turtles, and hermit crabs. The shell of a squid is restricted to a stiff, thin strip called a **pen** that serves to stabilize the squid while swimming. Octopi have no remnants of a shell (Figs. 19.24 and 19.25).

Respiration occurs primarily through gills. Cephalopods have a closed circulatory system with a heart and vessels. The nervous system is highly developed. The brain is lobed and is the largest of any invertebrate. With the exception of nautiloids, the sense organs are well developed. Octopi, as ambush hunters, are capable of learning, and they demonstrate complex behaviors.

Many cephalopods secrete dark sepia through ink glands that serves as a smoke screen during escape behaviors. Octopi, squid, and cuttlefish can change colors for seduction, warning, camouflage, communication, and attraction of prey. Chromatophores, which contain pigment granules, are responsible for color and pattern changes. Cephalopods are dioecious, and juveniles hatch directly from eggs.

FIGURE **19.22** Example cephalopods: (**A**) cuttlefish, *Sepiidae* sp.; (**B**) nautilus, *Nautilus pompilius*; (**C**) giant octopus, *Enteroctopus* sp.

FIGURE **19.23** (**A**) Nautilus, *Nautilus* sp., a cephalopod; (**B**) cross section showing gas-filled chambers within its shell that regulate buoyancy.

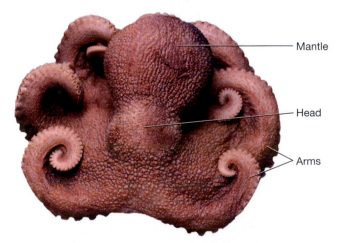

Mantle

Head

Arms

FIGURE **19.24** Dorsal view of an octopus, *Octopus* sp., collected in the Sea of Cortez, San Carlos, Mexico.

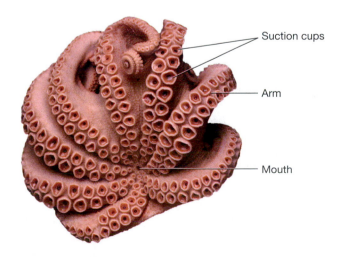

Suction cups

Arm

Mouth

FIGURE **19.25** Ventral view of an octopus, *Octopus* sp.

19

Class Bivalvia

Class Bivalvia consists of molluscs that feature two separate shells (valves) joined by a ligament called the **hinge** (Fig. 19.26). The oldest part of the shell, the **umbo**, looks like a large hump on the anterior end of the dorsal side of each valve. Powerful adductor muscles hold the valves of the shell together. Most bivalves have a posterior and an anterior adductor muscle. Scallops have just one.

Have you ever eaten fried clams or grilled scallops (*Pecten* spp.)? If so, you were eating adductor muscles! The bivalves also are called hatchet-footed animals because of their obvious hatchet-shaped muscular foot attached to the visceral mass of the organism. Examples of bivalves are clams, mussels, scallops, and oysters.

They lack a head and radula. They vary in size from seed shells about 1 millimeter in length to the giant clam, which can be longer than a meter and weigh 225 kilograms. Bivalves live in marine as well as freshwater environments. Most are **filter feeders.** They take in water and nutrients in the **incurrent siphon** and eliminate by the **excurrent siphon**. Gaseous exchange occurs through the mantle and gills. Bivalves are mostly dioecious. In freshwater clams the trochophore larvae develop into specialized veliger larvae known as glochidium larvae that attach to fishes to complete their development.

FIGURE **19.26** Examples of bivalves: (**A**) clam, *Tridacna derasa*; (**B**) mussel, *Mytilus californianus*; (**C**) scallop, *Pecten* sp.

Procedure 1

Macroanatomy of Classes Gastropoda, Cephalopoda, and Bivalvia

Materials

- ❏ Dissecting microscope or hand lens
- ❏ Petri dish
- ❏ Dissecting tray
- ❏ Probe
- ❏ Preserved specimens and shells of selected gastropods, including a snail, a slug, a whelk, a conch, an abalone, a limpet, a nudibranch, and others
- ❏ Preserved specimens and shells of select bivalves in jars, including an oyster, a freshwater clam or mussel, a scallop, a quahog, a shipworm, and others
- ❏ Preserved specimens and shells of select cephalopods, including an octopus, a squid, a cuttlefish, a chambered nautilus, a fossil ammonite, and others
- ❏ Colored pencils

1 Procure the needed equipment and specimens.

2 Examine the gastropod specimens in their jars with a dissecting microscope or hand lens or, if the instructor permits, place them in a dissecting tray or petri dish for examination (Fig. 19.27). Use a probe to move the specimen if necessary. Pay close attention to the structures discussed. Record your observations and labeled sketches in the space provided.

3 Examine the cephalopod specimens in their jars, or if the instructor permits, place them in a dissecting tray for examination. Pay particular attention to the structures discussed. Record your observations and labeled sketches in the space provided.

4 Examine the bivalve specimens in their jars, or if the instructor permits, place them in a dissecting tray for examination. Preserved shells should be placed on a tray. Pay particular attention to the structures discussed. Record your observations and labeled sketches in the space provided.

5 Clean your equipment and desktop thoroughly. Return the equipment and discard your specimens as directed.

19

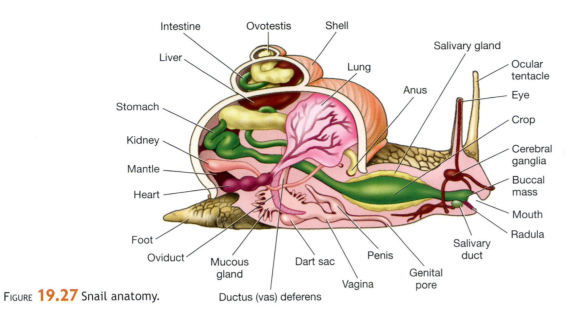

FIGURE **19.27** Snail anatomy.

Gastropod specimen 1 _____

Gastropod specimen 2 _____

Gastropod specimen 3 _____

Cephalopod specimen 1 _____

Cephalopod specimen 2 _____

Cephalopod specimen 3 _____

19

Bivalve specimen 1 _____

Bivalve specimen 2 _____

Bivalve specimen 3 _____

Procedure 2

Squid Dissection

Materials
- ❑ Dissecting tray
- ❑ Dissecting scissors
- ❑ Gloves
- ❑ Safety glasses
- ❑ Lab coat or apron
- ❑ Hand lens
- ❑ Water
- ❑ Squid
- ❑ Colored pencils

1 Procure the supplies and the squid.

2 Wash your specimen thoroughly under running water. Place the specimen on the dissecting tray.

3 Using a hand lens, observe the external features of your specimen (Fig. 19.28). Pay particular attention to the tentacles, mouth region, fin, anus, and eye. Record your observations and labeled sketch in the space provided.

WARNING

If preservative gets in your eyes, wash your eyes immediately, and contact the instructor.

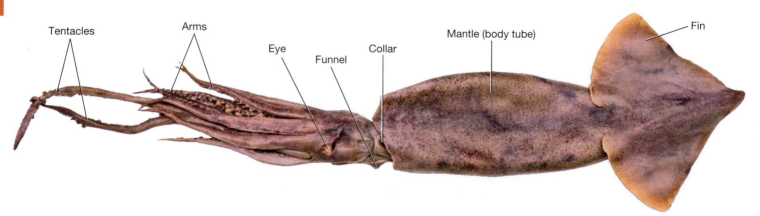

FIGURE **19.28** External anatomy of a squid, *Loligo* sp.

Squid dissection

Squid dissection

4 Keep the squid on the dissecting tray with the funnel facing upward. Carefully separate the two long tentacles from the eight shorter arms of the squid. Observe the tentacles and suckers. Using a pair of scissors, cut through the mantle from the anterior (near the collar) to the posterior end (fin) along the midline. Fold back the sides of the mantle and pin them down. Feel the mantle of the squid, and locate the pen. Remove the pen by carefully pulling it away from the mantle, examine it, and place it aside. Describe and sketch the pen in the space provided.

5 Examine the squid, and locate the structures found in Figures 19.29–19.31. Pay particular attention to the mouth region, gill, anus, ink sac, hearts, reproductive structures, stomach, kidney, and eye. Record your observations and labeled sketches in the space provided.

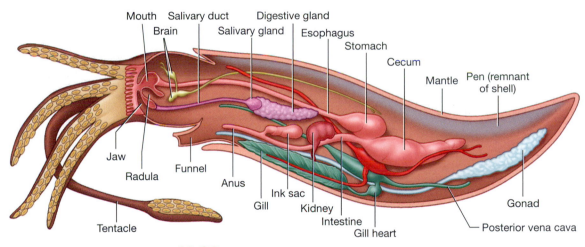

FIGURE **19.29** Internal anatomy of a squid, illustration.

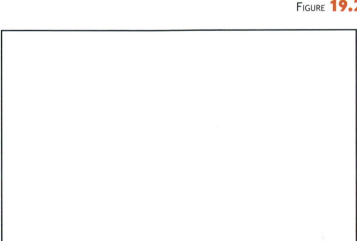

Squid dissection

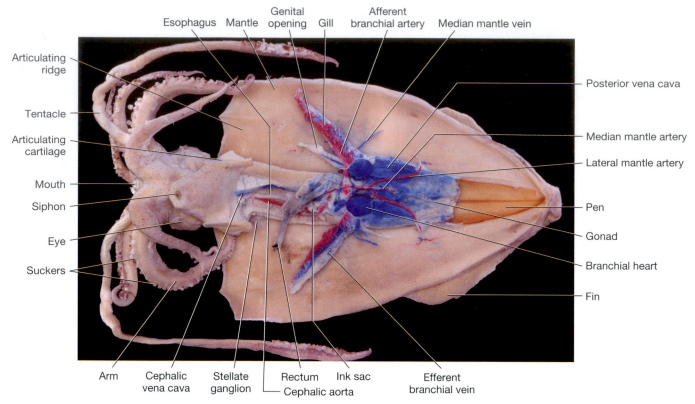

Articulating ridge
Tentacle
Articulating cartilage
Mouth
Siphon
Eye
Suckers

Esophagus   Mantle   Genital opening   Gill   Afferent branchial artery   Median mantle vein

Posterior vena cava
Median mantle artery
Lateral mantle artery
Pen
Gonad
Branchial heart
Fin

Arm   Cephalic vena cava   Stellate ganglion   Rectum   Cephalic aorta   Ink sac   Efferent branchial vein

FIGURE **19.30** Internal anatomy of a squid.

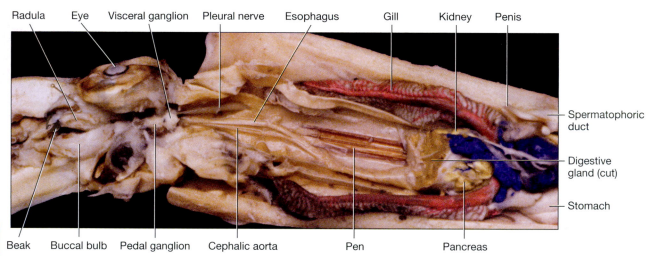

Radula   Eye   Visceral ganglion   Pleural nerve   Esophagus   Gill   Kidney   Penis

Spermatophoric duct
Digestive gland (cut)
Stomach

Beak   Buccal bulb   Pedal ganglion   Cephalic aorta   Pen   Pancreas

FIGURE **19.31** Internal anatomy of a squid, including head region.

6 Carefully remove the gills and ink sac, and examine them through a hand lens. Record your observations and sketches in the space provided.

7 Clean your equipment and desktop thoroughly. Return the equipment, and discard your specimen as directed.

19

Squid dissection

Squid dissection

✔ Check Your Understanding

3.1 Describe the major body regions of a typical mollusc.

3.2 In bivalves, the _____ muscles hold the two valves of the shell together.

3.3 List and describe the layers that make up the shell of a mollusc.

3.4 What is the mantle in a mollusc, and what is its function?

19

Phylum Annelida

Annelids are coelomates and exhibit bilateral symmetry. The body of an annelid is divided into similar rings, or **segments**. The annuli, the grooves dividing each segment, are readily visible on most species. Each segment is termed a **metamere**, or a **somite**. The repeated pattern of these metameres is termed *metamerism*. Internally, the segments are delimited by structures called **septa**.

Annelids, with the exception of leeches, possess tiny bristles known as **setae** on their somites that vary in size, form, and function in different species. Some setae, such as those in earthworms, serve as anchor mechanisms. These structures can be felt along the sides of an earthworm. Other setae are used in locomotion and respiration.

The typical annelid has a two-part head consisting of the **prostomium** and the **peristomium**. The prostomium, found in front of the mouth, usually is a small, liplike extension over the dorsal portion of the mouth. The peristomium contains the mouth. A segmented body follows the periostomium, and the posterior-most portion is called the pygidium. New segments form during development from the posterior end. Each segment typically contains the coelom and nervous, respiratory, circulatory, and excretory structures. With the exception of leeches, the coelom is filled with fluids, giving the animal a **hydrostatic skeleton.** The outer layer of annelids consists of a protective cuticle.

Annelids have extensive muscle systems consisting of both circular and longitudinal muscles. The muscles are involved in locomotion as well as peristalsis (movement of food through the digestive system). The digestive system of these animals is complete, beginning with a mouth and ending with an anus. Annelids undergo cutaneous respiration.

The circulatory system consists of ventral and dorsal longitudinal vessels in all annelids except leeches. Five pairs of pumping vessels serve as muscular hearts in annelids. The blood of many annelids contains the transport protein hemoglobin. The excretory system consists of a pair of **nephridia** that removes waste from each segment and discharges the waste through external pores. The nervous system is well developed, consisting of cerebral ganglia and a ventral nerve cord. The sense organs in annelids include eyes, chemoreceptors, and statocysts. Annelids may be monoecious like earthworms and leeches or dioecious like many sandworms. Some biologists suggest the oligochaetes and hirudineans should constitute a new class called Clitellata because both classes possess a **clitellum**, a glandular structure used during reproduction.

Phylum Annelida consists of three classes (Fig. 19.32):

1. Class Hirudinea (Fig. 19.33A) consists of leeches.

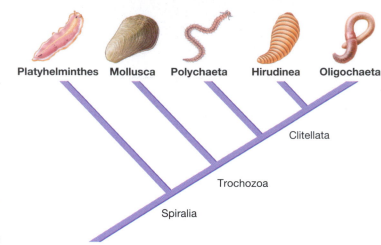

Platyhelminthes   Mollusca   Polychaeta   Hirudinea   Oligochaeta

Clitellata

Trochozoa

Spiralia

Figure **19.32** Phylogenetic relationships and classification of Annelida.

FIGURE **19.33** Examples of annelids: (**A**) leech, *Macrobdella* sp.; (**B**) earthworm, *Lumbricus* sp.; (**C**) bloodworm, *Glycera* sp.

2. Class Oligochaeta (Fig. 19.33B) contains about 10,000 species of marine and terrestrial worms. The earthworm is the best-known oligochaete.

3. Class Polychaeta (Fig. 19.33C) consists mostly of marine worms, such as sandworms and bloodworms.

Class Polychaeta

The polychaetes are the largest class of annelids, consisting of more than 10,000 species of mostly marine worms. These worms are commonly called paddle worms because of their **parapodia** (fleshy structures bearing setae) on each body segment. Polychaetes range in size from less than 1 millimeter to 3 meters. Many species of polychaetes are brightly colored. Polychaetes are dioecious animals. Gamete production occurs in individual segments or in specialized portions of the body. Most polychaetes do not copulate, instead releasing their gametes into the water for external fertilization.

Polychaetes most often are placed into two subclasses depending upon their mode of life. Errant polychaetes, subclass Errantia, crawl around stones, shells, and coral. They usually have a well-developed head with eyes and antennae. The parapodia are large and function like legs. The pharynx commonly contains teeth and powerful jaws. *Nereis* sp., the clam worm, is a typical errant polychaete. Its body can contain 200 segments and exceed 40 centimeters in length. Sedentary polychaetes, subclass Sedentaria, includes species that rarely expose more than their head from protective tubes and burrows. Sedentary burrowers construct a vertical burrow.

Class Hirudinea

Most leeches are inhabitants of freshwater habitats. Approximately 500 species of leeches have been described. Most leeches are between 2 and 6 centimeters in length, but one species of Amazonian leech, *Haementeria* sp., can reach 30 centimeters. Leeches usually are dorsoventrally flattened. Leeches are monoecious, having an obvious clitellum only in breeding season. The head of a leech usually is reduced, and setae are absent. In addition, leeches have no internal septa.

Contrary to popular belief, not all leeches are monstrous bloodsuckers. Approximately 25% of leeches are predaceous, feeding upon oligochaetes, snails, and insect larvae. The other 75% of leeches are **ectoparasites** on a variety of invertebrates and vertebrates. Blood-sucking leeches produce salivary secretions that contain an anesthetic, anticoagulants (hirudin), and a vasodilator. Many species of leeches can consume up to 11 times their weight in blood in a single 40-minute feeding period. Leeches have been used to remove blood in medical procedures for centuries. After water is removed from the blood, the digestive process may take up to 6 months.

Class Oligochaeta

The best-known oligochaete is the earthworm, or night crawler (*Lumbricus terrestris*). It burrows in moist, rich soil and usually emerges at night. During dry weather, these worms can burrow several feet below the surface and become dormant.

In wet weather, they stay near the surface, with their anus or mouth protruding through the burrow. When disturbed by too much water or certain stimuli, such as vibrations or chemicals, they will emerge from the burrow. Most earthworms are between 15 and 30 centimeters in length.

Earthworms are characterized by a prostomium on the anterior end and a pygostyle on the posterior end. In most earthworms, each segment contains four pairs of setae projecting from small pores in the cuticle to the outside. They feed primarily on decayed organic matter, bits of plant material, refuse, and animal matter. Food enters the mouth and, after leaving the esophagus, is stored in the **crop**. From the crop, food is passed to the **gizzard**, where it is ground into small pieces. Digestion and absorption take place in the intestine. Waste products are discharged through the anus. A pair of metanephridia is responsible for excretion.

Earthworms have no respiratory organs. Gaseous exchange takes place at the surface of the moist skin. The circulatory system is closed and characterized by the presence of five pairs of aortic arches. The nervous system consists of a central nervous system and peripheral nerves. Earthworms possess a brain and a ventral nerve cord that runs along the floor of the coelom to the last somite.

Earthworms are monoecious organisms, possessing both male and female organs in the same body, but they cannot undergo self-fertilization. The gonads are restricted to the anterior portion of the body. A distinct swelling called the clitellum can be seen on certain segments behind the genital pores. Along with the genital setae, the clitellum secretes mucus that holds mating earthworms together. Copulation involves the reciprocal transfer of sperm, which requires up to three hours. A few days after copulation, the clitellum secretes dense mucus that eventually will form the cocoon. Eggs from the female gonopores and sperm from the seminal receptacles are collected en route, and fertilization occurs within the cocoon. Embryonation takes place within the cocoon (Fig. 19.34). The young worm hatches, appearing similar to an adult, in two or three weeks.

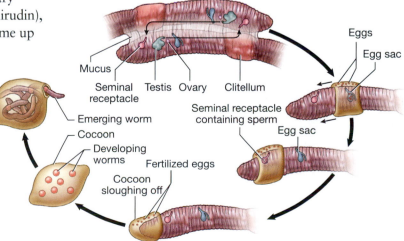

FIGURE **19.34** Earthworm copulation and formation of an egg cocoon.

Macroanatomy of Classes Polychaeta, Hirudinea, and Oligochaeta

Materials

- ❏ Dissecting microscope or hand lens
- ❏ Petri dish
- ❏ Probe
- ❏ Scalpel
- ❏ Preserved specimens of select polychaetes including *Nereis* spp., *Neanthes* spp., and others
- ❏ Preserved specimens of leeches
- ❏ Preserved specimens of select oligochaetes, including *Lumbricus terrestris*, tubiflex worms, and others
- ❏ Colored pencils

1 Procure the needed equipment and specimens.

2 Using the dissecting microscope, examine the polychaete specimens in their jars, or if the instructor permits, place them in a petri dish. Use the probe to move the specimen during your observations. Compare your observations with Figures 19.35 and 19.36. Pay close attention to the structures discussed. Record your observations and labeled sketches in the space provided.

3 Using the dissecting microscope, examine the leech specimens in their jars or, if the instructor permits, place them in a dissecting tray for examination (Figs. 19.37 and 19.38). Pay particular attention to the structures discussed in the figures. Record your observations and labeled sketches in the space provided.

FIGURE **19.35** Sandworm, *Nereis virens*.

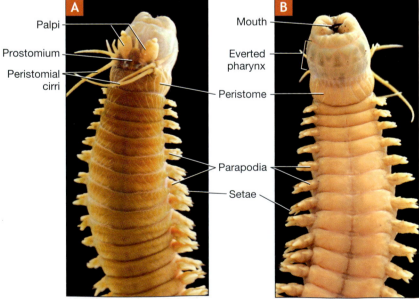

FIGURE **19.36** Anterior end of sandworm, *Nereis* sp.: (**A**) dorsal view; (**B**) ventral view.

Specimen 1 _____

Specimen 2 _____

19

Specimen 3 _____

_____

_____

Specimen 4 _____

_____

_____

FIGURE **19.37** Dorsal view of a leech, *Macrobdella* sp. (scale in mm).

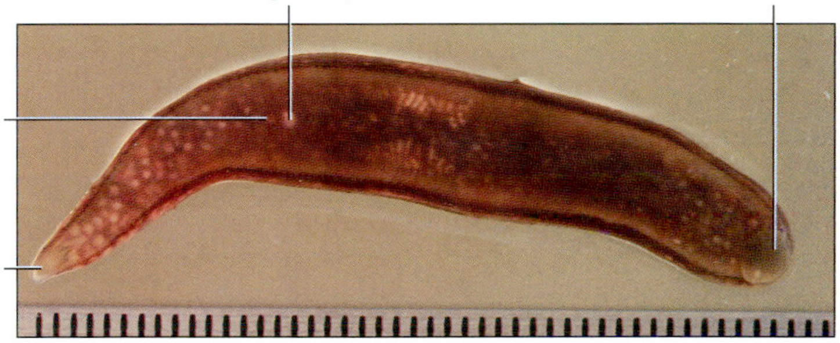

FIGURE **19.38** Ventral view of a leech, *Macrobdella* sp.

Specimen 5 _____

_____

_____

Specimen 6 _____

_____

_____

4 Using the dissecting microscope, examine the oligochaete specimens in their jars or, if the instructor permits, place them in a dissecting tray for examination. Compare your observations with Figures 19.39–19.41, paying particular attention to the structures discussed. Record your observations and labeled sketches in the space provided.

5 Clean your equipment and desktop thoroughly. Return the equipment, and discard your specimen as directed.

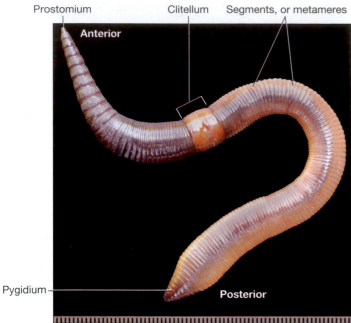

FIGURE **19.39** Dorsal view of an earthworm, *Lumbricus* sp. (scale in mm).

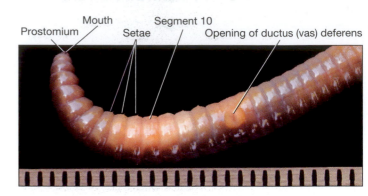

FIGURE **19.40** Anterior end of an earthworm, *Lumbricus* sp. (scale in mm).

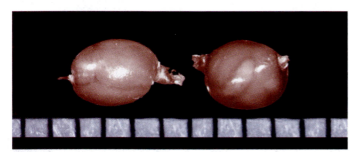

FIGURE **19.41** Earthworm cocoons (scale in mm).

Specimen 7 _____

Specimen 8 _____

Specimen 9 _____

19

Procedure 2

Earthworm Dissection

1 Procure the supplies and the earthworm.

2 Wash your specimen thoroughly under running water. Place the specimen on the dissecting tray.

3 Using a hand lens, observe the external features of your specimen. Notice that the dorsal side is more rounded and usually darker in color. Pay particular attention to the prostomium, clitellum, setae, genital pores, and anus. Record your observations and labeled sketch in the space provided on the following page.

4 Lay the earthworm on the dissecting tray with its dorsal side facing up. Begin the dissection about an inch posterior to the clitellum. Carefully lift up the integument, and snip an opening with a pair of sharp dissecting scissors. Insert the scissors into the opening, and cut in a straight line all the way up through the mouth. Take your time, and be sure to cut just the integument. If the cut is too deep, it may damage the internal organs. Using forceps and dissection pins, carefully pull apart the two flaps of skin, and pin them flat on the tray. The walls of some of the septa may have to be cut to pin down the earthworm properly.

5 Carefully examine the internal anatomy of the earthworm, comparing it with Figures 19.42–19.44. Pay particular attention to the pharynx, crop, gizzard, intestine, ventral nerve cord, reproductive structures, and aortic arches (hearts). Record your observations and labeled sketch in the space provided on the following page.

6 Clean your equipment and desktop thoroughly. Return the equipment, and discard your specimen as directed.

WARNING

If preservative gets in your eyes, wash your eyes immediately, and contact the instructor.

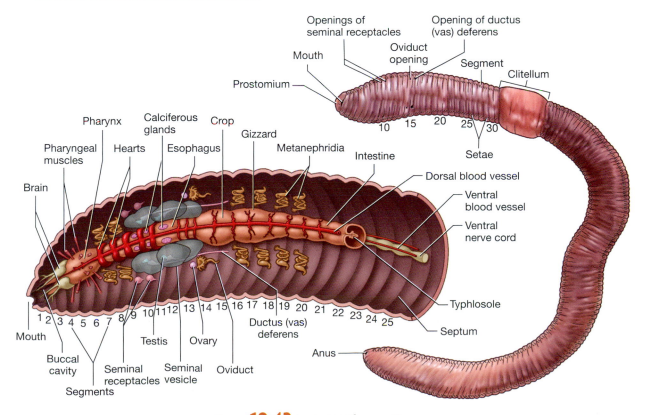

FIGURE **19.42** Anatomy of an earthworm.

19

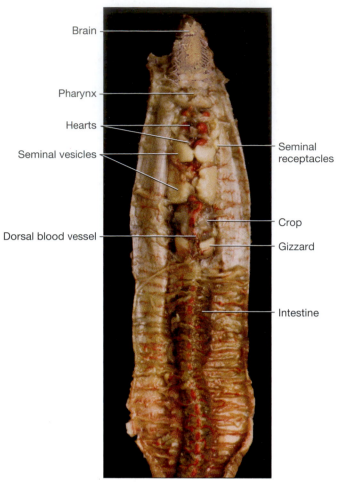

Figure **19.43** Internal anatomy of the anterior end of an earthworm, *Lumbricus* sp.

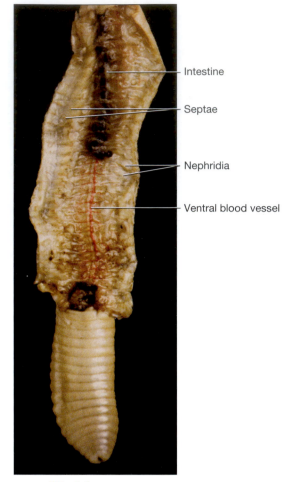

Figure **19.44** Internal anatomy of the posterior end of an earthworm, *Lumbricus* sp., with part of the intestine removed.

Lumbricus terrestris dissection

Lumbricus terrestris dissection

✓ Check Your Understanding

4.1 Label the descriptions with the correct class of the phylum Annelida, using the following key:

H = Hirudinea   O = Oligochaeta   P = Polychaeta

_____ These are marine species with paddle-like appendages (parapodia).

_____ These are dorsoventrally flattened with reduced heads and no setae.

_____ These are segmented with four pairs of setae per segment.

_____ *Nereis* sp. is an example.

_____ *Lumbricus terrestris* is an example.

_____ *Macrobdella* sp. and *Hirudo* sp. are examples.

4.2 What is the function of the clitellum in the earthworm?

_____

_____

_____

4.3 What is the role of hirudin in the leech?

_____

_____

_____

19

Animal Planet II
Understanding the Lophotrochozoans

MYTHBUSTING

From Coast to Coast

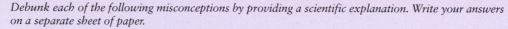

Debunk each of the following misconceptions by providing a scientific explanation. Write your answers on a separate sheet of paper.

1 A snail has a straight digestive tract.

2 Fleas have nothing to do with tapeworms.

3 You can get rid of a tapeworm by coaxing it out with a bowl of milk and cookies placed near your mouth—no kidding!

4 Squid and octopi do not change colors.

5 Oysters are always safe to eat.

1 What are the two defining characteristics found in organisms that are grouped in the clade Lophotrochozoa?

2 List five characteristics found in organisms placed in phylum Platyhelminthes.

3 Describe reproduction in tapeworms.

4 Name five characteristics of phylum Mollusca.

5 Sketch and label the basic external anatomy of a tapeworm in the space provided.

6 Why are cephalopods considered advanced molluscs?

_____

_____

_____

_____

7 Label Figure 19.45.

1. _____

2. _____

3. _____

4. _____

5. _____

6. _____

7. _____

8. _____

9. _____

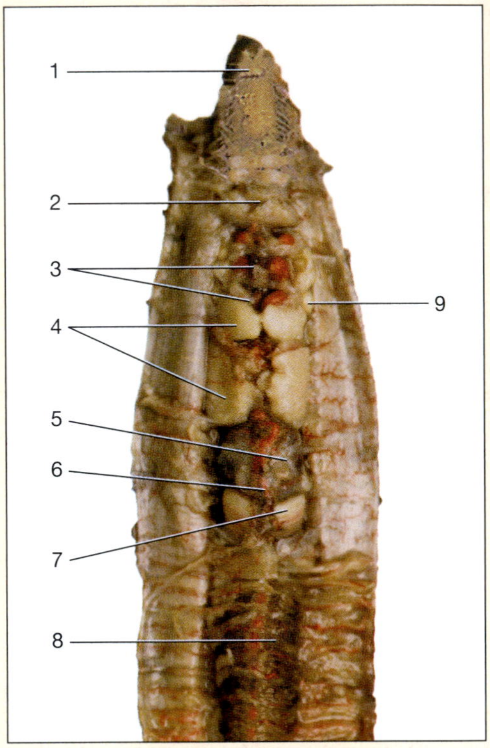

8 This mollusc structure is a protrusible, tonguelike organ with "teeth" used for rasping of algae or drilling through shells. (*Circle the correct answer.*)

   a. Operculum.

   b. Umbo.

   c. Labial palp.

   d. Radula.

   e. Girdle.

FIGURE **19.45** Anatomy of an earthworm.

9 Which of the following characteristics are not present in the phylum Annelida? (*Circle the correct answer.*)

   a. Coelomates exhibiting bilateral symmetry.

   b. Two-part head with a prostomium and periostomium.

   c. Segmented body with fairly well-developed nervous system.

   d. Visceral mass, which contains the circulatory, digestive, respiratory, and reproductive systems.

   e. Fresh-water, marine, and terrestrial species.

10 The _____ region in the phylum Mollusca contains the respiratory, circulatory, and reproductive systems.

11 Describe reproduction in earthworms.

_____

_____

_____

_____

19

Animal Planet III
Understanding the Ecdysozoans

20

Look closely at nature. Every species is a masterpiece, exquisitely adapted to the particular environment in which it has survived. Who are we to destroy or even diminish biodiversity?

— E.O. Wilson (1929–present)

Just wondering . . .

Consider the following questions prior to coming to lab, and record your answers on a separate piece of paper.

1 What was an arthropod called a eurypterid?

2 What is a tarantula "hair"?

3 What is *Loa loa*?

4 Why are medical scientists interested in horseshoe crabs?

5 Why did many arthropods in the Carboniferous become so large?

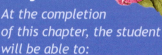

Objectives

At the completion of this chapter, the student will be able to:

1. Describe the taxonomical organization and characteristics of ecdysozoans.

2. Describe the characteristics and basic biology and provide examples of phylum Nematoda.

3. Discuss the fundamental characteristics and natural history and provide examples of phylum Arthropoda.

4. Discuss the basic biology and classification of phylum Arthropoda.

5. Compare and contrast the basic biology of subphyla Chelicerata, Myriapoda, Crustacea, and Hexapoda.

6. Dissect and describe the external and internal anatomy of a crayfish and a grasshopper.

I f you are an ecdysozoan, it's a hassle to grow because you must shed your protective cuticle, or **exoskeleton**. If you aren't careful, you may be eaten by a predator, or perhaps end up on someone's soft-shell crab platter. Look around: roundworms, spiders, insects, shrimp, and other ecdysozoans abound!

Ecdysis, or molting, is the basis for separation of the ecdysozoans from the lophotrochozoans (Fig. 20.1). Evolutionarily, the development of ecdysis influenced the further development of respiratory structures such as the trachea, gills, and lungs as well as internal fertilization and metamorphosis. As a result of these incredible innovations, the ecdysozoans are a large and diverse group with a great impact upon ecology, commerce, and medicine.

The ecdysozoans are represented by eight very different phyla of protostomes: **phylum Nematoda** (roundworms) and **phylum Nematomorpha** (horsehair worms), **phylum Kinorhyncha** (minute marine worms), **phylum Priapulida** (about 16 species of cold-water marine worms), **phylum Loricifera** (fewer than 100 species of tiny marine animals), **phylum Arthropoda** (insects and their relatives), **phylum Tardigrada** (water bears), and **phylum Onychophora** (velvet worms). Of these phyla that constitute the ecdysozoans, we will discuss Nematoda and Arthropoda in depth in this chapter.

Chapter Photo
Male blue-eyed hawker dragonfly, *Aeshna affinis*.

439

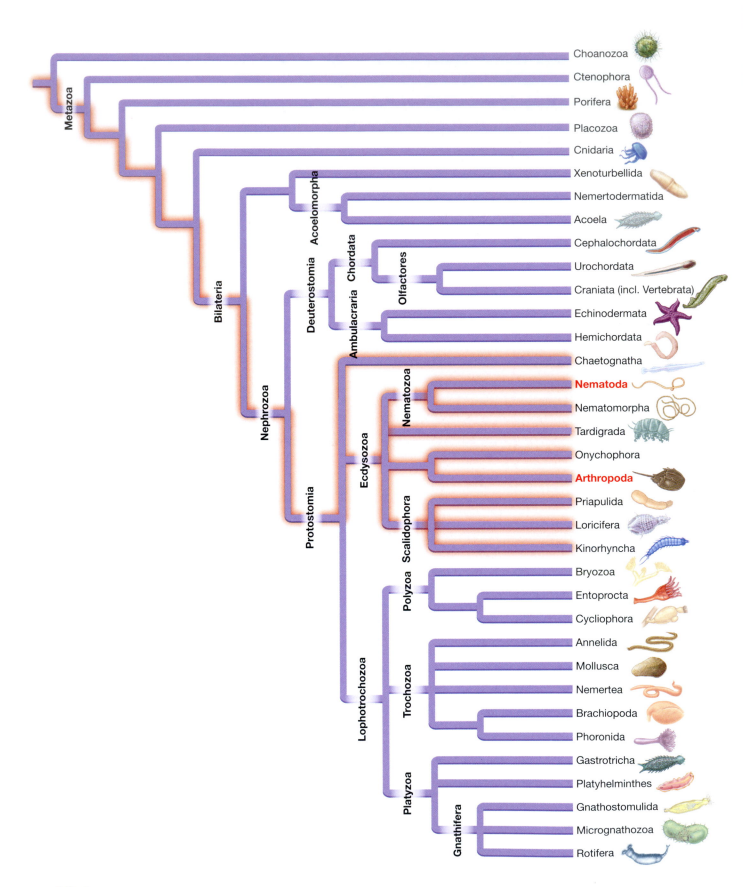

FIGURE **20.1** Phylogeny and classification of Metazoa (multicellular animals). The phyla shown in red are discussed in this chapter.

Phylum Nematomorpha

An old wives' tale warned that if a hair fell from a horse's mane or tail, it would turn into a worm. This is an excellent example of spontaneous generation. Today, we know better and realize the "worm" is actually a member of phylum Nematomorpha. The nematomorphans are still known as horsehair worms, or Gordian worms. Superficially, the worm resembles a brown skinny horsehair and is in some cases nearly a meter in length (Fig. 20.2). Approximately 350 species of nematomorphans have been identified.

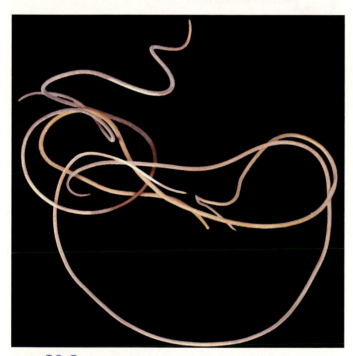

FIGURE **20.2** Horsehair worm, *Gordius aquaticus*.

In adult horsehair worms the digestive system is vestigial (a structure that has lost its function), and digestion occurs through absorption. Adult horsehair worms are free-living and can be found in damp environments, such as in puddles and watering troughs around a barn. The juvenile stage usually encysts on a plant eaten by a cricket or grasshopper. After being eaten, the juvenile develops in the host and eventually emerges.

Phylum Tardigrada

Under the microscope, members of phylum Tardigrada (water bears) are fascinating (Fig. 20.3). Like the velvet worms, water bears also possess a hemocoel lined by an extracellular matrix. Most of the 1,000 known species of tardigrades are microscopic. They can be found living in a film of water associated with moss, lichens, detritus, or moist plants. Marine and freshwater species exist. Tardigrades are considered polyextremophiles, surviving in both low and high temperatures, high radiation levels, dry conditions, and, as NASA determined, in the vacuum of space. Under extremely harsh conditions, tardigrades can enter a cryptobiotic (ametabolic) state.

150 ×

FIGURE **20.3** SEM of tardigrade, *Macrobiotus* sp.

Tardigrades have a stubby body with four pairs of short, unjointed legs terminating in four to eight claws. Tardigrades have a pair of stylets near their buccal tube that allows them to pierce plant tissues or animal body walls. Several species of water bears eat entire rotifers, other tardigrades, and small invertebrates. The tardigrade brain is large, and the nerve cord is associated with four ganglia that control the legs. Tardigrades are dioecious. ■

Phylum Nematoda

Phylum Nematoda consists of long, slender pseudocoelomates known as roundworms. Usually, when one thinks of roundworms, the following diabolical parasites come to mind: hookworms, pinworms, guinea worms, eye worms, heartworms, and whipworms (Fig. 20.4). It has been suggested that every animal on Earth has its own personal parasitic nematode, but just a small percentage of the 25,000 known species are "bad." Many species are harmless, existing in just about every environment on Earth. Rich topsoil can contain more than 1 billion nematodes per acre. Don't be surprised that the actual number of nematode species approaches between 500,000 and 1 million.

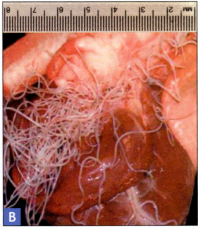

FIGURE **20.4** (**A**) Elephantiasis is caused by the nematode *Wuchereria bancrofti*. (**B**) Heartworm infections are caused by the nematode *Dirofilaria immitis*.

Anatomy of Nematodes

As a general rule, the bodies of nematodes are cylindrical and tapered at both ends. They vary in size from microscopic to longer than 30 centimeters. Nematodes possess a thick, protective, nonliving cuticle that grows between molts. The epidermis and subepidermal muscle can be found beneath the cuticle. In nematodes, the fluid-filled **pseudocoel** is well developed and serves as a **hydroskeleton**. Nematodes move in a thrashing motion produced by a layer of longitudinal muscle. This motion allows them to move between spaces in algae, sand, and soil particles.

The mouth of a nematode opens into a buccal cavity with teeth or a stylet (spear-like structure). The buccal cavity connects to the pharynx. Nematodes have a long, straight intestine that serves as a site of digestion and absorption. Nematodes lack protonephridia but possess glands and tubules that open through a **midventral pore** to the external environment.

The nervous system consists of nerve rings. Nematodes have sensory papillae, and nonparasitic species possess amphids (sensory organs) on each side of the head. Parasitic species have **phasmids** (sensory organs) near their posterior end. The majority of nematode species are dioecious, with the male being smaller than the female. Male nematodes usually possess copulatory spicules for internal fertilization. The female's fertilized eggs are stored in the uterus until deposition.

Free-living nematodes feed on bacteria, algae, yeast, fungi, small invertebrates, and other nematodes. Some free-living nematodes, called **coprozoic** nematodes, feed on fecal material, and still others may be **saprobes**, which feed on decaying organisms. Some species cause great agricultural damage by feeding on the juices of higher plants (usually roots). Perhaps the parasitic nematodes stimulate the most interest from humans. Parasitic forms are responsible for a variety of diseases in humans as well as other animals.

Nematodes are part of the food chain, eaten by insect larvae, mites, and even some species of nematode-capturing fungi. One species, *Caenorhabditis elegans*, has become a model organism in genetics, developmental biology, cell biology, and animal behavior research. The complete genetic sequence of *C. elegans* was finished in 2002, making it the first genome of a multicellular eukaryote in which every base is known.

Example of Nematodes

Ascaris lumbricoides

Ascaris lumbricoides is an excellent model of a typical nematode. Relatively common and large enough to dissect, it is one of the most common intestinal parasites in humans. Parasitologists estimate nearly 1.27 billion people are stricken with *Ascaris* infections worldwide. Other species of *Ascaris* are found in cats, horses, pigs, and a number of other vertebrates.

A typical female *Ascaris* is prolific, producing more than 200,000 eggs daily, which can remain viable for years (even found to be viable after being in preservative). The eggs pass in the host's feces and, given the proper soil conditions, can undergo embryonation. The eggs can remain viable in the soil for many months to perhaps 10 years. Eggs enter the body via uncooked vegetables, soiled fingers, etc. After the eggs enter a new host, they hatch in the small intestine. The juveniles burrow through the intestinal wall and into the veins and lymph vessels and are carried to the heart and lungs. Many times their presence in the lungs initiates serious pneumonia. From the lungs, the larvae make their way up the trachea. When the larvae reach the pharynx, they are swallowed to pass to the stomach and finally to the intestines, where the larvae mature and begin feeding on intestinal contents.

A Closer Look Other Nematodes

Trichinella spiralis

The painful and potentially deadly disease trichinosis, caused by *Trichinella spiralis*, can infect a variety of mammals including humans. *Trichinella*, the smallest nematode parasite in humans, is introduced to a host when the host eats muscle containing encysted juvenile parasites (such as infected pork). The parasite then develops in the host's intestine. Females produce living juveniles that penetrate blood vessels and eventually are carried to skeletal muscles, where they encyst (Fig. 20.5). On occasion, the larvae encyst in the heart or brain. General symptoms of trichinosis include vomiting, diarrhea, muscle soreness, fever, edema, and weakness. Mature females are approximately 3 millimeters in length and males 1.5 millimeters.

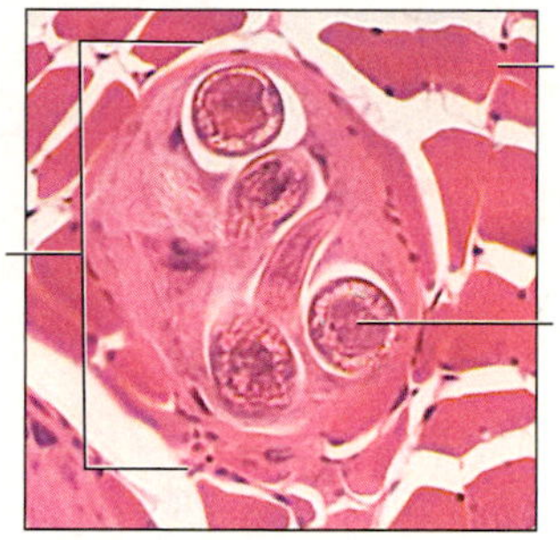

FIGURE **20.5** *Trichinella spiralis* can encyst in skeletal muscle.

Hookworms

Hookworms (*Necator americanus*) possess hook-like anterior ends (Fig. 20.6). Females can attain a length of 11 millimeters and males 9 millimeters. Hookworms infect a number of mammals including humans. Hookworm eggs pass in the feces of the infected host, and the juveniles hatch in the soil. Thus, going barefooted in an area where animals frequently defecate is not wise. The juveniles feed upon bacteria until they have the opportunity to burrow through the skin of a new host. The parasite travels to the lungs and, eventually, the intestine. The hookworm attaches to the intestinal mucosa, where it feeds on blood. Symptoms of a hookworm infection include anemia and protein deficiency and, in young children, the loss of proteins and iron. Left untreated, this loss can retard growth.

FIGURE **20.6** Hookworm, *Necator americanus*.

Pinworms

Pinworms (*Enterobius vermicularis*) are the most common intestinal parasite in the United States (Fig. 20.7). They have the uncanny ability to spread through an elementary school like wildfire. A female pinworm can exceed 12 millimeters in length. At night, the female pinworm migrates to the anal opening and lays her eggs. In the past, parents who suspected their child of having pinworms would place tape over the anus at bedtime to capture pinworms and diagnose an infection. The eggs contaminate bedding and clothing, and can be easily spread. One of the symptoms of pinworms is scratching the anal region. The eggs can get on the fingers and, with kids, there's no telling where they can end up, be swallowed, and start the cycle again.

FIGURE **20.7** Pinworm, *Enterobius vermicularis*.

Filarial Worms

Filarial worms get their name from their young, called microfilariae. In humans, a devastating filarial worm called *Wuchereria bancrofti* infects people in tropical countries in Africa and South America (Fig. 20.8). Females can exceed 10 millimeters in length and live in the lymphatic system. Females release microfilariae into the lymphatic system and blood. A mosquito bites the infected individual and is the vector for spreading the disease. *Wuchereria bancrofti* is responsible for the disfiguring disease elephantiasis (see Fig. 20.4). The most common filarial disease in the United States is heartworm, which occurs primarily in dogs and is caused by *Dirofilaria immitis*. Occasionally, a cat, a sea lion, or a human can acquire the infection. This disease can be deadly to dogs if left untreated. Mosquitoes are the vector for heartworms.

FIGURE **20.8** *Wuchereria bancrofti*.

Vinegar Eels

Vinegar eels (*Turbatrix aceti*) are free-living nematodes commonly found in fermented fruit juices and unpasteurized vinegar (Fig. 20.9). *Turbatrix* actively feed upon bacteria living in these liquids. Fortunately, the vinegar in your kitchen is pasteurized. Under the microscope, the 2 millimeters nematodes are noted for their thrashing movements.

FIGURE **20.9** Vinegar eel, *Turbatrix aceti*.

Symptoms of an *Ascaris* infection depend on the site and stage of the infection. In the intestines, *Ascaris* can cause malnutrition, blockage, and poor health. *Ascaris* has been known to block the bile duct, pancreatic duct, and appendix. Wandering worms can emerge from the throat and anus. The female is larger than the male. The male has a distinct crook in the tail. The female can attain a length of 30 centimeters or more.

Did you know . . .

Ultimate Survivor

In February 2003, the space shuttle Columbia was destroyed in a devastating accident. Aboard the shuttle were seven canisters containing the nematode *Caenorhabditis elegans*, used in zero-gravity experiments. Five of the canisters were recovered, and seven weeks after the crash, four canisters contained living *C. elegans*. Life in space, the fiery re-entry, and a 600-mile-per-hour impact had little effect on the ultimate survivors.

Procedure 1

Macroanatomy of *Ascaris*

Materials

- ❏ Dissecting microscope or hand lens
- ❏ Dissecting tray
- ❏ Scalpel
- ❏ Dissecting pins
- ❏ Safety glasses
- ❏ Gloves
- ❏ Lab coat or apron
- ❏ Running water
- ❏ Male and female specimens of *Ascaris lumbricoides* obtained from your instructor
- ❏ Colored pencils

1 Procure the needed equipment and specimens.

2 Carefully rinse a female and a male *Ascaris* and place them on a dissecting tray with running water (Figs. 20.10 and 20.11). (The water keeps the specimens from drying out.) Observe female and male *Ascaris* with a hand lens or a dissecting microscope. Record your observations and sketches in the space provided.

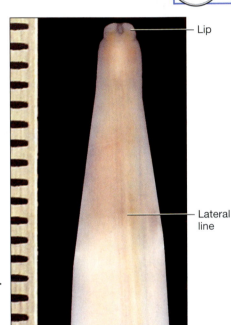

FIGURE **20.10** Head end of male *Ascaris* sp. (scale in mm).

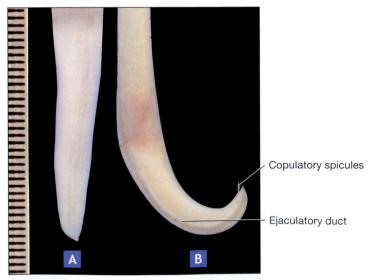

FIGURE **20.11** Posterior end of *Ascaris* sp. (scale in mm): (**A**) female; (**B**) male.

Ascaris lumbricoides

20

Ascaris lumbricoides

3 Position the specimens in the dissecting tray, ventral side down. Rotate the male to accommodate the curl of the tail. Gently grasp the female *Ascaris* with your thumb and forefinger of one hand. Use a scalpel or a pin to scrape a longitudinal mid-dorsal incision in the anterior third of the body through the cuticle and longitudinal muscles of the body wall. Perform the same procedure on the male *Ascaris*. Using the scalpel or pin, continue your incision to the mouth of both specimens. Carefully pin the cuticle of both specimens with dissecting pins into the dissecting tray, exposing the internal structures.

4 Carefully examine the internal structures of both the female and the male *Ascaris*, paying particular attention to those shown in Figures 20.12–20.14. Record your observations and labeled sketches in the space provided.

5 Clean your equipment and desktop thoroughly with the solutions provided by your instructor. Return the equipment, and discard your specimen as directed.

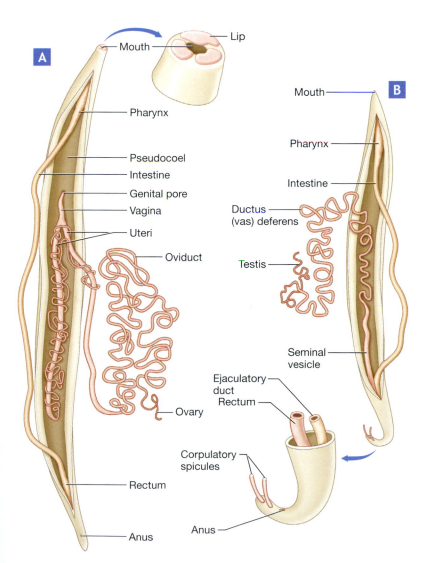

FIGURE **20.12** Internal anatomy of *Ascaris*: (**A**) female; (**B**) male.

Did you know . . . ?

Fit for a King (or Not)!

After intense research, in 2012 the battle-scarred remains of King Richard III of England were found beneath a parking lot in Leicester, England. The warrior king was brutally killed in the Battle for Bosworth Field in August 1485. His loyal soldiers buried him in a secret location near the church of Grey Friars to avoid further mutilation of his body. Richard III ruled England from 1482–1485 and was best known for his fierce disposition and severely misshapen and curved spine as the result of scoliosis.

Although the curved spine was a telltale sign that the body lying in the grave was Richard III, DNA testing proved his identity. To the researchers' surprise, throughout the soil where his pelvis once rested, vast numbers of eggs of the parasitic roundworm (*Ascaris lumbricoides*) were found. It is thought that the roundworms spread to Richard III by cooks who did not wash their hands after using the toilet, or by the use of human feces from towns to fertilize nearby fields, or by contaminated vegetables. Thus, even the body of a king can be taken over by ecdysozoan invaders.

FIGURE **20.13** Internal anatomy of male *Ascaris* sp. (scale in mm).

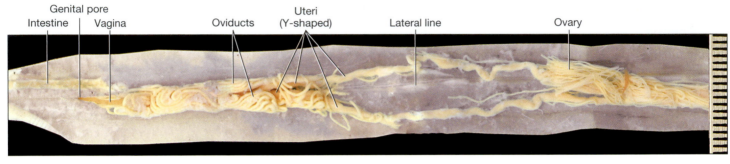

FIGURE **20.14** Internal anatomy of female *Ascaris* sp. (scale in mm).

Ascaris lumbricoides

Ascaris lumbricoides

✔ Check Your Understanding

1.1 Describe several characteristics of nematodes.

1.2 What two external characteristics did you observe that distinguish a female and male *Ascaris*?

1.3 It is always important to use good lab safety practices, but it is especially important when handling a female *Ascaris*. Why?

1.4 Why is *Caenorhabditis elegans* important to biology?

20

Phylum Arthropoda

Butterflies, fleas, centipedes, lobsters, barnacles, spiders, ticks, and fossil trilobites are a few examples of phylum Arthropoda, the joint-footed animals (Fig. 20.15). Presently, more than 1,000,000 species have been identified, and this number is expected to increase. No other phylum approaches the diversity or biomass of phylum Arthropoda (Fig. 20.16).

Arthropods have adapted successfully to every type of habitat and all modes of life. Living arthropods vary in size from 0.1 millimeters in length (*Stygotantulus stocki*, a tiny crustacean) up to the Japanese spider crab, *Macrocheira kaempferi*, which measures up to 4 meters in leg span. Eurypterids, a group of fossil chelicerates (described as giant carnivorous water scorpions), reached lengths exceeding 2 meters, and a giant dragonfly (*Meganeura monyi*) had a wingspan of nearly a meter. A relative of centipedes and millipedes (*Arthropleura*) grew to nearly 3 meters in length during the Carboniferous period. Without the arthropods, many ecosystems would literally collapse.

One of the most distinguishing characteristics of the arthropods is the presence of a **chitinous** exoskeleton. Some exoskeletons, such as those in a mosquito, are soft, and others are hard, such as those in crabs. The tough exoskeleton provides support and protection for the arthropods. Movement is possible because the exoskeleton usually is divided into plates over the body and cylinders around the appendages. To grow, arthropods must undergo ecdysis, or molting. The intervals between molts are termed **instars**.

Arthropods exhibit bilateral symmetry. Typically, they are divided into three body regions: **head, thorax**, and **abdomen**. Some species, such as spiders and shrimp, possess a **cephalothorax** and abdomen, and others have a head and trunk.

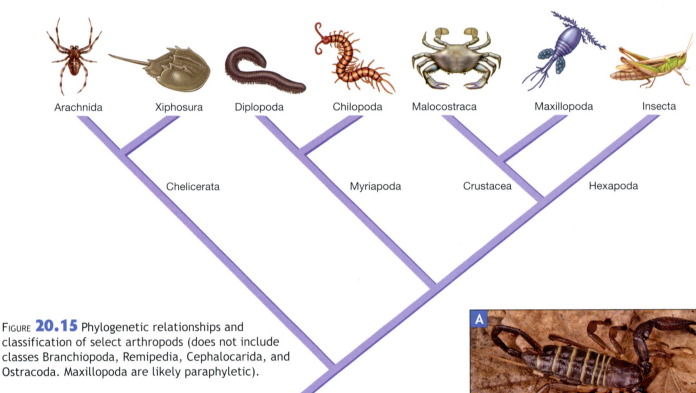

FIGURE **20.15** Phylogenetic relationships and classification of select arthropods (does not include classes Branchiopoda, Remipedia, Cephalocarida, and Ostracoda. Maxillopoda are likely paraphyletic).

FIGURE **20.16** Example arthropods include: (**A**) flat rock scorpion, *Hadogenes troglodytes*; (**B**) American giant millipede, *Narceus americanus*; (**C**) leafhopper, *Homalodisca vitripennis*.

Arthropods also exhibit segmentation, or metamerism; typically each somite has a pair of jointed appendages. In arthropods, segments may fuse, forming functional groups called **tagmata**. These appendages vary throughout the phylum and perform a variety of functions such as locomotion, food gathering, copulation, gas exchange, and egg brooding. Arthropods have complex muscular systems.

Typically, arthropods exhibit a great degree of cephalization. Also, they possess a variety of sense organs, such as **compound eyes** and **antennae**. The senses of touch, smell, hearing, balance, and chemical reception are generally well developed in arthropods. Arthropods are coelomates, although the coelom is reduced to portions of the reproductive and excretory systems. The major body cavity, the hemocoel, consists of blood-filled spaces within the tissues.

Arthropods have become adapted to a great number of dietary habits. Usually the appendages beside and just behind the mouth are used in feeding, and one pair of large appendages that flanks the mouth is used in biting and tearing. The digestive system is complete, consisting of a **foregut, midgut,** and **hindgut**. Osmoregulation in arthropods occurs in **Malpighian tubules** and glands. Respiration in arthropods takes place through the gills, tracheae, book lungs, and body surface.

Most arthropods have efficient **tracheae** (air tubes) that bring oxygen directly to the tissues and cells. The circulatory system is open; the heart consists of a pulsate vessel along the dorsal midline, and blood enters through the **ostia**. Generally, arthropods are dioecious and undergo internal fertilization. The females usually are **oviparous** (egg-laying). Many arthropods undergo **metamorphic changes**, including a larval form very different from the adult.

The arthropods have been divided into the following five distinct subphyla based on current research:

1. Subphylum Chelicerata includes horseshoe crabs, ticks, mites, sea spiders, spiders, scorpions, and extinct eurypterids.

2. Subphylum Myriapoda includes centipedes and millipedes.

3. Subphylum Trilobita includes extinct trilobites.

4. Subphylum Crustacea includes crabs, shrimp, lobsters, crayfish, barnacles, copepods, and pill bugs.

5. Subphylum Hexapoda includes insects, the largest subphylum.

Did you know . . .

What a Buzz!

Ever wonder why a mosquito buzzes by your ear? A butterfly is capable of four wing beats per second; a housefly and a bee are capable of 100 wing beats per second; the tiny fruitfly is capable of 300 wing beats per second; and, last but not least, the mosquito and midge are capable of wing beats in excess of 1,000 per second! That's why!

A Closer Look Trilobites

Like the dinosaurs, the trilobites, subphylum Trilobita, are icons of geologic time (Fig. 20.17). Trilobite-like organisms are perhaps the ancestors of all living arthropods. Trilobites first appeared in the Cambrian period approximately 540 million years ago, and eventually went extinct during the Permian period about 280 million years ago. Presently, the fossil record has yielded more than 5,000 known species, with many more yet to be named. Some fossil trilobites are actually shed exoskeletons. Trilobites ranged in size from less than a centimeter to more than 70 centimeters.

Trilobites at first showed little variation between the segments. In later arthropods, the segments tended to fuse and specialize. The term *trilobite* is derived from the pair of longitudinal grooves on the dorsal side, forming the left and right pleural lobes and the axial lobe. The body consisted of three tagmata: the cephalon (head region), the thorax (main body), and the pygidium (posterior fused segments). The appendages were biramous (having two distinct branches). Trilobites possessed a hypostome (hardened structure near the mouth used in feeding) instead of true mouthparts. Gills served as respiratory structures. Many trilobites could roll up similar to pill bugs today. ■

FIGURE **20.17** Examples of trilobites: (**A**) *Olenoides marjumensis;* (**B**) *Dicranurus elegans.*

Phylum Arthropoda, Subphylum Chelicerata

Members of subphylum Chelicerata are easily distinguished from other members of phylum Arthropoda. They are characterized by six pairs of appendages, including four pairs of **walking legs**, a pair of **chelicerae** (appendages behind the mouth used for feeding), and a pair of **pedipalps** (appendages that aid in chewing). Chelicerates have no antennae and no **mandibles**. The body has two regions: a cephalothorax and an abdomen. Three classes of chelicerates have been described in this manual: Merostomata and Arachnida.

Procedure 1

Macroanatomy of Class Merostomata

Materials

❏ Hand lens
❏ Dissecting tray
❏ Probe
❏ Horseshoe crab specimen provided by the instructor
❏ Colored pencils

Members of class Merostomata, commonly called the horseshoe crabs, have an ancient lineage not changed since the Triassic period (Fig. 20.18). Although intimidating at first look, these animals are harmless, living as omnivores and scavengers. These animals are the only gill-bearing chelicerates. Horseshoe crab blood contains a copper-rich blue molecule known as **hemocyanin**. Medical science uses horseshoe crab blood to detect and fight infections. Horseshoe crab larvae resemble trilobites. A common example is *Limulus polyphemus*, living in the Gulf of Mexico and the Atlantic Ocean.

1 Procure the needed equipment and specimen.

2 Place the horseshoe crab in a dissecting tray. Using a hand lens, observe the horseshoe crab, and record your observations and labeled sketch in the space provided.

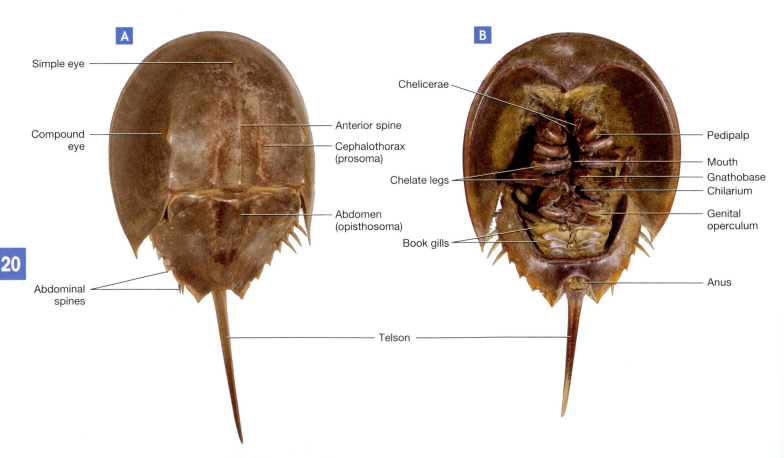

A
- Simple eye
- Compound eye
- Anterior spine
- Cephalothorax (prosoma)
- Abdomen (opisthosoma)
- Abdominal spines

B
- Chelicerae
- Pedipalp
- Mouth
- Gnathobase
- Chilarium
- Chelate legs
- Genital operculum
- Book gills
- Anus

Telson

FIGURE **20.18** Horseshoe crab, *Limulus* sp.: (**A**) dorsal view; (**B**) ventral view.

Limulus

Procedure 2

Macroanatomy of Class Arachnida

Materials

- ❏ Hand lens
- ❏ Dissecting tray
- ❏ Probe
- ❏ Harvestman specimens
- ❏ Spider specimens
- ❏ Scorpion specimens
- ❏ Tick specimens
- ❏ Colored pencils

Class Arachnida consists of 80,000 species of arthropods, such as spiders, scorpions, mites, ticks, and harvestmen ("daddy longlegs") (Fig. 20.19). Most arachnids are terrestrial. Many are predators and may possess fangs, claws, poison glands, and stingers. Generally, arachnids have sucking mouthparts with which they suck the fluids and soft tissues from the bodies of their prey.

Defining characteristics of arachnids include two **tagmata**, the **cephalothorax** and the **abdomen**. The cephalothorax consists of the head region and the thorax. The head usually has a pair of **pedipalps** (mandible-like structures), a pair of **chelicere** (maxilla-like structures), and four pair of **walking legs**. The claws, or **chelae** of scorpions are modified pedipalps. The abdomen consists of segments or **somites**.

1 Procure the needed equipment and specimen.

2 Your instructor will provide a number of arachnid specimens including: harvestmen, spiders, scorpions, and ticks.

3 Examine your specimens in a jar, or place it on a dissecting tray for study. Using a hand lens, record your observations and labeled sketch in the space provided, using Figure 20.20 as a reference.

FIGURE **20.19** Example arachnids: (**A**) jumping spider, *Phidippus regius*; (**B**) black widow, *Latrodectus hesperus*; (**C**) orb weaver, *Argiope trifasciata*; (**D**) cobalt tarantula, *Haplopelma lividum*; (**E**) tricolored scorpion, *Opistophthalmus ecristatus*; (**F**) harvestman, *Phalangium opilio*.

A

20

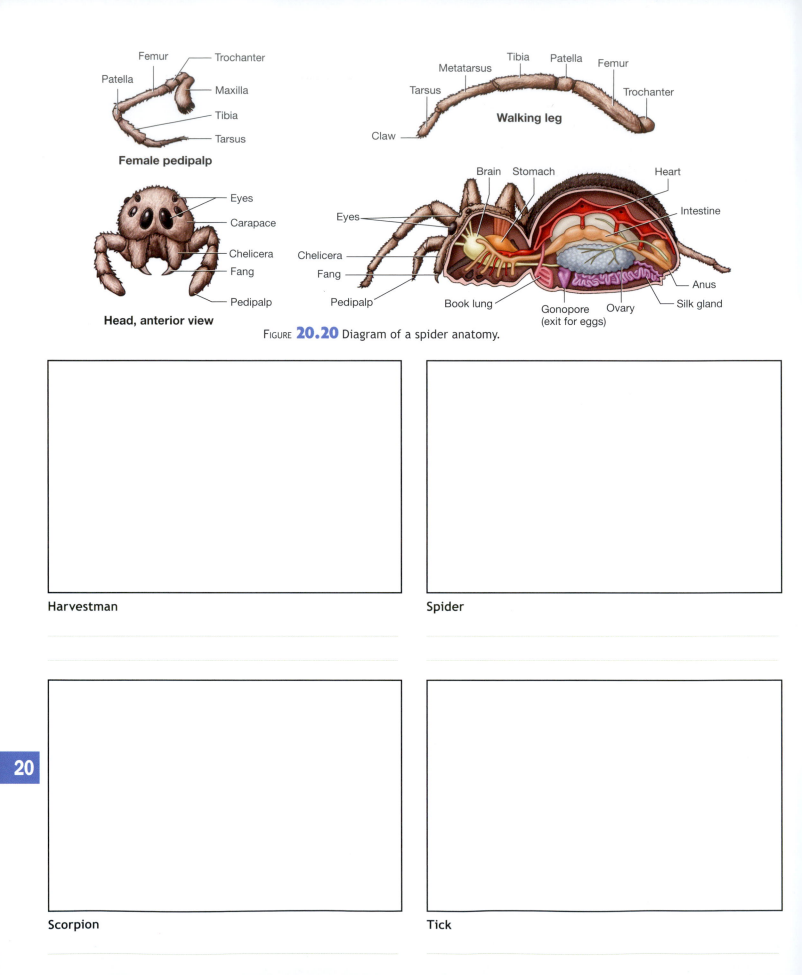

Female pedipalp

Femur
Trochanter
Patella
Maxilla
Tibia
Tarsus

Walking leg

Tibia
Patella
Metatarsus
Femur
Tarsus
Trochanter
Claw

Head, anterior view

Eyes
Carapace
Chelicera
Fang
Pedipalp

Brain   Stomach   Heart

Eyes
Chelicera
Fang
Pedipalp
Book lung
Intestine
Anus
Silk gland
Gonopore (exit for eggs)
Ovary

FIGURE **20.20** Diagram of a spider anatomy.

Harvestman

Spider

Scorpion

Tick

20

452   *Exploring Biology in the Laboratory: Core Concepts*

✔ Check Your Understanding

2.1 Members of the class Merostomata feature _____. (*Circle the correct answer.*)

  a. only gill-bearing chelicerates

  b. six pairs of appendages: walking legs, chelicerae, and telson

  c. presence of a heavy carapace on the dorsal surface

  d. both a and c

  e. all of the above

2.2 Match the following body regions and appendages with the correct subphyla or class.

_____ Trilobita

_____ Merostomata

_____ Arachnida

  A. Cephalothorax and abdomen, with a pair of chelicerae, a pair of sensory pedipalps, and four pairs of walking legs

  B. Cephalon, thorax, and pygidium, and biramous appendages

  C. Cephalothorax and abdomen, with a pair of chelicerae, a pair of pedipalps, and four pairs of walking legs (the last pair is modified with pincers)

2.3 Compare the anatomy and function of chelicerae and pedipalps of the horseshoe crab.

_____

_____

_____

_____

20

Order Araneae

Many people are terrified by arachnids, particularly spiders (Fig. 20.21). Despite their fearsome and perhaps creepy appearance, most spiders are harmless to humans. Presently, more than 40,000 species of spiders belong to order Araneae. Spiders can be found on every continent with the exception of Antarctica, and they live in a variety of habitats. These arachnids vary in size from 0.37 millimeters to the Goliath bird-eater spider (*Theraphosa leblondi*) with a leg span of more than 30 centimeters. By the way, it does not eat birds!

The body of a spider is compact, consisting of a cephalothorax joined to the abdomen by a pedicel. Spiders possess chelicerae modified to form fangs complete with venom sacs. They also have a pair of sensory pedipalps. Spiders do not have antennae. They have four pairs of walking legs. Most spiders liquefy the tissue of their prey, and then

FIGURE **20.21** Brown recluse, *Loxosceles apache*.

suck the liquid into the digestive system. Silk is produced by silk glands opening into two or three pairs of spinnerets. Some spiders weave complex and intricate webs. One interesting spider, *Mastophora dizzydeani* (named after the Baseball Hall of Fame pitcher Dizzy Dean), throws a sticky, ball-like web on the end of a thread to capture prey.

Book lungs, tracheae, or both structures are used in breathing. The heart of a spider is a long, slender tube in the abdomen that pumps the blood throughout the body. The blood returns to the heart through ostia from the hemocoel. In the excretory system, Malpighian tubules extract nitrogenous waste and uric acid from the hemocoel and release it into the cloaca. The nervous system is well developed. The majority of spiders have eight simple eyes located on the top anterior region of the cephalothorax. Spiders also have slit sensillae in the joints of their limbs that detect vibrations, especially when prey becomes entangled in a web.

Spiders are noted for their elaborate courtship behavior. Fertilized eggs develop in a cocoon that can be hidden or carried around by the female. In several species, the female devours the male shortly after mating. As a defense, tarantulas can release urticating hairs and deliver a painful bite.

Order Scorpiones

Approximately 2,000 species of scorpions have been classified and placed in order Scorpiones (Fig. 20.22). Many species can deliver a painful sting, and approximately 40 species of scorpions have a venomous sting potent enough to kill a human. Interestingly, some scorpions have fluorescent chemicals in their cuticle that fluoresce when exposed to black light. Most scorpions consume insects and are secretive and nocturnal, living as ambush hunters and stalkers.

FIGURE **20.22** Arizona desert hairy scorpion, *Hadrurus arizonensis*.

Order Acari

Order Acari consists of ticks, mites, and chiggers (red bugs). This order is large, with 40,000 described species and perhaps nearly 1 million total species (Fig. 20.23). The mites are especially hard to find because many are tiny and may live in obscure habitats. Members of order Acari can be found in a variety of habitats, including freshwater, marine, and many terrestrial. Many species are free-living, but several parasitic species exist.

This order is medically significant because many of its members are responsible for a host of diseases or serve as a vector for other diseases. Several species of minute mites known as house mites or dust mites, including *Dermatophagoides farina*, share our dwellings and cause allergies in some humans. These tiny mites consume dead skin cells, a major component of house dust.

Larval chiggers cause irritating dermatitis. The next time you fall victim to a chigger (*Trombicula* spp.), examine it with a dissecting microscope or a hand lens—they are strange-looking creatures! In humans, *Sarcoptes scabiei* can cause severe itching as they burrow beneath the skin, and they cause the dreaded disorder scabies. In dogs, *Demodex canis* is responsible for demodectic mange. Ticks can be an ectoparasite on a number of animals, including humans. Not only are ticks irritating, but they can be a vector for several diseases, including Lyme disease and Rocky Mountain spotted fever.

FIGURE **20.23** Wood tick, *Dermacentor* sp.

20

Phylum Arthropoda, Subphylum Myriapoda

Subphylum Myriapoda contains approximately 13,000 species of centipedes (Fig. 20.24), millipedes (Fig. 20.25), and their relatives. All members of this subphylum are terrestrial. The myriapods possess one pair of antennae and a pair of simple eyes. Similar to insects, myriapods breathe through their spiracles connected to the tracheal system. Malpighian tubules are the sites of excretion in the myriapods. The brain is poorly developed in this subphylum. Females lay eggs that hatch into young resembling the adults but with fewer segments and legs. The four classes of myriapods are:

1. Chilopoda (centipedes);

2. Diplopoda (millipedes);

3. Symphyla (pseudocentipedes; not covered in this manual); and

4. Pauropoda (soft-bodied myriapods; not covered in this manual).

FIGURE **20.24** Vietnamese centipede, *Scolopendra subspinipes*.

FIGURE **20.25** Sonoran desert millipede, *Orthoporus ornatus*.

Centipedes are commonly called "hundred-leggers" but the majority of species possess approximately 30 pair of legs. Centipedes have one pair of dominant legs per segment. These flattened, elongate arthropods have a chitinous exoskeleton and range in length from 4 millimeters to 30 centimeters. Of the 3,000 species most centipedes are dull brown in color and reside in the underbrush or perhaps your house. Centipedes are carnivorous and possess venomous claws, which actually are a pair of legs, modified to capture and kill prey.

Millipedes are commonly known as "thousand-leggers" and can have up to 200 pairs of legs. The millipede body is dome shaped and can consist of up to 100 segments with two pair of legs per segment. These arthropods range in size from 2 millimeters to 28 centimeters. Of the nearly 10,000 species the majority of millipedes live under logs and stones. Millipedes are herbivores or detritivores (consume dead plant and animal matter). When disturbed, millipedes may roll into a ball and emit a foul-smelling odor from repugnatorial glands. Larval millipedes resemble centipedes having one pair of legs per segment.

Comparison of Centipedes and Millipedes

Materials
- ❏ Dissecting microscope or hand lens
- ❏ Petri dish
- ❏ Probe
- ❏ Specimens of several species of centipedes and millipedes
- ❏ Colored pencils

1 Procure the needed equipment and specimens.

2 Observe your specimen in a jar, or place it on a petri dish. Thoroughly examine your specimens using a hand lens or dissecting microscope. Pay particular attention to the mouthparts and legs. Record your observations and labeled sketches in the space provided.

Centipedes

Millipedes

✔ Check Your Understanding

3.1 Label the following characteristics as either belonging to centipedes or millipedes, using the following key:

C = Centipedes   M = Millipedes

_____ Have one pair of legs per segment.

_____ Have dome-shaped body.

_____ Have two pairs of legs per segment.

_____ Have flattened, elongated bodies.

_____ Are herbivorous or detritivorous.

_____ Possess pair of venom claws to capture and kill prey.

3.2 Compare the dietary habits of centipedes and millipedes.

_____

_____

_____

_____

20

Phylum Arthropoda, Subphylum Crustacea

Fried shrimp, boiled crayfish, crab gumbo, steamed lobster—let's hear it for the crustaceans! Most of the 68,000 members of subphylum Crustacea are marine, but there are several freshwater and terrestrial forms (Fig. 20.26). Most crustaceans are free-living, although some parasitic species exist. Crustaceans possess a hard chitinous and calcified cuticle that must be shed (ecdysis) for the animal to grow. Most crustaceans have two tagmata—the cephalothorax and the abdomen. The number of body segments ranges from 6 to 32 segments, each with a pair of appendages. The more advanced forms are thought to have fewer segments.

Many crustaceans have a hardened carapace that covers the cephalothorax and may extend past the head in the form of a rostrum. Crustaceans are different from the other arthropods because they have two pairs of anterior antennae. The first pair is called the antennules and the second pair the antennae. The antennules are shorter and usually are associated with the sense of smell. Antennae are longer and are also sensory structures. Crustaceans have modified mouthparts, such as mandibles, **maxillae**, and **maxillipeds**, with important feeding and sensory functions.

In some species, such as crabs, crayfish, and lobsters, the first pair of walking legs, chelipeds, are modified to form powerful claws. Other walking legs are biramous (having two branches). In addition to walking legs, some crustaceans have feathery **swimmerets**, or **pleopods**, in the abdominal region. The tail of a crustacean consists of a nonsegmented telson flanked by fanlike **uropods**.

Crustaceans have an open circulatory system. If present, the heart resides in a blood-filled sinus and communicates by paired ostia (valves) to the body. In many crustaceans, food is broken down physically in the large **cardiac stomach** and also in the smaller **pyloric stomach**, part of the foregut. In the majority of crustaceans, excretion of nitrogenous wastes occurs by diffusion across the gills. Some species have **green glands**, which are excretory organs that rid the organism of water and are located near the base of the antennae.

Although several monoecious and parthenogenic species exist, most crustaceans are dioecious. Fertilized eggs may be released into water or carried by the female crustacean. Larval crustaceans are variable and may include the nauplius, zoea, and megalops stages.

Biologists recognize six classes of crustaceans. The four large classes are:

1. Class Branchiopoda (brine shrimp, tadpole shrimp, and fairy shrimp);
2. Class Ostracoda (ostracods and seed shrimp);
3. Class Maxillopoda (barnacles, tongue worms, fish lice, and copepods); and
4. Class Malacostraca (lobsters, crabs, shrimp, crayfish, and pill bugs).

Classes Remipedia and Cephalocarida include rather obscure crustaceans.

FIGURE **20.26** Examples of crustaceans: (**A**) peppermint shrimp, *Lysmata wurdemanni*; (**B**) hermit crab, *Coenobita clypeatus*; (**C**) blue crab, *Callinectes sapidus*; (**D**) Sally Lightfoot crab, *Grapsus grapsus*; (**E**) red reef lobster, *Enoplometopus* sp.

20

Class Branchiopoda

About 10,000 species of class Branchiopoda have been identified, of which fairy shrimp and brine shrimp are common members. These shrimp do not have a carapace. Brine shrimp (*Artemia salina*) are sold as novelty items packaged as sea monkeys. Brine shrimp eggs are metabolically inactive and in a state of cryptobiosis when they are purchased. After being placed in salt water, the brine shrimp hatch within a few hours, forming larvae called nauplii. The larvae soon mature and develop into those frolicking sea monkeys. The best-known branchiopod is the tiny water flea, *Daphnia* spp., commonly found in ponds and other freshwater environments (Fig. 20.27).

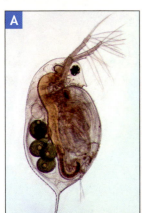

FIGURE **20.27** (**A**) Water flea, *Daphnia* sp.; (**B**) brine shrimp, *Artemia salina*; (**C**) tadpole shrimp, *Triops longicaudatus*.

Class Maxillopoda

Approximately 12,000 species of class Maxillopoda have been described. This diverse class includes the subclasses copepods and barnacles. As a general rule, its members have five cephalic segments, six thoracic segments, four abdominal segments, and a telson. Most maxillopods lack appendages. Some large barnacles (such as *Balanus nubilus*) can exceed 7.5 centimeters in length.

Subclass Copepoda

Copepods (subclass Copepoda) live in both freshwater and marine environments. Perhaps they have the largest animal biomass on Earth. Most copepods are less than a millimeter in length and have an elongated body that tapers toward the posterior. Early scientists referred to some copepods as cyclops because they have a single compound median eye in the adult form (Fig. 20.28). Copepods lack a carapace. In copepods, the antennules may be longer than the appendages. Copepods are a major component of zooplankton, important in the food chain. Several species of copepods are parasites on invertebrates, fishes, sharks, and marine mammals.

FIGURE **20.28** Cyclops copepod, *Abyssorum tatricus*.

Subclass Cirripedia

Charles Darwin devoted much of his life to studying barnacles, subclass Cirripedia. Biologists have described approximately 1,200 species of marine barnacles (Fig. 20.29). Barnacles have two larval stages. The nauplius larvae have a single median eye and are free-swimming. The cyprid larvae attach to a suitable substrate headfirst and develop into juvenile barnacles.

As adults, barnacles are sessile, and some species may be attached to a substrate by a stalk (such as the gooseneck barnacle) or cemented to a substrate (such as the bay barnacle). Most barnacles are enclosed in calcareous plates. Barnacles have no abdomen and a reduced head. When feeding, the top plates open, and feathery cirri filter food from the water. Some barnacles are parasitic on crabs. If barnacles grow in great numbers on a ship's hull, they can slow the ship, and these fouling organisms have to be removed periodically.

FIGURE **20.29** Gooseneck barnacles, *Pollicipes polymerus*.

20

Procedure 1

Macroanatomy of Class Malacostraca

Materials
- Dissecting microscope or hand lens
- Dissecting tray
- Petri dish or sheet of paper
- Preserved specimens of shrimp, prawns, hermit crabs, blue crabs, fiddler crabs, lobsters, pill bugs and crayfish
- Colored pencils

More than 22,000 members of class Malacostraca have been described. They are found worldwide inhabiting terrestrial, freshwater, and marine environments. The malacostracans usually possess three tagmata: a head with five segments, a thorax with approximately eight segments, and an abdomen with usually six segments.

Members of class Malacostraca are diverse, and their classification is lengthy and presently undergoing reorganization. Examples of this class include: lobsters, shrimp, crabs, crayfish, and pill bugs. Pill bugs (*Armadillidum* sp.) are a common representative of Class Malacostraca (Fig. 20.30). These curious animals are often found under stones and hidden in leaf matter and roll into a ball when threatened.

1 Procure the needed equipment and specimens.

2 Examine the preserved specimens in their jars and, if possible, place them on a dissecting tray for closer examination. After examining the specimens closely with a dissecting microscope and hand lens, place your observations and sketches in the spaces provided.

FIGURE **20.30** (A) Pill bug, *Armadillidum* sp.; (B) sea slater, *Ligia italica*.

Specimen 1

Specimen 2

Specimen 3

Specimen 4

Specimen 5

3 Place a living pill bug in a petri dish or a sheet of paper and record its locomotion and behavior.

Description of pill bug locomotion.

Description of pill bug behavior.

Procedure 2

Crayfish Dissection

Materials

- ❏ Dissecting tray
- ❏ Hand lens
- ❏ Safety glasses
- ❏ Gloves
- ❏ Lab coat or apron
- ❏ Scalpel
- ❏ Forceps
- ❏ Water
- ❏ Preserved crayfish
- ❏ Colored pencils

1 Procure the specimen and the needed equipment.

2 Rinse the preservative off of the crayfish with running water. Place the crayfish on its ventral side in the dissecting tray. Locate the external features of the crayfish shown in Figures 20.31–20.38. Record your observations and detailed sketch in the space provided.

3 With your hand lens, closely examine the mouthparts and appendages. Include your descriptions and detailed sketch in the space provided.

4 Determine the sex of the crayfish. Compare it with the specimen of the other sex that may be found at another lab station. Compare your male and female crayfish with Figure 20.38. Record your observations and sketch in the space provided.

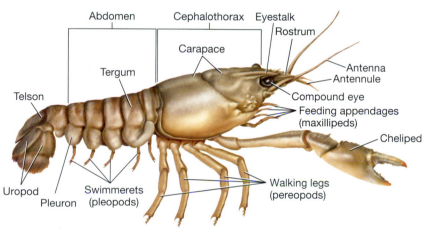

FIGURE **20.31** Crayfish.

Crayfish dissection

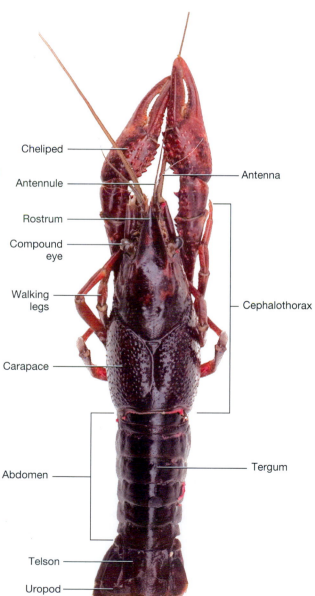

FIGURE **20.32** Dorsal view of crayfish, *Procambarus* sp.

20

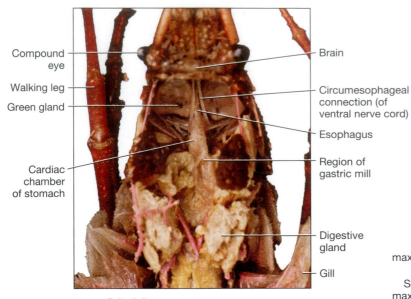

FIGURE **20.33** Dorsal view of oral region of crayfish, *Procambarus* sp.

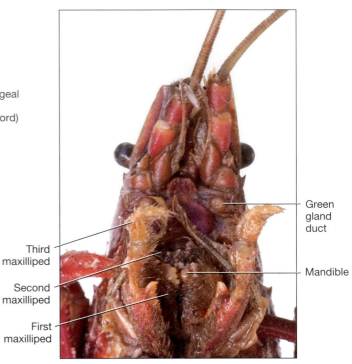

FIGURE **20.34** Ventral view of oral region of crayfish, *Procambarus* sp.

Crayfish dissection

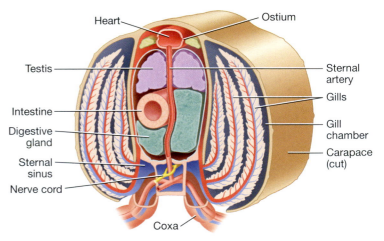

FIGURE **20.35** Transverse section of adult male crayfish.

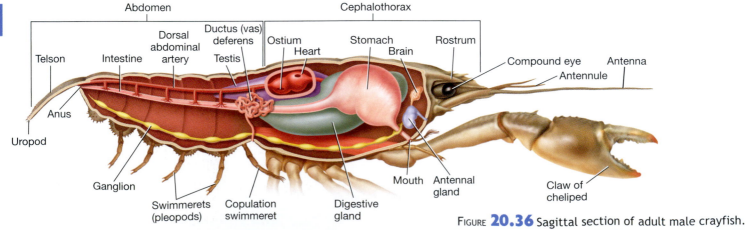

FIGURE **20.36** Sagittal section of adult male crayfish.

5 With a scalpel and forceps, carefully remove the carapace from the rostrum to the abdomen. Also remove the dorsal portion of three abdominal segments. Using Figure 20.38, find the major internal structures of the crayfish. Detach one of the walking legs of the crayfish, and notice the small gill attached to the leg. Place your observations and sketch in the space provided.

6 Dispose of all crayfish body parts in the waste container provided by the instructor. Clean the equipment and station thoroughly. Return the clean and dry equipment to the proper location.

Crayfish dissection

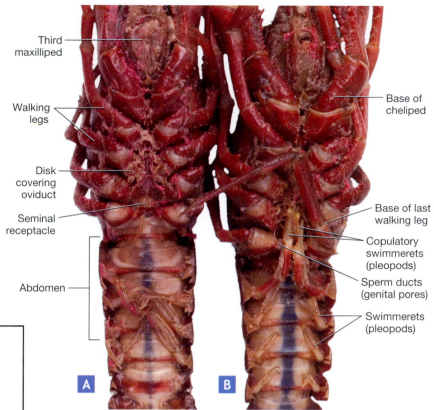

FIGURE **20.37** Ventral views of crayfish, *Procambarus* sp.: (**A**) female; (**B**) male.

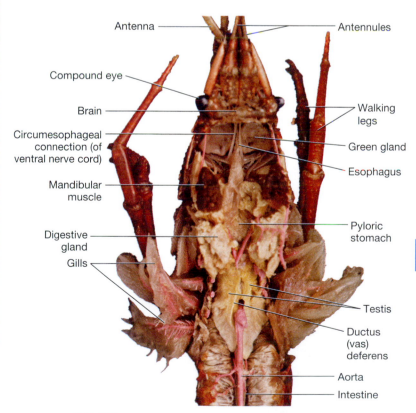

FIGURE **20.38** Dorsal view of anatomy of crayfish, *Procambarus* sp.

Crayfish dissection

_____

_____

_____

_____

✔ Check Your Understanding _____

4.1 Describe several general characteristics of malacostracans.

_____

_____

_____

4.2 Provide five examples of malacostracans.

_____

_____

_____

_____

_____

4.3 After observing your crayfish specimen, what external differences did you note between the male and female?

_____

_____

4.4 What structures are present in the circulatory system of the crayfish that would indicate an open circulatory system?

_____

_____

_____

4.5 Crayfish use chelipeds for defense and food gathering. (*Circle the correct answer.*)

True / False

Phylum Arthropoda, Subphylum Hexapoda

Nearly one-third of all animals on Earth are members of subphylum Hexapoda, which has two classes:

1. Class Entognatha, which includes order Collembolla (springtails); and

2. Class Insecta (ants, moths, fleas, etc.) Hexapods have six pairs of uniramous legs.

They also have three tagmata: a head, a thorax, and an abdomen.

The study of insects is called entomology. Class Insecta is the most diverse and abundant of all of the arthropods (Fig. 20.39). This class also represents the largest group of animals on Earth, with more than 1 million recognized species, and is estimated to increase manyfold in the future. Beetle species alone number more than 350,000.

The insects have undergone great adaptive radiation and have occupied virtually every type of habitat. Basically, insects differ from other arthropods by having three pairs of legs and, in the majority of species, two pairs of wings in the thoracic region. Insects vary in body shape, color, and size. Perhaps the smallest insect (a fairy fly from Costa Rica) measures about 0.13 millimeters in length, and one of the largest insects, the Titan beetle (from the Amazon), measures more than 17 centimeters in length.

The insect tagmata include the head, thorax, and abdomen (Fig. 20.40). The cuticle of each body segment is made of four plates, sclerites, connected to each other by a flexible hinge joint. The heads of insects usually have a pair of large compound eyes, three ocelli, and two antennae. The mouthparts are highly variable in insects. The thorax usually is divided into the **prothorax**, **mesothorax**, and **metathorax** (each with a pair of legs), which vary among insects. In most insects, the mesothorax and metathorax have a pair of wings. The abdomen has 9 to 11 segments. The end of the abdomen bears the external sex organs.

The digestive system consists of a foregut (mouth, salivary gland, esophagus, crop, and gizzard), a midgut (stomach and gastric caeca), and a hindgut (intestine, rectum, and anus). The feeding habits of insects vary from predaceous (praying mantis) to phytophagous

(grasshopper), to saprophagous (dung beetles) to parasitic (fleas). In insects, a tubular heart located in the pericardial cavity pumps hemolymph through the dorsal aorta. (The hemolymph has little to do with oxygen transport.) The tracheal system in insects is an efficient respiratory apparatus used primarily for breathing air. Insects, like spiders, possess a unique excretory system consisting of Malpighian tubules that operate in conjunction with specialized glands in the wall of the rectum for excretion.

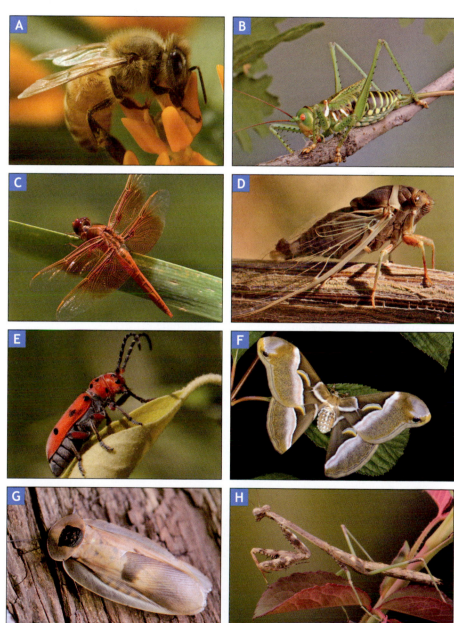

FIGURE **20.39** Examples of insects: (**A**) common honeybee, *Apis mellifera*; (**B**) eastern lubber grasshopper, *Romalea microptera*; (**C**) flame-skimmer dragonfly, *Libellula saturata*; (**D**) cicada, *Diceroprocta apache*; (**E**) milkweed beetle, *Tetraopes tetraophthalmus*; (**F**) cynthia moth, *Samia cynthia*; (**G**) giant cockroach, *Blaberus giganteus*; (**H**) Carolina mantis, *Stagmomantis carolina*.

20

A

Abdomen   Thorax   Head

— Antenna

— Compound eye

— Three pairs of legs

FIGURE **20.40** Basic insect anatomy.

The nervous system of insects is similar to the nervous systems of crustaceans. Insects have a variety of keen sense organs, primarily microscopic, located in the body wall. Sensilla are responsible for auditory reception. Interestingly some insects, such as grasshoppers, have tympanic organs in their legs. Many insects have keen chemoreception. Although chemoreceptors usually are located on mouthparts, they may appear on antennae (as in ants and bees) or even legs (as in butterflies and moths). Insect eyes follow two general plans: simple eyes are found in larvae, nymphs, and some adult insects. Compound eyes are found in the majority of species of insects. The compound eye aids insects to see simultaneously in nearly all directions. Insects exhibit a variety of complex behaviors and communication skills.

Insects are dioecious. Many species practice internal fertilization. Most insects are oviparous (egg-laying), but some **viviparous** (live-bearing) species exist. Female insects may possess an ovipositor that lays eggs. The stinger of insects is a modified ovipositor associated with venom glands. The insect egg exhibits early development. The hatching young insect escapes the egg through a variety of means. In gradual (incomplete) metamorphosis the young resemble the adults, but they are smaller and have different body proportions. In complete metamorphosis, the larvae look different from the adult and generally have different food requirements. Approximately 88% of insects undergo a complete metamorphosis that consists of three stages: the larva, the **pupa**, and the adult (Figs. 20.41 and 20.42).

There are more than 30 orders of insects, and entomologists have developed several means of classification. The orders listed in Table 20.1 are considered the user-friendly versions of some of the common orders, along with examples of each.

B

C

D

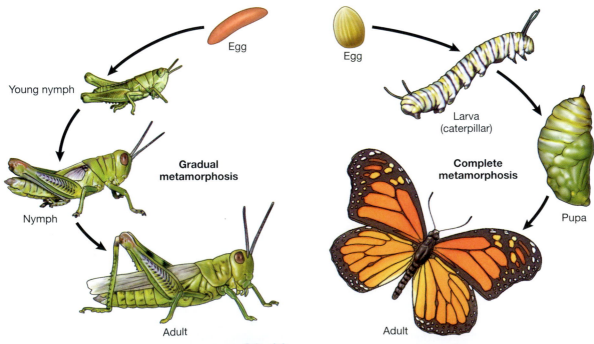

Egg

Young nymph

Gradual metamorphosis

Nymph

Adult

Egg

Larva (caterpillar)

Complete metamorphosis

Pupa

Adult

FIGURE **20.41** Insect development.

FIGURE **20.42** Developmental stages of the monarch butterfly, *Danaus plexippus*, include: (**A**) egg; (**B**) larval stage; (**C**) chrysalis; (**D**) adult.

20

TABLE **20.1** Major Orders of Insects

Order		Examples
Diptera		Housefly, crane fly, deerfly, fruit fly, gnat, mosquito, love bug, robber fly
Orthoptera		Grasshopper, cricket, mole cricket, locust
Coleoptera		Beetle, ladybug
Siphonaptera		Flea
Lepidoptera		Butterfly, moth
Dermaptera		Earwig
Odonata		Dragonfly, damselfly
Isoptera		Termite
Hymenoptera		Bee, wasp, hornet, ant, cicada killer
Blattodea		Roach
Mantodea		Praying mantis
Phasmatodea		Walking stick
Neuroptera		Lacewing, ant lion, dobsonfly
Homoptera		Cicadas, leafhopper, plant-hopper, scale insects, aphids, mealybug, spittlebug
Hemiptera		Stinkbug, giant water bug, assassin bug, water strider
Thysanura		Silverfish, bristletails
Ephemeroptera		Mayfly
Phthiraptera		Head louse, body louse, pubic louse, bird louse, poultry louse
Plecoptera		Stonefly

20

Materials

- ❏ Dissecting microscope
- ❏ Hand lens
- ❏ Dissecting tray
- ❏ Preserved and living specimens of various insects
- ❏ Colored pencils
- ❏ Camera

1 Procure the needed equipment and specimens. Consider photographing this activity.

2 Examine the preserved and living specimens of the sample insects. After examining the specimens closely with a dissecting microscope or hand lens, place your observations and sketches in the space provided below. Attempt to place them in their proper order as in Table 20.1.

Specimen 1 _____

Specimen 2 _____

Specimen 3 _____

Specimen 4 _____

Specimen 5 _____

Specimen 6 _____

20

Grasshopper Dissection

Materials

- ❏ Dissecting tray
- ❏ Hand lens
- ❏ Safety glasses
- ❏ Gloves
- ❏ Lab coat or apron
- ❏ Scalpel
- ❏ Forceps
- ❏ Scissors
- ❏ Water
- ❏ Preserved grasshopper
- ❏ Colored pencils

1 Procure the needed equipment and specimen.

2 Rinse the preservative off of the grasshopper with running water. Place the specimen on its ventral side in the dissecting tray. Using a hand lens, locate the external features of the grasshopper shown in Figures 20.43 and 20.44.

FIGURE **20.43** Anatomy of the grasshopper: (**A**) male; (**B**) female.

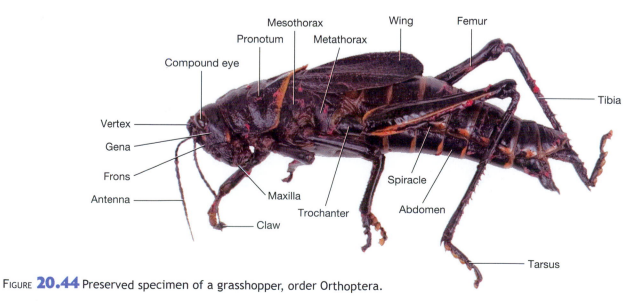

FIGURE **20.44** Preserved specimen of a grasshopper, order Orthoptera.

3 Using a hand lens, closely examine the head, mouthparts, and appendages of the grasshopper. Compare your specimen with the head region illustrated in Figure 20.45.

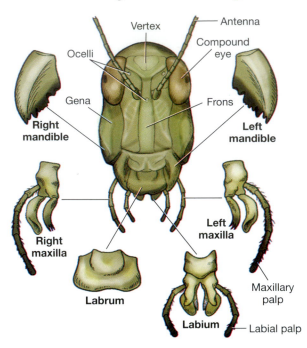

FIGURE **20.45** Head and mouthparts of a grasshopper.

4 Referring to Figure 20.46, determine the sex of the grasshopper. Compare it with a specimen of the opposite sex that may be at another lab station. Record your observations and sketches in the space provided.

5 With a scalpel and forceps, carefully remove the carapace from the rostrum to the abdomen. Also remove the dorsal portion of three abdominal segments. Referring to Figures 20.46 and 20.47, find the major internal structures of the grasshopper. Place your observations and sketch in the space provided.

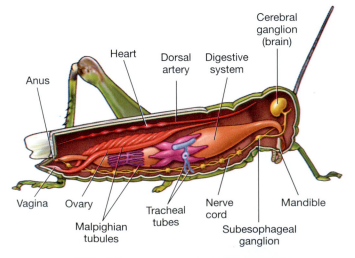

FIGURE **20.46** Internal anatomy of a grasshopper.

Grasshopper dissection

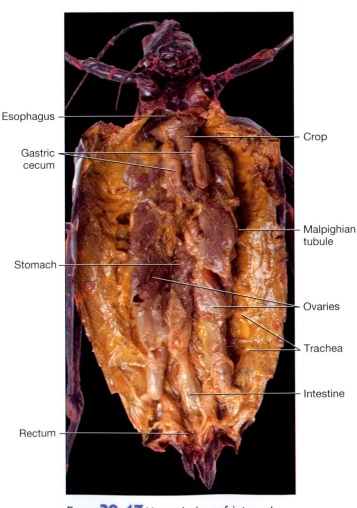

FIGURE **20.47** Ventral view of internal anatomy of grasshopper.

6 Place the grasshopper in the dissecting tray, ventral side up. Using scissors, cut through the exoskeleton on the ventral side from the head to the posterior-most end of the abdomen. Using the forceps, pull the sides apart. Pin each side of the insect to the dissecting pan. Locate the internal structures shown in Figures 20.46 and 20.47. Record your observations and sketch in the space provided.

7 Dispose of all grasshopper body parts in the waste container provided by the instructor. Clean the equipment and station thoroughly. Return the clean and dry equipment to the proper location.

Grasshopper dissection

Grasshopper dissection

✔ Check Your Understanding

5.1 What are three characteristics of the class Insecta?

5.2 How do body regions of the grasshopper compare with the crayfish?

5.3 List in order the structures found in the foregut, the midgut, and the hindgut of the grasshopper.

Did you know . . .

So That's the Answer!

Have you ever been called a nitpicker? What's a nitpicker anyway? A "nit" is the egg of a louse attached to the shaft of the hair of the host mammal. The term nitpicking commonly refers to the detailed and meticulous effort that has to be undertaken to locate the nit and remove it from the hair. Tiny combs were found in the wreckage of the Civil War ironclad, the USS Cairo, in Vicksburg, Mississippi. Archaeologists determined these combs were used to remove lice from the beards of sailors.

Did you know . . .

A Game of Deception

The flickering of fireflies on a quiet summer evening seems so peaceful. Well—maybe not. A game of deception may be going on beneath your eyes. Male fireflies fly around at night using specific blinking patterns to gain the attention of female fireflies residing on the ground. Females return a sequence of flashes that signal acceptance of the male. Unfortunately, some species of female fireflies can imitate the acceptance pattern of other females. When the unsuspecting male approaches the deceptive female, he is eaten by that female.

20

MYTHBUSTING

Crawling Critters

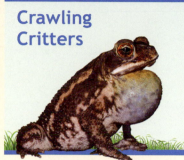

Debunk each of the following misconceptions by providing a scientific explanation. Write your answers on a separate sheet of paper.

1 Earwigs crawl in your ear and eat your brain.

2 Beware! Spiders can lay their eggs under the skin of humans.

3 People cannot get worms from cats and dogs.

4 Camel spiders can be over a foot long and run 25 mph.

5 All scorpions are deadly.

1 Which of the following are **not** found in the ecdysozoans? (*Circle the correct answer.*)

a. Growth requires shedding of the protective cuticle, or exoskeleton.

b. Gills, trachea, and lungs are present in various species.

c. All have an open circulatory system.

d. Fertilization occurs internally.

e. Both c and d.

2 Explain the taxonomical organization of the ecdysozoans.

3 Which of the following are characteristics found in the phylum Nematoda? (*Circle the correct answer.*)

a. Long cylindrical body with a thick, protective, nonliving cuticle.

b. Incomplete digestive tract with a long straight intestine, which serves as a site for digestion and absorption.

c. Fluid-filled pseudocoel, which serves as a hydroskeleton.

d. Both a and c.

e. All of the above.

4 Name five characteristics of phylum Arthropoda.

5 Describe and provide specific examples of the subphylum Chelicerata.

6 Label Figure 20.48.

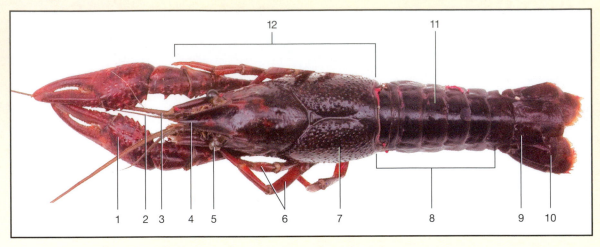

FIGURE **20.48**
Crayfish, dorsal view.

1. _____
2. _____
3. _____
4. _____
5. _____
6. _____

7. _____
8. _____
9. _____
10. _____
11. _____
12. _____

7 Explain the body organization of a typical crustacean.

_____

_____

8 Label Figure 20.49.

1. _____
2. _____
3. _____
4. _____
5. _____
6. _____
7. _____
8. _____

9. _____
10. _____
11. _____

FIGURE **20.49** Male grasshopper anatomy.

Animal Planet IV
Understanding the Deuterostomes

21

I looked into the jar; I don't know what they are; Some have scales and some lay eggs;
and some look pretty bizarre; I still don't know what they are; they must be vertebrates;
With the same subphylum trait; The farther I move from left; the closer I get to myself.

— Robert Everett, "Bio-Notes Vertebrate Songs" (1954–present)

Just wondering...

Consider the following questions prior to coming to lab, and record your answers on a separate piece of paper.

1 How did birds attain flight?
2 How do sharks detect minute movements in the water?
3 What is the most recent research on dolphin intelligence?
4 Why does Australia have such cool monotremes and marsupials?
5 What is the link between dinosaurs and birds?

Objectives

At the completion of this chapter, the student will be able to:

1. Provide examples of and describe the characteristics of deuterostomes and compare with protostomes.
2. Describe the basic characteristics and distinguish among the classes of phylum Echinodermata.
3. Dissect a sea star and identify specific external and internal structures.
4. Describe the defining characteristics shared by all organisms in phylum Chordata.
5. Provide examples of, describe, and compare and contrast the basic characteristics of subphylum Vertebrata (Craniata).
6. Identify the skeletal features of a bird, a frog, a snake, and a cat.

Perhaps we are more familiar with the deuterostomes than with any other group of animals. They include, among others, sea stars, sand dollars, sharks, frogs, turtles, birds, and humans (Fig. 21.1). All deuterostomes are triploblastic coelomates that undergo radial cleavage, and during embryological development, their blastopore forms the anal opening. The three phyla of deuterostomes are phylum Echinodermata (sea stars), phylum Hemichordata (acorn worms), and phylum Chordata (fishes and **tetrapods**). Figure 21.2 is a cladogram illustrating the relationship of the deuterostomes to the other animal phyla.

Phylum Echinodermata consists of nearly 7,000 species of organisms called the "spiny-skinned" animals. Echinoderms range in size from brittle stars measuring less than 1 centimeter in length or diameter to large sea cucumbers exceeding 2 meters in length. Except for some rare, brackish-water species, echinoderms inhabit marine environments.

Any study of the deuterostomes should mention phylum Hemichordata. The hemichordates (half chordates) are marine organisms previously placed in phylum Chordata, but now they reside in their own phylum. A classic example of a member of this phylum is the acorn worm.

Phylum Chordata consists of nearly 57,000 species. Three distinct subphyla of phylum Chordata have been recognized. Subphylum Urochordata is a nonvertebrate subphylum that includes the tunicates (sea squirts) and salps. Another nonvertebrate subphylum is Cephalochordata, which includes the sea lancelets. The largest subphylum of chordates is Vertebrata (Craniata), which includes fishes, amphibians, reptiles, birds, and mammals.

Chapter Photo
Leafy seadragon, *Phycodurus eques.*

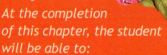

Figure **21.1** Representative deuterostomes: (**A**) red serpent star, or brittle star, *Ophioderma squamoisissinum*; (**B**) sea lancelet, or amphioxus, *Branchiostoma* sp; (**C**) coho or silver salmon, *Oncorhynchus kisutch*; (**D**) spotted salamander, *Ambystoma maculatum*; (**E**) Galapagos tortoise, *Chelonoidis nigra*; (**F**) hummingbird, *Eugenes spectabilis*; (**G**) polar bear, *Ursus maritimus*.

21

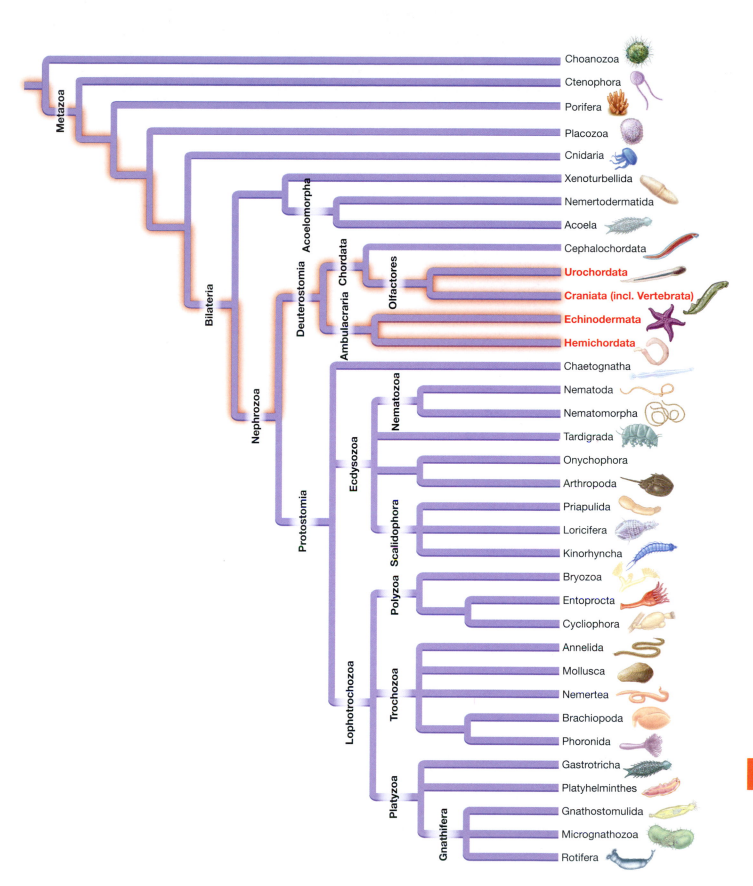

FIGURE **21.2** Phylogeny and classification of the Metazoans (multicellular animals). The phyla and subphyla shown in red are discussed in this chapter.

Phylum Echinodermata

The echinoderms are organisms with **pentamerous** (five-pointed) **radial symmetry** (Fig. 21.3). Echinoderms have a body wall containing an **endoskeleton** of small, calcareous **ossicles** that include surface **spines**. The spines can have jaw-like pincers, or **pedicellariae**, that discourage barnacles and other fouling organisms from settling on their surfaces. In some species, such as sea urchins, the spines are associated with poison glands.

The echinoderms possess a unique **water vascular system** of canals and appendages that function in locomotion, feeding, sensory reception, and gas exchange. The **tube feet** of sea stars are powered by this hydraulic system. Adult echinoderms lack a head, a brain, and segmentation. The digestive system of most echinoderms is complete. The circulatory system is reduced and radiates in five directions, circulating colorless blood. Minute **gills** that protrude from the coelom are responsible for respiration in some species. Some echinoderms, such as the sea cucumbers, use cloacal structures called **cloacal trees** for respiration.

The nervous system basically consists of nerves in a ring around the mouth extending radially outward from inside the body. Echinoderms do not have excretory organs. Echinoderms are dioecious. After fertilization, the zygote becomes a bilaterally symmetrical ciliated larva, or **bipinnaria**, that passes through several stages before becoming a radially symmetrical adult. Many species, such as the sea star, have great regenerative powers. Some species can purposely detach a limb to escape a predator, a process known as **autotomy**.

Phylum Echinodermata consists of five distinct living classes: Crinoidea, Asteroidea, Ophiuroidea, Echinoidea, and Holothuroidea.

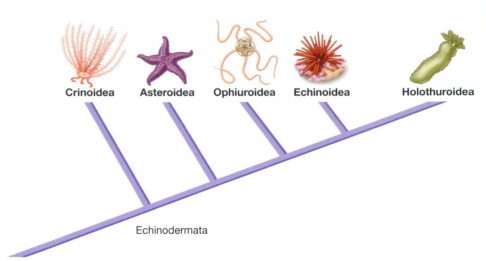

Crinoidea   Asteroidea   Ophiuroidea   Echinoidea   Holothuroidea

Echinodermata

FIGURE **21.3** Phylogenetic relationships and classification of Echinodermata.

called **pinnules**. The arms and the calyx constitute the **crown** of a crinoid. Sessile species possess a **stalk** of ring-shaped ossicles and branches called **cirri**.

Class Asteroidea

The sea stars are the representative members of class Asteroidea. Sea stars range in size from a few centimeters up to a meter in diameter. Sea stars are distinguished from other echinoderms because their bodies are gradually drawn out into five arms (rays) radiating from a **central disk**. Some species, however, can have six or more arms. The body of a sea star appears flattened and is covered by a thin epidermis that may contain calcareous spines. Most sea stars are dioecious, and fertilization is external. Fertilized eggs develop into planktonic bipinnaria larvae and, in some species, into **brachiolaria** (having armlike structures) larvae. The larvae are bilaterally symmetrical, unlike adult sea stars. Regeneration is common in sea stars.

Class Ophiuroidea

Class Ophiuroidea consists of more than 2,000 known species commonly called brittle stars, or basket stars. The arms of brittle stars are slender and sharply set off from the central disk. Their **ambulacral grooves** are covered with ossicles, and they do not possess pedicellariae. The tube feet of brittle stars do not have suckers. In these animals, the **madreporite** (sieve plate that connects the internal water vascular system to the exterior) is located on the oral surface. Five movable plates that serve as jaws surround the mouths of brittle stars. These animals do not possess an intestine or an anus. The visceral organs are confined to the central disk.

Class Crinoidea

Class Crinoidea is composed of numerous fossil species and living sea lilies and feather stars. Members of class Crinoidea have been present in Earth's oceans since Cambrian times, and about 700 species still thrive today. The oral region of a sea lily faces upward and resembles an upside-down sea star. The opposite side is attached to a stalk that attaches to a substrate by means of its lower portion, the **holdfast**. The body of a crinoid, called the **calyx**, is covered by a leathery **tegmen**. The five **arms** are composed of feathery branches

21

Class Echinoidea

Class Echinoidea includes sea urchins, sand dollars, and sea biscuits. Many beachcombers enjoy collecting the **test** formed by the fused ossicles of a sand dollar. Echinoids live in the deep ocean as well as the intertidal zone. Sea urchins seem to prefer rocky areas, and sand dollars prefer to burrow into sediment. Echinoids are dioecious and may produce planktonic larvae. Echinoids lack arms, but their pentamerous arrangement of parts is apparent on the aboral side of the test. The armlike extensions are called petalloids. Sea urchins feature a skeletal structure known as **Aristotle's lantern**. It has five hard "teeth" moved by complex struts and muscles and used for grinding food. Sea urchins possess large spines that can penetrate human skin.

Class Holothuroidea

Members of class Holothuroidea, sea cucumbers, are shaped like a cucumber or a sausage link and live on the surface of various substrates or burrow into sediments. Most holothurians are dioecious, although a few hermaphroditic species exist. Fertilization is external, and the zygote develops into planktonic larvae. When threatened, sea cucumbers can undergo **evisceration**, in which the entire digestive system and other organs, including the gonads, can be shot out of the mouth or anus.

Sea cucumbers have fleshy bodies, and their skeletons are reduced to isolated ossicles embedded in a muscular body wall. Sea cucumbers lie on their sides by means of three ambulacra called the **sole**. **Tentacles** surround their mouths and vary in number from 8 to 30. In sea cucumbers, the tube feet aid in locomotion. The fluid-filled coelom serves as a hydrostatic skeleton. The complete digestive system empties into a muscular cloaca. When threatened, sea cucumbers can rupture the hindgut, extruding their **Cuverian threads**, which can wrap around a threatening adversary. In sea cucumbers, the unique respiratory tree is both a respiratory and an excretory structure.

Procedure 1

Macroanatomy of Classes Crinoidea, Asteroidea, Echinoidea, and Holothuroidea

Materials

- ❏ Dissecting microscope or hand lens
- ❏ Dissecting tray
- ❏ Probe
- ❏ Select specimens of preserved crinoids, including fossil crinoids
- ❏ Sea star
- ❏ Select specimens of sea urchins, sand dollars, sea biscuits, and specimens of sea cucumbers
- ❏ Colored pencils

1 Procure the needed equipment and specimens.

2 Using a hand lens or dissecting microscope, observe the crinoid specimens in their jars (Figs. 21.4 and 21.5) and the fossil crinoids. Use a probe to move the specimens. If permitted, remove the crinoids from the jars and place them on a dissecting tray for closer examination.

3 Note their external anatomy. Is there much anatomical difference between the preserved and fossil crinoids? Record your observations and labeled sketches in the space provided.

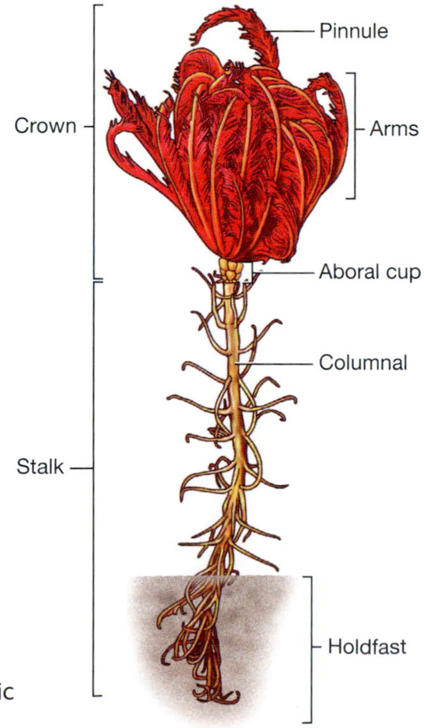

FIGURE **21.4** Basic crinoid anatomy.

Labels: Pinnule, Crown, Arms, Aboral cup, Columnal, Stalk, Holdfast

FIGURE **21.5** Red-stalked crinoid, class Crinoidea.

Crinoid

Crinoid fossil

4 Using a hand lens or dissecting microscope, observe the following echinoderms and compare them with the referenced figures: sea star (Fig. 21.6), sea cucumber (Fig. 21.7) sea urchin (Fig. 21.8C), sand dollar (Fig. 21.8D), and sea biscuit (Fig. 21.8E). If permitted, remove them from the jars and place them on the dissecting tray for closer examination. Note their external anatomy. Record your observations and labeled sketches in the spaces provided.

5 Clean the equipment and station thoroughly. Return the equipment to the proper location.

Cardiac stomach

FIGURE 21.6 Oral view of a sea star, *Asterias* sp.: (A) showing the cardiac stomach extended through mouth; (B) after retracting the stomach.

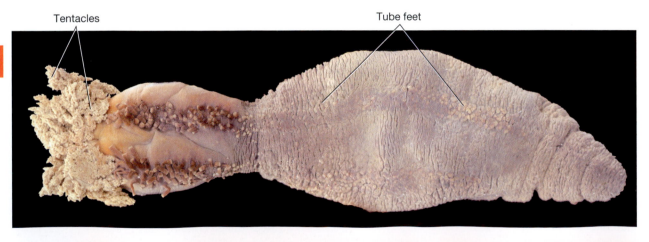

Tentacles

Tube feet

FIGURE 21.7 Sea cucumber, *Cucumaria* sp.

FIGURE **21.8** Examples of echinoderms: (**A**) chocolate chip star, *Nidorellia armata*, class Asteroidea; (**B**) green brittle star, *Ophiarachna incrassata*, class Ophiuroidea; (**C**) green sea urchin, *Strongylocentrotus droebachiensis*; (**D**) common sand dollars, *Echinarachnius parma*; (**E**) sea biscuit, *Clypeaster* sp., skeleton.

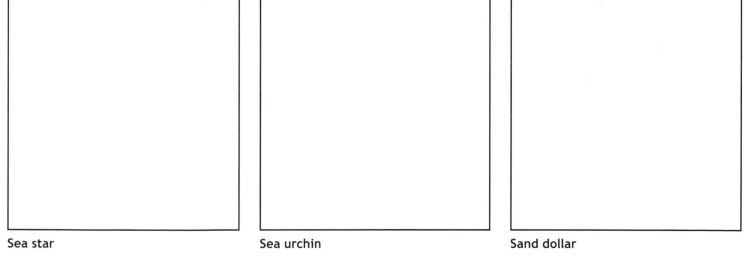

Sea star

Sea urchin

Sand dollar

Sea biscuit

Sea cucumber

Specimen _____

Procedure 2

Sea Star Dissection

Materials

- ❏ Hand lens or dissecting microscope
- ❏ Dissecting tray
- ❏ Dissecting pins
- ❏ Scalpel
- ❏ Safety glasses
- ❏ Lab coat or lab apron
- ❏ Gloves
- ❏ Water
- ❏ Sea star provided by your instructor
- ❏ Colored pencils

In this procedure, you will dissect a sea star. The mouth of a sea star is located on the oral (ventral) side. An ambulacral area runs along the oral side of each arm to the tip of the arm. The ambulacral groove is found in the center of this area. The groove is bordered by tube feet, or podia. Protective, movable spines appear near the tube edges of the tube feet. A distinct radial nerve is found in the center of each ambulacral groove.

The **aboral** (dorsal) surface may be smooth, granular, or covered with spines. Beneath the epidermis, a thick dermis layer secretes the endoskeleton of small ossicles, perforated by irregular canals filled with cells. A well-developed muscle layer allows the ossicle lattice to be flexible. Near the base of the spines are groups of minute pedicellariae (pincerlike structures) used to free the surface of debris and encrusting organisms. Small, fingerlike projections of the coelomic cavity, or papulae, cover the epidermis of the sea star. The papulae function in gas exchange and excretion.

In most sea stars, a distinct madreporite (external opening of the water vascular system) can be seen on one side of the central disk. In sea stars, a water vascular system is

responsible for movement, food gathering, respiration, and excretion. Primarily, the digestive system consists of a short esophagus, large stomach, small intestine, and inconspicuous anus (Fig. 21.9).

In sea stars, the sense organs are not well developed. Tactile organs and other sensory structures are scattered over the surface of the animal. Light-sensitive ocelli are present at the tip of each arm.

1 Procure the needed equipment and specimen.

2 Thoroughly rinse the sea star in running water.

3 Place the sea star in a dissecting tray with its aboral side facing upward. Pin the sea star down using dissecting pins.

4 Cut off the tip of one arm with a scalpel. Proceed to make two long, parallel incisions from the tip of the arm to the central disk. Carefully peel back the top layer of tissue, and locate the anatomical structures featured in Figure 21.9. Observe the sea star using a hand lens or dissecting microscope if desired. Record your observations and labeled sketch in the space provided.

5 Gently cut the tissue away from the central disk to find the internal structures featured in Figure 21.9. Record your observations and labeled sketch in the space provided.

21

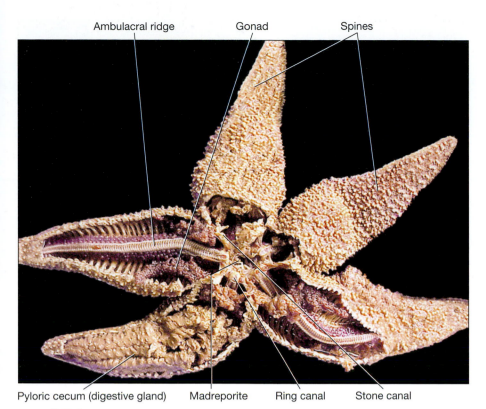

Ambulacral ridge   Gonad   Spines

Pyloric cecum (digestive gland)   Madreporite   Ring canal   Stone canal

FIGURE **21.9** Aboral view of the internal anatomy of a sea star, *Asterias* sp.

6 Thoroughly clean your laboratory station and equipment. Return the dry equipment, and discard the sea star as indicated by the instructor.

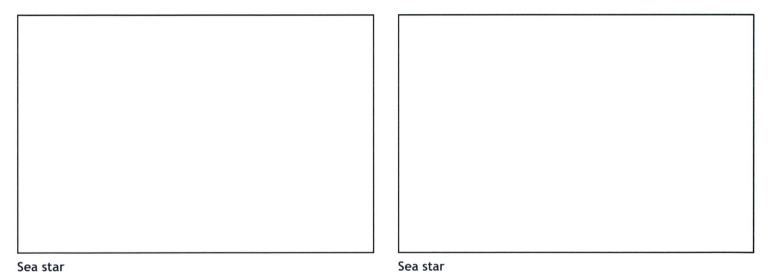

Sea star

Sea star

✔ Check Your Understanding

1.1 What is the function of the madreporite and the water vascular system?

_____

_____

1.2 What is unique about the respiratory tree in sea cucumbers?

_____

_____

1.3 Echinoderms vary in their anatomy, and differences are seen as the presence or absence of arms. Match the description with the correct echinoderm class.

_____ Asteroidea   A. Five arms composed of feathery branches, or pinnules

_____ Echinoidea   B. Five arms radiating from a central disk

_____ Ophiuroidea   C. No arms present; fused ossicles make up a test

_____ Crinoidea   D. No arms present; tentacles surrounding a mouth

_____ Holothuroidea   E. Five slender arms sharply set off from the central disk

Phylum Chordata, Classes Myxini and Cephalospidomorphi

All vertebrates follow the general body plan of the chordates, which includes a notochord, pharyngeal pouches, a dorsal hollow nerve chord, and a postanal tail (Table 21.1). Figure 21.10 is a cladogram of the vertebrate chordates. In the vertebrates (except for some fishes), the notochord is found only in embryos, where it guides development of the vertebrae. In the majority of vertebrates, the vertebral column surrounds or replaces the notochord. Another vertebrate characteristic is the presence of a dorsal hollow (tubular) nerve cord, found along the midline dorsal to the vertebrae. The spinal cord forms from the dorsal hollow nerve cord. In vertebrates, it is encased in the neural arches of the vertebrae.

The brain forms at the anterior end of the nerve cord and is encased in the cranium. The vertebrates are characterized by this vertebral column, a chain of structures that passes along the dorsal side from the head to the tail. All members of subphylum Vertebrata (Craniata) have a skull or cranium. Technically, some jawless fish, such as hagfish, lack vertebrae and keep their notochord. The craniates also have a **neural crest**, a group of embryonic cells that contribute to forming the cranium, jaws, teeth, and some nerves.

The vertebrates, or craniates, are metabolically more active than the protochordates and possess a more complex muscular system. Vertebrates feature a bony or cartilaginous endoskeleton consisting of a cranium, limb girdles, and two pairs of appendages. To keep the metabolic rate higher, all members of this group have a multichambered heart (two to four chambers) and hemoglobin-rich blood. Vertebrates also possess a number of specialized organs, including the liver

Did you know . . .

Big as a Whale's Heart

Blue whales are the largest animals ever known to have lived on Earth. The heart of a blue whale is about the size of a Volkswagen Beetle, and the tongue is as big as an elephant. A human actually can crawl through the aorta of a blue whale. A blue whale baby when born ranks as one of the largest animals on Earth, weighing in at more than 3 tons and more than 25 feet in length. It gains more than 200 pounds a day for the first year of its life.

TABLE **21.1** Representatives of the Subphylum Vertebrata

Taxa and Representative Kinds	Characteristics
Superclass Cyclostomata	Eel-like and aquatic; sucking mouth (some parasitic); lack jaws and paired appendages
Class Myxini—hagfishes	Terminal mouth with buccal funnel absent; nasal sac connected to pharynx; four pairs of tentacles; five to ten pairs of pharyngeal pouches
Class Cephalaspidomorphi (Petromyzontida)—lampreys	Suctorial mouth with rasping teeth; nasal sac not connected to buccal cavity; seven pairs of pharyngeal pouches
Superclass Gnathostomata	Jawed vertebrates; most with paired appendages
Class Chondrichthyes—sharks, rays, and skates	Cartilaginous skeleton; placoid scales; most have spiracle; spiral valve in digestive tract
Class Osteichthyes	Bony fishes; gills covered by bony operculum; most have swim bladder
Subclass Sarcopterygii	Bony skeleton; lobe-finned; paired pectoral and pelvic fins
Subclass Actinopterygii	Bony skeleton; most have dermal scales; ray-finned
Class Amphibia—salamanders, frogs, and toads	Larvae have gills and adults have lungs; scaleless skin (except apoda); an incomplete double circulation; three-chambered heart
Class Reptilia (Sauropsida)—turtles, snakes, and lizards	Amniotic egg; epidermal scales; three- or four-chambered heart; lungs
Class Aves—birds	Homeothermic (warm-blooded); feathers; toothless; air sacs; four-chambered heart with right aortic arch
Class Mammalia—mammals	Homeothermic; hair; mammary glands; most have seven cervical vertebrae; muscular diaphragm; three auditory ossicles; four-chambered heart with left aortic arch

21

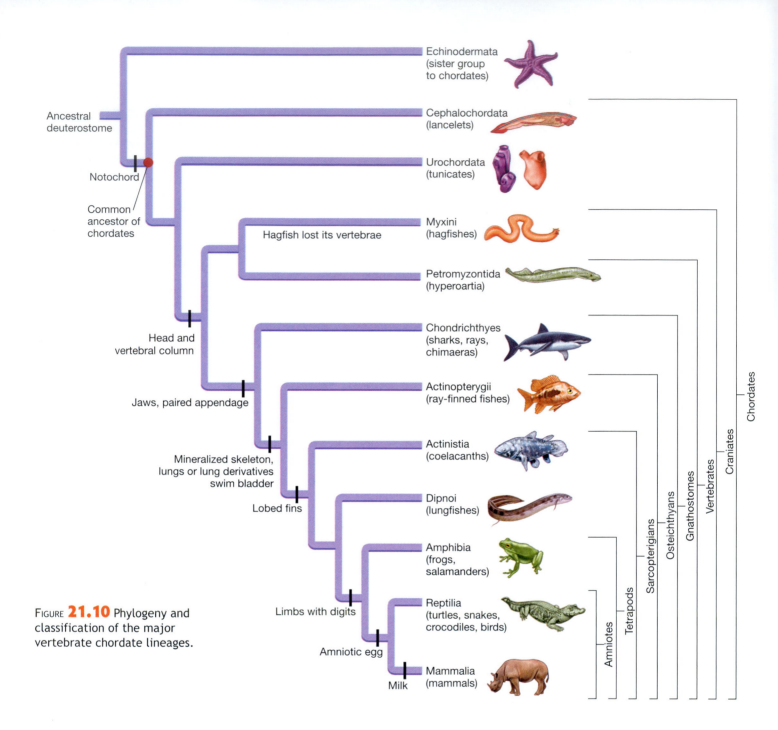

FIGURE **21.10** Phylogeny and classification of the major vertebrate chordate lineages.

and kidneys. The endocrine system is also well developed in vertebrates.

Subphylum Vertebrata comprises more than 55,000 species. It consists of two superclasses: Cyclostomata (Agnatha; jawless fishes) and Gnathostomata (jawed vertebrates). Superclass Cyclostomata consists of class Myxini (hagfishes) and class Cephalaspidomorphi (Petromyzontida; lampreys). These two classes consist of approximately 70 species of jawless fishes. Members of this group lack scales, internal ossification, and paired fins. Cyclostomata possess eel-like bodies with paired, pore-like gill openings.

Most animals with which we are familiar are members of superclass Gnathostomata. These organisms differ from the Cyclostomata in that they possess jaws and a variety of feeding devices designed to grasp, crush, shear, and chew. Superclass Gnathostomata consists of six classes: Chondrichthyes (cartilaginous fishes), Osteichthyes (bony fishes), Amphibia (amphibians), Reptilia (reptiles), Aves (birds), and Mammalia (mammals).

21

Classes Myxini and Cephalaspidomorphi

Class Myxini includes approximately 30 species of bottom-dwelling marine scavengers known as hagfishes. Hagfishes range in length from 18 centimeters to 1 meter. The mouth of a hagfish contains two keratinized plates with toothlike structures. Hagfishes have a small brain and eyes and highly developed senses of smell and taste. Lateral slime glands produce copious amounts of slime in self-defense.

Class Cephalaspidomorphi (Petromyzontida) includes approximately 41 species of freshwater and marine organisms called lampreys. Lampreys can vary in size from 15 centimeters to 1 meter in length. Nonparasitic lampreys do not feed after emerging as adults, because the alimentary canal degenerates. They usually spawn after reaching the adult stage and soon die. Marine lampreys are parasitic as adults, attaching to a fish with their sucker-like mouth and sharp horny teeth. They suck out body fluids, many times causing the death of the host. Marine lampreys such as *Petromyzon marinus* are **anadromous**, living in the ocean most of their lives and returning to freshwater to spawn.

Procedure 1

Macroanatomy of Hagfishes and Lampreys

Materials
- ❏ Dissecting microscope or hand lens
- ❏ Dissecting tray
- ❏ Probe
- ❏ Select specimens of *Myxine glutinosa* and *Petromyzon marinus*
- ❏ Colored pencils

1 Procure equipment and specimens.

2 Using a dissecting microscope or a hand lens, observe a specimen of a hagfish, *Myxine glutinosa* (Fig. 21.11) and a sea lamprey, *Petromyzon marinus* (Figs. 21.12–21.14) on a dissecting tray. Use a probe to move the specimen. Compare and contrast the external features of the two Cyclostomatas in the spaces provided, including observations and sketches.

3 Thoroughly clean your laboratory station and equipment. Return the dry equipment.

FIGURE **21.11** External anatomy of a Pacific hagfish, *Eptatretus* sp.

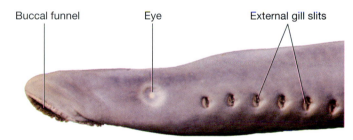

FIGURE **21.12** Lateral view of the anterior anatomy of a marine lamprey, *Petromyzon* sp.

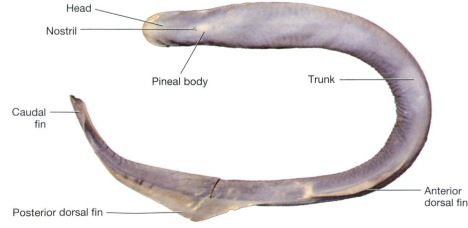

FIGURE **21.13** Dorsal view of the external anatomy of a marine lamprey, *Petromyzon* sp.

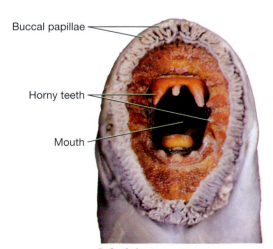

FIGURE **21.14** Oral region of a marine lamprey, *Petromyzon* sp.

21

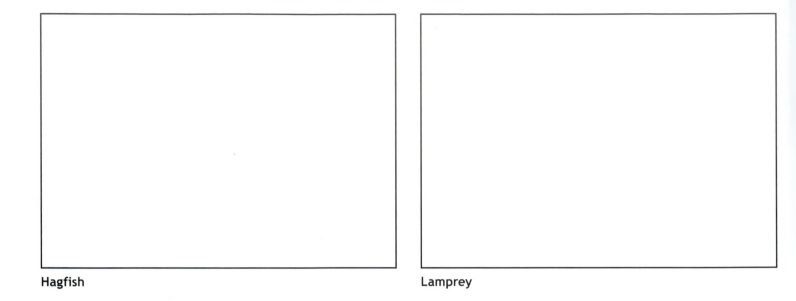

Hagfish

Lamprey

✔ Check Your Understanding

2.1 Organisms that are described as eel-like and aquatic, with sucking mouths, but lacking jaws and paired appendages are members of subphylum Vertebrata, _____ . (*Circle the correct answer.*)

a. superclass Gnathostomata

b. class Amphibia

c. class Chondrichthyes

d. superclass Cyclostomata

e. subclass Actinopterygii

21

Phylum Chordata, Classes Chondrichthyes and Osteichthyes

Class Chondrichthyes

Members of class Chondrichthyes are also known as cartilaginous fishes. There are nearly 1,000 living species, including sharks, rays, skates, sawfishes, and chimaeras (Fig. 21.15).

As gnathostomes, members of class Chondrichthyes possess a ventrally oriented mouth. The entire skeleton of members of class Chondrichthyes, including the skull, is cartilaginous. These animals feature paired pectoral and pelvic fins and two dorsal median fins. In males, the pelvic fins are modified to aid in sperm transfer and are called **claspers**. The skin of cartilaginous fishes usually has placoid scales (denticles, or small teeth) and mucous glands. The teeth of sharks and other members of the class are actually placoid scales (Figs. 21.16 and 21.17).

Members of class Chondrichthyes have a complete digestive system and a closed circulatory system with a two-chambered heart. Cartilaginous fishes also possess five to seven pairs of gill slits. **Spiracles**, found just below the eyes in some species, such as sawfishes and rays, aid in drawing water into the gills. Many members of class Chondrichthyes have well-developed senses of smell and hearing.

Rays are bottom-dwelling fishes that feature a slender, whiplike tail of serrated spines with venom glands at their base. Most rays are **ovoviviparous** species (eggs develop in the female without placental attachment and hatch immediately before birth). The young develop in the mother and are nourished by a yolk sac until birth. Some sharks are ovoviviparous as well, but several species are viviparous (rather than producing eggs, producing an embryo that develops inside the female prior to birth); the embryo receives nourishment from the mother's bloodstream from a placenta-like structure.

Presently, 200 species of skates have been named. Skates have a flattened body with a fleshy tail that lacks spines.

FIGURE **21.15** Examples of class Chondrichthyes: (**A**) black-tip reef shark, *Carcharhinus melanopterus*; (**B**) blue-spotted stingray, *Taeniura lymma*; (**C**) chimaera, *Hydrolagus colliei*.

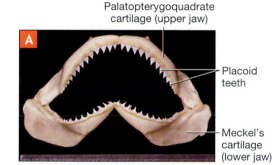

FIGURE **21.16** (**A**) Shark jaws; (**B**) detailed view showing rows of replacement teeth.

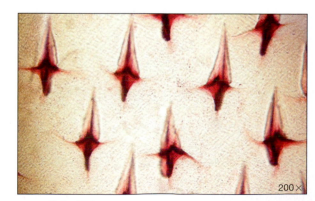

FIGURE **21.17** Placoid scales.

Skates can be distinguished from rays by the presence of a prominent dorsal fin. Skates possess small teeth as opposed to the grinding plates of rays. Skates are oviparous (egg-laying), producing a hard rectangular egg case called a "mermaid's purse" (Fig. 21.18).

The critically endangered marine sawfishes possess a unique, sawlike rostrum covered with motion-sensitive and electrosensitive pores (ampullae of Lorenzini), as sharks, rays, and skates do, that allow them to detect movement of prey. The rostrum may be used to dig up prey buried in the ocean floor or as a deadly slashing instrument. Chimaeras, sometimes called ratfish or ghostfish, are ancient cartilaginous fishes that usually inhabit deep marine waters. They are called chimaeras because of their appearance, which resembles a patchwork of other animals.

FIGURE **21.18** Many members of class Chondrichthyes produce an egg case known as a mermaid's purse. Here a young shark can be seen in silhouette. The round shape on the right is the yolk.

Class Osteichthyes

Class Osteichthyes includes approximately 27,000 species of bony fishes, the largest and most diverse chordate group (Fig. 21.19). In recent years, taxonomists have divided the osteichthyes into two distinct subclasses: subclass Sarcopterygii, or lobe-finned fishes; and subclass Actinopterygii, the ray-finned fishes. The sarcopterygians are well represented by many fossil species and the extant coelocanth. These fishes are the ancestor to the first amphibians.

The ray-finned fishes belonging to subclass Actinopterygii constitute nearly 95% of vertebrate species. Examples of this diverse class are goldfish, catfish, gars, tuna, seahorses,

FIGURE **21.19** Examples of class Osteichthyes: (**A**) preserved specimen of coelacanth, *Latimeria* sp.; (**B**) lungfish, *Neoceratodus forsteri*; (**C**) Christmas wrasse, *Biochoeres ornatissimus*; (**D**) leafy seadragon, *Phycodurus eques*; (**E**) rockfish, *Sebastes* sp.; (**F**) white sturgeon, *Acipenser transmontanus*.

21

pufferfish, swordfish, and trout. They are known as ray-finned fishes because their fins consist of webs of skin supported by bony spines. Actinopterygians feature paired pectoral and pelvic fins and skin with mucous glands usually embedded with dermal scales. Three types of scales can exist in these fishes:

1. Ganoid scales are flat, heavy scales composed of silvery ganoin on the top surface and bone on the bottom. They are shaped like an arrowhead, which is characteristic of gars (Fig. 21.20A).

2. Cycloid scales are thinner and more flexible scales than ganoids; overlap each other and feature growth rings. They are common in advanced bony fishes such as carp and salmon (Fig. 21.20B).

3. Ctenoid scales resemble cycloid scales but have comblike ridges on the exposed edge. They are found in fish such as bass and sunfish (Fig. 21.20C).

Most bony fishes have a fusiform (torpedo-shaped) body tapered at both ends. Evident segmentation of the muscles is present in the zigzag myomeres. Respiration in the actinopterygians occurs in the gills, covered with a protective flap, or **operculum**. Many species have a **swim bladder** that serves as a flotation device. Fishes possess a two-chambered heart and a closed circulatory system. Fishes are **ectotherms**, controlling their body heat through external sources. They have a complete digestive system supported by accessory organs, such as the pancreas and liver. The kidneys filter the wastes contained in the blood. Freshwater fishes have well-developed kidneys, and saltwater fishes have poorly developed kidneys.

The nervous system is well developed. The brain consists of a small cerebrum, olfactory lobes, a large cerebellum, and optic lobes. In many species, the eyes are acute, sounds can be detected in the inner ear, and olfaction (smell) is well developed. Also, in many fishes a **lateral line system** serves to detect vibrations. Most fishes are dioecious and oviparous.

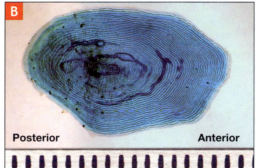
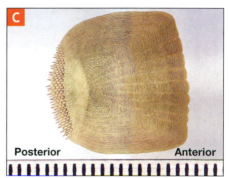

FIGURE **21.20** Fish scales: (**A**) ganoid scales; (**B**) cycloid scales; (**C**) ctenoid scales (scale in mm).

Procedure 1

Macroanatomy of Fishes

Materials

- ❏ Dissecting microscope or hand lens
- ❏ Dissecting tray
- ❏ Probe
- ❏ Select specimens of available members of class Chondrichthyes, including sharks, rays, skates, and a chimaera
- ❏ Examples of shark integument, shark teeth, a shark jaw, a sawfish blade, and a mermaid's purse
- ❏ Select specimens of available members of class Osteichthyes
- ❏ Examples of scale types and fish skeleton
- ❏ Colored pencils

1 Procure equipment and specimens.

2 Using a dissecting microscope or hand lens, observe the available Chondrichthyes specimens, paying particular attention to the shark and ray anatomy shown in Figures 21.21 and 21.22. Use a probe to move the specimen. Record your observations and labeled sketches in the space provided.

3 Using a hand lens, observe the available specimens of Osteichthyes on the dissecting tray. Pay particular attention to their external anatomy and their skeleton using Figures 21.23 and 21.24 as your references. Record your observations and labeled sketches in the space provided.

4 Observe the types of fish scales using the dissecting microscope. Or you may need a compound microscope for these observations. Record your observations and sketches in the space provided.

5 Thoroughly clean your laboratory station and equipment. Return the dry equipment.

21

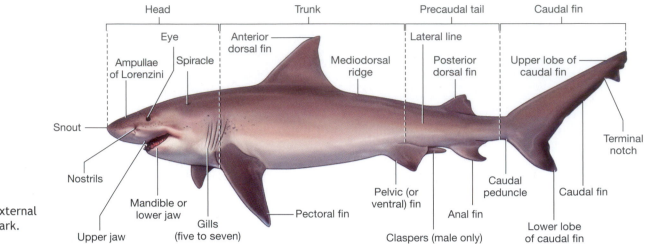

FIGURE **21.21** External anatomy of a shark.

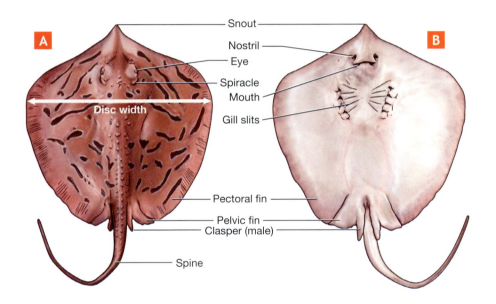

FIGURE **21.22** External anatomy of a stingray: (**A**) dorsal; (**B**) ventral.

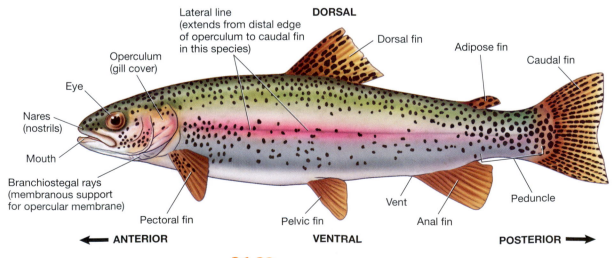

FIGURE **21.23** External anatomy of a fish.

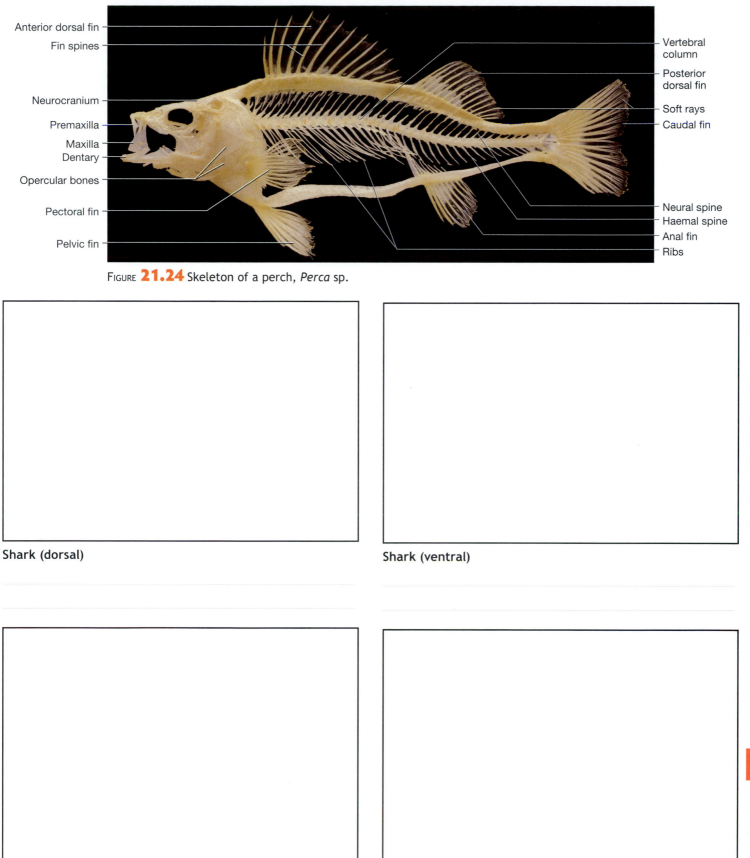

Figure **21.24** Skeleton of a perch, *Perca* sp.

Labels (clockwise from top left):
- Anterior dorsal fin
- Fin spines
- Neurocranium
- Premaxilla
- Maxilla
- Dentary
- Opercular bones
- Pectoral fin
- Pelvic fin
- Vertebral column
- Posterior dorsal fin
- Soft rays
- Caudal fin
- Neural spine
- Haemal spine
- Anal fin
- Ribs

Shark (dorsal)

Shark (ventral)

Stingray (dorsal)

Stingray (ventral)

21

[empty box]

Specimen 1

[empty box]

Specimen 2

[empty box]

Fish scales

✔ Check Your Understanding

3.1 Members of the subclass Actinopterygii are known as _____ fishes and have _____ within their fins. Fishes placed in the subclass Sarcopterygii are known as _____ fishes, which have _____ fins.

3.2 Compare and contrast the classes Chondrichthyes and Osteichthyes.

_____

_____

_____

_____

_____

_____

_____

Phylum Chordata, Class Amphibia

Class Amphibia is composed of nearly 6,000 species. Modern amphibians include frogs, toads, salamanders, amphiumas, sirens, newts, and caecilians. The amphibians evolved from sarcopterygian fishes during the Devonian period. To venture away from the waters and colonize the land, amphibians developed a protective **integument**, a means to breathe in the terrestrial environment, and specialized limbs. Despite these changes, amphibians did not evolve the ability to live their entire lives on land. They remained dependent on water to lay their eggs.

Members of class Amphibia are characterized by their bony skeletons with vertebrae, gills during development in the majority of species, and moist, glandular integument without external scales. In some amphibians the integument allows for taking in air through the skin, called cutaneous respiration. Pigment cells, or chromatophores, in the skin are responsible for a variety of colors.

The skull of an amphibian is short, broad, and incompletely ossified. The mouth is usually large, with small teeth. Two internal **nares**, or nostrils, open into the mouth cavity. Many species possess a protrusible muscular tongue attached to the front of the mouth. The circulatory system is closed, and the heart has three chambers, two atria and one ventricle. Amphibians eliminate nitrogenous waste primarily in the form of urea. Their nervous system is well developed. The brain consists of a forebrain, a midbrain, and a hindbrain. In many species the eyes have adapted to a terrestrial lifestyle.

The eyes are protected by an eyelid called the **nictitating membrane**. The auditory system contains a **tympanic membrane** for hearing.

Amphibians are dioecious. Frogs and toads undergo external fertilization, and salamanders primarily undergo internal fertilization. Amphibians are primarily oviparous. The eggs of amphibians are **mesolecithal**, having a large yolk and jellylike membranes. Larval forms, such as tadpoles, are aquatic and possess gills. Several species of salamanders are **neotenic**, retaining their gills in the adult form.

Class Amphibia is composed of three orders: Gymnophiona, Caudata, and Anura (Fig. 21.25). Order Gymnophiona consists of 173 species of elongated limbless apodans known as caecilians. At first glance, they may be mistaken for earthworms or snakes. Caecilians are typically blind and possess sensory tentacles between their eyes and nostrils. The body of a caecilian is arranged in rings, called annuli, giving caecilians an earthworm-like appearance.

Salamanders, newts, amphiumas, and sirens are members of order Caudata (Urodela), consisting of nearly 560 described species. The body of a typical salamander is lizard-like, characterized by a slender body, a short nose, and an elongated tail. Most salamanders have four toes on their front legs and five toes on their hindlegs, and they lack claws. Amphiumas and sirens have degenerate legs and an eel-like appearance. Some adult terrestrial salamanders have lungs, and some terrestrial species lack both lungs and gills. These

FIGURE **21.25** Examples of class Amphibia: (**A**) Cameroon caecilian, *Crotaphatrema bornmuelleri*, from the order Gymnophiona; (**B**) tiger salamander, *Ambystoma tigrinum*, from the order Caudata; (**C**) amphiuma, *Amphiuma means*, from the order Caudata; (**D**) axolotl, *Ambystoma mexicanum*, from the order Caudata; (**E**) blue-webbed gliding tree frog, *Rhacophorus reinwardtii*, from the order Anura; (**F**) toad, *Bufo bufo*, from the order Anura.

salamanders undergo cutaneous respiration. Gills are used in larval salamanders and some neotenic (paedomorphic) forms, such as sirens and axolotl.

Order Anura consists of 5,290 species of toads and frogs. Generally, toads possess a dry, bumpy integument, parotid glands behind the tympanic membrane, a blunt nose, no teeth, and no webs on their hind digits. Frogs have smooth, moist skin, teeth in the upper jaw, and webs between the hind toes. They possess extremely long, muscular hind legs and front legs that serve as shock absorbers. Adult frogs and toads are carnivorous, and the tadpoles are herbivorous.

Procedure 1

Macroanatomy of Amphibians

Materials

- ❏ Dissecting microscope or hand lens
- ❏ Dissecting tray
- ❏ Probe
- ❏ Select specimens of the following amphibians: Caecilian, salamanders, newts, amphiumas, sirens, toads, frogs
- ❏ Bullfrog skeleton
- ❏ Colored pencils

1 Procure equipment and specimens.

2 Using a hand lens or dissecting microscope, observe the available amphibian specimens. If permitted, remove the specimens from the jars and place them on a dissecting tray for closer examination. Use a probe to move the specimen. Record your observations and labeled sketches in the space provided. When possible, provide the scientific name of the specimen.

3 Using the hand lens, observe the skeleton of the bullfrog, comparing the bones with Figures 21.26 and 21.27.

Caecilian

Salamanders and newts

Amphiumas and sirens

Toads and frogs

21

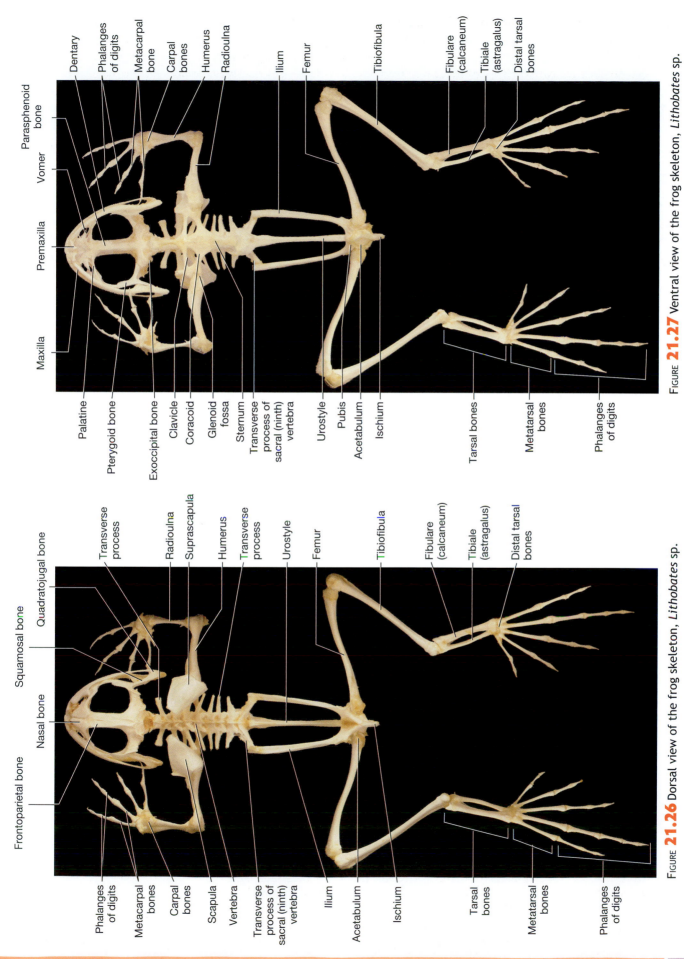

FIGURE **21.27** Ventral view of the frog skeleton, *Lithobates* sp.

Labels (figure 21.27, top):

Parasphenoid bone, Vomer, Premaxilla, Maxilla, Dentary, Phalanges of digits, Metacarpal bone, Carpal bones, Humerus, Radioulna, Ilium, Femur, Tibiofibula, Fibulare (calcaneum), Tibiale (astragalus), Distal tarsal bones, Palatine, Pterygoid bone, Exoccipital bone, Clavicle, Coracoid, Glenoid fossa, Sternum, Transverse process of sacral (ninth) vertebra, Urostyle, Pubis, Acetabulum, Ischium, Tarsal bones, Metatarsal bones, Phalanges of digits

FIGURE **21.26** Dorsal view of the frog skeleton, *Lithobates* sp.

Labels (figure 21.26, bottom):

Frontoparietal bone, Nasal bone, Squamosal bone, Quadratojugal bone, Transverse process, Radioulna, Suprascapula, Humerus, Transverse process, Urostyle, Femur, Tibiofibula, Fibulare (calcaneum), Tibiale (astragalus), Distal tarsal bones, Phalanges of digits, Metacarpal bones, Carpal bones, Scapula, Vertebra, Transverse process of sacral (ninth) vertebra, Ilium, Acetabulum, Ischium, Tarsal bones, Metatarsal bones, Phalanges of digits

21

CHAPTER **21** | **Animal Planet IV** Understanding the Deuterostomes **497**

4 Thoroughly clean your laboratory station and equipment. Return the dry equipment.

✔ Check Your Understanding

4.1 What adaptations allowed early amphibians to transition from water to land?

_____

_____

_____

4.2 This structure is found in amphibians and other animals, and serves to protect the eye. (*Circle the correct answer.*)

   a. Tympanic membrane.

   b. Nare.

   c. Nictitating membrane.

   d. Calamus membrane.

   e. None of the above.

4.3 Label each description as frogs, toads, or both using the following key:

          F = Frogs   T = Toads   F/T = Both frogs and toads

_____ Thin, smooth, moist integument

_____ Adults carnivorous; tadpoles herbivorous

_____ Blunt nose and no teeth

_____ No webbing between hind toes

_____ Dry, bumpy integument

_____ Very long muscular hindlegs with webs between hind toes

_____ Maxillary and vomerine teeth

Did you know . . . ?

Time Traveler

In June 2006, the biological world was saddened by the death of Harriet, the giant Galapagos land tortoise (*Geochelone nigra porteri*). It is believed that Harriet was collected by Charles Darwin in 1835 while visiting the Galapagos Islands on board the *HMS Beagle*. Records indicate that Harriet was brought to Australia by a former captain of the *Beagle*, John Wickham, in 1841. Harriet died in the Australia Zoo in 2006, the same year co-owner and TV personality Steve Irwin (Crocodile Hunter) died.

Phylum Chordata, Class Reptilia

With the evolution of reptiles, vertebrates finally conquered the land. Today, approximately 7,500 species of animals are in class Reptilia (Fig. 21.28). Presently, classification of reptiles is undergoing major revision, and several themes have become popular. In the future, terms such as "nonavian reptiles" may be used to describe turtles, lizards, snakes, tuataras, and crocodiles (Fig. 21.29).

Modern reptiles share several fundamental traits. Whether oviparous or ovoviviparous, reptiles produce amniotic eggs (Fig. 21.30) ideal for the transition to land. These eggs possess a yolk for nourishment and four distinct membranes important in development: the **yolk sac**, the **amnion**, the **chorion**, and the **allantois**. The yolk sac provides food for the embryo, the amnion encases and cushions the developing embryo in a fluid-filled cavity, the chorion allows for the exchange of vital respiratory gases, and the allantois collects metabolic waste.

The water-tight shell of the reptilian egg is highly protective. Fertilization occurs internally before the egg forms. The tough, dry, and scaly integument of reptiles protects them from desiccation and injury. The reptilian skeleton is well ossified and strong. The limbs are paired with five toes, with the exception of snakes, amphisbaenians, and some legless lizards. Usually two sacral vertebrae support the pelvic girdle. Reptiles possess well-developed lungs and undergo thoracic breathing, in which specialized muscles and ribs provide for the transport of copious amounts of air into and out of the lungs. With the exception of crocodilians, which have a four-chambered heart, reptiles have a three-chambered heart.

As ectotherms, reptiles depend upon the environment for thermoregulation. Reptiles have complete digestive systems with accessory structures, and they have well-developed kidneys. Urine is voided in a semisolid mass containing uric acid crystals. They possess a well-developed nervous system and various sensory structures. Reptiles are dioecious and have no larval stage.

Despite a rich history, only four living orders of reptiles exist today. Order Chelonia consists of turtles and tortoises, order Squamata includes lizards, snakes, and amphisbaenians, order Sphenodonta consists of tuataras, and order Crocodilia includes alligators and crocodiles.

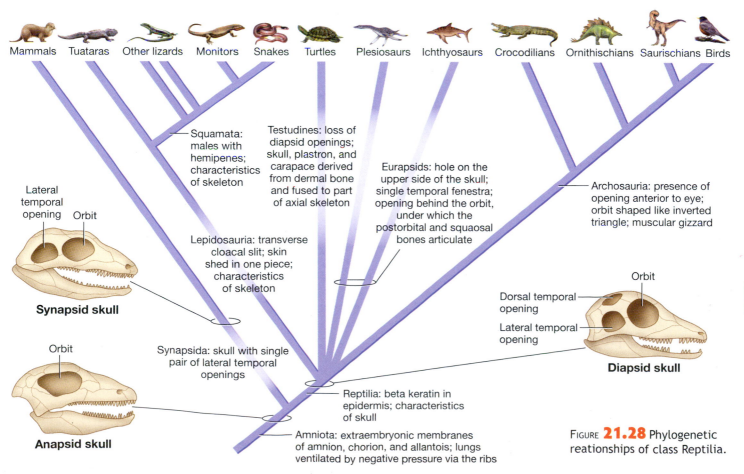

FIGURE **21.28** Phylogenetic reationships of class Reptilia.

FIGURE **21.29** Members of class Reptilia: (**A**) star tortoise, *Geochelone elegans*; (**B**) green basilisk, *Basiliscus plumifrons*; (**C**) Arizona mountain king snake, *Lampropeltis pyromelana*; (**D**) tuatara, *Sphenodon punctatus*; (**E**) American alligator, *Alligator mississippiensis*; (**F**) Gila monster, *Heloderma suspectum*.

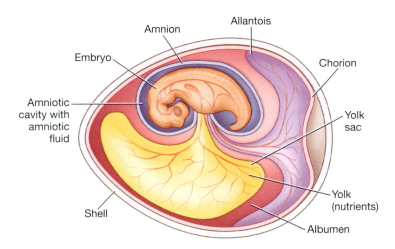

FIGURE **21.30** Amniotic egg.

Order Chelonia

Approximately 300 species of turtles and tortoises have been described. Although most turtles are aquatic, terrestrial and marine species exist. The most distinguishing feature of turtles is a protective shell that provides adequate defense. Many species can retract their heads and appendages into the shell. The box turtle, *Terrepene carolina*, possesses a **hinge** that actually closes up the head region. The dorsal portion of the shell is called the **carapace**, and the ventral portion is the **plastron.** The shell of a turtle is composed of hard, bony plates covered by corresponding keratinized **scutes.** Soft-shelled turtles have a pliable, leathery shell.

The ribs and body vertebrae are fused to the interior of the carapace in turtles. The shoulder and hip girdles of turtles are located within, instead of outside, the rib cage. Instead of teeth, turtles have a sharp, keratinized covering over their maxilla and mandible. Although the brain of a turtle is rather small, the sense of sight is well developed. Turtles are dioecious and oviparous. In several turtle families as well as in crocodiles and some lizards, the sex of turtles is determined by nest temperature. In temperature-dependent sex determination, high nest temperature results in females, and low nest temperature results in males.

Order Squamata

Order Squamata, the largest order of reptiles, encompasses nearly 7,000 species of lizards, snakes, and amphisbaenians. Squamates possess a skull with movable joints called a kinetic skull. The modified skull allows squamates to seize, hold, and swallow prey efficiently. Members of this order also possess distinct scales. Male squamates have a **hemipenis** used in copulation. Squamates also possess a Jacobson's organ (vomeronasal organ) that enhances the sense of smell. A snake flicking its tongue is carrying back molecules to the Jacobson's organ on the roof of the mouth for evaluation. Two living suborders of squamates are suborder Sauria (lizards and amphisbaenians) and suborder Serpentes (snakes).

Most of the nearly 4,000 species of lizards are terrestrial, but aquatic and marine forms exist. Lizards have an elongated body and, with the exception of the legless lizards, paired

21

appendages. In contrast to snakes, lizards have external auditory openings and a pair of eyelids. All lizards have the ability to lose their tail, and many can regenerate the lost tail. Some lizards have a colorful, flag-like structure, the dewlap, in the throat region.

Amphisbaenians also are known as "worm lizards" because they lack limbs and auditory openings and have underdeveloped eyes. The wormlike body has numerous independently moving rings used in locomotion. Approximately 160 species of amphisbaenians have been identified.

Of the 3,000 species of snakes, most are harmless, but rattlesnakes, coral snakes, cobras, and others keep a bad reputation going. Snakes first appeared in the early Cretaceous period, evolving from burrowing lizards. Constricting snakes such as boas and pythons appear to be the oldest group and still retain vestigial hindlimbs called spurs.

Snakes possess an elongated body with numerous vertebrae. The vertebrae (perhaps as many as 400) and their associated ribs are essential in the locomotion of snakes. Snakes have a complete digestive system. They can detect vibrations but do not have a good sense of hearing. Snakes lack eyelids and do not blink. In addition, they have a protective transparent membrane, the spectacle, over their eyes. Most snakes are oviparous, although some ovoviviparous species exist.

Order Crocodilia

Members of order Crocodilia, such as alligators, crocodiles, and caimans, along with the birds, are the only living descendents of the archosaurian lineage (the group that includes the now-extinct dinosaurs). Crocodilians possess an elongated, heavy skull with a robust mandible and maxilla housing large teeth in bony sockets. Members of order Crocodilia possess a secondary bony palate in their mouths that allows them to breathe air when they are partially submerged and their mouth is full of water.

Crocodilians have a four-chambered heart, a complete digestive system, and a well-developed nervous system. Crocodilians have excellent vision, including color vision. The presence of a light-reflecting layer, the tapetum, behind the retina greatly increases their ability to see at night. The tapetum also is responsible for their eyes seeming to glow at night. Nictitating membranes cover the eyes when the animal is underwater. Numerous sensory pits line the jaws of crocodilians and serve as a type of lateral line system. Crocodilians are oviparous, and some species show parental care.

Procedure 1

Macroanatomy of Reptiles

Materials
☐ Dissecting microscope or hand lens
☐ Dissecting tray
☐ Probe
☐ Select specimens of the following reptiles: turtles, lizards, amphisbaenians, snakes, and a crocodilian
☐ Snake skeleton
☐ Colored pencils

1 Procure equipment and specimens.

2 Using a dissecting microscope or hand lens, observe the available reptile specimens placed on a dissecting tray. Compare your observations with Figures 21.31 and 21.32. If permitted, remove the specimens from the jars for closer examination. Pay close attention to the previous description of the organisms. Use a probe to move the specimen. Record your observations and labeled sketches in the space provided. When possible, provide the scientific name of the specimen.

3 Using the hand lens, observe the skeleton of the snake, comparing the bones with Figures 21.33 and 21.34.

4 Thoroughly clean your laboratory station and equipment. Return the dry equipment.

21

FIGURE **21.31** Dorsal view of a turtle, *Trachemys* sp.

FIGURE **21.32** Ventral view of a turtle, *Trachemys* sp.

Dorsal view labels:
Eye
Nostril
Head
Pentadactyl foot
Nuchal scale
Vertebral scales
Costal scales
Marginal scales (encircle the carapace)

Ventral view labels:
Gular scales
Humeral scales
Pectoral scales
Abdominal scales
Femoral scales
Anal scales
Tail

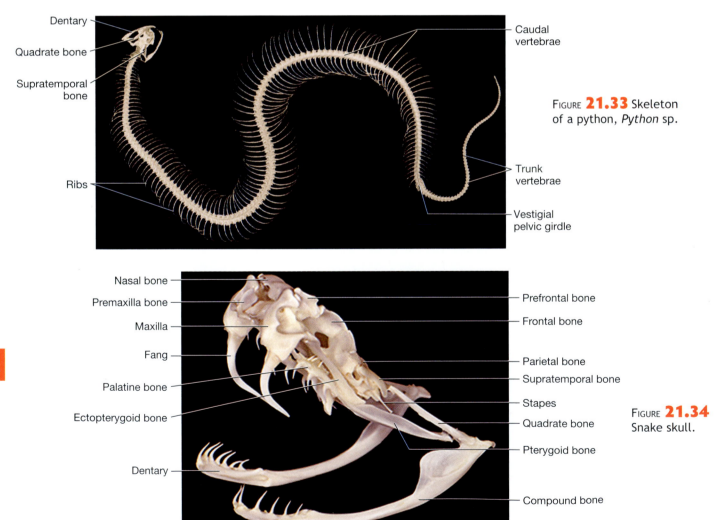

Dentary
Quadrate bone
Supratemporal bone
Ribs
Caudal vertebrae
Trunk vertebrae
Vestigial pelvic girdle

FIGURE **21.33** Skeleton of a python, *Python* sp.

Nasal bone
Premaxilla bone
Maxilla
Fang
Palatine bone
Ectopterygoid bone
Dentary
Prefrontal bone
Frontal bone
Parietal bone
Supratemporal bone
Stapes
Quadrate bone
Pterygoid bone
Compound bone

FIGURE **21.34** Snake skull.

Turtles

Lizards

Amphisbaenians

Snakes

Crocodilians

21

✔ Check Your Understanding

5.1 What is the function of the Jacobson's organ in snakes?

_____

_____

_____

5.2 What adaptation(s) allowed early reptiles to move further inland? (*Circle the correct answer.*)

   a. Integument composed of protective scales.

   b. Integument that is moist and highly vascularized.

   c. Amniotic egg.

   d. Anamniotic egg.

   e. Both a and c.

Phylum Chordata, Class Aves

The age-old question, "Where have all the dinosaurs gone?" can be answered today as, "They are the birds!" (Remember their ancestor, the velociraptor?) Approximately 9,700 extant species of class Aves, birds, have been described (Fig. 21.35).

Extant birds have been divided into two superorders. The superorder Paleognathae includes flightless birds (ratites) such as the ostrich. The ratites have no keel on their sternum, thus making flight impossible. Superorder Neognathae comprises 27 orders of modern birds. These birds possess a keeled sternum, but some are flightless.

Birds are bipedal, oviparous, endothermic, winged tetrapods. Their most distinguishing characteristic is the presence of feathers. The specific arrangement and appearance of feathers are called the bird's **plumage**. Plumage may vary within a bird species with respect to age and sex. Feathers are arranged on the bird's body in specific tracts, or **pterylae**. A typical bird has several types of feathers:

- Contour feathers provide the bird's shape and aid its flight.

- Down feathers help insulate a bird.

- Natal down feathers make a duckling or chick appear fluffy.

- Filoplumes are simple, hairlike feathers that provide sensory feedback on contour feather activity.

- Bristles are sensory and protective feathers found near the mouth of some birds, such as flycatchers.

Feathers require constant maintenance, and birds preen or groom daily. Birds apply an oily substance from the **uropygial gland**, located near the base of the tail, for waterproofing and to discourage microbial growth. A typical contour feather consists of a smooth, nonpigmented base extending beneath the skin, called the **calamus** (quill). The **inferior umbilicus** at the base of the calamus appears as a small hole. The shaft above the skin is called the **rachis**. The **vane** is the flat structure on each side of the feather composed of filaments called **barbs**. The barbs, in turn, consist of **barbules** connected by **hooklets** (Fig. 21.36).

Birds have a light but sturdy skeleton. Bird bones have numerous air cavities that make them lightweight. To make flight possible, birds' vertebrae are fused with the exception

FIGURE **21.35** Examples of class Aves: (**A**) brown pelican, *Pelecanus occidentalis*; (**B**) roseate spoonbill, *Ajaja ajaja*; (**C**) Magellanic penguin, *Spheniscus magellanicus*; (**D**) snow goose, *Chen caerulescens*; (**E**) red-tailed hawk (juvenile), *Buteo jamaicensis*; (**F**) emu, *Dromaius novaehollandiae*.

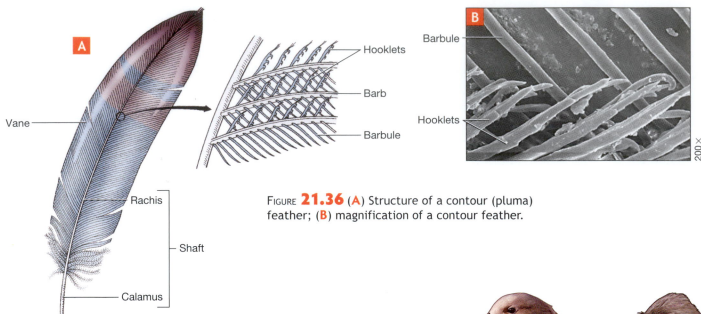

Vane, Hooklets, Barb, Barbule, Rachis, Shaft, Calamus, Barbule, Hooklets

200 ×

FIGURE **21.36** (**A**) Structure of a contour (pluma) feather; (**B**) magnification of a contour feather.

of the neck (cervical) vertebrae. The tail, or caudal, vertebrae are fused, forming a **pygostyle**. The sturdy pelvic girdle allows birds to walk and perch. In many birds, the sternum is keeled, allowing for the attachment of flight muscles. The "wishbone," or **furcula**, of a bird consists of fused clavicles that allow for flight. Bird skulls are highly fused and feature large orbits to accommodate the eyes. Birds have a single **occipital condyle.** A ring of bones called **sclerotic bones** encircles and supports the eyes of birds. Flight muscles are massive in birds of flight. Bird **talons** (claws), such as those of eagles, can produce great force.

The circulatory system of birds features a large, strong, four-chambered heart. Metabolism and heart rate are heightened. A chicken may have a resting heart rate of 250 beats per minute (bpm) and an active hummingbird more than 1,200 bpm. Birds have an advanced immune system. Their respiratory system is complex. Much of the air inhaled goes into a system of nine air sacs located primarily in the thorax and the abdomen. The **air sacs,** coupled with the lungs, keep the bird richly supplied with air to fuel their demanding oxygen requirement. A **syrinx**, located at the junction where the trachea forks into the lungs, allows birds to produce myriad sounds.

Birds exhibit a variety of feeding preferences and habits. The keratinized, toothless beaks of birds are highly specialized for specific diets (Fig. 21.37).

The digestive system is complete; the crop is an enlargement of the esophagus that serves as a storage chamber. The stomach has two unique compartments: the **proventriculus** and the **gizzard**. The proventriculus secretes gastric juices to help chemically break down food, and the muscular gizzard holds pebbles that help in the mechanical breakdown of food via grinding. The combined actions of these compartments is crucial for digestion, because birds do not have teeth. Solid wastes, urine, and reproductive cells exit the body by means of the **cloaca.** Birds excrete nitrogenous wastes as uric acid crystals. Marine birds have salt glands above each eye to eliminate excess salt.

Generalist, Insect catching, Grain eating, Nut cracking, Nectar feeding, Fruit eating, Chiseling, Dip netting, Surface skimming, Filter feeding, Probing, Pursuit fishing, Scavenging, Flesh eating

FIGURE **21.37** Variation in bird beaks.

21

In birds, the nervous system is well developed. The three major portions of the brain are:

1. The cerebrum, which primarily controls complex behavior patterns, migration, navigation, mating behavior, and nest building;

2. A fine-tuned cerebellum that controls muscular coordination and flight-related matters; and

3. Large optic lobes responsible for acute vision and association.

Birds have good color vision and excellent monocular and binocular vision. A nictitating membrane lubricates and protects the eye. With the exception of flightless birds, waterfowl, and several other groups, the senses of smell and taste are poorly developed in birds. Hearing in birds, like vision, is highly developed.

All birds are oviparous. In most male species, the **testes** become active only during breeding season. Most male bird species lack a penis. In females, only the left **ovary** and **oviduct** develop. To reproduce, most birds must bring their cloacal surfaces together. Fertilization occurs in the upper region of the oviduct before the albumen (egg white) and shell are added to the egg. Eggs usually are laid in a nest and are incubated by one or both parents. Upon hatching, some young, such as chicks and ducklings, are **precocial**—ready to run or swim. Other birds, such as the mockingbird and canary, are **altricial** (helpless), requiring parental care for a period of time.

Procedure 1

Macroanatomy of Birds

Materials

- ❏ Dissecting microscope or hand lens
- ❏ Compound microscope
- ❏ Dissecting tray
- ❏ Probe
- ❏ Variety of bird feathers
- ❏ Microscope slides of bird feathers
- ❏ Skeleton of a pigeon (*Columba* spp.)
- ❏ Select specimens of representative birds
- ❏ Colored pencils

1 Procure equipment and specimens.

2 Observe select specimens of representative birds placed on a dissecting **tray.** Record your observations and labeled sketches in the spaces provided. Use a probe to move the specimen. Along with the common name, when possible, provide the order and the scientific name of the specimen.

Specimen 1

Specimen 2

Specimen 3

3 Observe the available feather specimens with a dissecting microscope or hand lens. Record your observations and labeled sketch in the space provided.

4 Observe a microscope slide of a feather, using a compound microscope noting the ultrastructure of the feather. Record your observations and labeled sketches in the space provided below.

5 Using the hand lens, observe the skeleton of the pigeon, and compare it with Figure 21.38. Pay particular attention to the special features discussed in the procedure introduction.

6 Thoroughly clean your laboratory station and equipment. Return the dry equipment.

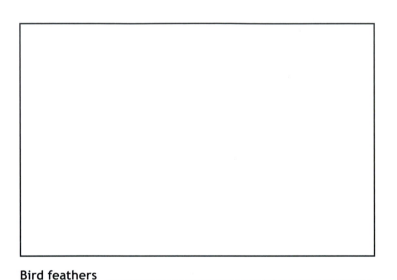

Bird feathers

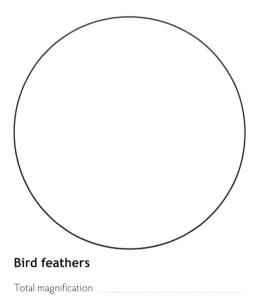

Bird feathers

Total magnification

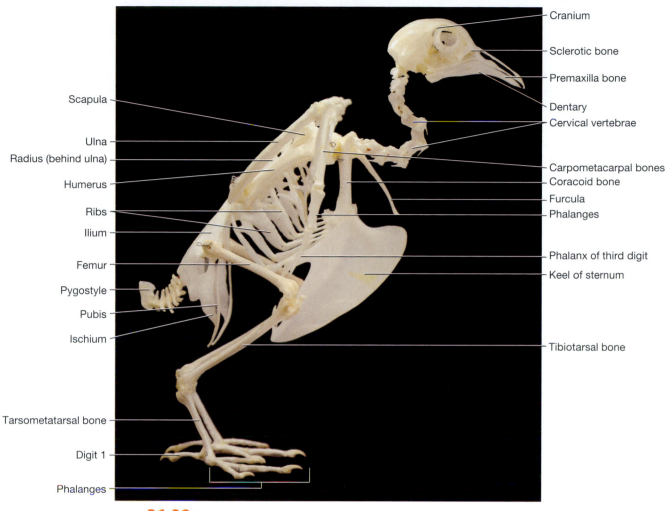

Labels on figure: Cranium, Sclerotic bone, Premaxilla bone, Dentary, Cervical vertebrae, Carpometacarpal bones, Coracoid bone, Furcula, Phalanges, Phalanx of third digit, Keel of sternum, Tibiotarsal bone

Scapula, Ulna, Radius (behind ulna), Humerus, Ribs, Ilium, Femur, Pygostyle, Pubis, Ischium, Tarsometatarsal bone, Digit 1, Phalanges

Figure **21.38** Skeleton of a pigeon, *Columba* sp.

✔ Check Your Understanding

6.1 What adaptations are found in birds that allow for flight?

6.2 Describe the process birds use to maintain their feathers.

Phylum Chordata, Class Mammalia

Members of class Mammalia have conquered marine, freshwater, aerial, and terrestrial habitats (Fig. 21.39). Presently, nearly 5,000 species of mammals have been described, constituting 26 orders. The monotremes, or prototherians, are represented by one order; the marsupials, or metatherians, are represented by seven orders; and the placentals, or eutherians, are represented by 18 orders.

Monotremes are egg-laying mammals. The oviparous duck-billed platypus and echidna are the vestiges of this ancient group. About 120 million years ago, the marsupial (having a **marsupium**, or pouch) mammals appeared. In marsupials, pregnant females develop a modified yolk sac in the womb, which provides the embryo nutrients, and they give birth at an early stage of embryological development. After birth, the newborn marsupial crawls to the pouch and attaches itself to a nipple for nourishment.

Shortly after the marsupials, another branch, the **placental** mammals, now the majority, appeared. In placental mammals the embryo remained in the uterus, receiving vital nutrients and oxygen from the mother for an extended time, allowing for further development of the brain. Mammals are dioecious animals and undergo internal fertilization. The placenta serves as the interface between the embryo and the mother. The young are born altricial (such as kittens and mice) or precocial (such as calves and dolphins).

The most obvious characteristic of mammals is the presence of hair. Hair can provide insulation and protection, serve as camouflage or warning, and give sensory feedback. The coat of a mammal is called its pelage. Vibrissae, or whiskers, are long, coarse hairs used in gathering tactile information. Mammals also possess a number of glands in the integument, including sweat glands, scent glands, sebaceous (oil) glands, and mammary glands. Members of class Mammalia are endothermic, maintaining their own body temperature. Mammals have a four-chambered heart and a respiratory system driven by a muscular diaphragm. Mammals can be carnivorous, herbivorous, or omnivorous, each with specific modifications to their complete digestive systems. Mammals exhibit diverse dentition based on their diet, from toothless anteaters to the conical, fish-eating homodont teeth of dolphins. Most mammals exhibit a number of tooth types (heterodont), including incisors, canines, premolars, and molars.

In mammals, the lower jaw is a single bone. The middle ear of mammals contains three **ossicles** (bones): the stapes (stirrup), incus (anvil), and malleus (hammer). With several exceptions, all mammals from mice to giraffes have seven cervical (neck) vertebrae. The nervous system of mammals is well developed and features a large cerebrum responsible for the majority of complex behaviors and learning. The senses are developed in various mammals depending upon their role in nature. The special sense of echolocation is extremely well developed in bats and dolphins.

FIGURE **21.39** Examples of class Mammalia: (**A**) echidna, *Tachyglossus aculeatus*; (**B**) eastern grey kangaroo, *Macropus giganteus*; (**C**) killer whale, *Orcinus orca*; (**D**) lion, *Panthera leo*; (**E**) meerkat, *Suricata suricatta*; and (**F**) gorilla, *Gorilla gorilla*.

Procedure 1

Macroanatomy of Mammals

Materials
- ❏ Dissecting microscope or hand lens
- ❏ Dissecting tray
- ❏ Probe
- ❏ Select specimens of representative mammals
- ❏ Cat skeleton
- ❏ Colored pencils

1 Procure equipment and specimens.

2 Observe the available mammal specimens using a dissecting microscope or hand lens. If permitted, remove the specimens from the jars and place them on the dissecting trays for closer examination. Use a probe to move the specimen. Pay close attention to the previous description of the organisms. Record your observations and labeled sketches in the spaces provided. Along with the common name, when possible, provide the order and the scientific name of the specimen.

3 Using the hand lens, observe the skeleton of the cat, and compare it with Figures 21.40 and 21.41. Pay particular attention to the special features discussed in the procedure introduction.

Specimen 1

Specimen 2

Specimen 3

Specimen 4

Specimen 5

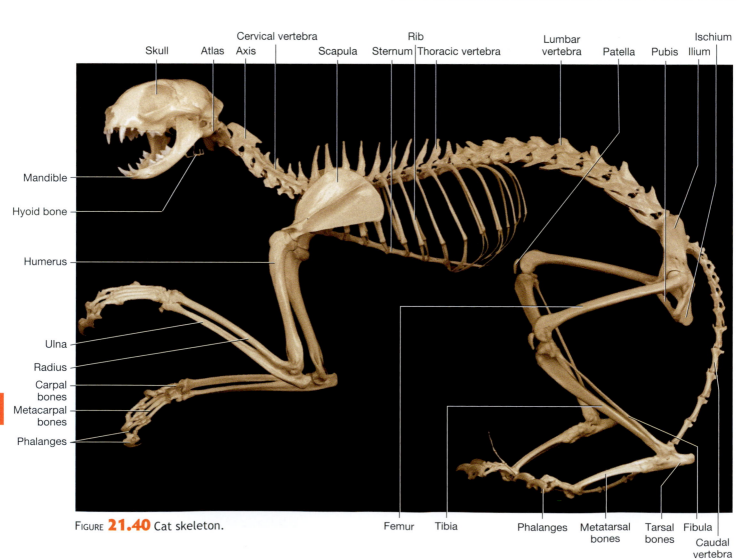

FIGURE **21.40** Cat skeleton.

Labels: Skull, Atlas, Axis, Cervical vertebra, Scapula, Sternum, Rib, Thoracic vertebra, Lumbar vertebra, Patella, Pubis, Ilium, Ischium, Mandible, Hyoid bone, Humerus, Ulna, Radius, Carpal bones, Metacarpal bones, Phalanges, Femur, Tibia, Phalanges, Metatarsal bones, Tarsal bones, Fibula, Caudal vertebra

21

Exploring Biology in the Laboratory: Core Concepts

4 Thoroughly clean your laboratory station and equipment. Return the dry equipment.

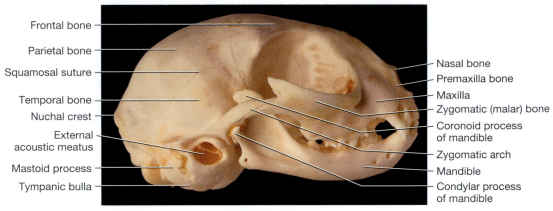

Frontal bone
Parietal bone
Squamosal suture
Temporal bone
Nuchal crest
External acoustic meatus
Mastoid process
Tympanic bulla

Nasal bone
Premaxilla bone
Maxilla
Zygomatic (malar) bone
Coronoid process of mandible
Zygomatic arch
Mandible
Condylar process of mandible

FIGURE **21.41** Cat skull, lateral view.

✔ Check Your Understanding

7.1 Describe three traits common to mammals.

7.2 List the three major kinds of mammals, and give two examples of each.

Mammal	Example 1	Example 2

21

Did you know . . .

Really?

Many perfumes have a unique and gross component: whale vomit! Ambergris, sometimes called whale vomit, is a foul smelling secretion from the bile ducts and intestines of sperm whales. It is thought to help sperm whales ease the passage of sharp objects such as squid beaks. Once it ages, ambergris has a sweet, earthy scent. Historically, it has been used as a fixative to allow perfumes to keep their fragrance longer. Today it has been replaced in most perfumes by ambriene, a plant product.

21

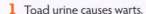

Animal Planet IV
Understanding the Deuterostomes

MYTHBUSTING

A Little Bird Told Me

Debunk each of the following misconceptions by providing a scientific explanation. Write your answers on a separate sheet of paper.

1 Toad urine causes warts.

2 Skinks, sometimes called scorpions, are very venomous.

3 If a skunk or raccoon comes out in the day, they have rabies.

4 If a snapping turtle bites you, it will not let go until it thunders.

5 Snakes have cold, slimy skin.

6 Don't touch that baby bird. The mother will pick up your scent and abandon her young!

1 In _____, the blastopore gives rise to the mouth, and in _____, the blastopore gives rise to the anus.

2 Sketch, label, and describe the parts that make up a typical contour bird feather.

3 Draw and label the internal anatomy of a typical sea star.

4 Which of the following are characteristics not found in the phylum Echinodermata? (*Circle the correct answer.*)

a. Pentamerous radial symmetry.

b. Water vascular system.

c. Complete digestive system.

d. Some members can eviscerate.

e. Members all have a body wall containing an endoskeleton of calcareous ossicles with surface spines.

5 Which of the following are found in the class Chondrichthyes? (*Circle the correct answer.*)

a. Gills covered with operculum.

b. 5–7 gill slits.

c. Placoid scales.

d. Both b and c.

e. Both a and c.

6 Give two examples of a deuterostome and two examples of a protostome.

_____

_____

7 Members of the subclass Actinopterygii make up 95% of vertebrate species and have the following characteristics: _____. (_Circle the correct answer._)

   a. It is a diverse class with examples such as goldfish, tuna, seahorses, and trout

   b. Organisms possess skin with mucous glands embedded with dermal scales

   c. Gills are covered with an operculum

   d. Organisms possess swim bladders and two-chambered hearts

   e. All of the above

8 Label the external anatomy of the fish in Figure 21.42.

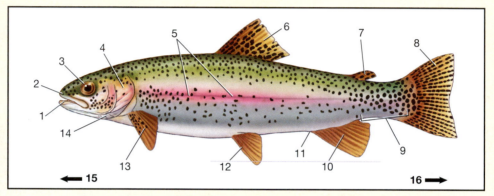

FIGURE **21.42** Fish, external anatomy.

1. _____	9. _____
2. _____	10. _____
3. _____	11. _____
4. _____	12. _____
5. _____	13. _____
6. _____	14. _____
7. _____	15. _____
8. _____	16. _____

21 **9** List the major classes of subphylum Vertebrata (Craniata), and give an example of each.

_____

_____

_____

_____

_____

The Cutting Edge
Understanding Vertebrate Dissections

I have destroyed almost the whole race of frogs ... For in this (frog anatomy) owing to the simplicity of the structure, and the almost complete transparency of the vessels, which admits the eye into the interior, things are more clearly shown so that they will bring the light to other more obscure matters.

— Marcello Malpighi (1628–1694)

Just wondering . . .

Consider the following questions prior to coming to lab, and record your answers on a separate piece of paper.

1 Are frogs capable of cutaneous respiration? If so, why?
2 What is the function of vacuities and vomerine teeth in frogs?
3 What is the significance of the attachment of the tongue in frogs?
4 In frogs, what is the urostyle and its function?
5 What is the value of dissecting in biology labs?

Objectives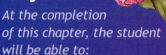

At the completion of this chapter, the student will be able to:

1. Observe and label the external anatomy of the bullfrog.
2. Dissect a bullfrog and identify specific muscles and internal structures.
3. Observe and label the external anatomy of the fetal pig.
4. Dissect a fetal pig and identify specific muscles and internal structures.
5. Relate the external and internal anatomy of the bullfrog and fetal pig to humans.

One thing that scientists have learned over the centuries is that vertebrate anatomy is a variation on a theme. Since the vertebrates, especially the tetrapods, including frogs, pigs, and humans, share common ancestry, they are built from the same basic components. Bone for bone, muscle for muscle, and internal structure for internal structure, there are no major significant differences between the frog, pig, and human.

Upon initial observation there seems to be a multitude of differences between various vertebrates. Color, shape, integument, lifestyle, and specialized features paint a diverse picture. After the initial descriptions of these animals, an anatomical theme begins to appear. This variation on a theme holds true and is tweaked by evolution.

Charles Darwin (1809–1882) was one of the first scientists to notice the similarity between the skeletons of the tetrapods and who commented that a pattern of skeletal development had been adapted over time by different environmental pressures to perform different functions. The comparative anatomist Richard Owen (1804–1892) noted that in the tetrapod limbs a pattern of one bone, two bones, and many bones was a reoccurring theme. With the discovery of *Tiktaalik*, Neil Shubin (1960–present) and colleagues showed that this common pattern emerged in the Devonian period about 375 million years ago. The same variation of a theme also holds true for the muscles and internal organs.

Note:
When you dissect these animals, remember that they have given their all for your learning experience. Treat them with respect and be awed by their incredible anatomy.

Chapter Photo
Domesticated pig, *Sus scrofa.*

Dissection of Common Chordates

Procedure 1

Bullfrog Dissection

Materials

- ❑ Dissecting microscope or hand lens
- ❑ Dissecting tray
- ❑ Probe
- ❑ Dissecting scissors
- ❑ Scalpel
- ❑ Water
- ❑ Safety glasses
- ❑ Lab coat or lab apron
- ❑ Gloves
- ❑ Paper towels
- ❑ Doubly injected preserved bullfrog
- ❑ Colored pencils

One way to distinguish male from female frogs is that during breeding season male frogs have thickened thumb pads known as nuptial tuberosities. They are used to grip the female during reproduction. Consider photographing this activity (Fig. 22.1).

1. Procure the doubly injected preserved bullfrog, and thoroughly rinse it in running water.

2. Place the bullfrog in a dissecting tray. Locate the dorsal and ventral external anatomical features labeled in Figure 22.2A. Label the external features of the bullfrog in the frog outline provided (Fig. 22.3).

3. Using a probe, pry open the mouth as wide as you can. Locate the anatomical features labeled in Figure 22.2B. Press on the eyes, and notice the movement of the vacuities. When a frog swallows, it blinks its eyes, and the vacuities help push the food into the esophagus. Record your observations and labeled sketch in the space provided.

FIGURE **22.1** American bullfrog, *Lithobates catesbeiana* (= *Rana catesbeiana*).

4. With a pair of sharp scissors carefully make a cut to include just the skin completely around the waist (about where a pair of pants would fit). Remove the top half of the skin (the "sweater") and the bottom half of the skin (the "pants"). Use a scalpel or a pair of scissors as an aid. After the skin is completely removed, begin comparing the musculature of the specimen with

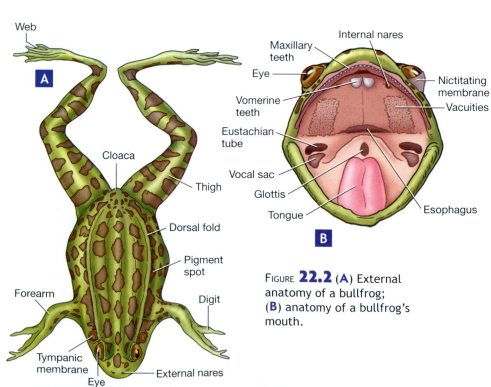

FIGURE **22.2** (**A**) External anatomy of a bullfrog; (**B**) anatomy of a bullfrog's mouth.

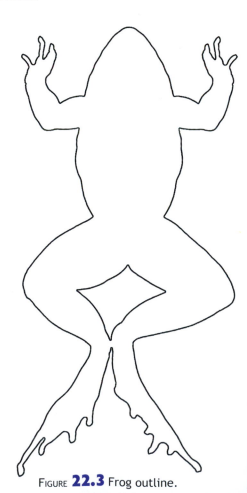

FIGURE **22.3** Frog outline.

22

Frog mouth

Figures 22.4 and 22.5. Draw and label either the dorsal muscles or ventral muscles of the bullfrog in the frog outline provided (Fig. 22.6), and label your illustration to indicate the view of your sketch.

5 Using sharp scissors, make two shallow incisions just through the muscles from the junction of the hindlegs upward to the bottom of the jaw just a few millimeters to each side of the linea alba. You will have to cut through the pectoral girdle.

6 Using the scalpel, separate the blue ventral abdominal vein from the muscles. Proceed to move the muscular flap. Using the scissors, cut laterally through the external oblique muscle on each side from the "shoulder" to the "waist." Remove the muscle, or pin it in place.

7 Determine the sex of the bullfrog. Locate the anatomical features in Figures 22.7–22.9. Observe the anatomy of a frog of the opposite sex from another laboratory station. Draw and label the external and internal features of the bullfrog on the two frog outlines provided (Fig. 22.10).

8 Thoroughly clean your laboratory station and equipment. Return the dry equipment, and discard the frog as indicated by the instructor.

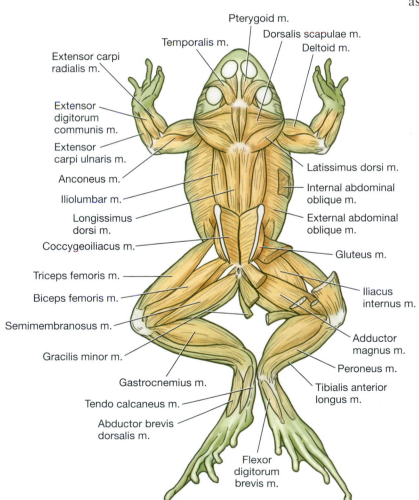

FIGURE **22.4** Diagram of the dorsal frog musculature.

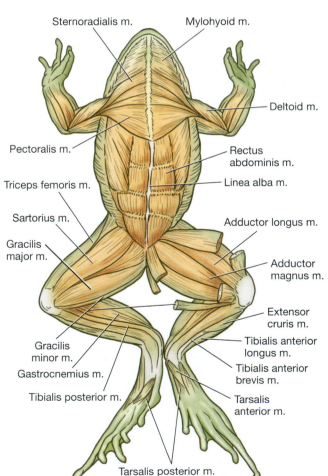

FIGURE **22.5** Diagram of the ventral frog musculature.

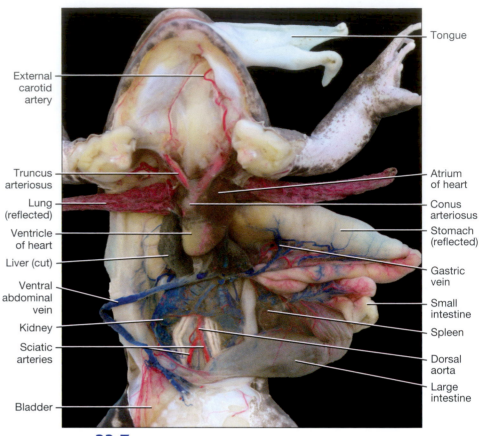

FIGURE **22.6** Frog outline for sketching muscles.

FIGURE **22.7** Internal anatomy of the frog.

Labels in Figure 22.7: Tongue, External carotid artery, Truncus arteriosus, Atrium of heart, Lung (reflected), Conus arteriosus, Ventricle of heart, Stomach (reflected), Liver (cut), Gastric vein, Ventral abdominal vein, Small intestine, Kidney, Spleen, Sciatic arteries, Dorsal aorta, Large intestine, Bladder

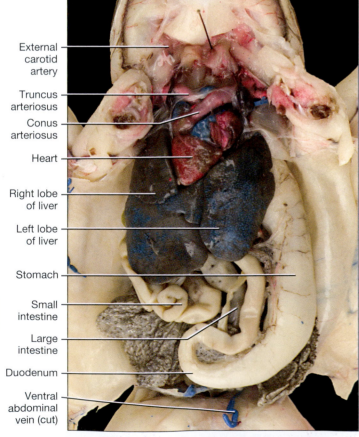

FIGURE **22.8** Ventral view of the frog viscera.

Labels in Figure 22.8: External carotid artery, Truncus arteriosus, Conus arteriosus, Heart, Right lobe of liver, Left lobe of liver, Stomach, Small intestine, Large intestine, Duodenum, Ventral abdominal vein (cut)

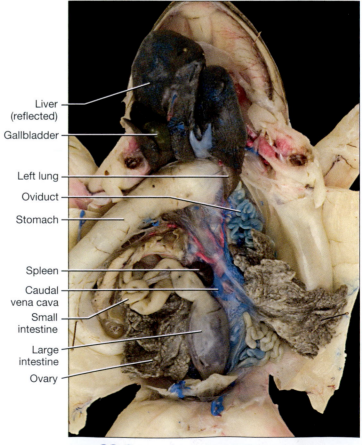

FIGURE **22.9** Deep view of the frog viscera.

Labels in Figure 22.9: Liver (reflected), Gallbladder, Left lung, Oviduct, Stomach, Spleen, Caudal vena cava, Small intestine, Large intestine, Ovary

22

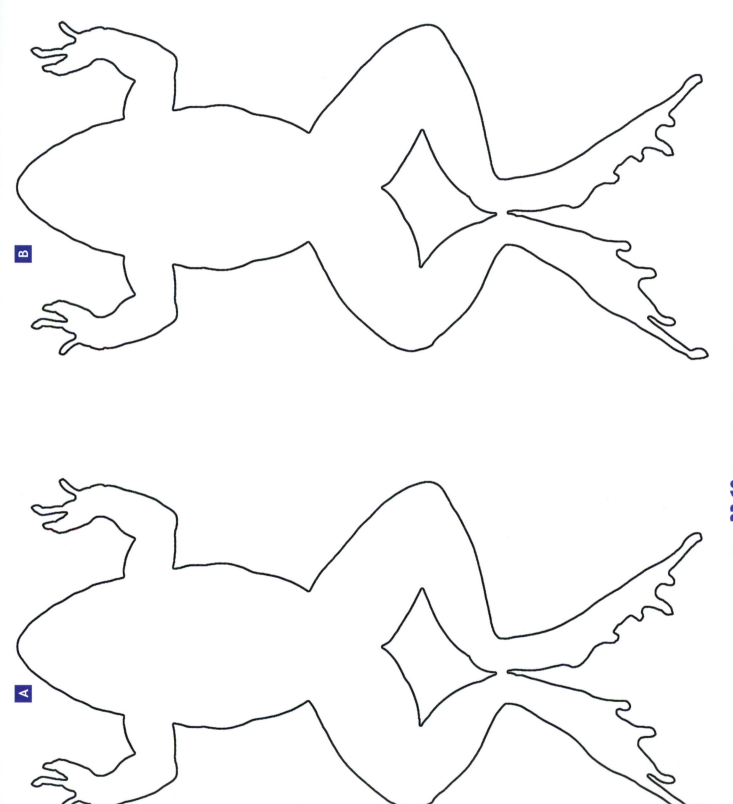

FIGURE **22.10** Frog outlines for sketching viscera.

Procedure 2

Fetal Pig Dissection

Materials

- ❏ Dissecting microscope or hand lens
- ❏ Dissecting tray
- ❏ Scalpel
- ❏ Safety glasses
- ❏ Lab coat or lab apron
- ❏ Gloves
- ❏ Paper towels
- ❏ Preserved fetal pig (*Sus scrofa*)
- ❏ Colored pencils

Consider photographing this activity.

1 Procure the needed equipment and specimen.

2 Thoroughly rinse the fetal pig in running water.

3 Examine the external anatomy of the fetal pig, comparing it with Figures 22.11 and 22.12.

4 Place your specimen on a dissecting tray, ventral side up. Using a sharp scalpel, make a shallow incision through the skin, extending from the chin caudally to the umbilical cord. Carefully continue your cut around one side of the umbilical cord. If your specimen is a male, make a diagonal cut from the umbilical cord to the scrotum. If it is female, continue a midventral incision from the umbilical cord to the genital papilla. Make an incision around the genitalia and tail.

5 From the midventral incision, extend an incision down the medial surfaces of the front legs to the hooves, and then do the same for the skin of the hindlegs. Make circular incisions around each of the hooves. Following the ventral borders of the lower jaws, make extended cuts from the chin dorsolaterally to just below the ears.

6 Grasp the cut edge of the skin, and carefully remove it from your specimen. If the skin is difficult to remove,

grasp the cut edge of the skin with one hand, and push on the muscle with the thumb of the other hand.

7 After the specimen is skinned, the internal anatomy can be seen more easily if the moisture is sponged away with a paper towel (Figs. 22.13–22.27). The muscles of a fetal pig are extremely delicate, and as you proceed to dissect your specimen, make certain you separate the muscles along their natural boundaries. When transection of a muscle is necessary, carefully isolate the muscle from its attached connective tissue and make a clean cut across the belly of the muscle, leaving the origin and insertion intact.

8 Record your observations and sketch and label the external and internal anatomy of the pig fetus in the two pig outlines provided (Fig. 22.28).

9 Thoroughly clean your laboratory station and equipment. Use the paper towel in the clean up. Return the dry equipment, and discard the fetal pig as indicated by the instructor.

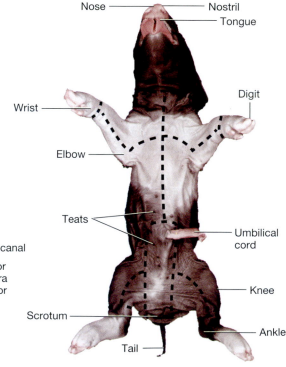

FIGURE **22.12** Ventral view of the surface anatomy of the male fetal pig, *Sus* sp. Dashed lines show suggested incisions for opening the pig.

FIGURE **22.11** Directional terminology and superficial structures in a male fetal pig, *Sus* sp. (quadrupedal vertebrate).

22

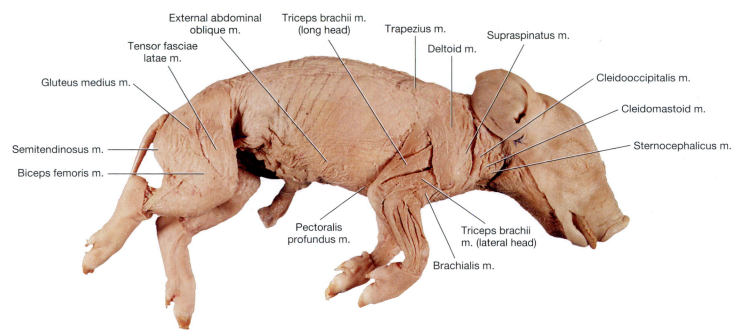

External abdominal oblique m.
Tensor fasciae latae m.
Gluteus medius m.
Triceps brachii m. (long head)
Trapezius m.
Deltoid m.
Supraspinatus m.
Cleidooccipitalis m.
Cleidomastoid m.
Semitendinosus m.
Biceps femoris m.
Sternocephalicus m.
Pectoralis profundus m.
Triceps brachii m. (lateral head)
Brachialis m.

FIGURE **22.13** Lateral view of superficial musculature of the fetal pig, *Sus* sp.

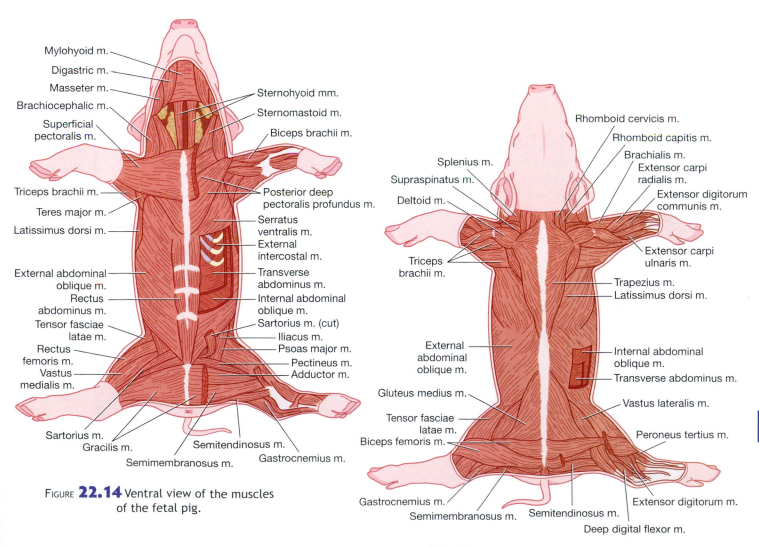

Mylohyoid m.
Digastric m.
Masseter m.
Brachiocephalic m.
Superficial pectoralis m.
Sternohyoid mm.
Sternomastoid m.
Biceps brachii m.
Triceps brachii m.
Teres major m.
Latissimus dorsi m.
Posterior deep pectoralis profundus m.
Serratus ventralis m.
External intercostal m.
External abdominal oblique m.
Transverse abdominus m.
Rectus abdominus m.
Internal abdominal oblique m.
Tensor fasciae latae m.
Sartorius m. (cut)
Iliacus m.
Psoas major m.
Rectus femoris m.
Pectineus m.
Vastus medialis m.
Adductor m.
Sartorius m.
Gracilis m.
Semitendinosus m.
Semimembranosus m.
Gastrocnemius m.

FIGURE **22.14** Ventral view of the muscles of the fetal pig.

Splenius m.
Supraspinatus m.
Deltoid m.
Rhomboid cervicis m.
Rhomboid capitis m.
Brachialis m.
Extensor carpi radialis m.
Extensor digitorum communis m.
Triceps brachii m.
Extensor carpi ulnaris m.
Trapezius m.
Latissimus dorsi m.
External abdominal oblique m.
Internal abdominal oblique m.
Transverse abdominus m.
Gluteus medius m.
Tensor fasciae latae m.
Vastus lateralis m.
Biceps femoris m.
Peroneus tertius m.
Gastrocnemius m.
Semimembranosus m.
Semitendinosus m.
Extensor digitorum m.
Deep digital flexor m.

FIGURE **22.15** Dorsal view of the muscles of the fetal pig.

22

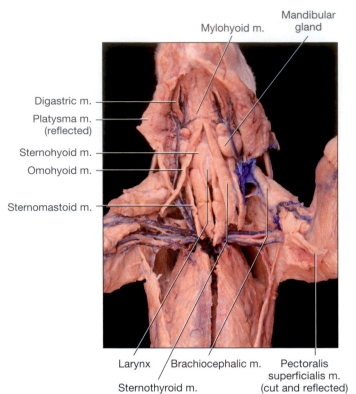

Mylohyoid m.
Mandibular gland
Digastric m.
Platysma m. (reflected)
Sternohyoid m.
Omohyoid m.
Sternomastoid m.
Larynx
Brachiocephalic m.
Sternothyroid m.
Pectoralis superficialis m. (cut and reflected)

FIGURE **22.16** Ventral view of superficial muscles of the neck and upper torso.

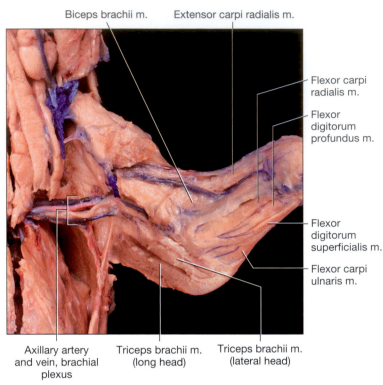

Biceps brachii m.
Extensor carpi radialis m.
Flexor carpi radialis m.
Flexor digitorum profundus m.
Flexor digitorum superficialis m.
Flexor carpi ulnaris m.
Axillary artery and vein, brachial plexus
Triceps brachii m. (long head)
Triceps brachii m. (lateral head)

FIGURE **22.17** Superficial medial muscles of the forelimb.

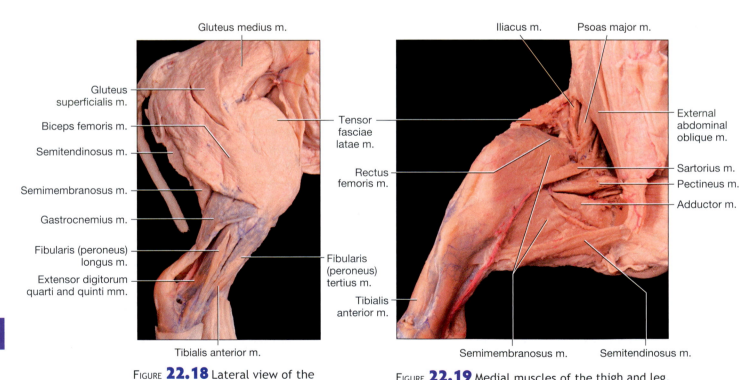

Gluteus medius m.
Gluteus superficialis m.
Biceps femoris m.
Semitendinosus m.
Semimembranosus m.
Gastrocnemius m.
Fibularis (peroneus) longus m.
Extensor digitorum quarti and quinti mm.
Tensor fasciae latae m.
Rectus femoris m.
Fibularis (peroneus) tertius m.
Tibialis anterior m.
Tibialis anterior m.

FIGURE **22.18** Lateral view of the superficial thigh and leg.

Iliacus m.
Psoas major m.
External abdominal oblique m.
Sartorius m.
Pectineus m.
Adductor m.
Semimembranosus m.
Semitendinosus m.

FIGURE **22.19** Medial muscles of the thigh and leg.

22

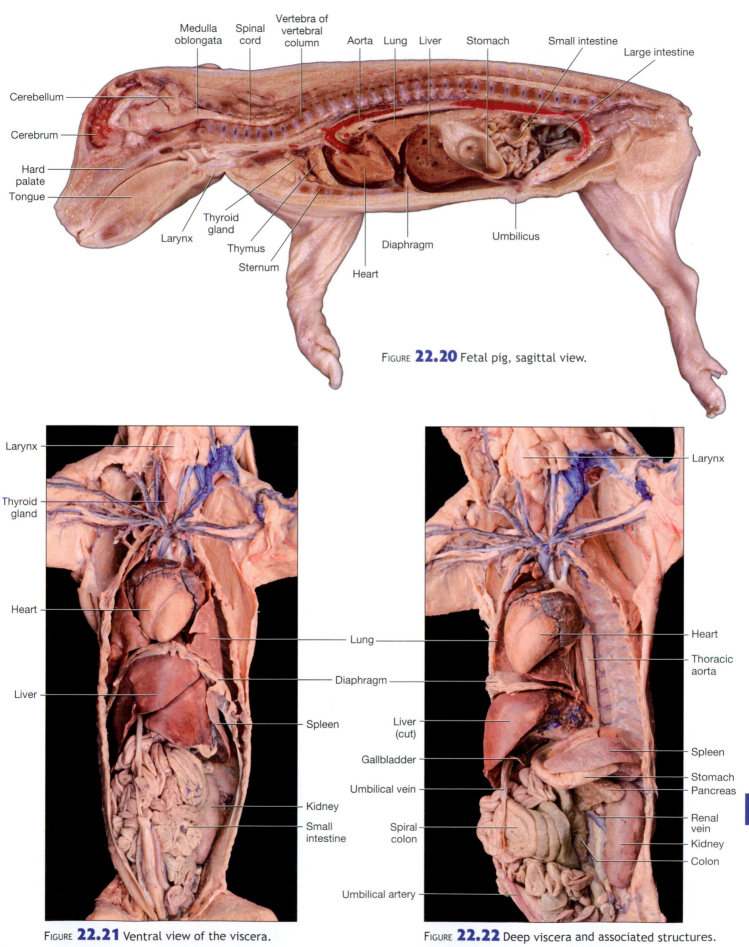

FIGURE **22.20** Fetal pig, sagittal view.

Labels in Figure 22.20: Cerebellum, Cerebrum, Hard palate, Tongue, Medulla oblongata, Spinal cord, Vertebra of vertebral column, Aorta, Lung, Liver, Stomach, Small intestine, Large intestine, Larynx, Thyroid gland, Thymus, Sternum, Heart, Diaphragm, Umbilicus

FIGURE **22.21** Ventral view of the viscera.

Labels in Figure 22.21: Larynx, Thyroid gland, Heart, Liver, Lung, Diaphragm, Spleen, Kidney, Small intestine

FIGURE **22.22** Deep viscera and associated structures.

Labels in Figure 22.22: Larynx, Heart, Thoracic aorta, Liver (cut), Gallbladder, Umbilical vein, Spiral colon, Umbilical artery, Spleen, Stomach, Pancreas, Renal vein, Kidney, Colon

22

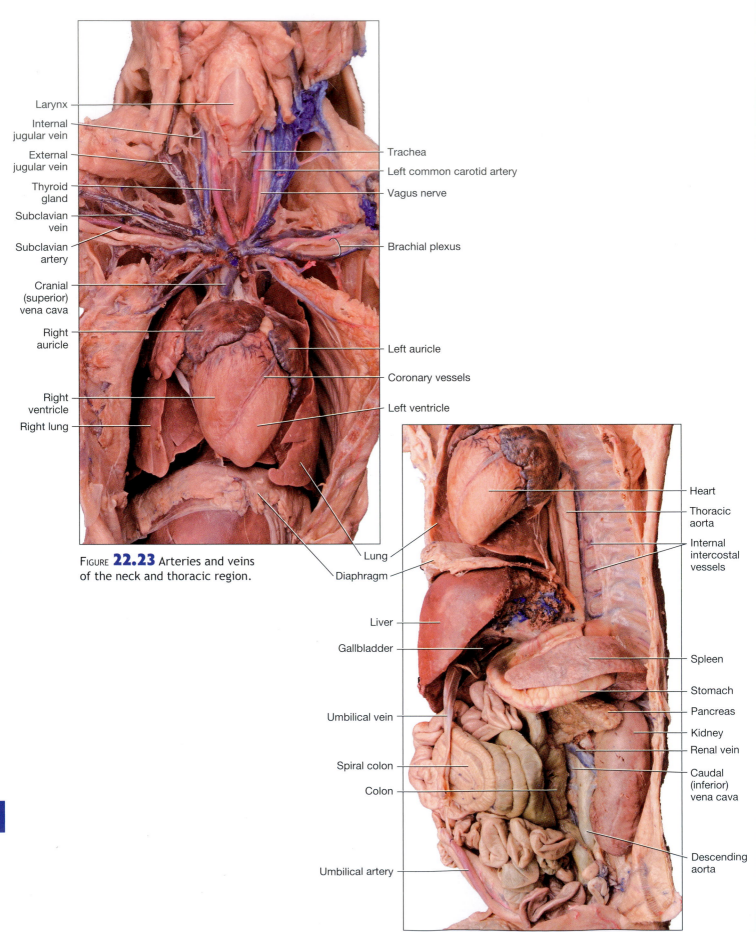

Larynx
Internal jugular vein
External jugular vein
Thyroid gland
Subclavian vein
Subclavian artery
Cranial (superior) vena cava
Right auricle
Right ventricle
Right lung

Trachea
Left common carotid artery
Vagus nerve
Brachial plexus
Left auricle
Coronary vessels
Left ventricle

FIGURE **22.23** Arteries and veins of the neck and thoracic region.

Lung
Diaphragm

Heart
Thoracic aorta
Internal intercostal vessels

Liver
Gallbladder

Umbilical vein

Spiral colon
Colon

Umbilical artery

Spleen
Stomach
Pancreas
Kidney
Renal vein
Caudal (inferior) vena cava

Descending aorta

FIGURE **22.24** Structures of the abdomen and lower extremities.

22

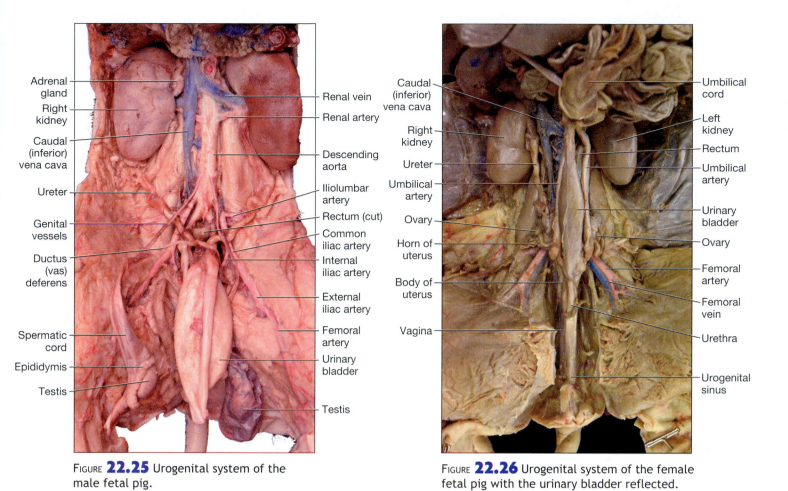

Adrenal gland
Right kidney
Caudal (inferior) vena cava
Ureter
Genital vessels
Ductus (vas) deferens
Spermatic cord
Epididymis
Testis

Renal vein
Renal artery
Descending aorta
Iliolumbar artery
Rectum (cut)
Common iliac artery
Internal iliac artery
External iliac artery
Femoral artery
Urinary bladder
Testis

FIGURE **22.25** Urogenital system of the male fetal pig.

Caudal (inferior) vena cava
Right kidney
Ureter
Umbilical artery
Ovary
Horn of uterus
Body of uterus
Vagina

Umbilical cord
Left kidney
Rectum
Umbilical artery
Urinary bladder
Ovary
Femoral artery
Femoral vein
Urethra
Urogenital sinus

FIGURE **22.26** Urogenital system of the female fetal pig with the urinary bladder reflected.

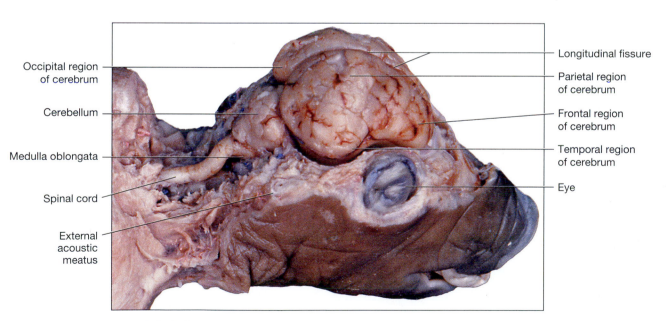

Occipital region of cerebrum
Cerebellum
Medulla oblongata
Spinal cord
External acoustic meatus

Longitudinal fissure
Parietal region of cerebrum
Frontal region of cerebrum
Temporal region of cerebrum
Eye

FIGURE **22.27** General structures of the fetal pig brain. Because the cerebrum is less defined in pigs, the regions are not known as lobes as they are in humans.

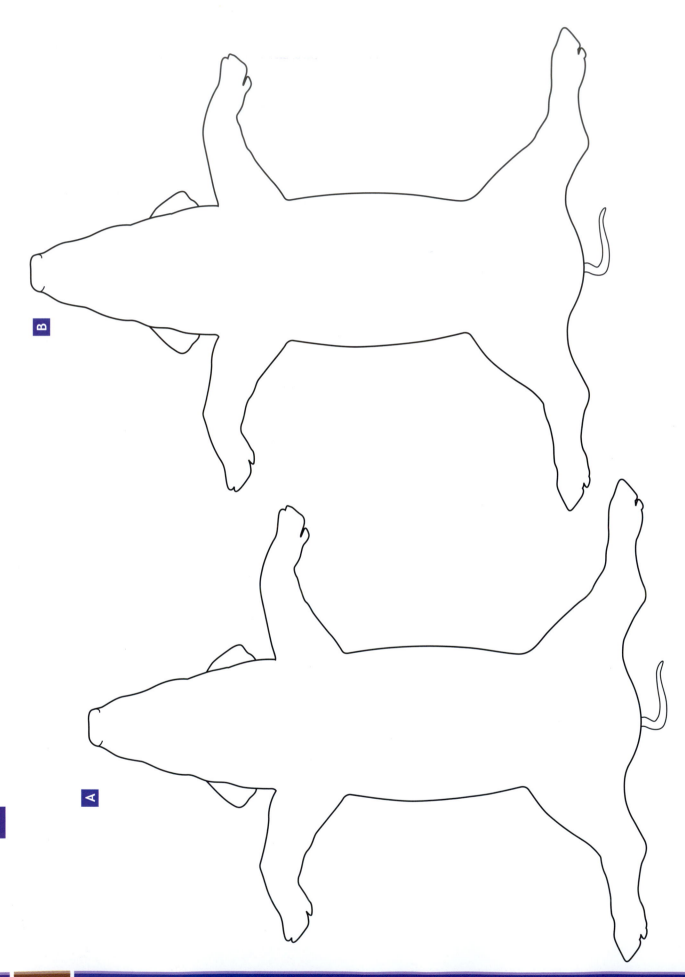

FIGURE **22.28** Pig outlines for sketching external and internal anatomy.

✔ Check Your Understanding

1.1 Name the two types of teeth that are found in the mouth of the bullfrog. What is their function?

1.2 Describe the skin of the frog. What is the importance of the skin to the frog and its transition to a terrestrial environment?

1.3 What is the "dome" of tissue found in the pig that separates the thoracic cavity from the abdominal cavity. What is its function in the mammal?

1.4 What structures did you note at the caudal end of your pig specimen?

1.5 Note similarities and differences in the external anatomy of the bullfrog and fetal pig.

1.6 Note similarities and differences in the internal anatomy of the bullfrog and fetal pig.

22

MYTHBUSTING

In a Pig's Eye

Debunk each of the following misconceptions by providing a scientific explanation. Write your answers on a separate sheet of paper.

1 Frogs have teeth in the maxilla and mandible.

2 The front appendages of frogs are short because this allows them to climb.

3 Fetal pigs have no anatomical similarity to humans. After all, a pig is a pig.

4 Frogs have internal fertilization.

5 That translucent covering of a frog's eye (nictitating membrane) serves as a magnifier.

1 Label the muscles in the ventral view of the bullfrog (Fig. 22.29).

1. _____
2. _____
3. _____
4. _____
5. _____
6. _____
7. _____
8. _____
9. _____
10. _____
11. _____
12. _____
13. _____
14. _____
15. _____
16. _____
17. _____
18. _____
19. _____

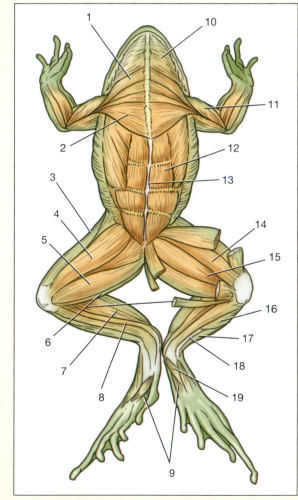

FIGURE **22.29** Frog musculature.

2 In some mammals, the _____ serves as the interface between the embryo and the mother.

3 Amphibians moved to a terrestrial environment; however, they did not evolve the ability to live their entire lives on land. Explain the reason that they are still dependent on an aquatic environment.

4 Describe how the tongues of bullfrogs (and other frogs) function. Where is the tongue attached in the bullfrogs compared to fetal pigs and humans?

5 Label the internal organs in the ventral view of the bullfrog (Fig. 22.30).

1. _____
2. _____
3. _____
4. _____
5. _____
6. _____
7. _____
8. _____
9. _____
10. _____
11. _____
12. _____

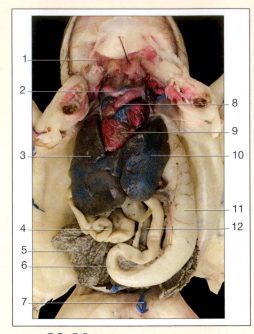

FIGURE **22.30** Bullfrog internal organs

6 Label the internal organs in the ventral view of the pig (Fig. 22.31).

1. _____
2. _____
3. _____
4. _____
5. _____
6. _____
7. _____
8. _____
9. _____

7 Why can we say that comparative vertebrate anatomy is a variation on a theme?

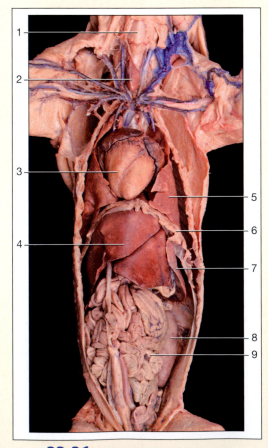

FIGURE **22.31** Viscera of the pig.

22

Homo sapiens I
Understanding Human Body Structure

23

Each one of you has something no one else has, or has ever had: your fingerprints, your brain, your heart. Be an individual. Be unique. Stand out. Make noise. Make someone notice. That's the power of individuals.

— Jon Bon Jovi (1962–present)

Just wondering...

Consider the following questions prior to coming to lab, and record your answers on a separate piece of paper.

1 What happens in Parkinson's disease?
2 What makes my muscles sore after working out for the first time in a long time?
3 Are carrots the most important food for good eye health?
4 What is the function of the gyri, or convolutions, in the brain?
5 Is there something to left-brain and right-brain research?

In 1871 when Charles Darwin (1809–1882) speculated about the origins of humans in his work *The Descent of Man and Selection in Relation to Sex*, he had no idea that more than a century later, the fossil record would yield a plethora of hominid fossils that would substantiate his ideas. Since the humble emergence of our species, we have come a long way in understanding our world and ourselves.

Like all other organisms, humans possess certain traits that make them unique members of the living world. *Homo sapiens* is a **generalist species**, being highly adaptable to a number of environments and ways of life. Human body type varies tremendously. The average size of a human is 1.5–1.8 meters tall and 54–83 kilograms in mass. Being **bipedal** (walking on two legs) frees up people's upper limbs to engage in a number of activities from operating a computer to catching a baseball. The **manipulative hands,** coupled with an **opposable thumb,** allow humans to display a **power grip** for grasping and a **technical grip** for fine activities such as writing.

Humans are **omnivorous,** consuming both plant and animal products. Humans have **stereoscopic color vision**. The **brain** is highly developed, capable of abstract reasoning, problem solving, tool-making, consequential thinking, and introspection. **Broca's area** of the brain in humans is well developed, allowing humans to have complex **language** skills. Humans are eutherians who usually give birth to one offspring at a time. The young have an extended **preadult** period in which to grow both physically and intellectually.

The human body is an incredible machine consisting of nearly 100 trillion cells working together to sustain life. These cells are organized into tissues, such as those discussed in Chapter 18. An organ is composed of at least two different types of tissue and has discrete boundaries, and an organ system is composed of the organs that work together to perform a specific function. Humans consist of several organ systems, outlined in Table 23.1.

You may want to review the basic anatomical orientation terms provided in Chapter 18 on page 381.

Objectives

At the completion of this chapter, the student will be able to:

1. List and describe characteristics that make us human.

2. Discuss the basic components of the integumentary system and their functions.

3. Identify the major bones of the skeletal system, compare and contrast the axial and appendicular skeletons, and discuss their functions.

4. Describe the basic anatomy and functions of the olfactory and gustatory structures.

5. Identify, describe, and discuss the structures and functions of the ear.

6. Identify, describe, and discuss the structures and functions of the eye.

7. List and describe the major components and functions of the digestive system.

Chapter Photo
Light micrograph of skeletal muscle fibers.

533

TABLE **23.1** Organ Systems of the Human Body

Organ System	Principal Organs	Basic Functions
Integumentary system	Skin, hair, nails, cutaneous glands	Protection, defense, thermoregulation, vitamin D synthesis, electrolytic balance, tactile information
Skeletal system	Bones, cartilage, ligaments	Movement, protection, support, blood formation, calcium storage
Muscular system	Muscles, tendons, aponeuroses (tendinous sheets)	Movement, support, production of heat
Nervous system	Brain, spinal cord, nerves, sense organs	Coordination, control, communications, sensory input, response, reflexes, cognition
Cardiovascular system	Heart, blood vessels, blood	Distribution of oxygen, nutrients, and hormones, removal of wastes
Respiratory system	Nose, sinuses, pharynx, larynx, trachea, bronchi, lungs, alveoli	Ventilation of lungs, provides oxygen to blood cells, removal of carbon dioxide, acid-base balance, generation of sounds
Lymphatic system	Lymph, lymphatic vessels, lymph nodes, spleen, thymus, tonsils, red bone marrow	Defense against pathogens, returns fluids to bloodstream, absorption of dietary lipids
Endocrine system	Pituitary gland, thyroid gland, parathyroid glands, hypothalamus, pineal gland, thymus, adrenal glands, pancreas, ovaries, testes	Coordination, control, communications, metabolic activities, regulation of sex cell production
Digestive system	Teeth, tongue, salivary glands, pharynx, esophagus, stomach, small intestine, large intestine, liver, gallbladder, pancreas	Ingestion, chemical and mechanical breakdown of food, digestion, absorption, elimination of wastes
Urinary system	Kidneys, ureters, urinary bladder, urethra	Storage and removal of urine and other wastes, regulation of blood pressure and blood volume, electrolytic balance, water balance, acid-base balance
Female reproductive system	Ovaries, uterine tubes, uterus, vagina, mammary glands	Production of the egg, site of fertilization, site of fetal development, nourishment of fetus, production of sex hormones, production of milk
Male reproductive system	Testes, epididymis, seminal vesicles, prostate gland, penis	Production and delivery of sperm, production of sex hormones

23

EXERCISE 23.1

Integumentary System

Surprisingly, the **integumentary system,** including the skin, hair, nails, and associated glands, is the largest organ system in the body, covering nearly 2 square meters. The integumentary system serves as a protective layer against ultraviolet (UV) light and harmful chemicals. It is the first line of defense against potentially harmful microbes. The system is involved in vitamin D synthesis, prevention of desiccation, regulation of body heat, excretion, and sensory reception.

The skin consists of a superficial layer known as the **epidermis,** a second layer called the **dermis,** and a deep subcutaneous layer called the **hypodermis.** The dermis is found beneath the epidermis. It may be thick in some regions of the body, such as the palms of the hands. The surface appears uneven because of the **dermal papillae,** which in the fingers can form fingerprints.

Procedure 1

Fingerprinting

Materials

- ❏ Printer's ink
- ❏ Ink roller, if necessary
- ❏ Stamp pad
- ❏ Hand lens
- ❏ Colored pencils
- ❏ Soap
- ❏ Paper towels

Fingerprint identification, or **dermatoglyphics,** is a method of personal identification using the impressions of fingertip ridges. No two persons (including identical twins) have exactly the same fingerprint patterns because patterns are based on the whorling patterns of the amniotic fluid, and the patterns remain unchanged throughout life. This procedure is designed to introduce the fascinating world of dermatoglyphics. The content and fingerprint types have been simplified. The true science of dermatoglyphics is meticulous and requires experience, a keen eye, and patience.

Work in groups of two to four students for this procedure.

1 Procure the needed equipment and supplies.

2 Make sure the ink is evenly distributed on the stamp pad. It can be recharged with printer's ink and spread evenly with an ink roller. Each member of the team should insert his or her fingerprints in the manual. Carefully place the medial edge of the thumb of the right hand against the stamp pad. Roll the thumb slowly and

naturally across the pad to the lateral edge of the thumb, covering the thumb in ink. Using the same movement, roll the thumb across the space in the space provided. One at a time, repeat for each finger.

3 Wash your hand thoroughly, and allow the ink enough time to dry.

4 Compare your fingerprints with Figure 23.1, and determine the specific pattern of your fingerprints with a hand lens Record your observations in the space provided, including any unusual patterns noted. More details on fingerprints can be found on websites such as http://www.encyclopedia.com/topic/fingerprint.aspx.

Right thumb	Index finger	Middle finger	Ring finger	Little finger

23

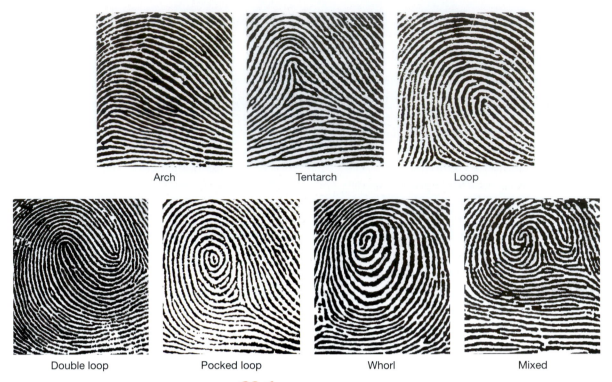

Arch Tentarch Loop

Double loop Pocked loop Whorl Mixed

FIGURE **23.1** Basic fingerprint types.

5 Compare your fingerprints with those of other members of your group. Record your observations in the space provided.

✔ Check Your Understanding

1.1 The dermal papillae within the epidermis are responsible for producing fingerprints. (*Circle the correct answer.*)

True / False

1.2 List the three fundamental skin layers and their location in relationship to each other.

1.3 How is the science of dermatoglyphics used today?

23

Skeletal System

The vertebrates, including humans, have an **endoskeleton** composed of cartilage or bone (as opposed to the chitinous **exoskeleton** of arthropods). In humans, the **skeletal system** consists of bones, cartilage, ligaments, and associated connective tissues. The bones protect internal organs; allow for movement and leverage; support the body; provide sites of attachment for muscles, ligaments, and tendons; allow for growth; store energy; provide a reservoir for calcium; aid in electrolytic and acid-base balance; and give rise to red blood cells in **hematopoiesis**.

Amazingly, at birth a human may have more than 270 bones, whereas the average adult human has approximately 206 bones, depending upon genetic and developmental factors. Distinct anatomical variation between individuals may result in the formation of more or fewer bones. The skeleton is divided into the axial skeleton and the appendicular skeleton (Table 23.2).

The **axial skeleton** consists of 80 bones including the skull, ossicles of the ear, hyoid bone, ribs, vertebrae, and sternum. These bones form the axis of the body, providing support to the body, attachment for muscles, stabilization of the appendicular skeleton, movements associated with respiration, and protection for the brain and internal organs.

The **appendicular skeleton** consists of 126 bones that include the pectoral girdle, the arms, the hands, the pelvic girdle, the legs, and the feet. The pectoral and pelvic girdles attach the appendicular skeleton to the axial skeleton, aid in locomotion, and allow for manipulation of objects.

TABLE **23.2** Skeletal System Terms of the Human Body

Bone	Description
Axial Skeleton	

Cranial Bones
    Frontal bone (1)
    Parietal bone (2)
    Temporal bone (2)
    Occipital bone (1)
    Sphenoid bone (1)
    Ethmoid bone (1)
	Bones in the skull that encase the brain

Facial Bones
    Mandible (1)
    Maxillae (2)
    Lacrimal bones (2)
    Nasal bones (2)
    Vomer (1)
    Inferior nasal
       conchae (2)
    Zygomatic bones (2)
    Palatine bones (2)
	Bones in the skull that form the face and house the sense organs

Hyoid	Horse-shoe shaped bone in the anterior neck region
Suture	An immovable joint between cranial bones
Vertebrae	24 bones that make up the vertebral column
Ribs	Bowed, flat bones that attach to the thoracic vertebrae and house the lungs and heart. 24 bones
Sternum	Flat bone that lies in the midline of the thorax
Appendicular Skeleton	

Pectoral Girdle
    Scapula
    Clavicle
	2 bones that frame the each shoulder

Humerus	Single bone of the upper arm
Radius	Lateral bone of the forearm
Ulna	Medial bone of the forearm
Carpals	8 bones found in the wrist
Metacarpals	5 long bones of the hand
Phalanges	14 long bones in the fingers

Pelvic Girdle
    Ilium
    Ischium
    Pubis
    Sacrum
    Coccyx
	Bones that attach the lower limbs to the axial skeleton

Femur	Single bone of the upper leg
Tibia	Single bone of anterior portion of lower leg
Fibula	Single bone of posterior portion of lower leg
Patella	Knee cap
Tarsals	7 bones forming the ankle
Metatarsals	5 long bones of the foot
Phalanges	14 long bones in the toes

Procedure 1

Labeling the Human Skeleton

Materials
- ❏ Articulated human skeletons
- ❏ Skull
- ❏ Colored pencils

1 Procure a skeleton and skull. Referencing Figures 23.2–23.5 and Table 23.2, observe the skeleton and skull, noting all anatomical features.

2 Label Figures 23.6–23.9 with the bones of the skeleton and skull listed in Table 23.2.

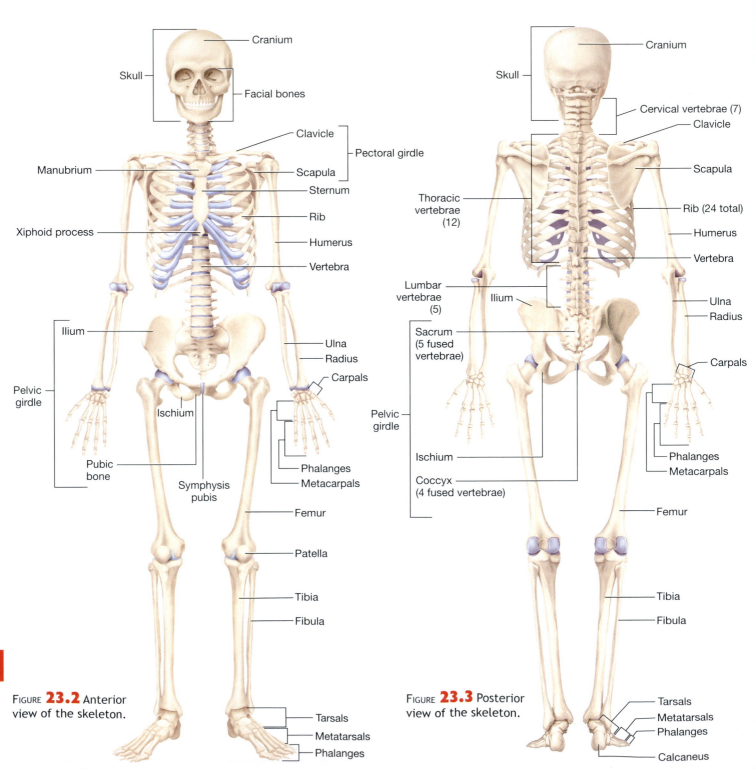

FIGURE **23.2** Anterior view of the skeleton.

FIGURE **23.3** Posterior view of the skeleton.

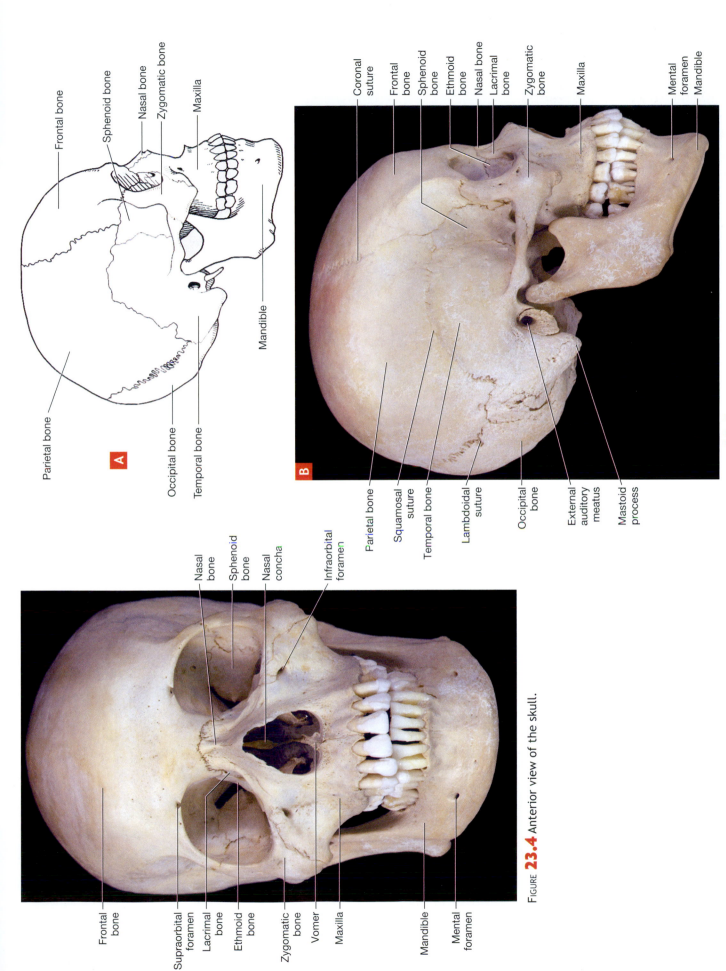

A

Frontal bone
Sphenoid bone
Nasal bone
Zygomatic bone
Maxilla
Mandible
Parietal bone
Occipital bone
Temporal bone

B

Coronal suture
Frontal bone
Sphenoid bone
Ethmoid bone
Nasal bone
Lacrimal bone
Zygomatic bone
Maxilla
Mental foramen
Mandible
Parietal bone
Squamosal suture
Temporal bone
Lambdoidal suture
Occipital bone
External auditory meatus
Mastoid process

FIGURE **23.5** (**A**) Illustration of human skull; (**B**) lateral view of the skull.

Nasal bone
Sphenoid bone
Nasal concha
Infraorbital foramen
Frontal bone
Supraorbital foramen
Lacrimal bone
Ethmoid bone
Zygomatic bone
Vomer
Maxilla
Mandible
Mental foramen

FIGURE **23.4** Anterior view of the skull.

23

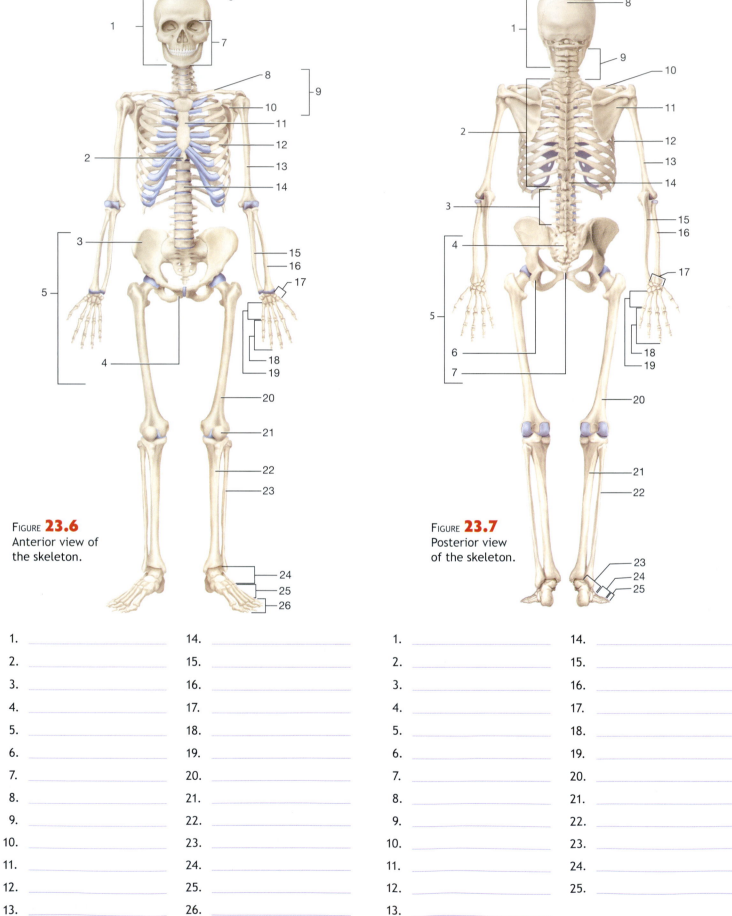

FIGURE **23.6**
Anterior view of
the skeleton.

FIGURE **23.7**
Posterior view
of the skeleton.

1. _____   14. _____   1. _____   14. _____
2. _____   15. _____   2. _____   15. _____
3. _____   16. _____   3. _____   16. _____
4. _____   17. _____   4. _____   17. _____
5. _____   18. _____   5. _____   18. _____
6. _____   19. _____   6. _____   19. _____
7. _____   20. _____   7. _____   20. _____
8. _____   21. _____   8. _____   21. _____
9. _____   22. _____   9. _____   22. _____
10. _____ 23. _____   10. _____   23. _____
11. _____ 24. _____   11. _____   24. _____
12. _____ 25. _____   12. _____   25. _____
13. _____ 26. _____   13. _____

23

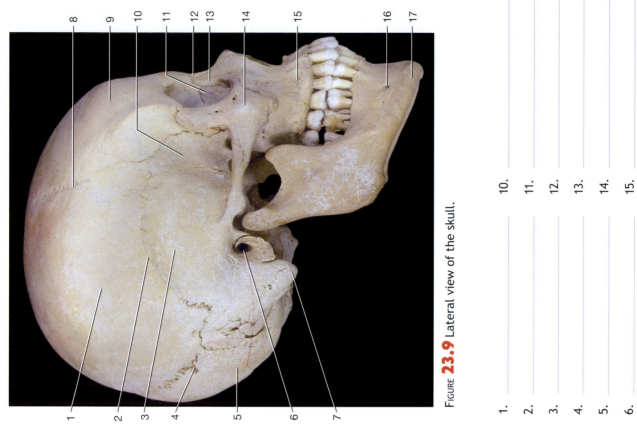

FIGURE **23.9** Lateral view of the skull.

1. _____
2. _____
3. _____
4. _____
5. _____
6. _____
7. _____
8. _____
9. _____
10. _____
11. _____
12. _____
13. _____
14. _____
15. _____
16. _____
17. _____

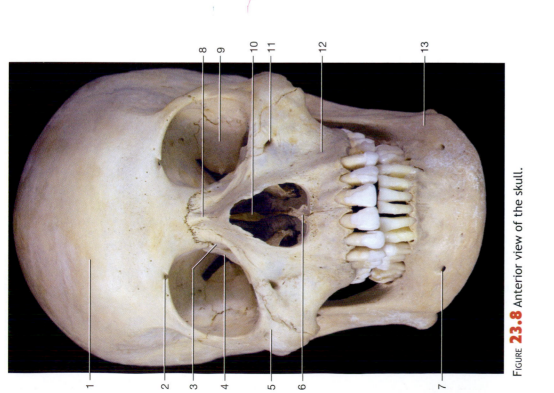

FIGURE **23.8** Anterior view of the skull.

1. _____
2. _____
3. _____
4. _____
5. _____
6. _____
7. _____
8. _____
9. _____
10. _____
11. _____
12. _____
13. _____

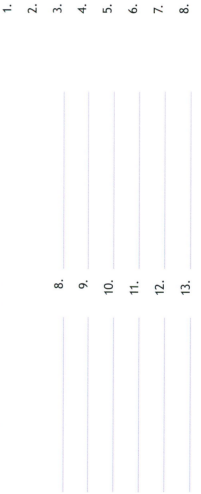

3 See if you can locate the following openings in your skull: supraorbital foramen, infraorbital foramen, mental foramen, external auditory meatus. What gives the mental foramen its name? Explain what each foramen is for.

4 See if you can locate the following sutures: frontal suture, squamosal suture, coronal suture, lambdoidal suture. Frontal sutures are present in children and allow for flexibility in the skull during the birthing process. This feature may persist into adulthood, or it may fuse together into one bone. Can you find it on the skull provided to you in lab?

5 Finally, look for the mastoid process, which is the site of attachment for several muscles.

✔ Check Your Understanding

2.1 Identify the following as characteristics of the axial skeleton or the appendicular skeleton:

_____ Bones included: skull, ossicles of ear, ribs, vertebrae, and hyoid bone

_____ Allows for manipulation of objects

_____ Bones included: pectoral girdle, arms, hands, pelvic girdle, legs, and feet

_____ Provides support to body, attachment for muscles

_____ Aids in locomotion

_____ Protects the brain and internal organs

2.2 Describe five functions of the skeletal system.

2.3 The largest organ system in the human body is the _____ . (*Circle the correct answer.*)
   a. integumentary system
   b. female reproductive system
   c. respiratory system
   d. cardiovascular system
   e. digestive system

23

Articulations

An **articulation** is a joint and refers to a point where two bones meet. Humans have approximately 147 articulations, which allow for movement and provide strength. Obvious examples of articulations include movable joints (**diarthroses, or synovial joints**), such as the knee, hip, ankle, elbow, and shoulder. Immovable joints (**synarthroses**) include the sutures of the skull and the gomphosis between the teeth and jaw. An **amphiarthrosis** is a slightly movable fibrous or cartilaginous joint. Examples include the junction between the pubic bones at the pubic symphysis, the connection between the tibia and fibula, and the articulation between the vertebrae at the intervertebral disks.

Procedure 1

Joint Movements

Materials

❏ Figure 23.10

1 The joints allow for a number of movements as indicated in Figure 23.10. Referring to Figure 23.10, review the joints, and with your lab group practice the various types of movement (almost like a version of Simon Says).

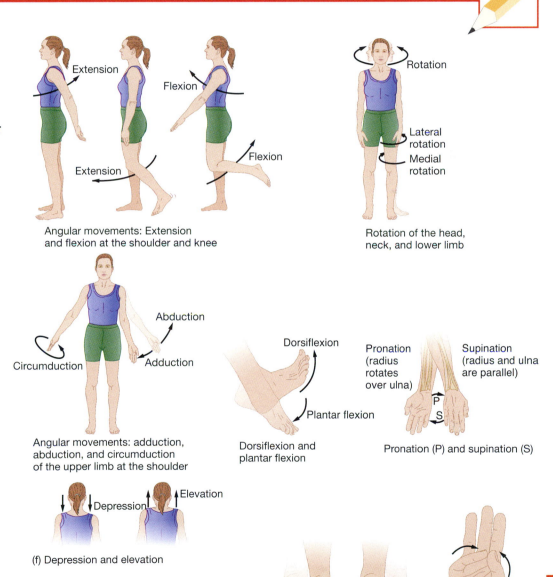

FIGURE **23.10** Motions of synovial joints.

2 Describe an activity that requires the following joint movements:

a. Flexion of the elbow

b. Extension of the knee

c. Opposition of the thumb

d. Eversion of the ankle

e. Adduction of the arm

f. Elevation of the shoulder

g. Rotation of the neck

h. Circumduction of the arm

✔ Check Your Understanding

3.1 Define articulation.

3.2 Define diarthroses and provide three examples.

23

EXERCISE 23.4

Muscular System

The three types of muscle tissue were discussed in Chapter 18. Review this information, particularly the information related to skeletal muscle. **Skeletal muscle** is responsible for movement, maintenance of posture, generation of body heat, and support of soft tissues. It also serves to guard and regulate bodily entrances and exits. A human has approximately 600 skeletal muscles, accounting for about 40% of body mass. Some major muscles of the human body are shown in Table 23.3.

Most skeletal muscles are attached to bone at both ends by tendons. A **tendon** attaches muscle to bone, and a **ligament** attaches bone to bone. An example of a tendon is the Achilles tendon (tendo calcaneous) of the lower leg, and an example of a ligament is the cruciate ligament of the knee. **Fascia**, a band or sheath of fibrous connective tissue, serves to cover muscles, and attaches muscle to skin. The **origin** of a muscle refers to the more stationary attachment, and the **insertion** refers to the more movable attachment (Fig. 23.11). The movement provided by a muscle is called **muscle action**. Skeletal muscles usually work in groups to provide movement.

TABLE **23.3** Major Muscles of the Human Body

Muscle Region	Description
Head and Neck	
Sternocleidomastoid	Lateral neck muscle
Trapezius	Triangular muscle in the back
Frontalis	Muscle of the forehead
Zygomaticus	Muscle of the upper cheek
Orbicularis oculi	Circular muscles around the eyes
Orbicularis oris	Circular muscle around the mouth
Masseter	Muscle of the jaw
Thorax, Abdomen, and Back	
Rectus abdominis	Flat muscle over the abdomen
External oblique	Diagonal muscle on either side of the torso
Deltoid	Triangular muscle over the shoulder
Pectoralis major	Triangular chest muscle
Latissimus dorsi	Flat muscle on the back
Infraspinatus	Triangular muscle underneath the scapula
Upper Limb	
Biceps brachii	Anterior arm muscle
Triceps brachii	Posterior arm muscle
Brachioradialis	Deep anterior forearm muscle
Brachialis	Lateral muscle of upper arm
Wrist/digit flexors (e.g., flexor carpi radialis)	Anterior forearm muscles
Wrist/digit extensors (e.g., extensor carpi ulnaris)	Posterior forearm muscles
Lower Limb	
Gluteus maximus	Muscles over the buttock
Sartorius	Muscle that crosses anterior thigh
Quadriceps femoris (includes rectus femoris, vastus lateralis, vastus medialis, and vastus intermedius; latter not shown)	Anterior thigh muscles
Adductor group (includes the adductor longus, adductor magnus, and adductor brevis; latter not shown)	Muscles on the medial thigh
Pectineus	Deep anterior thigh muscle
Gracilis	Medial thigh muscle
Hamstring group (includes biceps femoris, semitendinosus, and semimembranosus)	Posterior thigh muscles
Gastrocnemius	Posterior calf muscle
Tibialis anterior	Anterior calf muscle
Soleus	Deep posterior calf muscle

23

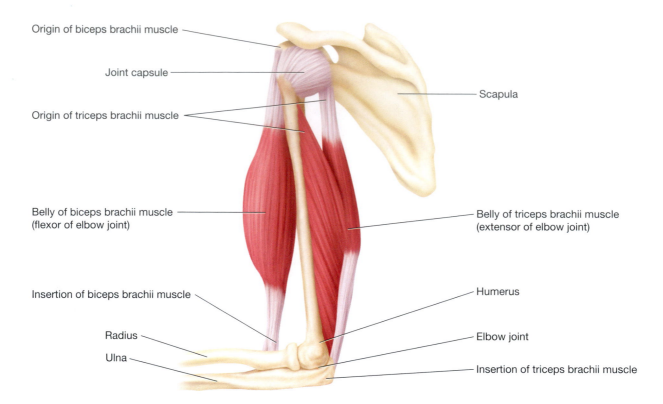

Origin of biceps brachii muscle

Joint capsule

Origin of triceps brachii muscle

Scapula

Belly of biceps brachii muscle
(flexor of elbow joint)

Belly of triceps brachii muscle
(extensor of elbow joint)

Insertion of biceps brachii muscle

Humerus

Radius

Ulna

Elbow joint

Insertion of triceps brachii muscle

FIGURE **23.11** Origin and insertion of biceps brachii muscle.

Procedure 1

Labeling the Superficial Muscles of the Human Body

Materials

- ❏ Figures 23.12–23.13
- ❏ Model of human superficial muscles, if available

1 Referencing Figures 23.12 and 23.13 and Table 23.3, locate and identify the muscles.

2 Label the muscles on Figures 23.14 and 23.15.

23

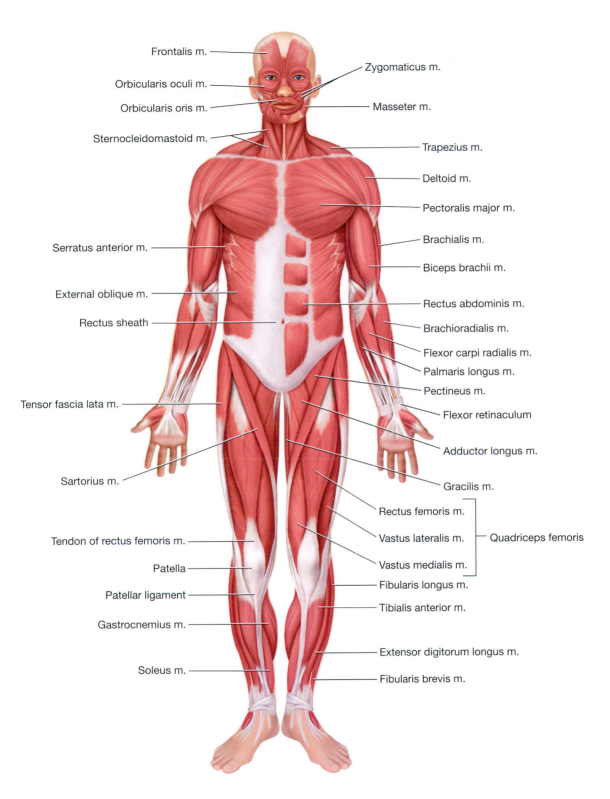

Frontalis m.

Zygomaticus m.

Orbicularis oculi m.

Orbicularis oris m.

Masseter m.

Sternocleidomastoid m.

Trapezius m.

Deltoid m.

Pectoralis major m.

Serratus anterior m.

Brachialis m.

Biceps brachii m.

External oblique m.

Rectus abdominis m.

Rectus sheath

Brachioradialis m.

Flexor carpi radialis m.

Palmaris longus m.

Pectineus m.

Tensor fascia lata m.

Flexor retinaculum

Adductor longus m.

Sartorius m.

Gracilis m.

Rectus femoris m.

Vastus lateralis m.

Quadriceps femoris

Tendon of rectus femoris m.

Vastus medialis m.

Patella

Fibularis longus m.

Patellar ligament

Tibialis anterior m.

Gastrocnemius m.

Extensor digitorum longus m.

Soleus m.

Fibularis brevis m.

FIGURE **23.12** Anterior view of human musculature (m. = muscle).

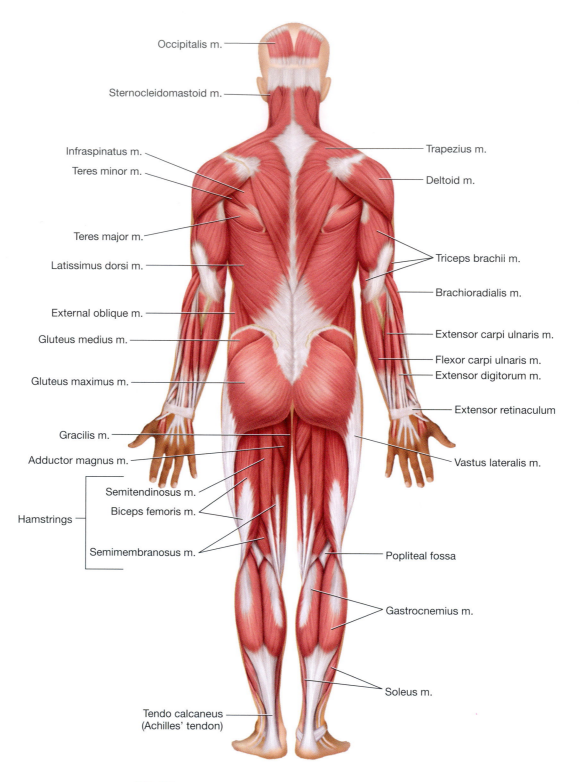

Occipitalis m.

Sternocleidomastoid m.

Infraspinatus m.

Teres minor m.

Teres major m.

Latissimus dorsi m.

External oblique m.

Gluteus medius m.

Gluteus maximus m.

Gracilis m.

Adductor magnus m.

Semitendinosus m.

Biceps femoris m.

Hamstrings

Semimembranosus m.

Tendo calcaneus
(Achilles' tendon)

Trapezius m.

Deltoid m.

Triceps brachii m.

Brachioradialis m.

Extensor carpi ulnaris m.

Flexor carpi ulnaris m.

Extensor digitorum m.

Extensor retinaculum

Vastus lateralis m.

Popliteal fossa

Gastrocnemius m.

Soleus m.

FIGURE **23.13** Posterior view of human musculature (m. = muscle).

23

Exploring Biology in the Laboratory: Core Concepts

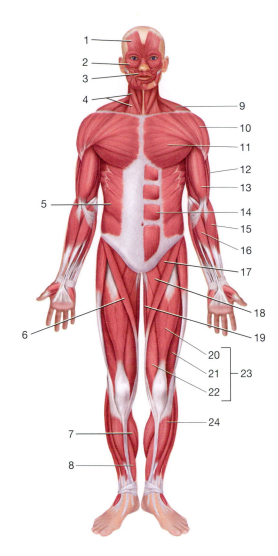

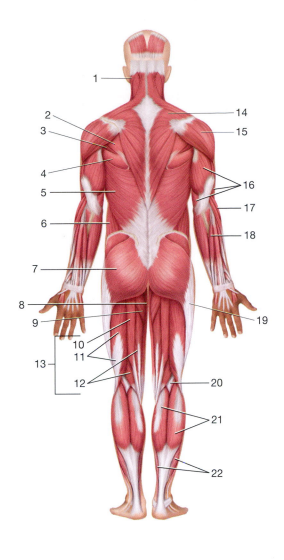

FIGURE **23.14** Anterior view of human musculature.

FIGURE **23.15** Posterior view of human musculature.

1. _____
2. _____
3. _____
4. _____
5. _____
6. _____
7. _____
8. _____
9. _____
10. _____
11. _____
12. _____

13. _____
14. _____
15. _____
16. _____
17. _____
18. _____
19. _____
20. _____
21. _____
22. _____
23. _____
24. _____

1. _____
2. _____
3. _____
4. _____
5. _____
6. _____
7. _____
8. _____
9. _____
10. _____
11. _____
12. _____

13. _____
14. _____
15. _____
16. _____
17. _____
18. _____
19. _____
20. _____
21. _____
22. _____

23

✔ Check Your Understanding

4.1 A _____ attaches muscle to bone and a _____ attaches bone to bone.

4.2 What is the difference between the origin and the insertion of a muscle?

_____

_____

_____

4.3 Match the description of the muscle with the type of muscle found in the head and neck region.

_____ Frontalis

_____ Trapezius

_____ Orbicularis oculi

_____ Masseter

A. Muscle of the jaw

B. Triangular muscle in the back

C. Circular muscle around the eyes

D. Muscle of the forehead

4.4 What is fascia, and what purpose(s) does it serve in the human body?

_____

_____

_____

4.5 Name the following muscles by the description of their location:

_____ Posterior calf muscle

_____ Lateral muscle of upper arm

_____ Lateral neck muscle

_____ Diagonal muscle on either side of the torso

_____ Muscle that crosses anterior thigh

_____ Anterior arm muscle

23

Nervous System

Your body is a fantastic machine comprised of a multitude of working parts from systems to individual cells hopefully working in harmony to maintain a state of homeostasis. One of the major functions of the nervous system is to coordinate and control these various parts. Many times we take these functions for granted until we are stricken with diseases such as amyotrophic lateral sclerosis (Lou Gehrig's disease) or Parkinson's disease. The nervous system consists of two major anatomical divisions: the **central nervous system (CNS)** and the **peripheral nervous system (PNS)**. The CNS consists of the brain and spinal cord, and the PNS consists of all nervous tissue outside of the brain and spinal cord, including cranial nerves, spinal nerves, and ganglia (Fig. 23.16).

The smell of a rose, the taste of your favorite candy, the sound of a soothing voice, and the sight of a majestic waterfall stimulate your senses. We develop an understanding of the world through a number of receptors, including those of the special senses. A **receptor** is a specialized structure that detects a given stimulus. Receptors are not limited to the special senses of smell

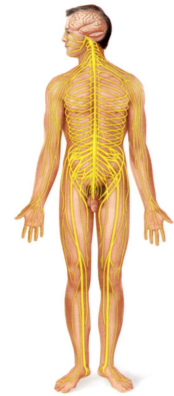

FIGURE **23.16** Nervous system.

(**olfaction**), taste (**gustation**), hearing, sound, and vision; receptors can also detect touch (**tactile**), pressure (**baroreceptors**), temperature (**thermoreceptors**), pain (**nociceptors**), and position (**proprioreceptors**). Through our senses and their intimate connection to the nervous system, we are able to interpret the world within and around us.

Regions of the Human Brain

The **brain** consists of billions of neurons and, by number, contains nearly 98% of the body's neural tissue (Table 23.4). Although brain measurements vary between individuals, the typical adult human brain weighs approximately 1.4 kilograms and has a volume of approximately 1,200 cubic centimeters.

The brain (Figs. 23.17 and 23.18) is covered by **meninges** that serve to protect and nourish it. The largest portion of the brain is the **cerebrum,** which interacts with all other parts of the brain and contains the centers for higher-level thought, learning, speech, and memory. It is divided into left and right **cerebral hemispheres** by a **longitudinal fissure**. The left hemisphere is more adept at analytical thought, mathematics, and language. The right hemisphere is more adept at artistic and musical skills, emotions, spatial relationships, and pattern recognition. **Sulci** divide each hemisphere into distinct lobes.

The **corpus callosum** is a thick collection of nerve fibers connecting the hemispheres. A superficial layer of **gray matter,** or the **cerebral cortex** (5 centimeters thick), covers the surface. Obvious ridges, the **gyri,** and shallow depressions, sulci, cover the surface. The deep depressions are called **fissures**. Gray matter contains the neuronal dendrites and cell bodies. The white matter, deep to the cortex, is largely composed of the axons of neurons surrounded by myelin, an insulating material that appears white because of its lipid content.

TABLE **23.4** Characteristics of the Human Brain

Region	Feature	Part	Description and Function
Hindbrain	Cerebellum		Motor coordination; monitoring sensory input, muscle movements, and muscle tone
	Pons		Connects cerebellum to other parts of the brain and spinal cord; contains sensory and motor nuclei of several cranial nerves and nuclei that involve sleep, respiration, swallowing, hearing, equilibrium, taste, eye movements, bladder control, and movements of the head
	Medulla oblongata		Center for autonomic regulation of heartbeat, breathing, constriction and relaxation of blood vessels, sneezing, coughing, gagging, vomiting, hiccupping, and swallowing
Midbrain	Superior colliculi		Vision
	Inferior colliculi		Hearing
	Red nucelus		Muscle tone and upper limb positioning
	Substantia nigra		Relaying of inhibitory signals to the thalamus
	Reticular formation		Cardiovascular control, pain modulation, visual tracking, consciousness and habituation
Forebrain	Diencephalon	Epithalamus	Contains pineal gland (an endocrine structure that secretes melatonin); circadian rhythms, day-night cycles, and regulation of reproductive functions
		Thalamus	Final relay point for ascending sensory information going to the primary sensory cortex; involved in emotion, motivation, touch, pain, temperature, position, and visual and auditory signals
		Hypothalamus	Thirst, eating, body temperature, and circadian rhythms
	Limbic system		Found along the border of the diencephalon and cerebrum; facilitates memory storage and retrieval, learning, establishing emotional states, and linking the conscious and unconscious functions of the cerebral cortex
	Cerebrum*	Frontal lobe	Memory, planning, emotion, speech, judgment, mood, voluntary control of skeletal muscle, and aggression
		Parietal lobe	Perception of touch, pressure, pain, taste, and temperature
		Temporal lobe	Hearing, smell, memory, emotional behavior, and visual recognition
		Occipital lobe	Visual centers of the brain

*There are five cerebral lobes. The four listed are superficial; the fifth lobe, the insula, is deep to the parietal and temporal lobes and is less well characterized than the superficial lobes.

23

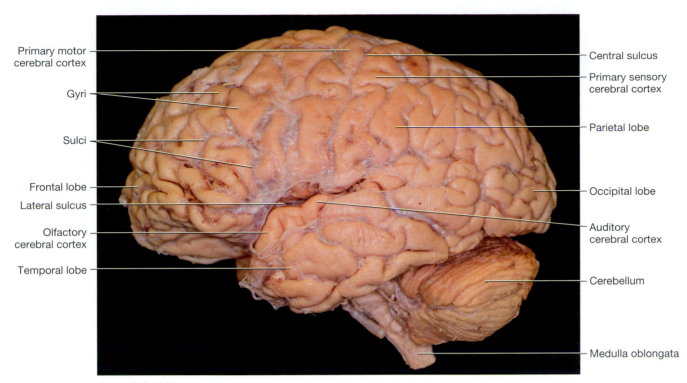

Primary motor cerebral cortex
Gyri
Sulci
Frontal lobe
Lateral sulcus
Olfactory cerebral cortex
Temporal lobe

Central sulcus
Primary sensory cerebral cortex
Parietal lobe
Occipital lobe
Auditory cerebral cortex
Cerebellum
Medulla oblongata

FIGURE **23.17** Lateral view of the brain.

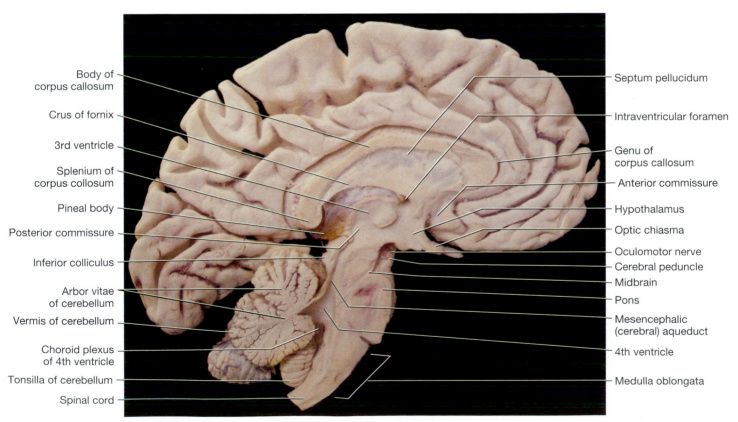

Body of corpus callosum
Crus of fornix
3rd ventricle
Splenium of corpus collosum
Pineal body
Posterior commissure
Inferior colliculus
Arbor vitae of cerebellum
Vermis of cerebellum
Choroid plexus of 4th ventricle
Tonsilla of cerebellum
Spinal cord

Septum pellucidum
Intraventricular foramen
Genu of corpus callosum
Anterior commissure
Hypothalamus
Optic chiasma
Oculomotor nerve
Cerebral peduncle
Midbrain
Pons
Mesencephalic (cerebral) aqueduct
4th ventricle
Medulla oblongata

FIGURE **23.18** Sagittal view of the brain.

23

Procedure 1

Labeling the Human Brain

Materials
- ❏ Brain model
- ❏ Colored pencils

1 Procure a brain model. Referencing Figures 23.17 and 23.18, observe the model, noting all anatomical features.

2 Label and color Figures 23.19 and 23.20 with the anatomical features of the brain listed in Table 23.4.

3 Return the brain model.

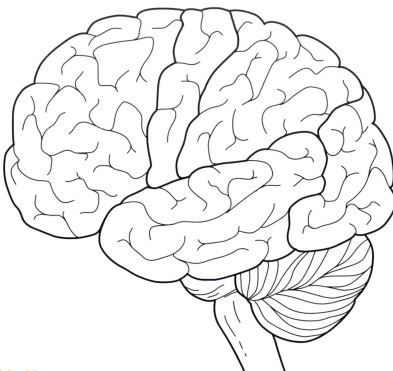

FIGURE **23.19** Lateral view of the brain.

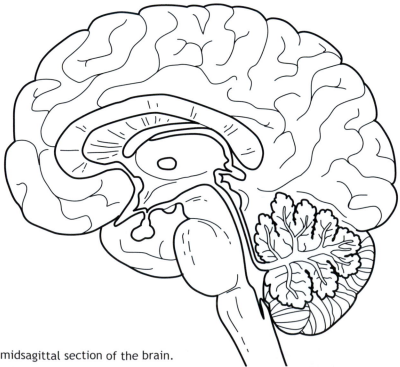

FIGURE **23.20** Medial view of the midsagittal section of the brain.

✔ Check Your Understanding

5.1 The _____ serves to protect and nourish the brain.

5.2 Describe the corpus callosum and its function.

_____

_____

_____

_____

5.3 The medulla oblongata _____. (*Circle the correct answer.*)
a. is part of the hindbrain
b. is part of the midbrain
c. serves as the center for autonomic regulation of heartbeat, breathing, and constriction and relaxation of blood vessels
d. serves as the center for perception of touch, pressure, and taste
e. both a and c

5.4 Identify the function described by the correct superficial lobe of the cerebrum, using the following key:

OL = Occipital lobe   FL = Frontal lobe   TL = Temporal lobe   PL = Parietal lobe

_____ Memory, emotion, speech, planning, and judgment

_____ Visual center of brain

_____ Perception of touch, pressure, and taste

_____ Hearing, smell, memory, emotional behavior, and visual recognition

_____ Perception of pain and temperature

_____ Aggression, voluntary control of skeletal muscle, and mood

Did you know . . . ?

Are You Habituated Yet?

The reticular formation includes neurons in diverse portions of the brain. It is involved in habituation, somatic motor control, sleep and consciousness, and pain modulation. In habituation, the brain learns to ignore repetitive, meaningless stimuli while remaining sensitive to other stimuli. Thus, a person can sleep through loud traffic in a large city, but can be awakened by an alarm. Habituation of the reticular formation allows many students to study with the iPod blaring or to ignore a noisy air conditioner in a lecture hall. For example, just record a lecture and notice all the sounds that you filter out such as people coughing and shuffling their feet.

23

Special Senses of Smell and Taste

Olfaction

The senses of smell (olfaction) and taste (gustation) involve special receptors classified as **chemoreceptors**. **Olfactory receptors** are located in the roof of the **nasal cavity**. The average human can distinguish between 2,000 and 4,000 distinct odors. One common classification of odors recognizes eight primary odors humans can detect:

1. **Putrid** (e.g., rotten eggs);

2. **Pungent** (e.g., vinegar);

3. **Minty** (e.g., peppermint);

4. **Floral** (e.g., rose);

5. **Musky** (e.g., ferret musk);

6. **Ethereal** (e.g., solvents);

7. **Camphoraceous** (e.g., eucalyptus oil); and

8. **Fishy** (fish).

These eight basic odors can be combined to produce myriad unique odors. As examples, the smell of a jasmine flower employs more than 90 odoriferous compounds, and the smell of cow manure is composed of 75 odoriferous compounds. In this activity, designate a test subject and do not allow them to review the test substance.

Procedure 1

Olfaction

Materials

❏ Small labeled bottles with removable caps containing a variety of substances provided by instructor. Examples may include but are not limited to: peppermint, wintergreen, freshly ground coffee, strawberry or other freshly cut fruit, garlic, vinegar, oregano, chocolate, vanilla, distilled water, rose oil, orange peel, cinnamon, eucalyptus oil, a pungent cheese, milk of magnesia, and fresh bread.

This procedure demonstrates the sense of olfaction using common substances. Work in groups of two to four for this procedure.

1 Each group will procure 5 of the 20 unknown bottles that you will use in this procedure.

2 Trial I: Instruct your partner to sit comfortably, close his or her eyes, and keep them closed when testing a substance. Remove the cap from the first bottle, and hold it approximately 5 cm from your partner's nose for about 3 seconds. Place the cap back on the bottle. Have your partner describe or identify the odor, and record his or her response in Table 23.5. Allow ample time for your partner to cleanse his or her palate.

3 Repeat this step for the other bottles in your possession. Do not let the test subject know the results until the activity is completed.

4 Trial II: Repeat steps 2 and 3 using the same bottles in a different order. Record the results in Table 23.5. Return the bottles, and discuss your results.

5 Tally the number of correct and incorrect responses offered by your partner for each trial. Compare your group's results with those of other groups.

a. Did you find any evidence that individuals vary in their ability to recognize odors?

_____

_____

b. Did you note any discrepancies between trials I and II of the olfaction test?

_____

_____

c. Which odors were the hardest to detect?

_____

_____

23

TABLE **23.5** Olfactory Identification of Substances

Substance	Identification or Description of Substance in Trial I	Identification or Description of Substance in Trial II	Correctly Identified Substance? (Y/N)
1.			
2.			
3.			
4.			
5.			

Taste

Taste receptors are associated with taste buds located on the tongue (Fig. 23.21), inside the cheeks, and on the soft palate, pharynx, and epiglottis. An average human has between 3,000 and 4,000 taste buds, most of which are on the tongue. The tongue has four kinds of observable bumps, or **lingual papillae**. The most common papillae are called **filiform papillae** covering the front two-thirds of the tongue. Filiform papillae do not contain taste buds. These structures are generally rough (as observed more easily on a cat's tongue) in order to create friction and sense the texture of food.

Foliate papillae, more common in young children, are found on the sides of the tongue. On average, humans have between 5 and 6 foliate papillae located on each side of the tongue each containing about 115 taste buds.

Circumvallate papillae are large and located in a "V" shape near the back of the tongue. An average person possesses from 7 to 12 of these structures with as many as 250 taste buds per papilla. Circumvallate papillae contain about half of all taste buds in humans.

Fungiform papillae are shaped like mushrooms and are located all over the tongue, especially at the tip and on the sides. They appear as red spots on the tongue because they are richly supplied with blood. Humans have about 200 of these papillae. Fungiform papillae house between 1 and 15 taste buds each depending upon their location.

Physiologists recognize five primary tastes:

1. **Sweet** (sugars);
2. **Salty** (salt);
3. **Sour** (citrus fruits and acids);
4. **Bitter** (alkaline substances and caffeine); and
5. **Umami** (meaty broth and amino acids, such as glutamic and aspartic acid).

Recently, the presence of water receptors in the pharynx has been described by physiologists. The senses of smell and taste are important in helping us enjoy favorable stimulants and detect dangerous substances.

The primary taste sensations can be detected all over the tongue; however, specific regions of the tongue seem more sensitive than others to these sensations. The tip of the tongue is generally more sensitive to sweet tastes. The frontal lateral margins of the tongue are generally more sensitive to salty tastes, and the midlateral margins are more sensitive to sour tastes. Generally, the rear of the tongue is more sensitive to bitter tastes.

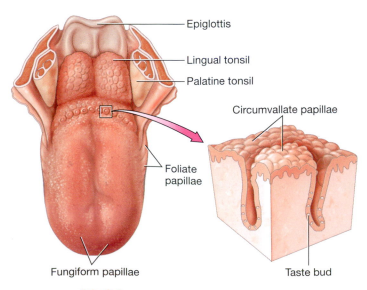

FIGURE **23.21** Generalized overview of the tongue.

Procedure 2

Gustation

Materials
- ❑ Paper cups
- ❑ Drinking water
- ❑ Clean paper towels
- ❑ Clean cotton swabs
- ❑ Small capped bottles with the following solutions:
  - 5% sucrose
  - vinegar
  - 5% NaCl solution
  - 0.5% quinine-sulfate solution

1 Procure the needed materials.

2 Have one member of the group thoroughly rinse out his or her mouth with water.

3 Using a clean, dry paper towel, have the student blot his or her tongue. You may have to do this multiple times. Dispose of the paper towel.

4 Dip a clean cotton swab into the 5% sucrose solution. Remove as much excess liquid from the swab as possible by pressing it against the side of the bottle.

5 If your lab partner permits (he or she may want to do this independently because it can trigger the gag reflex), gently touch specific regions of the tongue, cheek, gums, and roof of the mouth with the moistened tip of the swab. What regions of the tongue seemed more sensitive to sweet sensations?

6 Wait 5 minutes, and then repeat steps 2–5, using a vinegar solution. What regions of the tongue seemed more sensitive to sour sensations?

7 Wait 5 minutes, and then repeat steps 2–5, using a 5% NaCl solution. What regions of the tongue seemed more sensitive to salty sensations?

8 Wait 5 minutes, and then repeat steps 2–5, using a 0.5% quinine-sulfate solution. What regions of the tongue seemed more sensitive to bitter sensations?

9 Properly dispose of the cotton swabs and paper towels.

10 Repeat the same series of steps with another member of the group.

11 Based upon your results, briefly sketch the tongue and denote specific regions that were sensitive to certain tastes.

Tongue

Did you know . . . ?

Try This!

The senses of taste and smell are intimately connected and provide us with information on the presence of chemicals in our foods. As anyone with a cold can attest, food tastes different when the sense of smell is impaired. To illustrate this relationship, hold your nose when tasting a piece of flavored candy or even a small piece of an onion. What happens?

23

✔ Check Your Understanding

6.1 List the eight basic types of odors and give an example of each.

6.2 Describe any pattern observed in the location of the reception of the various taste sensations on the tongue.

6.3 Match the characteristics of the four types of lingual papillae with its location on the tongue. (Some will have more than one answer.)

_____ Filiform papillae

_____ Fungiform papillae

_____ Circumvallate papillae

_____ Foliate papillae

A. Sides of the tongue
B. Surface of the tongue (most common type)
C. All over the tongue, but more focused at tip and sides
D. "V" near back of the tongue
E. Papillae not involved with taste
F. Mushroom-shaped, with three tastebuds each
G. Large, contain about half of all taste buds
H. Rough; creates friction and senses the texture of food

6.4 List five basic types of taste.

23

Special Senses of Hearing and Equilibrium

The two major functions of the ear are **equilibrium** and **hearing**. Equilibrium involves rotation, linear acceleration, and gravity, allowing us to determine the position of the head with respect to the environment. Hearing (audition) detects and interprets sounds in the environment. The range of frequencies detected by various animals varies. In humans, the general audible range of frequencies is from 20 Hz (cycles per second) to 20,000 Hz, although individuals vary considerably with respect to genetics, sex, age, and occupation. (To gain a better understanding of frequencies, play with the equalizer on a stereo.)

The ear has three major regions:

1. The **outer ear** consists of the conspicuous **auricle**, or **pinna,** and the **external auditory meatus,** or canal (Fig. 23.22). The canal is lined with **ceruminous glands** that produce earwax, or **cerumen.**

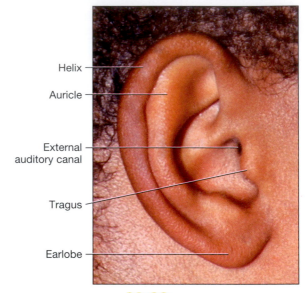

FIGURE **23.22** Surface anatomy of the auricle.

Did you know . . .

Oh, That Terrible Feeling!

Have you ever been seasick? If so, you know it's a miserable feeling. Seasickness happens when the brain receives conflicting messages about motion and body position. The conflicting information is delivered from the inner ear, eyes, and perhaps skin and muscle receptors. The symptoms of seasickness may include dizziness, profuse sweating, nausea, and vomiting.

2. The middle ear is primarily a cavity filled with air (Fig. 23.23). The middle ear contains the three **auditory ossicles**—the **malleus, stapes,** and **incus**—as well as the **tympanic membrane** (eardrum) and the **eustachian tube.**

3. The inner ear consists of the **vestibule** and the **semicircular canals** (both involved in equilibrium) and the **cochlea** (organ of hearing).

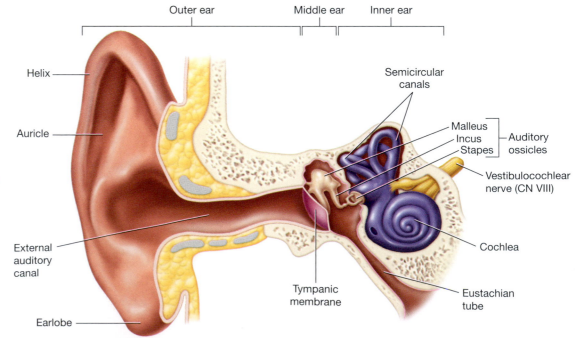

FIGURE **23.23** Generalized overview of the ear.

Procedure 1

Hearing Acuity Test

Materials

- ❏ Analog clock (one that ticks)
- ❏ Meterstick
- ❏ Sterile cotton
- ❏ Blindfold

Hearing acuity refers to the sharpness of hearing or the ability to detect a sound with respect to intensity and distance. This simple test measures hearing acuity.

1 Have your lab partner sit comfortably on a lab stool or chair.

2 Have your partner keep his or her eyes closed, or blindfold your partner during testing. Instruct your partner to keep his or her head stationary during testing.

3 Gently place a sterile cotton plug into your partner's left ear.

4 Hold a ticking clock approximately 0.5 m from your partner's right ear, and slowly move the sound source away from the ear in a straight line. Hold a meterstick in the other hand. Have your partner indicate when he or she can no longer hear the ticking sound.

5 Using the meterstick, measure the distance to where your partner lost the sound of the ticking clock. Convert the distance to centimeters, and record it in Table 23.6.

TABLE **23.6** Hearing Acuity Test

Student	Distance (Right Ear)	Distance (Left Ear)

6 Repeat the procedure using a new sterile cotton plug, but place it in your partner's right ear. Record your results in Table 23.6.

7 Exchange places with your partner and repeat the above steps.

Procedure 2

Equilibrium Test

Materials

None required

1 Work in groups of four for this procedure. Have your laboratory partner stand erect and as still as possible with both feet placed closely together for 2 minutes. Observe how he or she maintains balance. Record your observations in Table 23.7.

2 Have your laboratory partner repeat the procedure with his or her eyes closed. Observe how the test subject maintains his or her balance, and record your observations in Table 23.7.

3 Have your partner stand erect on one foot for 1 minute with his or her eyes open. Observe how the test subject maintains balance, and record your observations in Table 23.7.

4 Have your laboratory partner repeat the procedure with his or her eyes closed. Observe how the test subject maintains balance, and record your observations in Table 23.7.

WARNING

For safety during this activity, a lab partner should stand on each side of the subject.

23

TABLE **23.7** Equilibrium Test

Procedure	Observations of Response
Standing on two feet with both eyes open	
Standing on two feet with both eyes closed	
Standing on one foot with both eyes open	
Standing on one foot with both eyes closed	

✔Check Your Understanding

7.1 Was the hearing acuity the same for the right and left ears for both you and your partner?

7.2 How important was the relationship of vision to equilibrium during Procedure 2?

7.3 The inner ear consists of the vestibule, the semicircular canals, and the cochlea. Which is involved with equilibrium and which with hearing?

Special Sense of Vision

Humans are highly dependent upon vision, and the eyes are the structures responsible for vision. The eye refracts and focuses incoming light waves onto sensitive **photoreceptors,** the **rods** and **cones,** located in the **retina.** Nerve impulses from the stimulated photoreceptors are conveyed along visual pathways to the **occipital lobe** of the cerebrum of the brain, where vision is perceived.

The human eyeball (Fig. 23.24) is an irregular, sphere-shaped structure about the size of a ping-pong ball, consisting of three layers called the **tunics.** The **fibrous,** or **outer tunic,** consists of the **sclera** and the **cornea.** The sclera, or the "white" of the eye, is comprised of fibrous tissue. A thin layer called the **conjunctiva** covers the sclera. The cornea is transparent and is continuous with the sclera. The fibrous tunic provides support and a site of attachment for muscles and aids in focusing.

The **vascular,** or middle, **tunic** contains vessels, lymphatics, and some eye muscles; it regulates light entering the eye, secretes and absorbs aqueous humor, and controls the shape of the eye. The iris, choroid, and ciliary body make up the vascular tunic. The **iris,** the colored portion of the eye, regulates the diameter of the **pupil.** The pupil, like the aperture of a camera, allows light into the eye. Behind the retina is the **choroid,** a highly vascularized and pigmented portion of the eye. The ciliary body is a thickened portion of the choroid that forms a ring around the **lens.**

The lens helps to form images and divides the eye into two chambers, or compartments. The **anterior chamber,** between the cornea and lens, is filled with **aqueous humor,** a liquid that provides nutrients, carries away wastes, and cushions the eye. The **posterior chamber,** behind the eye, contains gelatinous **vitreous humor** that helps to give the eyeball its shape and supports the retina.

The neural, or inner, tunic contains the retina, and the retina has photoreceptors. Rods are responsible for vision in dim light, such as twilight, and cones are responsible for color vision in bright light. Cones provide sharper images than rods do. The highest concentration of cones is in the **fovea centralis** of the **macula lutea** (a region with no rods). The fovea centralis is the region of sharpest vision. The **optic disc,** the blind spot, is a region where the **optic nerve** connects to the wall of the eye.

Did you know . . .

Glowing Eyes

Have you ever noticed that cat eyes and the eyes of other animals tend to glow (eyeshine) when shined with a light at night? They possess a structure called a tapetum lucidum. Humans do not have a tapetum lucidum. Although many photos of your friends have red-eye, it occurs when a camera captures light reflecting from the retina at the back of your subject's eye when a flash is used at night and in dim lighting.

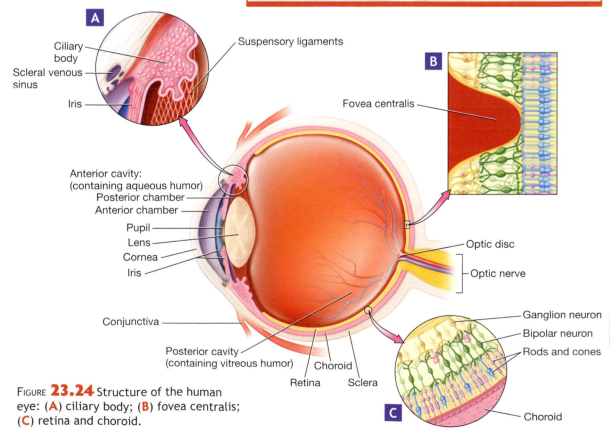

FIGURE **23.24** Structure of the human eye: (**A**) ciliary body; (**B**) fovea centralis; (**C**) retina and choroid.

Procedure 1

Snellen Test of Visual Acuity

Materials
- ❏ Masking tape
- ❏ White index cards
- ❏ Snellen chart
- ❏ Meterstick or yardstick

The **Snellen eye chart** is commonly used to measure visual acuity (Fig. 23.25). The chart has several sets of letters in different sizes printed on a white poster. The larger letters are at the top of the chart and the smallest letters at the bottom. Each set of letters is associated with an acuity value such as 20/20. Normal eyes can see these letters at a distance of 20 feet. If a person has a visual acuity of 20/50, that person can see detail from 20 feet away the same as a person with normal eyesight would see it from 50 feet away. People with a visual acuity of 20/50 have less than normal acuity.

1 Hang a Snellen chart in a quiet, well-lit area in the laboratory.

2 Measure a distance on the floor of 20 feet from the chart, and place a piece of masking tape on the floor at this spot.

3 Remove your eyeglasses or contact lenses if you wear them.

4 Face the Snellen chart, and place an index card over your left eye. Read the letters on the fifth line of the chart aloud so your partner can hear your response. If you cannot read the letters at this level, go up one level at a time until you can accurately read the line. If you can accurately read the fifth line, proceed to the line with the smallest letters you can read accurately. Which line could you accurately read with your right eye? What was the visual acuity value?

5 Repeat steps 3 and 4, with the index card covering your right eye. Which line could you accurately read with your left eye? What was the visual acuity value?

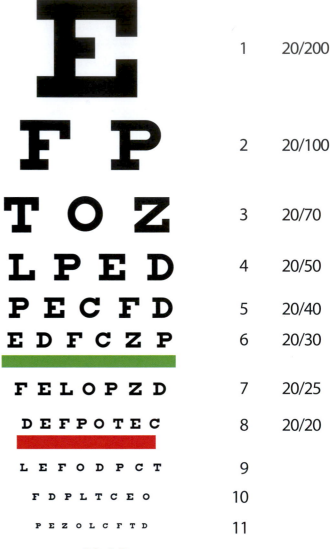

FIGURE **23.25** Example of a Snellen chart.

6 If you removed your eyeglasses or contact lenses, use them and repeat steps 3 and 4. Which line could you accurately read with your right eye? What was the visual acuity value? Which line could you accurately read with your left eye?

23

Procedure 2

Color Vision Test

Materials
- ❏ *Ishihara's Tests for Colour Blindness* book or Figure 23.26
- ❏ Meterstick or tape measure

The term **color blind** is commonly used to describe individuals who cannot distinguish certain colors; however, the term **color deficient** is more appropriate. Although the primary cause of color deficiency is genetic, certain types of eye, nerve, or brain damage can also cause color deficiency. The most common form of color deficiency is red/green color deficiency. The test for evaluating color deficiency is based on color plates developed by Shinobu Ishihara (1879–1963).

1 Examine the plates in *Ishihara's Tests for Colour Blindness*, if the book is available. If not, the three plates in Figure 23.26 will serve as example plates.

2 Administer this test in bright light. Place the plate about 76 cm (30 in.) in front of your lab partner. Your partner should respond by telling you a number within the plate or not describe a number at all within 5 seconds. Record his or her response.

3 Test other members of the group as well as yourself.

Lab partner 1's response _____

Lab partner 2's response _____

Lab partner 3's response _____

Your response _____

4 Compare your group's responses with the correct response, and discuss the results.

a. Did the class results for the males reflect the expected results? Why?

_____

_____

b. Did the class results for the females reflect the expected results? Why?

_____

_____

_____

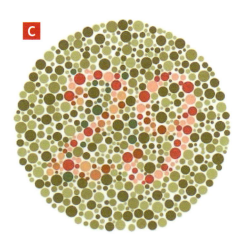

Did you know . . .

Do You Have Daltonism?

The English scientist John Dalton (1766-1844), famous for his work on atomic theory and gas laws, was color deficient. He attempted to study color deficiency, and for many years color deficiency was called Daltonism. He requested that, after his death, his eyes be removed and studied. His physician, Joseph Ransome, removed the eyes, and they were scientifically examined. The examination was normal, meaning that the cause of his color deficiency was not the result of abnormal eye anatomy. Today, one of his eyes is on display in Dalton Hall in Manchester, England.

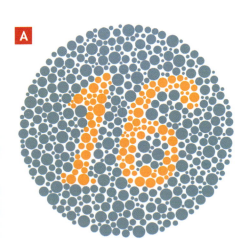

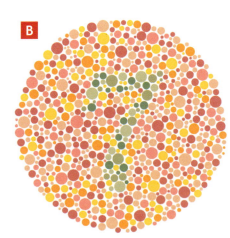

FIGURE **23.26** Color-blindness tests. These are only examples and not meant for diagnosis. (**A**) Everyone should see the number 16. (**B**) A person with normal color vision would see a 7 here. (**C**) A person with normal color vision would see a 29 here. A person with red/green color deficiency would see the number 70 or no number.

Procedure 3

Afterimage Test

Materials
- ❑ Figure 23.27
- ❑ Meterstick or tape measure
- ❑ Timing device

You have seen an **afterimage** after staring at a light or seeing a ghost image on a TV screen when it goes blank. An afterimage results when an image continues to appear after exposure to the original image has stopped. Negative afterimages are caused when photoreceptors in the eyes are overstimulated and lose their sensitivity. When the eyes are diverted to a blank wall or closed, the colors are muted, and the brain interprets it as the complementary color.

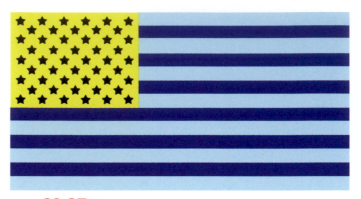

FIGURE **23.27** Afterimage test.

1 Hold the page with the flag (Fig. 23.27) about 30 cm from your face. Intensely fixate on the bottom right star for 1 minute. After 1 minute, look away toward the ceiling or a blank wall, or close your eyes.

2 Describe the flag.

Procedure 4

Blind-Spot Test

Materials
- ❑ Figure 23.28
- ❑ Meterstick or tape measure

The eye has a blind spot where the optic nerve exits the rear of the eye. This region, termed the optic disc, contains no receptor cells. Your blind spot can be easily detected by the following test.

1 Close your left eye, and keep it closed.

2 Hold the page with the blind-spot test (Fig. 23.28) about 30 cm from your face.

3 Fixate on the plus sign with your right eye, and start moving the page toward your eye until the dot disappears. When the image of the dot disappears, have your lab partner measure the distance from the eye to the dot in centimeters. The dot disappeared because the image of the dot passed over the optic disc.

Distance _____

4 Repeat steps 1 through 3, closing the right eye and testing the left eye.

Distance _____

Did you note any differences between the right and left eye?

FIGURE **23.28** Blind-spot test.

✔ Check Your Understanding

8.1 The _____ is the thin layer of the eye that covers the sclera, or "white," of the eye.

8.2 Name the material found within the eye that gives the eye its shape and provides support to the retina. (*Circle the correct answer.*)
   a. Conjunctiva.
   b. Choroid.
   c. Vitreous humor.
   d. Aqueous humor.
   e. Tunic.

8.3 Compare and contrast the structure and function of rods and cones.

8.4 The optic disc, or the blind spot, is a region where the optic nerve connects to the wall of the eye. (*Circle the correct answer.*)

True / False

8.5 An optometrist tells you that you have 20/20 vision; what does this mean?

MYTHBUSTING

Use Your Eyes and Ears

Debunk each of the following misconceptions by providing a scientific explanation. Write your answers on a separate sheet of paper.

1 What did you say? Playing my stereo loud will not affect my hearing!

2 Osteoporosis is always painful.

3 Don't worry about it. The sun does not hurt your eyes.

4 The smarter you are, the more the convolutions in your brain.

5 Brain cancer involves the nervous tissue of the brain in adults.

1 Describe three anatomical characteristics that make us human and differentiate us from other animals.

2 What is/are the basic function(s) of the integumentary system? (*Circle the correct answer.*)
   a. Vitamin D synthesis.
   b. Temperature regulation, excretion, and sensory reception.
   c. Protection and defense against harmful microrganisms.
   d. Both b and c.
   e. All of the above.

3 What two factors in humans make fingerprint patterns useful in the field of criminalistics?

4 Describe three types of articulations, and give an example of each.

5 Describe the four basic functions of muscle.

6 What are the eight major functions of bone?

_____

_____

_____

_____

_____

_____

_____

_____

7 Label the bones of the arm in Figure 23.29.

1. _____

2. _____

3. _____

4. _____

5. _____

6. _____

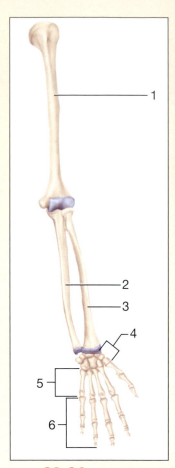

FIGURE **23.29** Bones of the arm.

8 Name the muscles of the leg in Figure 23.30.

1. _____

2. _____

3. _____

4. _____

5. _____

6. _____

7. _____

8. _____

9 Compare and contrast the composition and function of the central and peripheral nervous systems.

_____

_____

_____

_____

_____

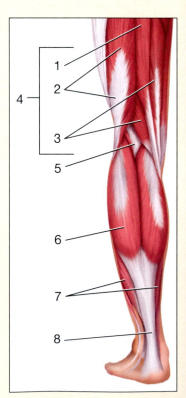

FIGURE **23.30** Muscles of the leg.

23

10 The term *color blind*, or more appropriately, *color deficient*, describes people who cannot distinguish certain colors. Which of the following are causes of this condition? (*Circle the correct answer.*)

a. Eye or brain damage.

b. Genetic.

c. Nerve damage.

d. Both a and b.

e. All of the above.

11 The sharpness of hearing, or the ability to detect a sound with respect to intensity and distance, is known as

_____ .

12 Label the human ear (Fig. 23.31).

1. _____
2. _____
3. _____
4. _____
5. _____
6. _____
7. _____
8. _____
9. _____
10. _____
11. _____

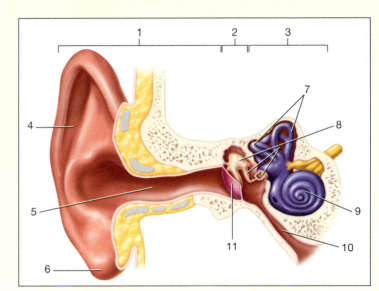

FIGURE **23.31** Generalized overview of the ear.

13 Label the human eye (Fig. 23.32).

1. _____
2. _____
3. _____
4. _____
5. _____
6. _____
7. _____
8. _____
9. _____
10. _____
11. _____
12. _____

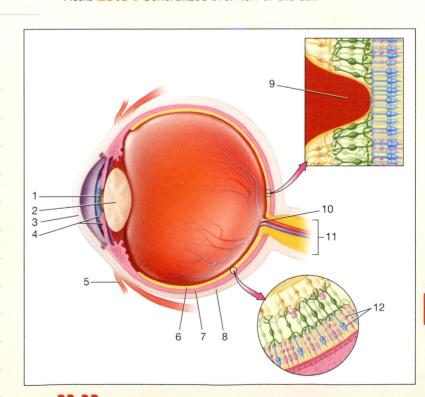

FIGURE **23.32** Structure of the human eye.

14 Describe the function of the following:

a. Medulla oblongata

b. Cerebrum

c. Pons

d. Reticular formation

e. Cerebellum

f. Thalamus

g. Hypothalamus

h. Frontal lobe

i. Temporal lobe

j. Parietal lobe

23

Homo sapiens II
Understanding Human Body Systems

*I… learn… anatomy not from books but from dissections,
not from tenets of philosophers but from the fabric of nature.*

— William Harvey (1578–1657)

Just wondering…

Consider the following questions prior to coming to lab, and record your answers on a separate piece of paper.

1 What is the difference between type I and type II diabetes?
2 Who was Dr. William Beaumont?
3 What is the impact of cystic fibrosis on the lungs?
4 President John F. Kennedy had Addison's disease. What is this disorder?
5 Why are urinary tract infections more common in females?

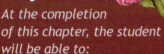

Objectives

At the completion of this chapter, the student will be able to:

1. Describe the major components of the circulatory, respiratory, and lymphatic systems along with their basic functions.
2. Identify and differentiate structures of the heart.
3. Trace the route of blood through the heart.
4. Distinguish between the pulmonary and systemic circuits.
5. Describe the functions and list the glands of the endocrine system, and describe the function of each hormone it produces.
6. List and describe the major components and functions of the digestive system.
7. Describe the major function of the urinary system, and list its major components.
8. List and describe the functions of the major components of the female and male reproductive systems.

Chapter Photo
SEM of an artery with red blood cells.

The cardiovascular, respiratory, and lymphatic systems are essential components of the human body, serving primarily as delivery and defense systems. The **cardiovascular system** consists of the heart and blood vessels (arteries, capillaries, and veins)—most of the components of the **circulatory system**, which also includes blood. The function of the heart and vessels is to transport various substances throughout the body (Figs. 24.1 and 24.2). These substances include cells, oxygen, nutrients, waste products, and solutes. Several common diseases are associated with the circulatory system, including: arteriosclerosis, atherosclerosis, hypertension, cardiomyopathy, heart valve defects, heart attacks, angina, and congestive heart failure.

The **respiratory system** is made up of the airways that carry air to and from the lungs (nose, sinuses, pharynx, larynx, trachea, bronchi, bronchioles, and alveoli). The functions of the respiratory system are to provide oxygen to cells (so they can produce ATP) and to remove carbon dioxide. Diseases such as cystic fibrosis, asthma, pneumonia, and emphysema can impair lung function and can be life threatening.

The **lymphatic system** consists of lymph, lymphatic vessels, lymph nodes, lymphatic tissues, and lymphatic organs, such as the thymus and spleen. It is an important part of the body's defense system against microbes and other harmful agents. It also preserves the body's fluid homeostasis. Diseases of the lymphatic system include lymphedema, Castleman disease, Hodgkin's lymphoma, and lymphoid leukemias.

The **endocrine system** consists of a variety of small glands, such as the pituitary, thyroid, and adrenal glands, essential to our survival. Like the nervous system, the endocrine system controls bodily processes. However, the endocrine system uses chemical messengers to regulate the metabolic activities of every system and cell in the body. This action is long term (e.g., maintenance of the reproductive system) in contrast to, for instance, a nervous system action that may be short term (e.g., reaction to a stimulus). Diseases of the endocrine system include: Addison's disease, Cushing syndrome, and Grave's disease.

The **digestive system** is responsible for taking in food, breaking down food into nutrients that can be used by the body, absorbing those nutrients into the bloodstream, and getting rid of indigestible substances. We will look at the structures of the digestive system, including the **gastrointestinal tract** and the **accessory digestive organs and structures**. Noteworthy digestive system disorders include: ulcers, Crohn's disease, irritable bowel syndrome, appendicitis, and gastroesophageal refux disease (GERD).

The **urinary system** function with which most people are familiar is the elimination of waste products from the bloodstream. This system also plays a role in blood-cell formation, helps the liver detoxify some substances, makes glucose when necessary, and regulates the balance of fluids and electrolytes. We will become familiar with the structures of the urinary system, including the kidneys, ureters, urinary bladder, and urethra. Nephrosis, kidney stones, polycystic kidney disease, and urinary tract infections are common diseases of the urinary system.

The female and male reproductive systems are important because they allow for the perpetuation of the species. They do this by means of gametogenesis, the process of forming new gametes. The **female reproductive system** contains the ovaries, uterus, oviducts, vagina, external genitalia, and

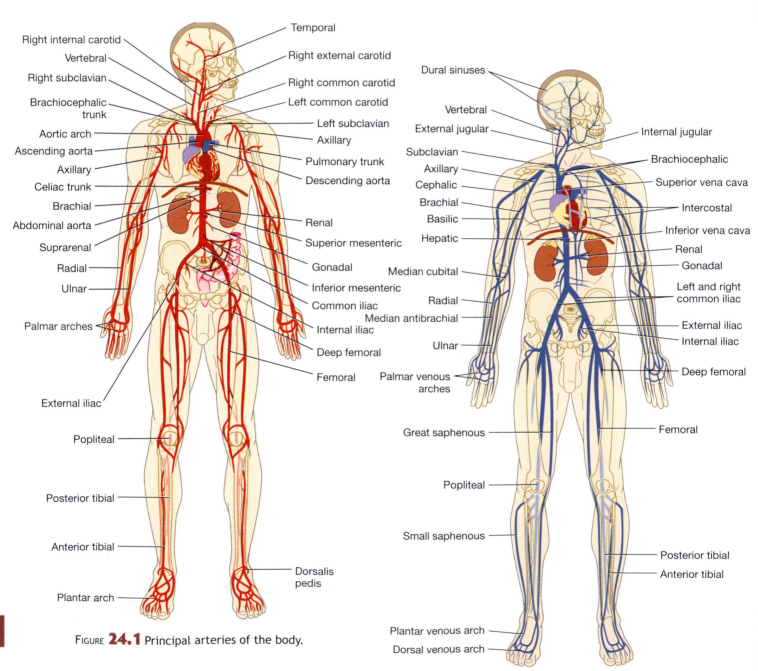

FIGURE **24.1** Principal arteries of the body.

FIGURE **24.2** Principal veins of the body.

24

mammary glands. The **male reproductive system** includes the testes, scrotum, seminal vesicles, prostate gland, bulbourethral gland, and penis. Disorders of the reproductive system include: prostate cancer, abruptio placentae, genital herpes, syphilis, gonorrhea, chlamydia, papilloma virus, and infertility.

Cardiovascular System

The **heart** is a muscular organ about the size of a clenched fist. The heart lies in the mediastinum, the region of the body bordered superiorly by the base of the neck, inferiorly by the diaphragm, anteriorly by the sternum, posteriorly by the thoracic vertebrae, and laterally by the pleural cavities, where the lungs reside. The distal end, or **apex**, of the heart is blunt-shaped and points to the left.

Several large vessels attach to the top (base) of the heart (Figs. 24.3 and 24.4). The vessels make up the **pulmonary**

circuit, which carries blood to and from the lungs, and the **systemic circuit,** which carries blood to and from the rest of the body. The heart is enclosed in the **pericardium,** a double-walled sac. The fibrous outer portion of the pericardium is called the **parietal pericardium,** and the inner serous layer is called the **visceral pericardium,** or **epicardium**. The epicardium covers the surface of the heart (Fig. 24.5).

A **pericardial cavity** filled with **pericardial fluid** exists between the parietal and visceral pericardia. The fluid reduces friction between the parietal pericardium and beating heart. The **endocardium** is a thin layer of epithelial tissue that lines the inside of the heart and blood vessels. The muscular **myocardium** exists between the epicardium and endocardium. The myocardium, the thickest region of the heart wall, is composed of cardiac muscle.

Within the heart are four chambers consisting of the **atria** and the **ventricles** (Figs. 24.6 and 24.7). The **left** and **right atria** receive blood, and the **left** and **right ventricles** pump blood. The atria are separated by a thin-walled **interatrial septum,** and the ventricles are separated by a muscular **interventricular septum**. The **right atrium,** a part of the systemic circuit, receives deoxygenated blood from the **superior vena cava,** the **inferior vena cava,** and the **coronary sinus**.

Deoxygenated blood flows from the right atrium to the **right ventricle** through the **right atrioventricular,** or **tricuspid, valve**. The right ventricle pumps blood into the **pulmonary trunk,** which divides into the **left** and **right pulmonary arteries** (see Fig. 24.1). The pulmonary arteries carry blood to the lungs, where gas exchange occurs. Blood returns to the **left**

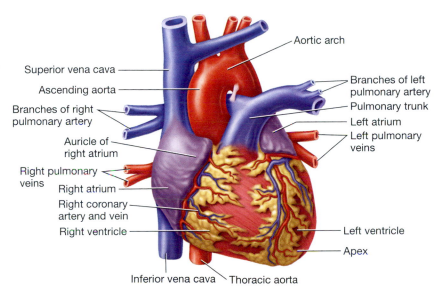

FIGURE **24.3** Ventral view of the structure of the heart.

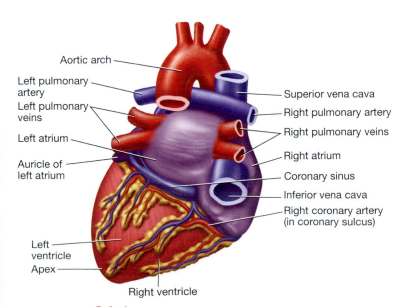

Figure **24.4** Dorsal view of the structure of the heart.

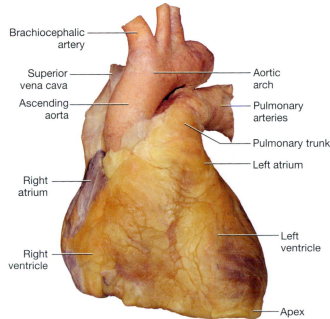

Figure **24.5** Ventral view of the heart.

atrium by way of the left and right pulmonary veins (see Fig. 24.2). Note that arteries always carry blood away from the heart, and veins always return blood to the heart.

The opening between the left atrium and ventricle is guarded by the **left atrioventricular,** or **bicuspid,** or **mitral, valve. Chordae tendineae,** also called the "strings of the heart," connect the papillary muscles in the ventricles to the tricuspid and bicuspid valves (see Figs. 24.6 and 24.7). The **left ventricle** has a thick wall because it is responsible for pumping oxygenated blood through the systemic circuit to the rest of the body.

Blood leaves the left ventricle through the **aortic valve,** or **semilunar valve,** to the **aorta,** the largest artery in the body (Fig. 24.8). From the aorta, blood is distributed throughout the body.

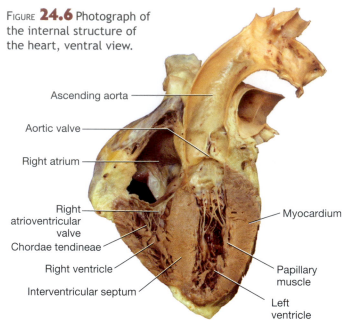

FIGURE **24.6** Photograph of the internal structure of the heart, ventral view.

Labels: Ascending aorta, Aortic valve, Right atrium, Right atrioventricular valve, Chordae tendineae, Right ventricle, Interventricular septum, Myocardium, Papillary muscle, Left ventricle

Labels: Aortic arch, To lungs, Pulmonary valve, Interatrial septum, Left atrium, From lungs, Aortic valve, Right atrium, Mitral (bicuspid) valve, Left ventricle, From lungs, Chordae tendinae, Tricuspid valve, Papillary muscle, Interventricular septum, Right ventricle, Endocardium, Myocardium, Visceral pericardium

FIGURE **24.7** Illustration of the internal structure of the heart, ventral view.

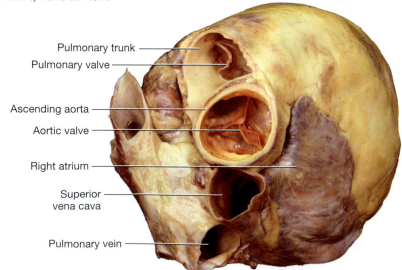

Labels: Pulmonary trunk, Pulmonary valve, Ascending aorta, Aortic valve, Right atrium, Superior vena cava, Pulmonary vein

FIGURE **24.8** Superior view of the great vessels of the heart and aortic valve.

Cardiovascular System in Action

Cardiovascular physiology attempts to explain how the heart works. Knowledge of this discipline is essential to health-care providers. It can be as simple as measuring pulse and listening to heart sounds or as complex as describing the action of the heart and its electrophysiology.

Procedure 1

Listening to the Heart

Materials
- ❏ Stethoscope
- ❏ Alcohol pads
- ❏ Timing device

Auscultation is the process of listening to sounds generated by the body. Auscultation of the heart can provide a medical professional vital information about the heart and its health. An instrument called the **stethoscope** is used to listen to heart sounds; each cardiac cycle generates two, or perhaps three, of these sounds. When listening to heart sounds, the first S_1 and second S_2 are responsible for the typical "lub" "dub" sounds. The S_1 sound is louder and longer than S_2 and represents the closing of the tricuspid and bicuspid valves and the beginning of ventricular contraction. The S_2 sound occurs when the ventricles start to fill after the semilunar valves close.

In individuals under age 30 and trained atheletes, a third heart sound, S_3, sometimes can be heard. If so, it is called a *gallup* or *triple rhythm*. In these individuals it can be considered normal and should disappear before middle age. Later in life, it is considered abnormal and could indicate serious heart problems such as congestive heart failure. S_3 results from the oscillation of blood back and forth between the walls of the ventricles and is initiated by the inflow of blood from the atria.

Heart sounds are best heard in the four regions indicated in Figure 24.9. The normal resting heart rate is about 65 to 80 beats per minute. Athletes may have slower heart rates.

1 Procure the stethoscope and alcohol pads. Thoroughly clean the earpieces and diaphragm of the stethoscope with an alcohol pad. Place the earpieces in your ears, and gently tap on the diaphragm to ensure the diaphragm is oriented properly.

2 You may listen to your own heart sounds or your partner's. You or your partner should sit quietly, and place the diaphragm in the aortic area under the shirt or blouse. Your partner may wish to place the diaphragm of the stethoscope on his or her own chest. Listen for cardiac cycles, and note the characteristic "lub" "dub." Count the heart rate for 15 seconds, and multiply this number by 4 to determine the number of beats per minute. Record the number of beats in the space provided.

Number of heartbeats in 15 seconds: _____

Number of heartbeats in 1 minute: _____

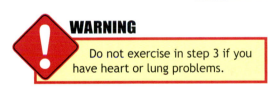

WARNING

Do not exercise in step 3 if you have heart or lung problems.

3 Move your stethoscope to the pulmonic, tricuspid, and mitral areas of the heart. Jog in place for 2 minutes, and measure the heart rate at the mitral area immediately afterward. Record the number of heartbeats in the space provided.

Number of heartbeats in 15 seconds: _____

Number of heartbeats in 1 minute: _____

4 Discuss in the space provided below the difference in heartbeat at rest and after exercise.

_____

_____

_____

_____

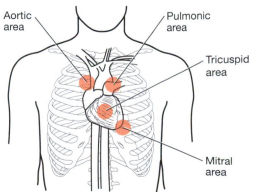

Figure **24.9** Common auscultation areas.

5 Clean the earpiece and diaphragm of the stethoscope with another alcohol pad, and return it to the proper place.

Procedure 2

Measuring the Pulse

Materials
- Sanitizing soap
- Timing device

As blood surges through the arteries, it creates the **pulse**. Monitoring the pulse rate with the fingertips is common and is called **pulse palpitation**. **Pulse rate** represents the heartbeats per minute. Typical pulse rates range from 65 to 80 beats per minute. The body has several common pulse points (Fig. 24.10), but the most common one used by medical professionals is the **radial pulse**.

1 Thoroughly wash your hands, and have your partner thoroughly wash his or her hands and wrist area with sanitizing soap. Have the person being measured sit down. Lightly place your index and middle finger on the thumb side of the inner wrist to locate the radial artery of your partner, and press down slightly. After you have located the pulse, count the pulse rate for 15 seconds. Multiply this number by 4 to determine the pulse rate per minute.

2 Repeat step 1, but this time monitor the carotid artery.

Pulse rate in 15 seconds: _____

Pulse rate for 1 minute: _____

3 Have your partner hold his or her breath for 20 seconds, and measure the radial pulse.

Pulse rate in 15 seconds: _____

Pulse rate for 1 minute: _____

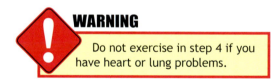

WARNING

Do not exercise in step 4 if you have heart or lung problems.

4 Ask your partner to vigorously jog in place for 2 minutes, and then sit down. Measure the radial pulse immediately.

Pulse rate in 15 seconds: _____

Pulse rate for 1 minute: _____

5 Measure the radial pulse of your partner 2 and 5 minutes after exercise.

2 minutes after exercise.

Pulse rate in 15 seconds: _____

Pulse rate for 1 minute: _____

5 minutes after exercise.

Pulse rate in 15 seconds: _____

Pulse rate for 1 minute: _____

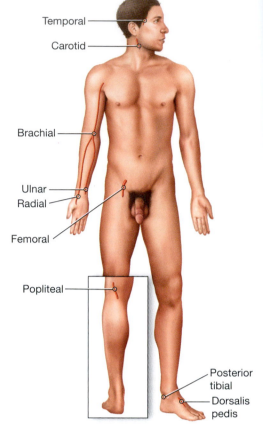

FIGURE **24.10** Common pulse points.

6 Thoroughly wash your hands, and have your partner thoroughly wash his or her hands and wrist area with sanitizing soap.

7 Fill in Table 24.1. Account for deviation of pulse rate from rest for each measurement.

TABLE **24.1** Pulse Rates

Activity	Pulse Rate per Minute
At rest	
Holding breath	
Immediately after exercise	
2 minutes after exercise	
5 minutes after exercise	

Procedure 3

Measuring Blood Pressure

Materials
- ❏ Sphygmomanometer
- ❏ Stethoscope
- ❏ Alcohol pads
- ❏ Timing device
- ❏ Bucket of ice water

Basically, **blood pressure** is a measure of the pressure exerted upon the surface of the vessels by the blood. This pressure serves to circulate the blood throughout the body. The blood leaving the heart generates the **systolic pressure,** and the pressure when the heart relaxes is the **diastolic pressure**. Blood pressure is measured in millimeters of mercury (mm Hg). The systolic pressure is the larger pressure and should average between 100 and 120 mm Hg. Diastolic pressure averages between 60 and 80 mm Hg.

An instrument called a **sphygmomanometer** is used to measure blood pressure. Partners may be switched after each major measurement, but be sure to note these changes in Table 24.2.

> ## Note:
> **Several types of sphygmomanometers are available. Listen to your instructor to learn how to correctly use the shygmomanometer you will use in this procedure.**

1 Procure the sphygmomanometer, stethoscope, and alcohol pads. Thoroughly clean the earpieces and diaphragm of the stethoscope with an alcohol pad. Have your partner sit comfortably in a chair next to a lab table and rest his or her arm on the table.

2 If your partner is not wearing short sleeves, have him or her roll up one sleeve. Wrap the cuff of the sphygmomanometer apparatus around the arm (your choice) of your partner. The cuff should sit about 3.5 cm above the antecubital fossa (the depression at the end of the elbow). Do not make it tight—just tight enough to stay in place. If you are uncomfortable with the degree of tightness, have your instructor check your setup.

3 Clean the arm to be tested near the brachial artery with an alcohol pad. Familiarize yourself with the gauge of the sphygmomanometer. Place the diaphragm of the stethoscope over the brachial artery. Place the earpieces in your ears. No sounds should be heard.

4 Close the screw on the bulb of the sphygmomanometer by tightening it in a clockwise direction. Watch the pressure gauge, and begin to gently pump the bulb. Pump the bulb up to about 190 mm Hg. Your partner may feel some discomfort.

5 Keeping the stethoscope above the brachial artery and listening closely, watch the pressure gauge, and slowly release the pressure on the cuff by turning the screw in a counterclockwise direction. Eventually you will see the needle on the gauge jump and hear the pulse in the brachial artery.

6 Note the value when you hear the pulse; this is the systolic pressure. Continue releasing the pressure with the screw until you can no longer hear the pulse. Note the value at this point. This is the diastolic pressure. Record the two pressures in the space provided.

Systolic pressure: _____

Diastolic pressure: _____

TABLE **24.2** Blood Pressure Measurements

Activity	Blood Pressure
At rest	
At rest (2nd reading)	
Immediately after standing	
5 minutes while standing	
10 minutes while standing	
After resting 3 minutes	
During water immersion	
After water immersion	
Prior to exercise at rest	
Immediately after exercise	
2 minutes after exercise	
5 minutes after exercise	

7 Record the blood pressure as a fraction with the systolic pressure on top.

Blood pressure: _____

8 Repeat steps 2–7 now that you understand the process.

Blood pressure: _____

9 Repeat steps 2–7 with your partner standing. Take a reading immediately after he or she stands. Then take readings 5 minutes and 10 minutes after your partner stands up.

Blood pressure immediately after standing: _____

Blood pressure after standing for 5 minutes: _____

Blood pressure after standing for 10 minutes: _____

10 Have your partner sit and relax for 3 minutes. Measure your partner's blood pressure at rest.

Blood pressure resting: _____

11 Have your partner immerse his or her free hand in a bucket of ice water. With the hand in the water, quickly measure the blood pressure.

Blood pressure during water immersion: _____

12 Have your partner remove his or her hand from the bucket and relax. Measure the blood pressure 5 minutes after the hand has been removed from the ice water.

Blood pressure after water immersion: _____

13 Take a resting blood pressure of your partner 10 minutes after the hand has been removed from the ice water.

Blood pressure resting: _____

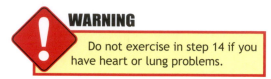

WARNING

Do not exercise in step 14 if you have heart or lung problems.

14 Have your partner rigorously jog in place for 3 minutes. Immediately take the blood pressure. Take the blood pressure while sitting 2 and 5 minutes after exercising.

Blood pressure immediately after exercise: _____

Blood pressure 2 minutes after exercise: _____

Blood pressure 5 minutes after exercise: _____

15 Clean the earpiece and diaphragm of the stethoscope with another alcohol swab, and return the materials to the proper place.

16 Fill in Table 24.2.

Did you know . . .

How Long?

Our bodies are constantly bombarded by a variety of microscopic pathogens, such as viruses, bacteria, protists, and fungi. The lymphatic system, consisting of the tonsils, thymus, lymph nodes, spleen, and cisterna chyli, serves to protect our body from pathogens as well as internal dangers, such as cancer. The lymphatic system has several important functions, including production, distribution, and maintenance of white blood cells known as lymphocytes, filtering out microorganisms, collection and circulation of extracellular fluids, return of plasma proteins from extracellular fluids to the blood, and transport of dietary lipids from the intestines to the bloodstream (all functions of lymph).

24

✔ Check Your Understanding

1.1 Match the heart valve with the structures separated by each valve.

_____ Tricuspid valve

_____ Pulmonary valve

_____ Aortic valve

_____ Mitral valve

A. Separates the left atrium and the left ventricle
B. Separates the left ventricle and the aorta
C. Separates the right atrium and the right ventricle
D. Separates the right ventricle and the pulmonary artery

1.2 The heart is enclosed in a double-walled sac, or _____ ; the _____ is a thin layer of epithelial tissue that lines the inside of the heart and blood vessels. The inner serous layer of the pericardium that covers the heart is known as the visceral pericardium, or _____ .

1.3 What is the difference between pulmonary and systemic circulation?

1.4 Explain any differences between the four areas of the heart during auscultation.

1.5 Did you witness an extra heartbeat or extra sounds in Procedure 1?

1.6 Blood pressure _____ . (*Circle the correct answer.*)
   a. is the measure of pressure exerted upon the surface of the vessels by the blood
   b. is the pressure when blood is circulated throughout the body
   c. is the combined measure of diastolic and systolic pressures; blood leaving the heart generates the diastolic pressure while the pressure when the heart relaxes is systolic pressure
   d. both a and b
   e. all of the above

1.7 Discuss and account for differences in blood pressure readings in Procedure 3.

24

Respiratory System

The **respiratory system**, working in conjunction with the cardiovascular system, supplies oxygen to the cells of the body so they can undergo aerobic respiration and remove the waste product of cellular respiration—carbon dioxide—from the body. In addition, the respiratory system is involved in vocalizations, the sense of smell, and controlling the pH of the body (Fig. 24.11).

The respiratory tract is divided into two distinct regions:

1. The **upper respiratory tract** consists of the structures found outside of the thoracic cavity and includes the nose, nasal cavity, paranasal sinuses, and pharynx (nasopharynx, oropharynx, and laryngopharynx). These structures filter, warm, and humidify the incoming air and protect the respiratory tract.

2. The **lower respiratory tract** (passages within the thoracic cavity) consists of the larynx (voice box), trachea (windpipe), bronchi, bronchioles, alveoli, and lungs.

The **conducting portion** of the respiratory system consists of the structures essential in movement of air. The **respiratory portion** consists of structures involved in gas exchange, such as the respiratory bronchioles and alveoli.

The **lungs** are the largest organs in the respiratory tract. In humans, the right lung has three lobes, and the left lung has two lobes. Air enters the lungs by way of the **left** and **right primary bronchi**. The **bronchial tree** has several divisions within the lungs ending in **terminal bronchioles** or respiratory bronchioles (which have one or more alveoli attached and are thus sites of gas exchange). The respiratory rate in an average adult human is 12 times per minute, or nearly 17,500 times a day.

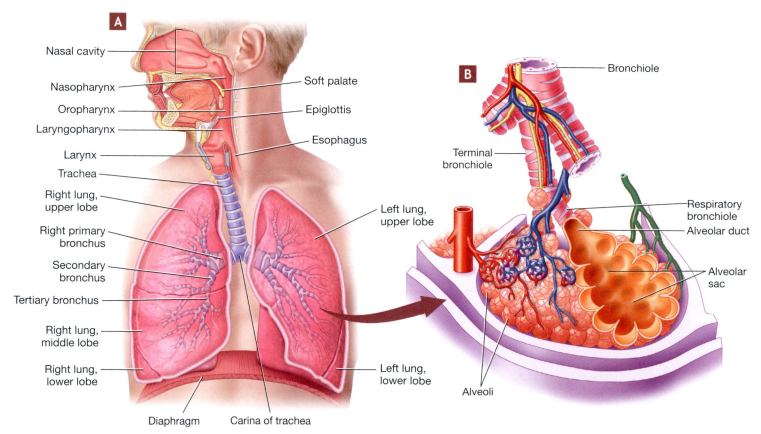

FIGURE **24.11** Structures of the respiratory system: (**A**) overview; (**B**) detailed view of the lung.

Procedure 1

Labeling the Respiratory System

Materials
- ❏ Model of respiratory system
- ❏ Colored pencils

1 Procure the respiratory system model.

2 Examine the respiratory system model, locating the structures labeled in Figure 24.11.

3 Label Figure 24.12.

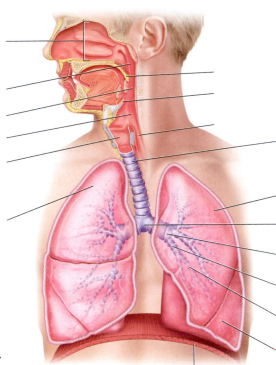

FIGURE **24.12** Structures of the respiratory system.

Procedure 2

Listening to the Lungs

Materials
- ❏ Stethoscope
- ❏ Alcohol pads
- ❏ Timing device

Auscultation of the lungs can provide a medical professional with vital information about the lungs and their health. A stethoscope is used to listen to the lungs. Your partner's size and weight can affect the transmission of lung sounds. The more you practice auscultating lung sounds, the easier it will be to identify normal and abnormal sounds. A number of internet sources provide audio clips for various lung sounds. Humans take between 12 and 20 breaths per minute.

Tracheal breath sounds are heard in the neck and trachea regions. These distinct sounds are louder and have a higher pitch than other lung sounds. These sounds can be heard during both inspiration and expiration. **Bronchial** sounds are very high-pitched, coarse sounds heard over the trachea and manubrium. The bronchial expiratory sounds last longer than inspiratory. Pneumonia can cause abnormal bronchial

sounds (consolidation) because the infection causes the alveoli in the lungs to fill with fluids, leukocytes, and erythrocytes. **Vesicular breath** sounds are heard across the lung surface. These sounds have a lower-pitched, rustling sound with higher intensity during inspiration.

Abnormal sounds include wheezing, stridor, and crackles. Wheezing can occur during any condition that narrows the airways. Asthma and chronic bronchitis are common causes of wheezing. In stridor, you hear a loud, high-pitched wheeze in the upper airways. Stridor indicates some degree of upper airway obstruction and becomes more intense during the process of inspiration. This kind of obstruction can lead to respiratory arrest. Crackles are clicking, rattling, or crackling noises heard during auscultation of the lung caused by the popping open of small airways and alveoli collapsed by fluid. Crackles may be associated with asthma, pulmonary edema, heart failure, tuberculosis, bronchitis, and cystic fibrosis.

1 Procure the stethoscope and alcohol pads. Thoroughly clean the earpieces and diaphragm of the stethoscope with an alcohol pad. Place the earpieces in your ears, and gently tap on the diaphragm to ensure the diaphragm is oriented properly.

2 When listening to lung sounds, start with the back, and, if possible, have your partner sit up, lean forward, and cross their arms in front of their chest. Apply firm pressure and auscultate in the intercostal spaces so you are not listening directly over bone. Begin at the top of the back and work down the back. Listen in the same area on both sides of the spine before moving down to the next intercostal space. Describe the sounds and record the number of breaths from one of your readings below.

3 Auscultate your partner in the anterior tracheal and neck regions, and describe these sounds below.

4 Clean the earpiece and diaphragm of the stethoscope with another alcohol pad, and return it to the proper place.

✔ Check Your Understanding

2.1 Using the following key, indicate which of the following make up the upper and which the lower respiratory tract.

UR = Upper respiratory tract   LR = Lower respiratory tract

_____ Nose and nasal cavity

_____ Larynx and trachea

_____ Alveoli and lungs

_____ Paranasal sinuses and pharynx

_____ Bronchi and bronchioles

2.2 The respiratory system _____. (*Circle the correct answer.*)
   a. supplies oxygen to the cells of the body
   b. is involved in vocalizations, sense of smell, and controls pH of the body
   c. supplies oxygen that enables cells to undergo aerobic respiration and remove carbon dioxide
   d. both a and c
   e. all of the above

2.3 The _____ are the sites of gas exchange in the lungs.

2.4 Describe tracheal, bronchial, and vesicular lung sounds.

24

Endocrine System

The **endocrine system** is a major control system of the body, and like the nervous system, helps maintain homeostasis. The endocrine system is responsible for producing a variety of **hormones**. Hormones reach their **target** cells—those with receptors for a given hormone—by traveling through the bloodstream. Hormones are essential for every bodily activity, including the processes of digestion, metabolism, growth, reproduction, and mood.

The endocrine system comprises ductless glands that produce and release hormones. In addition to these glands, structures in other systems, such as the heart, liver, and kidneys, also have endocrine functions. Even adipose tissue has some endocrine function. Table 24.3 provides an overview of several endocrine structures, the hormones they produce, and the primary function, or action, of these hormones.

TABLE **24.3** Endocrine Glands, Associated Hormones, and Their Functions

Endocrine Gland or Structure	Hormone	Target	Action
Hypothalamus	Releasing hormones	Anterior pituitary	Stimulates release of hormones of the anterior pituitary
	Inhibiting hormones	Anterior pituitary	Inhibits release of hormones of the anterior pituitary
Anterior pituitary (adenohypophysis)	Follicle-stimulating hormone (FSH)	Ovaries and testes	Stimulates spermatogenesis and development of ovarian follicle
	Luteinizing hormone (LH)	Ovaries and testes	Stimulates ovulation and testosterone secretion
	Thyroid-stimulating hormone (TSH)	Thyroid gland	Stimulates the thyroid gland and metabolic rate
	Adrenocorticotrophic hormone (ACTH)	Adrenal cortex	Stimulates adrenal cortex
	Prolactin (PRL)	Mammary glands	Stimulates production of milk
	Growth hormone (somatotropin or GH)	Connective tissue, internal organs, and soft tissues	Promotes cell division, protein synthesis, and growth
Posterior pituitary* (neurohypophysis)	Antidiuretic hormone (ADH)	Kidneys	Stimulates water resorption
	Oxytocin (OT)	Uterus and mammary glands	Stimulates uterine contraction and release of milk
Thyroid	Thyroxine (T_4) and triiodothyronine (T_3)	All tissues	Stimulates metabolic rate, and regulates growth and development
	Calcitonin	Bone	Reduces blood calcium level and promotes ossification
Parathyroid	Parathyroid hormone (PTH)	Kidneys, bone, and digestive tract	Raises blood calcium levels and activates vitamin D
Adrenal glands Adrenal cortex	Glucocorticoids (cortisol)	All tissues	Raises blood glucose level, stimulates protein breakdown, and mobilizes fat
	Mineralocorticoids (aldosterone)	Kidneys	Promotes regulation of mineral homeostasis in the blood by stimulating the reabsorption of sodium and excretion of potassium by the kidneys
	Sex steroids (androgen, estrogen)	Gonads, skin, bones, and muscles	Stimulates the development of sex characteristics and reproductive organs
Adrenal glands Adrenal medulla	Epinephrine (adrenaline) and norepinephrine (noradrenaline)	Muscles, heart, and most tissues	Raises heart rate, blood glucose levels, and metabolic rate; dilates blood vessels and mobilizes fat reserves

(continues)

24

TABLE **24.3** Endocrine Glands, Associated Hormones, and Their Functions (*continued*)

Endocrine Gland or Structure	Hormone	Target	Action
Pancreas	Insulin	Liver, skeletal muscle, and adipose tissue	Lowers blood glucose levels; promotes fat, protein, and glycogen synthesis
	Glucagon	Liver, skeletal muscle, and adipose tissue	Raises blood glucose levels; stimulates the liver's breakdown of glycogen
Thymus	Thymosins	T lymphocytes	Promotes the development of T lymphocytes
Pineal gland	Melatonin	Gonads, brain, and pigment cells	Controls biological rhythms such as circadian rhythms; perhaps controls sex organ maturation
Testes	Androgens (testosterone)	Gonads, muscles, bones, and skin	Stimulate the development of male secondary sex characteristics, spermatogenesis, and red blood cell synthesis
	Inhibin	Anterior pituitary	Inhibits secretion of FSH
Ovaries	Estrogen	Female reproductive tract, skin, muscles, and bones	Stimulates the development of female secondary sex characteristics and oogenesis and prepares endometrium for pregnancy
	Progesterone	Uterus and mammary glands	Stimulates development of mammary glands and completes preparation for pregnancy
	Inhibin	Anterior pituitary	Inhibits secretion of FSH

*The hormones released by the posterior pituitary are made in the hypothalamus.

Procedure 1

Labeling the Endocrine System

Materials

- ❏ Model of the endocrine structures
- ❏ Colored pencils

1 Procure the model of the endocrine structures.

2 Locate and be able to briefly describe the functions of the endocrine organs included in Figure 24.13.

3 Label Figure 24.14, and write next to the label the name of the hormone each structure synthesizes.

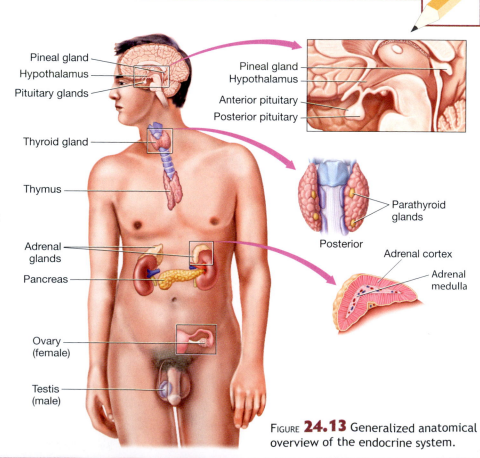

FIGURE **24.13** Generalized anatomical overview of the endocrine system.

24

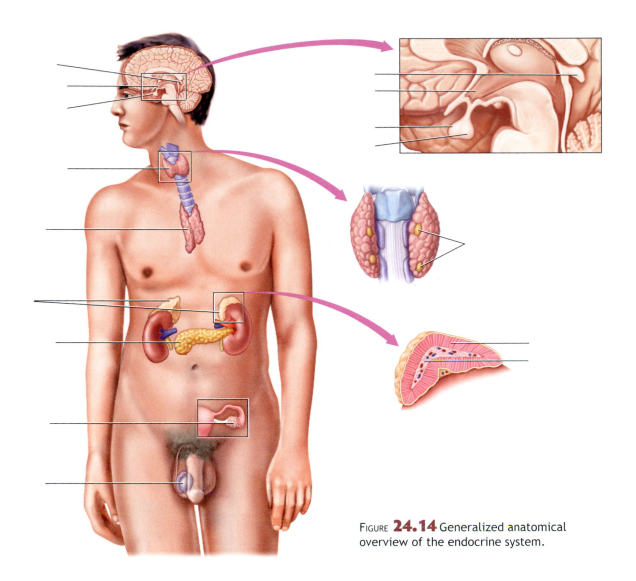

FIGURE **24.14** Generalized anatomical overview of the endocrine system.

Check Your Understanding

3.1 The endocrine system in the human body _____. (*Circle the correct answer.*)
  a. is the major control system
  b. is composed of ductless glands that produce and release hormones
  c. helps to maintain homeostasis
  d. both a and b
  e. all of the above

3.2 Hormones reach their target cells, which are those with receptors for a given hormone, by traveling through the bloodstream. (*Circle the correct answer.*)

  True / False

3.3 The _____ gland, which is found in the neck, stimulates the metabolic rate and regulates growth and development in the human body.

Digestive System

The alimentary canal is basically a long tube extending from the mouth to the anal opening (Fig. 24.15). Accessory digestive structures are found along the tract to aid in the breakdown of food. The digestive system begins with the oral cavity, or opening of the mouth, where the food is chewed (**mastication**) and mechanically broken down by the **teeth**. The **lips, cheeks, palate,** and **tongue** help contain and manipulate the **bolus,** or food mass. The **salivary glands** release **saliva** (containing salivary amylase) that initiates the digestion of carbohydrates. After the bolus is prepared, swallowing (**deglutition**) occurs.

The **pharynx** receives the bolus from the oral cavity and passes it to the esophagus. The pharynx is shared by the respiratory and digestive systems. The **esophagus** is a tube approximately 25 centimeters in length and 2 centimeters at its widest point. Its function is to allow the passage of the bolus to the stomach through peristalsis, a wave of smooth muscle contractions.

The stomach is a J-shaped organ that performs the following functions:

- Serves as a storage structure for ingested food;
- Continues the mechanical breakdown of food;
- Chemically breaks down food; and
- Produces a mixture of partially digested food called **chyme,** passed to the **duodenum** of the small intestine.

The next two sections of the small intestine are the **jejunum** and the **ileum**. The small intestine is responsible for most of the chemical digestion and nutrient absorption in the body. The small intestine averages 6 meters in length and has a diameter of about 1 centimeter.

Secretions from the liver and pancreas are essential to proper functioning of the small intestine. The **liver,** an accessory digestive structure, consists of four lobes. The liver is the largest internal organ, weighing about 1.5 kilograms. The digestive function of the liver is bile production. Other functions of the liver include:

- Storage of iron and copper;
- Conversion of glucose to glycogen;
- Storage of glycogen;
- Storage of vitamins (particularly fat-soluble vitamins); and
- Detoxification of harmful substances.

The **bile** produced in the liver is stored in the **gallbladder,** a pear-shaped structure located inferior to the right lobe of the liver. Bile is released into the duodenum, where it acts as an emulsifier of fats.

The **pancreas** is a pinkish, elongated structure of about 15 centimeters that extends laterally from the duodenum toward the spleen. It is responsible for secretion of pancreatic juice containing acid-neutralizing bicarbonate and digestive enzymes, and production of the hormones insulin (lowers the concentration of glucose in the bloodstream) and glucagon (raises the concentration of glucose in the bloodstream).

The **large intestine** receives undigested wastes from the small intestine; absorbs water, salts, and vitamins; and stores indigestible material until it is eliminated. The large intestine does not produce any digestive enzymes. The large intestine is about 1.5 meters in length and has a diameter of 7.5 centimeters.

The ileum of the small intestine empties into the **cecum** of the large intestine. The **vermiform appendix,** attached to the cecum, has some lymphatic function. The **colon,** the largest section of the large intestine, is shaped like a horseshoe and consists of the **ascending colon,** the **transverse colon,** the **descending colon,** and the **sigmoid colon.** The **rectum,** which lies inferior to the sigmoid colon, forms the last 15 centimeters of the alimentary canal. It serves as a temporary storage structure for feces. The last portion of the rectum terminates in the **anus.**

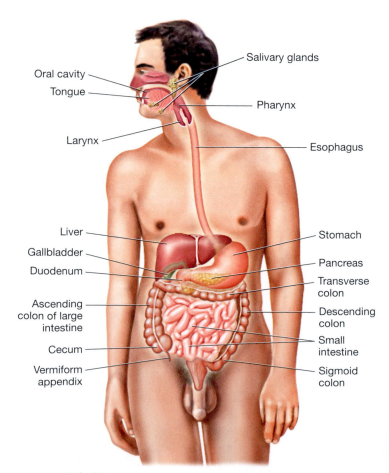

FIGURE **24.15** Basic anatomy of the digestive system.

24

Procedure 1

Labeling the Digestive System

1 Procure the model of the digestive system. Consider photographing the model.

2 Locate and be able to briefly describe the functions of the digestive structures included in Figure 24.15.

3 Label Figure 24.16, and write next to the label the function of the structure.

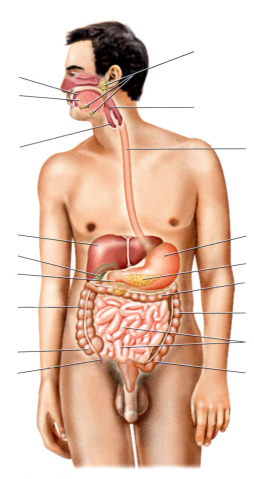

FIGURE **24.16** Basic anatomy of the digestive system.

Did you know . . .

Oh, How Embarrassing!

Have you ever sat in silent a classroom, a church, or an office and your "stomach" begins to growl or even roar? This embarrassing sound is called borborygmus. It is basically caused by gas moving through the intestines and is rarely a sign of disease.

24

Procedure 2

Amylase in Action in the Digestive System

Materials

- ❏ Clean test tubes (6)
- ❏ Test-tube rack
- ❏ 0.5% amylase solution
- ❏ 0.5% starch solution
- ❏ Iodine-potassium-iodide solution
- ❏ Benedict's solution
- ❏ Water
- ❏ Wax pencil or Sharpie
- ❏ Hot plate
- ❏ 500 mL beaker
- ❏ Thermometer
- ❏ Timing device
- ❏ Medicine droppers (4)
- ❏ Test-tube clamps (3)
- ❏ Reaction wells
- ❏ Blank typing paper
- ❏ Toothpicks
- ❏ Gloves
- ❏ Safety goggles
- ❏ 25 mm graduated cylinder (3)

In this procedure, you will investigate the action of the enzyme salivary amylase, a digestive enzyme found in saliva that initiates the digestion of starch, a complex carbohydrate.

WARNING

Benedict's reagent is corrosive. Wear gloves. If it splashes onto your skin, wash the area immediately with soap and water.

1 Procure three test tubes and other needed supplies.

2 With a wax pencil or Sharpie, write "Amylase," "Starch," and "Amylase + Starch," one each on three separate test tubes.

3 Using a graduated cylinder, add 6 mL of amylase solution to the test tube marked "Amylase," add 6 mL of starch solution to the test tube marked "Starch," and add 1 mL of amylase solution and 5 mL of starch solution to the test tube marked "Amylase + Starch." Gently shake the test tubes for 10 seconds.

4 Prepare a water bath using the hot plate and a 500 mL beaker filled three-quarters full of water. Stabilize the temperature to 37°C (same as body temperature, 98.6°F). Carefully place the three test tubes in the beaker of water, and let them sit in the water bath for 10 minutes.

5 Place a reaction well over a piece of white typing paper. Using a medicine dropper, drop 1 mL of the amylase solution in one reaction well, and label its position. Using a second medicine dropper, drop 1 mL of the starch solution in another well, and label its position. Using a third medicine dropper, drop 1 mL of the amylase and starch solution in one well, and label its position with a wax pencil.

6 Add one drop of the iodine-potassium-iodide solution with a clean medicine dropper to each well, and stir with

a toothpick. If the solution turns a bluish black, starch is present. Record and label your results from the reaction wells, especially any changes in color, in the space provided.

Amylase: _____

Starch: _____

Amylase + Starch: _____

7 With a wax pencil or Sharpie, write "Amylase," "Starch," and "Amylase + Starch," respectively, on another set of three test tubes.

8 Add 1 mL of amylase solution to the test tube marked "Amylase," add 1 mL of starch solution to the test tube marked "Starch," and add 1 mL of amylase and starch solution to the test tube marked "Amylase + Starch." Add 1 mL of Benedict's solution to each of the three test tubes.

9 Place a 500 mL beaker filled three-quarters full of water on the hot plate, and bring the water to a boil. Using a test-tube clamp, place one test tube at a time in the boiling water for 2 minutes. Keep in mind that the color blue indicates no sugar is present, red indicates the greatest amount of sugar, green indicates a minimal amount of sugar, and yellow or orange represents a significant amount of sugar.

10 In the space provided, record any changes in color.

Amylase: _____

Starch: _____

Amylase + Starch: _____

11 Discard the solutions as directed, clean the glassware thoroughly, and return the materials.

12 Complete Table 24.4, and interpret the test results in the space provided.

_____

_____

_____

_____

_____

24

TABLE **24.4** Amylase and Starch Test Results

Container	Starch Test	Sugar Test
Amylase solution		
Starch solution		
Amylase + starch solution		

✔ Check Your Understanding

4.1 In order, list the organs through which a piece of food passes in its journey from the mouth to the anus.

4.2 Match the structures making up the digestive system with their major functions.

_____ Small intestine

_____ Gallbladder

_____ Pancreas

_____ Esophagus

_____ Stomach

_____ Liver

_____ Large intestine

A. Production of insulin and glucagon

B. Production of bile

C. Chemical breakdown of food

D. Absorption of water, salts, vitamins, and storage of undigested waste until elimination

E. Passage of bolus to the stomach through peristalsis

F. Storage of bile

G. Chemical digestion and nutrient absorption

4.3 What is the role of the enzyme salivary amylase in the digestive process?

Urinary System

The **urinary system** is perhaps one of the most underappreciated systems in the human body. The system consists of a pair of kidneys, two ureters, the urinary bladder, and the urethra. In the male, the urinary system is associated with the reproductive system because the urethra serves as a passage for both urine and sperm. The term **urogenital system** refers to both the urinary and reproductive systems.

The **kidneys** are two organs about 10 centimeters long, 5 centimeters wide, and 2.5 centimeters thick (each about the size of a bar of soap) that lie against the dorsal abdominal wall at about the height of the 12th rib. A single kidney is composed of 1.2 million functional units called **nephrons**. The primary function of a nephron is to filter the blood to regulate the concentration of water and several soluble substances, primarily sodium. Needed substances are reabsorbed, and waste is excreted as **urine**.

From the kidneys, urine enters the ureters, a pair of tubes (about 27 centimeters in length and 3.5 millimeters in diameter) that aid in propelling it to the urinary bladder.

The human bladder resembles a pear and can hold about 400–600 milliliters of urine. Urine exits the bladder via the urethra, which carries the urine out of the body. The urethra is longer in males (20 centimeters) than in females (4 centimeters) because it passes through the penis.

Urine is a pale-yellow to amber-colored solution composed of water, metabolic waste such as urea, salts, and other organic compounds. The characteristic smell of urine is from ammonia (did you ever smell a kitty litter box ignored for a while?), a product from the breakdown of urea. Other smells, such as those from ketones, also may be detected in urine. For centuries, medical practitioners have analyzed urine (odor, color, consistency) to determine the health of individuals. At one time, urine was even tasted to determine if the person had diabetes. As in the past, diagnostically helpful attributes of urine, such as pH, specific gravity, and the presence of nitrates, bilirubin, glucose, proteins, ketones, blood, and leukocytes are ascertained through **urinalysis**.

Procedure 1

Labeling the Urinary System

Materials
❑ Model of the human urinary system
❑ Colored pencils

1 Procure the model of the human urinary system.

2 Examine and locate the structures in Figure 24.17.

3 Label Figure 24.18, and write the basic function of each structure next to the label.

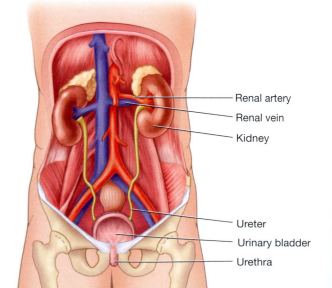

FIGURE **24.17** Organs of and structures associated with the urinary system.

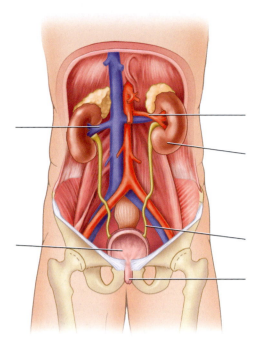

FIGURE **24.18** Organs of the urinary system.

✔ Check Your Understanding ──────────────────

5.1 The main function of the urinary bladder is to collect urine excreted by the kidneys and to filter the urine before it is moved to the outside of the body. (*Circle the correct answer.*)

True / False

5.2 What is the role of the 1.2 million nephrons that are found in a typical kidney?

_____

5.3 The _____ connects the urinary bladder to the outside of the body, and the _____ are tubes of smooth muscles that move urine from the kidney to the urinary bladder.

24

EXERCISE 24.6

Reproductive Systems

In humans, the reproductive system consists of **primary sex structures** and **secondary sex structures**. The primary sex structures are responsible for the production of sex cells, or **gametes**. In females the **ovaries** produce ova, or eggs, and in males the **testes** produce sperm. Secondary sex structures function to ensure that the gametes reach their intended destinations: in males, this includes structures that ensure the maturation and conveyance of sperm, and in females, a place for fertilization and for development of the fetus. Male secondary sex structures are the seminal vesicles, prostate gland, bulbourethral gland, urethra, and penis. Female secondary sex structures include the uterine tubes, uterus, vagina, external genitalia, and mammary glands.

Female Reproductive System

The **female reproductive system** is more complex than the male reproductive system, simply because the former has many more duties. With the exception of the ovaries, the female reproductive system exists in the pelvic cavity. The **ovaries** are a pair of almond-shaped organs that lie in the **ovarian fossa** of the dorsal portion of the pelvic wall. The ovaries are responsible for the production of the egg, or ova, and several hormones including estrogen, progesterone, and inhibin. The ovaries are held in place by ligaments.

Oocytes develop within follicles in the ovary. During **ovulation**, a **secondary oocyte** is released from the **Graafian** (mature) **follicle** into the pelvic cavity, where fingerlike projections called **fimbriae** direct the oocyte to the **uterine tube** (**fallopian tube;** oviduct). The uterine tube is about 10 centimeters long.

Fertilization occurs in the uterine tube, and if fertilization does not occur, unfertilized eggs degenerate in the uterine tubes or the **uterus**. About 6 days after fertilization, the developmental stage called the blastocyst enters the uterus. If the embryo continues to develop in the uterine tube, it is called an **ectopic** (tubal) **pregnancy**.

The uterus is a muscular chamber about 7.5 centimeters long, 5 centimeters in diameter, and 2.5 centimeters thick. In pregnancy, the uterus can exceed 30 centimeters in length. Anatomically, the uterus consists of the upper **fundus**, the large middle region known as the **body** (where the fetus is housed), and the lower **cervix** extending into the **vagina**. The uterus consists of three distinct layers: an outer layer called the **perimetrium**, a middle muscular layer called the **myometrium**, and an inner layer known as the **endometrium**. After the blastocyst leaves the uterine tubes, it implants in the endometrial lining. The endometrial lining becomes part of the **placenta** and is called the **decidua**. If there is no implantation of the blastocyst, the endometrium is shed during menstruation.

The **vagina** is a muscular tube that extends from the cervix to the **vaginal orifice** (opening). The external genitalia of the female, collectively called the **vulva**, consist primarily of the **urethral orifice, clitoris,** a number of glands, the **labia majora** (singular = labium majus), and the **labia minora** (singular = labium minus). The **mons pubis** is a bulge of adipose tissue that lies superior to the vulva.

The paired breasts lie above the pectoralis major muscle. The breasts consist of the **body, adipose tissue, axillary tail** (extension toward the armpit—unfortunately, sometimes a route for metastasis of breast cancer), **nipple,** and **areola.** The nipple is a conical structure where the ducts from the **mammary glands** open to the surface of the skin. A darker, circular region, the **areola,** surrounds the nipple. Although not true reproductive organs, the mammary glands are an essential part of the female reproductive system. The function of the mammary glands is to produce milk (**lactation**) to nourish the infant. The mammary glands are controlled by hormones of the reproductive system and placenta.

During pregnancy, the mammary glands develop 15 to 20 **lobes** that contain milk-producing **alveoli** within **lobules**. The milk travels from the alveoli to an **alveolar duct** and eventually to a chamber called the **lactiferous sinus**. Milk is released from 15 to 20 **lactiferous ducts** that open at the surface of the nipple.

Procedure 1

Labeling the Female Reproductive System

Materials
❏ Model of the human female reproductive system
❏ Colored pencils

1 Procure the model of the human female reproductive system.

2 Examine and locate the structures in Figure 24.19.

3 Label Figure 24.20, and write the basic function of each structure next to the label.

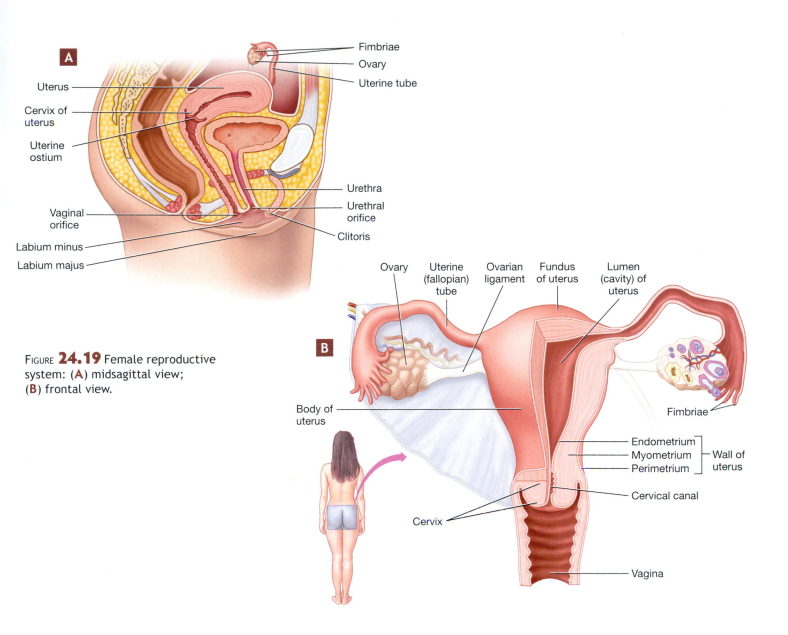

A
Fimbriae
Ovary
Uterine tube
Uterus
Cervix of uterus
Uterine ostium
Urethra
Urethral orifice
Clitoris
Vaginal orifice
Labium minus
Labium majus

FIGURE 24.19 Female reproductive system: (A) midsagittal view; (B) frontal view.

B
Ovary
Uterine (fallopian) tube
Ovarian ligament
Fundus of uterus
Lumen (cavity) of uterus
Fimbriae
Body of uterus
Endometrium
Myometrium
Perimetrium
Wall of uterus
Cervical canal
Cervix
Vagina

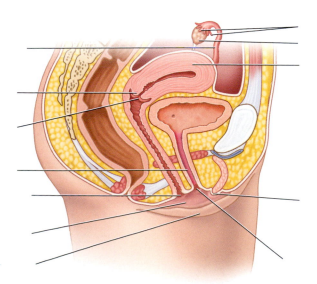

FIGURE 24.20 Reproductive organs of the female reproductive system.

24

Male Reproductive System

The major functions of the **male reproductive system** are to produce and store spermatozoa or sperm, produce male sex hormones (androgens, such as testosterone), and ejaculate semen. The gamete-producing structures in the male are the **testes,** housed in an external sac, the **scrotum**. Sperm cannot develop at 37°C (body temperature). Thus, the scrotum is located outside the body cavity to allow sperm to develop in a cooler environment.

A single testis is an oval structure approximately 4 centimeters long and 2.5 centimeters wide. The testis is divided into more than 250 **lobules**. Each lobule is composed of several tightly packed **seminiferous tubules,** the sites of sperm production. **Sertoli cells** in the seminiferous tubules protect and aid the development of sperm.

Located between the seminiferous tubules are **Leydig cells** that produce testosterone. The seminiferous tubules lead to the **rete testes**, where sperm partially mature. **Efferent ductules** lead to the **epididymis,** where sperm finish maturation. Mature sperm can be stored in the epididymis and associated ductus (vas) deferens for up to 60 days before being reabsorbed. The ductus (vas) deferens, testicular nerves, testicular artery, and testicular venous structure pass through the body wall within a structure called the **spermatic** cord. Posterior to the urinary bladder, each ductus deferens joins with a gland called the **seminal vesicle,** forming the **ejaculatory ducts**.

The ejaculatory duct connects to the **urethra**. Originating at the bladder and terminating at the end of the **penis,** the urethra carries urine and **semen**. Semen consists of 95% secretions and 5% sperm cells. The volume of semen ejaculated by human males averages between 2 and 5 milliliters and contains between 20 million and 130 million sperm per milliliter. The seminal vesicles secrete fructose to nourish the sperm and factors to enhance the motility of sperm. The **prostate gland** releases an alkaline substance that buffers seminal and vaginal acidity in order to activate sperm. The **bulbourethral gland** adds more alkaline substances to the semen.

The penis has the dual function of eliminating urine and depositing semen during intercourse. The **corpora cavernosa** (anterior) and the **corpus spongiosum** (surrounding the penile urethra) serve as erectile tissue during penile erection. The **root** of the penis is interior and connects the penis to the body wall. The **shaft** of the penis is the tubular external portion of the penis. The **glans penis** is the expanded distal end of the penis. The **external urethral meatus** is the opening of the penis.

Procedure 2

Labeling the Male Reproductive System

Materials
- ❏ Model of the human male reproductive system
- ❏ Colored pencils

1. Procure the model of the human male reproductive system.

2. Examine and locate the structures in Figures 24.21 and 24.22.

3. Label Figure 24.23, and write the basic function of each structure next to the label.

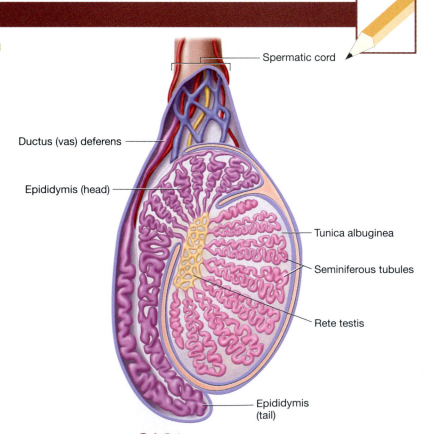

FIGURE **24.21** Midsagittal section through the testis.

24

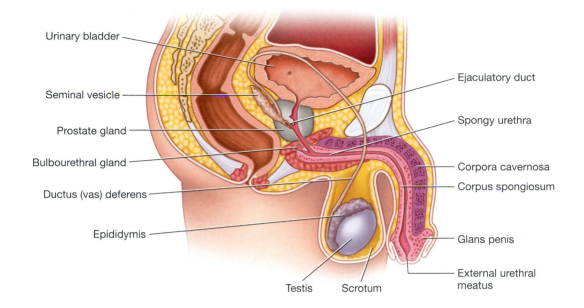

Urinary bladder

Seminal vesicle

Prostate gland

Bulbourethral gland

Ductus (vas) deferens

Epididymis

Ejaculatory duct

Spongy urethra

Corpora cavernosa

Corpus spongiosum

Glans penis

External urethral meatus

Testis   Scrotum

FIGURE **24.22** Male reproductive system.

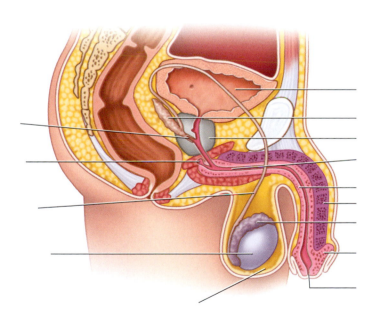

FIGURE **24.23** Male reproductive system.

24

✔ Check Your Understanding

6.1 List five secondary sex structures in females and males.

6.2 In the male reproductive system, the _____ is/are the site(s) of sperm production. (*Circle the correct answer.*)

a. Leydig cells

b. seminiferous tubules

c. scrotum

d. rete testes

e. efferent ductules

6.3 Name the three layers that make up the uterine wall. Describe their orientation within the wall.

24

MYTHBUSTING

Take Heart

Debunk each of the following misconceptions by providing a scientific explanation. Write your answers on a separate sheet of paper.

1 I am too young to worry about heart disease.

2 Watch it. Spicy foods cause ulcers!

3 People living with Crohn's disease only have to take their medicine when they feel bad.

4 There is no danger in taking growth hormone.

5 Diabetes has no impact upon the heart.

---

1 Follow the path of oxygen as it travels through the lower respiratory tract. (*Circle the correct answer.*)

a. Larynx, alveoli, bronchi, trachea, bronchioles.

b. Alveoli, bronchioles, bronchi, trachea, larynx.

c. Larynx, trachea, bronchi, bronchioles, alveoli.

d. Larynx, bronchi, trachea, bronchioles, alveoli.

e. None of the above.

2 Label the external structures of the heart (Fig. 24.24).

1. _____
2. _____
3. _____
4. _____
5. _____
6. _____
7. _____
8. _____
9. _____
10. _____
11. _____
12. _____
13. _____

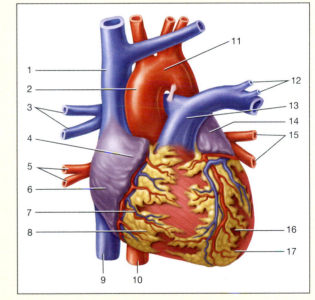

FIGURE **24.24** External structure of the heart, ventral view.

14. _____
15. _____
16. _____
17. _____

3 _____ is produced in the liver, stored in the gallbladder, and is released into the duodenum where it acts to emulsify fats.

4 Label the internal structures of the heart (Fig. 24.25).

1. _____
2. _____
3. _____
4. _____
5. _____
6. _____
7. _____
8. _____
9. _____
10. _____

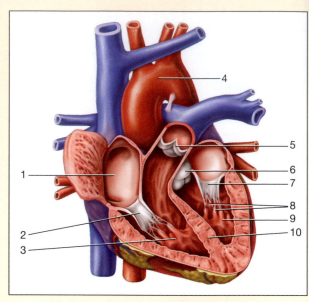

FIGURE **24.25** Internal structure of the heart, ventral view.

5 Trace the route of blood through the heart.

_____
_____
_____
_____
_____

6 Structures associated with the lymphatic system include the _____ . (*Circle the correct answer.*)
a. spleen and cisterna chyli
b. tonsils
c. thymus, lymph nodes, and thoracic duct
d. both a and b
e. all of the above

7 Label the components of the respiratory system (Fig. 24.26).

1. _____
2. _____
3. _____
4. _____
5. _____
6. _____
7. _____
8. _____
9. _____
10. _____

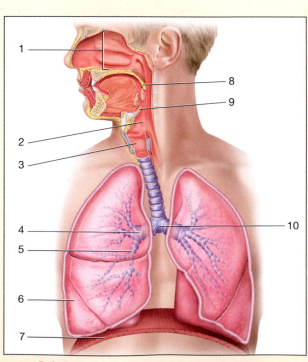

FIGURE **24.26** Structures of the respiratory system.

24

8 Label the figure of the endocrine system (Fig. 24.27).

1. _____
2. _____
3. _____
4. _____
5. _____
6. _____
7. _____
8. _____
9. _____

9 Label parts of the digestive system (Fig. 24.28).

1. _____
2. _____
3. _____
4. _____
5. _____
6. _____
7. _____
8. _____
9. _____
10. _____
11. _____
12. _____
13. _____
14. _____
15. _____
16. _____
17. _____
18. _____
19. _____
20. _____
21. _____
22. _____

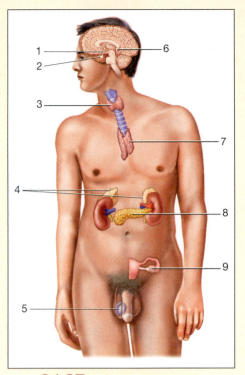

FIGURE **24.27** Generalized anatomical overview of the endocrine system.

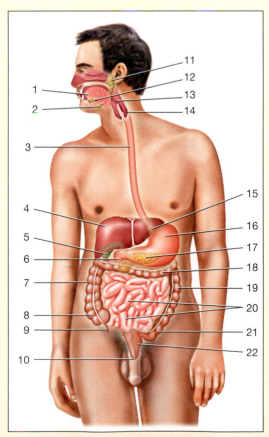

FIGURE **24.28** Basic anatomy of the digestive system.

10 Which gland, located on the thyroid gland, produces a hormone that targets the kidneys, bone, and digestive tract, raises blood calcium levels, and activates vitamin D? (*Circle the correct answer.*)

 a. Adrenal gland.

 b. Pineal gland.

 c. Pancreas.

 d. Hypothalamus.

 e. Parathyroid gland.

11 Which endocrine gland produces hormones that promote the development of T lymphocytes? (*Circle the correct answer.*)

 a. Pancreas.

 b. Hypothalamus.

 c. Parathyroid.

 d. Thymus.

 e. Pineal gland.

12 Label the general anatomy of the urinary system (Fig. 24.29).

 1. _____

 2. _____

 3. _____

 4. _____

 5. _____

 6. _____

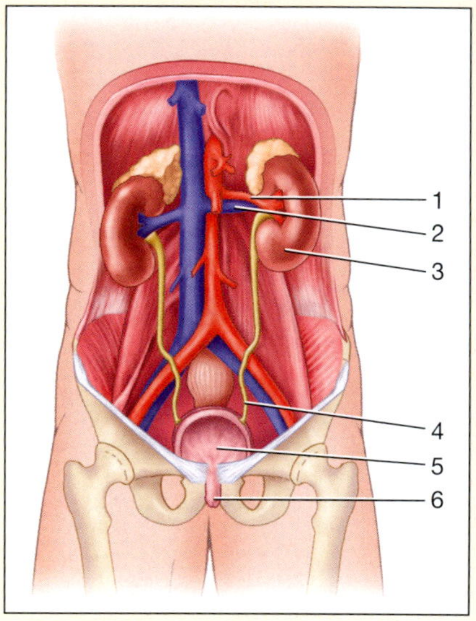

FIGURE **24.29** Organs of the urinary system.

13 Describe the important role of the lymphatic system in the human body.

14 Discuss the role of the urethra in the male urogenital system.

15 What is the difference between primary and secondary sex structures?

16 What is the function of the prostate and bulbourethral glands?

24

But One Earth
Understanding Basic Ecology

25

However fragmented the world, however intense the national rivalries, it is an inexorable fact that we become more interdependent every day. I believe that national sovereignties will shrink in the face of universal interdependence. The sea, the great unifier, is man's only hope. Now, as never before, the old phrase has a literal meaning: We are all in the same boat.

— Jacques Yves Cousteau (1910–1997)

Just wondering . . .

Consider the following questions prior to coming to lab, and record your answers on a separate piece of paper.

1 What is the ICUN Red List?
2 What can I do to help threatened and endangered species?
3 How can I determine if an organism is an invasive species?
4 What is the greatest threat to biodiversity?
5 What has been the biological and economic impact of the BP oil spill so far?

Objectives

At the completion of this chapter, the student will be able to:

1. Define and differentiate ecology and environmental science.

2. Define and describe population, community, ecosystem, biome, and aquatic zone, and investigate the biotic and abiotic factors that contribute to ecosystems in your area.

3. Construct and describe a food chain, a food web, and an ecological pyramid based on trophic levels.

4. Define and investigate primary and secondary succession, pioneer species, and climax communities in your area.

5. Define and investigate several keystone, endangered, threatened, and invasive species in your area.

6. Define pollution, and describe several types.

7. Discuss the effects of oil as a pollutant.

John Muir (1838–1914) wisely stated, "When one tugs at a single thing in nature, he finds it attached to the rest of the world." We must develop an understanding and appreciation of Earth, its composition, cycles, life, and connections that enable us to survive. **Ecology,** the study of the interrelationship between organisms and their environment, is a broad discipline encompassing principles of the physical sciences and the biological sciences. Among the many subdisciplines of ecology are molecular, evolutionary, organismal (physiological and behavioral), population, community, ecosystem, conservation, and quantitative ecology.

Many times, the terms *ecology* and *environmental science* are used interchangeably. **Environmental science** is a multidisciplinary study including the scientific as well as social impact of humans upon Earth. Environmental science includes topics such as philosophy, politics, ethics, economics, sustainability, and stewardship.

A **population** of organisms consists of individuals of the same species living in a given area. Their home is called the **habitat,** and their **niche** is their role in the environment. These organisms interact with other species, forming a **community**. An **ecosystem** consists of the biological, or **biotic,** community and the nonliving, or **abiotic,** environment. Abiotic components include factors such as moisture or water, sunlight, weather or climate, nutrients, substrate, soil, pH, salinity, and oxygen availability. An ecosystem can consist of a small ditch or an entire swamp. Ecosystems can be natural, such as a forest, or artificial, such as an aquarium.

All of Earth's ecosystems make up the **biosphere** (Fig. 25.1). The terrestrial portions of the biosphere, or biomes, consist of large regions of land with distinct climates and specific species of organisms. Examples of biomes are deserts, coniferous forests, deciduous forests, evergreen forests, tropical forests, grasslands, savannas, and tundra (Fig. 25.2). The parts of the biosphere dominated by water constitute the aquatic life zones and include freshwater

Chapter Photo
Ecosystems, like the one on this beach, have living organisms and abiotic factors.

603

systems, marine environments, estuaries, and wetlands. Within a given ecosystem, **keystone species** play a critical role in maintaining a healthy ecosystem despite the fact that their biomass is disproportionate (inversely) to their value to the ecosystem. Examples of keystone species are bison on grasslands, alligators in a swamp, gopher tortoises in a desert, sea otters in a kelp forest, beavers near a lake, sea stars in a sound, and killer whales in the ocean. If a keystone species is removed from an ecosystem, the ecosystem will be prone to significant change.

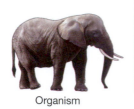
Organism

Population

Community

Ecosystem

Biosphere

FIGURE **25.1** Biological levels of organization.

FIGURE **25.2** Despite the apparent differences between these ecosystems, they have many characteristics in common: (**A**) desert; (**B**) coniferous forest; (**C**) swamp; (**D**) marsh; (**E**) kelp forest; (**F**) tropical forest; (**G**) rocky beach; (**H**) grassland.

25

Investigating Local Ecosystems

The organisms residing within an ecosystem can be assigned **trophic levels** based on their source of energy acquisition. **Producers** make up the fundamental trophic level and consist of autotrophs, such as photosynthetic organisms (some bacteria, algae, and plants) and chemosynthetic organisms (some bacteria, such as those living in hydrothermal vents). Photosynthetic organisms directly capture the radiant energy of the sun and turn it into energy-rich organic molecules, such as glucose, they and other organisms can use.

Consumers, such as caterpillars and owls, are heterotrophs, organisms that obtain their energy by consuming other organisms. Primary consumers, such as rabbits and deer, eat plant materials and are called **herbivores**. Secondary consumers are organisms that eat primary consumers. In turn, they may be eaten by tertiary consumers. Consumers that eat flesh are called **carnivores**, of which bass, bobcats, dolphins, and eagles are examples. **Omnivores** consume both plant and animal matter and include, among many others, bears, chickens, raccoons, and humans. **Scavengers** are animals that primarily consume the carcasses of dead animals, such as vultures and occasionally an opossum or coyote. **Detritivores**, such as earthworms, feed on decomposing organic matter.

Decomposers are the final trophic level, the ultimate recyclers. These organisms use nonliving organic matter as a source of energy and return inorganic materials to the environment when they die.

Trophic levels can be appreciated by constructing a simple **food chain** (Fig. 25.3). A **food web** (Fig. 25.4) is a more complete representation of the relationships in an ecosystem, illustrating the many trophic-level interactions in a community of organisms.

Energy flows through each trophic level. With each level, the amount of energy to the next trophic level decreases. The percentage of usable energy transferred through the trophic levels is termed the *ecological efficiency* of a system. Only about 10% of the energy is passed from one level to the next. As a result, the biomass (total weight of all organisms at a given trophic level) decreases from one level to the next. This explains why there are more grazing animals than top carnivores on grassland. An ecological pyramid illustrates energy relationships in an ecosystem (Fig. 25.5). The most biologically productive ecosystems on Earth are the swamps and marshes, estuaries, and tropical rainforests.

Did you know ...

Just for Thought!

How much biomass is there in a well-trimmed lawn? Calculating the biomass of grass clippings in a large lawn would be quite a task. Provided that the grass in the lawn is uniform, you could manage this task taking a square meter to study over the course of a year. On a regular schedule, you would clip all of the grass to a uniform height and save the clippings. Then you would determine the wet and dry weight of the clippings during the year.

Of what does most of the weight of grass consist? For a rough estimate of the biomass of the clippings, multiply the dry weight by the number of square meters of grass in the yard. Unless you have the task of cutting the lawn, this number may be surprising. It does not include the biomass of the remainder of the plant and other living things in the yard such as trees, insects, and other animals. WOW!

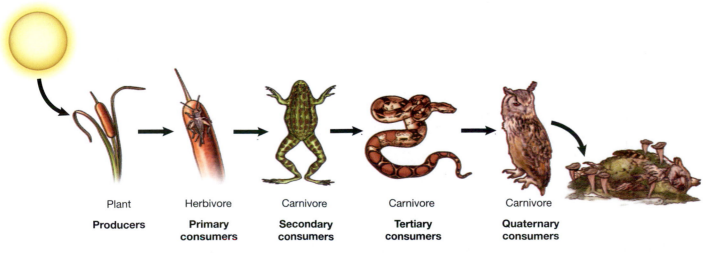

Plant	Herbivore	Carnivore	Carnivore	Carnivore
Producers	**Primary consumers**	**Secondary consumers**	**Tertiary consumers**	**Quaternary consumers**

FIGURE **25.3** Simple food chain.

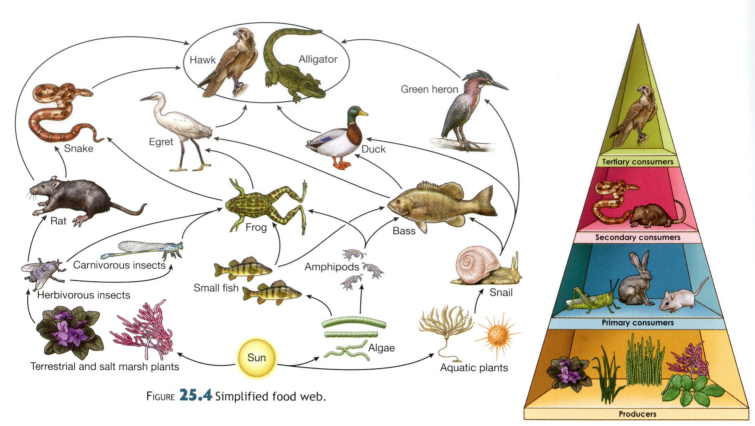

FIGURE **25.4** Simplified food web.

FIGURE **25.5** Simplified ecological pyramid.

Procedure 1

Local Ecosystems

Materials
- ❏ Appropriate clothes
- ❏ Camera
- ❏ Binoculars
- ❏ Field thermometer
- ❏ Small rake
- ❏ Net
- ❏ Field guides to common plants and animals
- ❏ Compound microscope
- ❏ Microscope slides and coverslips
- ❏ Eyedropper
- ❏ Collection jar
- ❏ Meterstick
- ❏ Hand lens
- ❏ Notebook
- ❏ Pencils

One of the things many people like about biology is that classroom concepts are best illustrated outdoors. You can apply your knowledge of ecosystems and succession by taking a local field trip. This procedure introduces you to several ecosystems near your campus. Preferably, at least two natural ecosystems (a terrestrial ecosystem and an aquatic ecosystem, if possible) and one artificial ecosystem will be observed. The instructor will obtain permission to explore and take photographs of these environments.

1 Discuss various ecosystems near your campus. Determine which ecosystems would be interesting to visit, and then take a field trip to these locations.

2 In a notebook, answer the following questions for each ecosystem, and take photographs during the field trip.

3 If the instructor requests, construct a PowerPoint presentation for each of the three ecosystems.

Terrestrial Ecosystem

Where is the ecosystem located?

When are you conducting this study (time of year and date)?

Measure the weather conditions. Take the temperature 2 meters above the ground and at the surface of the soil in both shade and sunlight. Also, if there is cement nearby, take the temperature 2 meters above the cement and at the cement level in both shade and in sunlight. Record your results in Table 25.1.

25

TABLE **25.1** Terrestrial Ecosystem

Location	Shade Temperature	Sunlight Temperature
2 meters above ground		
Soil surface		
2 meters above cement		
Cement surface		

Write a general overview of the study area, and take several general photographs.

_____

_____

_____

Describe the abiotic conditions of the study area.

_____

_____

_____

State the dominant soil type (sand, clay, etc.).

_____

What is the dominant plant ground cover?

_____

Is there leaf litter? If so, how deep and rich is the litter? Do any animals inhabit the litter? If so, what are they? Do any animals live in the soil? If so, what are they?

_____

_____

_____

Describe the most obvious animals, both invertebrate and vertebrate, in the study area.

_____

_____

Using scientific and common names, identify several organisms and their place in the food web.

_____

Aquatic Ecosystem

Where is the ecosystem located?

_____

When are you conducting this study (time of year and day)?

_____

Measure the weather conditions. Take the temperature at the shoreline 2 meters above the ground and at the surface of the ground in both shade and sunlight. What is the temperature at the surface of the water? If possible, take a temperature reading at various depths beneath the surface of the water and 2 meters above the surface of the water. Record these temperatures in Table 25.2.

Write a general overview of the study area, and take several general photographs. Is the water flowing, still, or stagnant?

_____

_____

_____

Describe the abiotic conditions of this aquatic ecosystem.

_____

_____

Describe the type of bottom, if possible (sand, clay, silt).

_____

Describe the condition of the water as related to light penetration (muddy, murky, clear).

_____

_____

TABLE **25.2** Aquatic Ecosystem

Location	Shade Temperature	Sunlight Temperature
2 meters above ground		
Soil surface		
2 meters above water		
Below water surface		

25

What are the dominant plants on the shoreline? Are any plants living in the water?

Take a sample of the water, if possible, with the collection jar. Bring the jar to the lab as soon as possible, and examine the water with a compound microscope. Describe and identify, if possible, the organisms found in the water. Discard the water and slide as instructed.

Name the most obvious animals, both invertebrate and vertebrate, in your study area.

Using both scientific and common names, identify several organisms and their place in the food web.

Artificial Ecosystem

Where is the ecosystem located (aquarium, wildlife park, etc.)?

When are you conducting this study (time of year and date)?

Describe the abiotic conditions or weather conditions. If possible, take temperature readings and record them in the space provided.

Write a general overview of the study area, and take several general photographs.

Describe the abiotic conditions of the ecosystem.

Describe the environment.

Describe the most obvious plants and animals in the ecosystem.

Using both the scientific and the common names, identify several organisms and their place in the food web.

Describe the apparent health of each ecosystem.

25

✔ Check Your Understanding

1.1 List several producers and several consumers in the ecosystem you observed.

1.2 List the trophic levels within an ecosystem and their source of energy.

Example

Producer: consists of autotrophs, photosynthetic organisms, and chemosynthetic organisms that directly capture radiant energy from the sun.

1.3 What is a keystone species? Does the loss of a keystone species have an effect on the ecosystem? Describe a keystone species in the ecosystem you observed.

1.4 The percentage of usable energy transferred through trophic levels is known as ecological efficiency of a system. How much energy is passed from one trophic level to the next level? (*Circle the correct answer.*)

a. 90%.

b. 25%.

c. 10%.

d. 75%.

e. 100%.

1.5 Explain how population, community, and ecosystem are different.

25

Species Decline

One of the major factors decreasing **biodiversity** is the alarming extent of species decline. In recent years, scientists introduced the acronym HIPPO to describe five major issues affecting life on Earth (Table 25.3).

Background extinction is a naturally occurring process in the biological world when organisms no longer can succeed in their natural environment. Mass extinction involves many species going extinct over a relatively short period of time, such as the extinction event at the end of the Cretaceous period.

Human activities are endangering and threatening many species, driving some to the threshold of extinction. Endangered species are small populations in immediate jeopardy of going extinct. Threatened (vulnerable) species have declining numbers and could become endangered in the future.

TABLE **25.3** Issues Affecting Species Decline

Letter	Description
H	habitat loss from destruction, degradation, and fragmentation. Today, habitat loss is the greatest threat to the success of organisms. It can be brought about by deforestation, destruction of wetlands, degradation of coral reefs, unwise land-use management, and fragmentation of existing habitats.
I	invasive species, introduced accidentally or deliberately, such as kudzu and zebra mussels, which are overcompeting and replacing native populations.
P	pollution of the air, water, and soil, which can destroy organisms as well as their habitat.
P	population growth: the alarming exponential growth of the human population that has pushed many species to extinction and many others to the brink of extinction.
O	overexploitation of plants and animals by humans. It is thought the illegal trade of wildlife products is worth up to $10 billion annually to poachers and smugglers. As examples, a wild rhinoceros horn is worth up to $13,000 a pound, and an imperial Amazon macaw may sell for $30,000.

Procedure 1

Endangered and Threatened Species

Materials
- ❏ Appropriate clothes
- ❏ Camera
- ❏ Notebook
- ❏ Pencils

In this procedure, you will visit a game preserve, zoo, or botanical garden and observe endangered and threatened species (Figs. 25.6 and 25.7). If these sources are not available, reference the internet.

1 In a group, discuss extinction, and name several animals and plants extinguished in the past 200 years. State the difference between endangered and threatened species. Discuss several endangered and threatened species on the worldwide level, in the United States, and in your region.

2 Visit a game preserve, botanical garden, or zoo as arranged by your instructor, and observe several endangered and threatened species. Fill in the following data, and use a separate notebook to record your observations.

   a. Place visited: _____

   b. Time of year: _____

   c. Weather conditions: _____

   d. Species observed: _____

   e. Condition of the species and how the organisms were housed or displayed:

3 Prepare a brief report on the organism, and provide photographs. In addition, your instructor may ask you to develop a PowerPoint presentation.

25

FIGURE **25.6** Threatened species: (**A**) bison, *Bison bison*; (**B**) mountain plover, *Charadrius montanus*; (**C**) Utah prairie dog, *Cynomys parvidens*.

FIGURE **25.7** Endangered species: (**A**) Amur leopard, *Panthera pardus orientalis*; (**B**) Haleakala silversword, *Argyroxiphium sandwicense*; (**C**) gorilla, *Gorilla gorilla*.

Procedure 2

Invasive Species

Materials

❏ Appropriate clothes
❏ Camera
❏ Field guides to common plants and animals
❏ Hand lens
❏ Notebook
❏ Pencils

Approximately 50,000 non-native or exotic species of plants and animals live in the United States. Through the years many non-native plant and animal species have been a source of food, medicine, aesthetics, and other beneficial applications. One in seven non-native species, however, is considered an **invasive species** because it is harmful to the ecosystem (Fig. 25.8).

One of the most significant threats to ecosystems is the introduction of invasive species. Many of these organisms have no natural predators, competitors, pathogens, or parasites that can keep their populations in check. As a result, they over-compete with the native species, disrupting the natural ecosystem. In the United States alone, more than 7,000 invasive species have been introduced either accidentally or deliberately into various ecosystems.

This procedure introduces you to non-native and invasive species. Before engaging in this activity, you should conduct some research on non-native and invasive species in the region. The instructor will obtain permission for the group to explore these environments and photograph the species in them.

1 After determining some non-native and invasive species in your geographic region, list and describe several species and discuss their impact on local life.

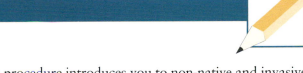

FIGURE 25.8 Invasive species: (A) nutria, *Myocaster coypus*; (B) water hyacinth, *Eichhornia crassipes*; (C) gold dust day gecko, *Phelsuma laticauda*; (D) European starling, *Starnus vulgaris*.

2 Develop a plan of action to find several of these organisms.

_____

_____

_____

_____

_____

3 As arranged by your instructor, visit areas where you would expect to find these organisms and try to photograph them.

Did you know . . .

They must be Crazy!

A new invasive species that originates in South America has spread across Texas and several Gulf Coast states. It is known as the Rasberry crazy ant, or tawny ant (*Nylanderia fulva*), because of its fast, unpredictable movements; it seems to wildly scurry from place to place. The ant is called Rasberry after Tom Rasberry, the exterminator who first identified them. They are small, reddish ants that feed on minute insects and plant material. When threatened the ant can bite but not sting and excretes formic acid through a hairy circle at the end of the abdomen causing temporary discomfort to its victim. Interestingly, the formic acid can also serve as antidote against fire ant venom. These curious little ants have an affinity to electrical equipment, causing short circuits, overheating, and destruction of insulation.

✔ Check Your Understanding

2.1 Small populations that are in immediate jeopardy of extinction are known as _____; whereas, vulnerable or _____ species are those declining in numbers and may become endangered in the future.

2.2 Describe several endangered and threatened species in your region.

_____

_____

_____

_____

2.3 One in seven non-native plant and animal species introduced into the United States is considered to be an invasive species. Why do these species become a threat to the environment?

_____

_____

25

Effects of Oil as a Pollutant

Pollution is defined as any physical or chemical entity that degrades the environment. When we think about pollution, some types that come to mind are foul air, toxic wastes, acid rain, radiation leaks, oil spills, and intolerable noise. Also, pollution conjures up images of death, filth, and despair that haunt our thoughts. Most pollution results from human activities, such as burning fossil fuels; however, natural sources of pollution, such as volcanoes, also affect the environment. The two main sources of human pollutants are:

1. Point-source pollutants, such as oil spills, exhaust from vehicles, and noise from a jet engine (single-source, localized, and easily recognized forms of pollution); and

2. Non-point-source pollutants, such as urban runoff, pesticide runoff, and fertilizer runoff (more difficult to identify).

Pollutants disrupt life-support systems, such as food chains. They damage and destroy living things. They even alter lifestyles, as with the unpleasant noise, odors, and sights surrounding some industrial areas encroaching on neighborhoods.

On April 20, 2010, the Deepwater Horizon oil rig exploded in the Gulf of Mexico approximately 50 miles off the coast of Louisiana (Fig. 25.9). The resulting spill released tens of thousands of barrels of crude oil per day into the water, threatening the rich estuaries, fragile wetlands, and pristine beaches of Louisiana, Mississippi, Alabama, and Florida. The spill was responsible for the horrific destruction of wildlife and coastal plants, disruption of the tourist industry, and devastation of the seafood industry. It jeopardized a way of life in a region rich in tradition.

Initial use of unproven contemporary methods to stop the oil leak at its source failed, allowing an unprecedented amount of oil to leak into the Gulf of Mexico. Several strategies were attempted to contain and clean up the spill. These methods included letting the spill break down by natural means, using booms to channel and collect the oil, blocking the spill with barges, building sand berms, using biological agents to degrade the oil, burning off the oil, and using dispersants to break up the oil.

A common method to clean up oil spills is using dispersants to break up the oil and speed the natural biodegradation process. Dispersants are similar to emulsifying surfactants (or surface active agents), acting to reduce surface tension of water that prevents oil and water from mixing (Fig. 25.10). This method forms smaller droplets of oil that can be broken down more rapidly.

Unfortunately, many dispersants are toxic to living things. Dispersants are not appropriate for all oils and all locations. If the oil is dispersed through the water column, it can affect marine organisms that form the basis of the food chain or are important to the seafood industry.

A common dispersant safe for classroom use is dishwashing liquid. The detergent loosens grease and oil from dirty dishes by acting as a surfactant, which works at the interface between the oil or grease and the water. Surfactant molecules consist of a hydrophilic face (attracted to water) and a hydrophobic face (repelled by water).

In a cleaning solution, the hydrophobic face of the surfactant molecule orients itself toward the oil and grease. Then surfactant molecules disrupt the hydrophobic interactions between the molecules of oil and grease (breaking them up into small pieces), similar to the way in which a dispersant interacts with oil in an aquatic environment. In addition, the hydrophilic face of the surfactant molecules projects into the water, causing the oil and grease to become suspended in the cleaning solution.

FIGURE **25.9** The Deepwater Horizon incident was devastating to the environment, economy, and way of life in the northern Gulf of Mexico.

FIGURE **25.10** Spraying dispersant in the Gulf of Mexico.

Procedure 1

Oil and Feathers

Materials

- ❑ Gloves
- ❑ Lab coat or apron
- ❑ Eye protection
- ❑ Feathers (soaked in outboard motor oil), 6 per group
- ❑ 1% Dawn dishwashing liquid solution
- ❑ Small plastic trays, 3 per group
- ❑ Cotton balls
- ❑ Large plastic container
- ❑ Paper towels
- ❑ Disposal bin for oil and oily water
- ❑ Timing device
- ❑ Water

One of the most heartbreaking scenes from an oil spill is the oil-covered wildlife. Birds, such as pelicans, are particularly vulnerable to being exposed to oil. This experiment is designed to illustrate how dishwashing liquid can be used to clean bird feathers. Keep in mind the real-life process is highly coordinated and much more complex. Oiled animals should not be treated or cleaned without proper facilities and trained personnel.

Although valiant efforts have been made to clean oil-soaked birds, recent studies indicate only a small percentage of the birds actually survive, especially those inhabiting colder waters. Many die from hypothermia and shock as well as kidney- and liver-related complications. Also, detergents may damage a bird's natural waterproofing oil mechanisms. In addition, surviving and released birds may become disoriented and not adjust to new environments.

1 Procure the equipment for this experiment. Place the oil-soaked feathers in one of the small plastic trays.

2 Place three oil-soaked feathers in a small plastic tray containing only water.

3 Wearing gloves, a lab coat, and eye protection, gently scrub the feathers with cotton balls until the water becomes dirty with oil. When the water is dirty, move the feathers to another container containing only water, repeating until the feathers appear clean, or for a maximum of 5 minutes. Record your observations in the space provided.

4 Place three oil-soaked feathers in a small plastic tray containing 1% dishwashing solution.

5 Gently scrub the feathers in the 1% dishwashing solution with cotton balls until the water becomes dirty with oil. When the water is dirty, move the feathers to another container, repeating until the feathers appear clean, or for a maximum of 5 minutes. Rinse the feathers with water. Record your observations in the space provided.

6 Dispose of the feathers, water, and oil as indicated by your instructor.

7 Discuss your findings in the space provided.

Exploring Biology in the Laboratory: Core Concepts

Procedure 2
Oil and Radish Seeds

Materials
- ❏ Lab coat or apron
- ❏ Gloves
- ❏ Eye protection
- ❏ Radish seeds
- ❏ Radish seedlings growing on a moist paper towel in a petri dish
- ❏ Petri dishes
- ❏ Paper towels
- ❏ Shoebox or plastic container
- ❏ Ruler
- ❏ Hand lens
- ❏ Timing device
- ❏ Pipette
- ❏ Variety of graduated cylinders
- ❏ Outboard motor oil
- ❏ Mineral oil
- ❏ Water
- ❏ Oil-based paint
- ❏ Variety of beakers
- ❏ Thermometer
- ❏ Camera
- ❏ Applicator sticks
- ❏ Colored pencils
- ❏ Other materials students select

Along the marshes of the northern coast of the Gulf of Mexico, wetland plants were exposed to oil. The oil proved lethal to many plants, seeds, and seedlings, increasing the chances of erosion and the loss of valuable wetlands.

Working in groups, you will design and conduct a valid experiment that tests the consequences of exposure to oil on radish seeds. Communication within the group is important. You also will prepare a detailed lab report, including photographs and, if requested by the instructor, a PowerPoint presentation.

1 Decide to test either seed germination or seedlings. Decide to test either mineral oil or outboard motor oil. If you decide to use a solution of mineral oil and water, consider adding a bit of food coloring to the water.

2 Answer the following questions:

a. Discuss some background material.

b. List the negative control and the independent and dependent variables in your experiment.

c. State your hypothesis.

d. List the materials you decided to use.

e. Describe the procedure.

f. Describe your findings.

g. What is your means of safely disposing of the material at the end of the experiment?

25

✔ Check Your Understanding

3.1 Explain how dispersants work when used on oil spills.

3.2 Which of the following will result when dispersants are used following an oil spill? (*Circle the correct answer.*)
   a. Dispersants may damage marine organisms that form the basis of the food chain.
   b. Dispersants break oil down into smaller droplets that can break down more rapidly.
   c. Oils may be easier to collect.
   d. Both a and c.
   e. All of the above.

3.3 Which is more effective for removing oil from bird feathers: water or dilute dishwashing liquid?

3.4 List several problems a professional or a volunteer may encounter in cleaning birds.

3.5 Studies have shown that only a small percentage of oil-soaked birds that have had their feathers cleaned actually survive. What problems face the birds following the feather cleaning?

25

MYTHBUSTING

The Squeaky Wheel Gets the Grease

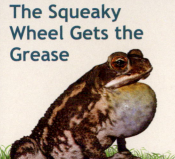

Debunk each of the following misconceptions by providing a scientific explanation. Write your answers on a separate sheet of paper.

1 Over time, oil will break down or sink and never be a problem again to the environment.

2 All invasive species have a negative impact upon their new environment.

3 Once an animal becomes endangered, it is not worth saving!

4 Who cares about the loss of this species or that? It has no impact on humans!

5 Most environmental concerns are "just made up."

1 What is the difference between the disciplines of ecology and environmental science? Provide several examples of each.

2 List several of the most bioproductive environments on Earth.

3 Explain the difference between primary succession and secondary succession.

4 Following an oil spill, most of us immediately are concerned about the birds and other marine life in the area. What is the effect of oil on the wetland plants in marshes and the long-term effects for the environment following an oil spill?

5 What does the acronym HIPPO stand for?

6 Match the following examples with the correct trophic level.

_____ Fungi, bacteria

_____ Grass

_____ Rat

_____ Snake

_____ Frog—carnivore

_____ Zooplankton

_____ Hawk—carnivore

_____ Rabbits, deer—herbivores

_____ Algae, some bacteria

_____ Grasshopper—herbivore

_____ Phytoplankton

_____ Raccoon

_____ Mosquito larvae

A. Producers

B. Primary consumers

C. Secondary consumers

D. Tertiary consumers

E. Quaternary consumer

F. Decomposers

7 Construct a simple marine food web.

8 What is the greatest threat to the success of organisms?

_____

_____

_____

9 Discuss several types of pollution in your region.

_____

25

_____

Index